Current Topics in Microbiology and Immunology

Volume 420

More information about this series at http://www.springer.com/series/82

Benjamin F. Cravatt · Ku-Lung Hsu ·
Eranthie Weerapana
Editors

Activity-Based Protein Profiling

Responsible series editor: Peter K. Vogt

Springer

Editors
Benjamin F. Cravatt
Department of Chemistry
The Scripps Research Institute
La Jolla, CA, USA

Ku-Lung Hsu
Department of Chemistry
University of Virginia
Charlottesville, VA, USA

Eranthie Weerapana
Department of Chemistry
Boston College
Chestnut Hill, MA, USA

ISSN 0070-217X ISSN 2196-9965 (electronic)
Current Topics in Microbiology and Immunology
ISBN 978-3-030-11142-7 ISBN 978-3-030-11143-4 (eBook)
https://doi.org/10.1007/978-3-030-11143-4

Library of Congress Control Number: 2018968084

This Springer imprint is published by the registered company Springer Nature Switzerland AG
The registered company address is: Gewerbestrasse 11, 6330 Cham, Switzerland

Preface

Activity-based protein profiling (ABPP) is a chemical proteomic technology that applies the fundamental principles of chemistry for the global analysis of protein function and small molecule interactions in native biological systems. A large fraction of the proteome remains functionally unannotated, and an even smaller portion has selective small molecule probes for pharmacological characterization. ABPP addresses these challenges by deploying broadly reactive chemical probes to interrogate functional and druggable sites on many proteins (100s–1000s) in parallel. These profiles can furnish novel insights into protein function, as well as hit compounds for selective chemical probe development against a wide array of proteins. The power of ABPP thus lies in its multidisciplinary nature that can seamlessly bridge basic protein function with the translational potential afforded by modern synthetic and analytical chemistry methods. This volume provides a focused collection of recent developments that highlight the merits and opportunities of using ABPP for biological discovery.

The first half of this volume focuses on using ABPP to illuminate functional alterations in bacterial proteomes during the formation of microbial communities and in response to host–pathogen interactions. In the first chapter, Wright and Whidbey describe how ABPP can be implemented to study alterations in protein biology as bacteria interact with each other and respond to environmental changes as exemplified by the microbiome. Carlson and colleagues follow with a detailed overview of chemical probe design, applications, and proof-of-concept examples for the application of ABPP in prokaryotic systems. Seeliger and colleagues describe ABPP's utility in the understanding of immune evasive strategies of mycobacteria, with emphasis on serine hydrolase and kinase activities involved in metabolic reprogramming of these bacteria during the transition toward a persistent dormant state. The next set of chapters shift focus toward changes in proteome activity related to host–pathogen interactions. Hatzios and colleagues provide a synopsis of several in vitro and animal microbial infection models where ABPP has the potential to discover new antibacterial targets and corresponding lead inhibitors. Hang and Peng describe the use of ABPP to exploit a specific metabolic vulnerability of pathogenic bacteria and fungi. Specifically, protein fatty acylation of both host and bacterial proteins is important for survival and pathogenesis, and inhibitors that block these fatty acylation pathways may lead to new antimicrobial and antifungal strategies. Ovaa and colleagues present another strategy for targeting bacterial and viral effector proteins through perturbation of ubiquitin signaling pathways that regulate protein homeostasis in pathogen and host cells. Pezacki and colleagues provide specific examples of applying ABPP for targeting host–virus interactions, while Child and Tate focus on applications for the study of parasites. Finally, Hsu and colleagues put forth a general overview of ABPP to study the regulation of T cells, a key immune cell subset for immunity against pathogens as well as cancer.

The second half of the volume focuses on new enabling tools for ABPP investigations of protein biology including posttranslation modifications (PTMs), protein–protein interactions, and protein–small molecule interactions. Cohen and colleagues summarize the use of ABPP to study and target poly-ADP-ribose polymerases (PARPs), which are key enzymes that regulate ADP-ribosylation signaling in cells. Thompson and colleagues describe ABPP to study protein citrullination, an unusual PTM resulting in the conversion of arginine residues into citrulline that is important for apoptosis, terminal differentiation, and transcriptional regulation of autoimmune disorders. Verhelst and colleagues shift from PTM analysis to enzyme subclasses under investigation by ABPP, specifically, the use of ABPP for the development of selective inhibitors against proteases that contain highly similar active sites. van der Stelt and colleagues tackle lipid enzymes, which present different challenges that can be addressed by ABPP including assaying integral membrane enzymes without the need for purification. Considering that proteins do not function in isolation, Kakeya and Ishikawa describe the application of ABPP for studies of large multidomain protein complexes known as non-ribosomal peptide synthetases (NRPS). NRPS systems are a rich resource for the discovery of new antibiotics, and ABPP can provide fresh insights into the protein machinery required to harness their synthetic potential. As an appropriate follow-up to the discussion of natural products, Nomura and Maimone describe mining natural product structures for electrophilic scaffolds that can be converted into new ABPP probes for protein and inhibitor discovery. Backus concludes the volume with a provocative overview of the growing number of functional and druggable sites on proteins being revealed by ABPP through, for instance, global assessments of cysteine reactivity in biological systems.

In summary, ABPP has emerged as a fundamental technology for the functional analysis of proteins in native biological systems. Ranging from studies of protein activity and posttranslational regulation to protein–protein and protein–small molecule interactions, ABPP is enabling chemists and biologists to address important basic and translational research problems that have been historically beyond the reach of more conventional methods. The chapters in this volume collectively exemplify the exciting diversity of approaches and applications of ABPP, with a common theme of enabling new opportunities for discovery in the fields of microbiology, immunology, and beyond.

La Jolla, USA Benjamin F. Cravatt
Charlottesville, USA Ku-Lung Hsu
Chestnut Hill, USA Eranthie Weerapana

Contents

Activity-Based Protein Profiling—Enabling Multimodal Functional Studies of Microbial Communities

Christopher Whidbey and Aaron T. Wright

Contents

By mistake only one of the two authors was marked as corresponding author in the original version of this chapter. This has now been corrected.

C. Whidbey (✉) · A. T. Wright (✉)
Chemical Biology and Exposure Sciences Group, Pacific Northwest National Laboratory, Richland, WA 99352, USA
e-mail: whidbeyc@seattleu.edu

A. T. Wright
e-mail: aaron.wright@pnnl.gov

C. Whidbey
Department of Chemistry, Seattle University, Seattle, WA 98122, USA

A. T. Wright
The Gene and Linda Voiland School of Chemical Engineering and Bioengineering, Washington State University, Pullman, WA 99163, USA

Current Topics in Microbiology and Immunology (2019) 420:1–22
DOI 10.1007/82_2018_128

Published Online: 09 November 2018

Abstract Microorganisms living in community are critical to life on Earth, playing numerous and profound roles in the environment and human and animal health. Though their essentiality to life is clear, the mechanistic underpinnings of community structure, interactions, and functions are largely unexplored and in need of function-dependent technologies to unravel the mysteries. Activity-based protein profiling offers unprecedented molecular-level characterization of functions within microbial communities and provides an avenue to determine how external exposures result in functional alterations to microbiomes. Herein, we illuminate the current state and prospective contributions of ABPP as it relates to microbial communities. We provide details on the design, development, and validation of probes, challenges associated with probing in complex microbial communities, provide some specific examples of the biological applications of ABPP in microbes and microbial communities, and highlight potential areas for development. The future of ABPP holds real promise for understanding and considerable impact in microbiome studies associated with personalized medicine, precision agriculture, veterinary health, environmental studies, and beyond.

Keywords Activity-based protein profiling · Microbiomes · Microbial communities · Activity-based probes

1 Introduction

Microorganisms impact nearly every facet of life on Earth. Human health, biogeochemical cycles, and the maintenance of a food supply all involve microbial processes. While microbiology is a well-established field, the majority of its history has been focused on the study of single (i.e., axenic) strains. Though the study of axenic cultures is important and can be informative, it misses the fact that the overwhelming majority of microbial life is spent in complex communities with other microbes. These complex microbial communities can dramatically differ from axenic cultures in regard to gene expression, protein production, and the metabolism they conduct. Understanding the mechanistic underpinnings of community structure, interactions, and functions will be critical for understanding more complex biological systems ranging from wetland and soil ecosystems to the human gut and resulting physiology (Gilbert et al. 2014; Biteen et al. 2016; Fierer 2017). Additionally, microbiomes are not static, and it will also be necessary to understand how microbiomes respond to various perturbations.

While the importance of determining the mechanisms that drive microbial communities is clear, the study of such communities is in its infancy due to its challenging nature. Ideally, one would have the ability to grow communities in a controlled fashion and use traditional genetic and biochemical approaches to test a specific hypothesis. In practice, difficulties isolating microbes from complex matrices, low biomass recovery, and the inability to grow many microbes in the laboratory have been major barriers to working with microbial communities. Furthermore, interindividual and site-specific variance in microbial community structure and function introduce more layers of complication.

In order to address these difficulties and study microbial communities, a new set of molecular tools is emerging (Arnold et al. 2016). Recent technological developments, such as high-throughput (previously described as 'next-generation') nucleic acid sequencing, have been a major step forward in this regard. High-throughput sequencing (HTS) has enabled culture-independent detection and identification of microbes, single-cell sequencing, and the ability to know the sequence of all the genes present in the complex community (the metagenome). The utility of HTS approaches can be extended to RNA as well, allowing detection of all of the actively transcribed genes within a sample (the metatranscriptome). The ability to study all of the proteins present within a sample (the metaproteome) has been slower to develop, for reasons discussed in Sect. 3.

The role of HTS techniques in opening up the study of microbial communities cannot be overstated. At the same time, nucleic acid sequencing has some important technical and epistemological caveats (Zarraonaindia et al. 2013). Amplification bias, contamination, undersampling, and a lack of an agreed-upon universal analysis framework have had an appreciable impact on the ability to draw conclusions from different studies (Pollock et al. 2018; Weiss et al. 2014; Kim et al. 2017). More fundamentally, the presence of a gene or transcript does not mean that a functionally active protein is produced. The limitations of gene annotation and presence of closely related organisms mean that metagenomic and metatranscriptomic studies can identify potential function but cannot, in themselves, draw a causal inference between function and gene or organism. Metaproteomics is likely closer to representing true function, but many enzymes require posttranslational modifications or cofactors to be active. Thus, there exists a clear need for tools capable of bridging the gap between the two big questions about any microbial community system—'who is there?' and 'what are they doing?'.

At present, activity-based protein profiling (ABPP) for microbial community applications is in its early stages, but there are unprecedented opportunities for function-dependent studies to reveal the mechanisms of microbial communities impacting our health and the world around us. ABPP also provides an excellent avenue to determine how external exposures, such as chemical insults, changing climate, or altered diets or physiology result in functional alterations to microbiomes. Herein, we illuminate the current state and prospective contributions of ABPP as it relates to microbial communities. We will discuss the design, development, and validation of probes, challenges associated with ABPP in complex microbial communities, provide some specific examples of the biological applications of ABPP in microbes and microbial communities, and highlight potential areas for development.

2 Probe Design

2.1 Activity- and Affinity-Based Probes for Multimodal Characterization of Microbiome Function

As described throughout this book and in numerous reviews, chemical probes for ABPP consist of three moieties: a reactive group that forms an irreversible covalent bond with a target protein (extracellular or intracellular), a binding group (e.g., protein substrate or metabolite) that biases the probes toward a target protein or protein family and may also impart cell permeability, and a reporter tag for rapid and sensitive measurement of labeled enzymes. In microbiomes, where so little is known about organism content or overall functional capacity, the potential power of probes to reveal the functional landscape is enormous and varying the reactive groups can yield probing of diverse functions.

The reactive group of the probe can be developed in two ways. First, probes can incorporate electrophilic moieties that enable selective targeting of intracellular catalytically active enzymes by a direct mechanism-based reaction between the enzyme and the probe (Sadler and Wright 2015; Cravatt et al. 2008; Sadaghiani et al. 2007). Probes of this nature can target hydrolases, kinases, oxidation and reduction catalyzing enzymes, and others in a direct activity-dependent manner. The second type of probe mimics a small molecule and preserves the physio-chemical properties of the small molecule but does not directly react with target proteins (Dubinsky et al. 2012); therefore, a photoaffinity moiety such as a diazirine or benzophenone must be incorporated to enable covalent probe-protein binding. These probes can enable characterization of microbial transporters and intracellular proteins involved in the metabolism of the small molecule or enzymes utilizing that small molecule as a cofactor (Romine et al. 2017; Anderson et al. 2016; Nair et al. 2017).

An advantage of ABPP is the ability to characterize the functional capacity of a microbiome at both cell and protein scales. Multimodal profiling of probe targets by altering the reactive group can enable sorting and imaging at the cell scale, and SDS-PAGE and chemoproteomics at the protein scale (Fig. 1). Once proteins are probe labeled, options exist for directly incorporating a reporter tag (e.g., fluorophore or biotin) in the probe or exploiting click chemistry (CC) reactions to attach a reporter to the probe after it has bound its target. The latter option maintains a smaller probe size which can minimize undesirable impacts on reactivity with the target protein or cell permeability. Furthermore, it permits the user to readily exchange the type of reporter that is applied based on the desired application and outcome of the study, and properties of the sample being assayed.

For microbiome characterization, probe-assisted fluorescence-activated cell sorting (FACS) is a powerful tool for characterizing the functional capacity of the microbiome at the cell scale. Choosing a fluorophore to enable FACS is dictated by the background fluorescence of microbiome samples, the fluorescent signal strength required to detect the desired functional cell type, permeability of the cells to the fluorophore, and compatibility with other fluorescent reagents. Activity-based probe-enabled cell sorting enables the isolation of functional guilds of microbial cells from complex microbiomes. Additionally, probe-enabled sorting can provide reduced sample complexity to assist with proteomics analyses and partitioning of cells displaying an active function for subsequent DNA sequencing. Due to microbiome complexity, as well as redundancy when looking at a specific enzyme type, cell sorting can help yield a reduced organism search space for complementing proteomic studies (see Sect. 3). Finally, in the future probes may be used

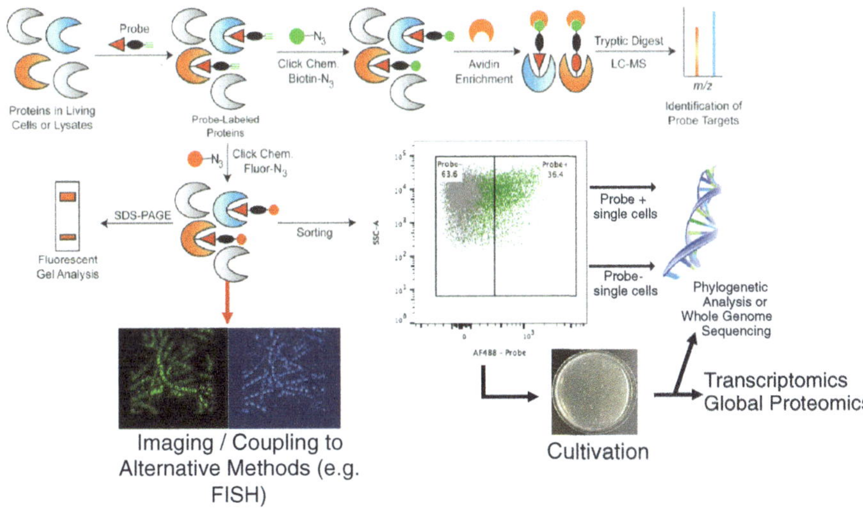

Fig. 1 Multimodal analyses of microbial communities by ABPP. Probes can be used for enrichment and proteomics identification and quantification of target proteins in complex microbiomes or via attachment of a fluorophore spatial resolution of functional cells and proteins can be determined and/or cells can be isolated by cell sorting and sequenced. The future may also see function-dependent live cell sorting for subsequent cultivation and enhanced characterization of active cells from microbiomes

to isolate functional groups of living organisms from microbiomes for subsequent cultivation and analytical studies providing an unprecedented view of microbiome function (Fig. 1).

2.2 *Click Chemistry and Sample Analysis*

One of the biggest challenges in microbial community research is the complexity of the sample matrix. For both environmental and host-associated communities, samples from the native environment are more complex than those from laboratory media. As will be discussed in Sect. 3, this complicates recovery of biomaterial for analysis. However, it also complicates the ability to perform otherwise straightforward chemistry in the sample. The application of CC approaches to ABPP has been especially useful where the incorporation of a full reporter tag would impede probe function and/or in cases where multimodal analyses are desired, e.g., chemoproteomics and fluorescence imaging. The most common CC reactions in ABPP are strain promoted azide–alkyne cycloaddition (SPAAC) and copper-catalyzed azide–alkyne cycloaddition (CuAAC) (Kolb et al. 2001; Speers et al. 2003; Agard et al. 2006). Unfortunately, both exhibit a large degree of non-specific labeling in samples of microbial mats, cecal content, fecal material, and soil (unpublished observations). CuAAC, in addition, has dramatically reduced efficacy in microbial cell lysate from fecal material or cecal content. This can be overcome by increasing the concentration of copper, at the expense of increasing Cu(I)-induced biomolecule damage. To address this, it is possible to 'pre-click' alkyne and azide probes to their reporter group prior to probing the sample or to utilize probes that were synthesized with reporter groups attached. The introduction of a large and (in the case of fluorophores) charged group to the probe may cause a reduction in probe labeling. The trade-off between background and possible loss of signal is likely specific to the probe, sample, and question being addressed.

The complex sample matrix also impacts the ability to analyze samples after probe labeling. SDS-PAGE of fluorescently labeled sample, the traditional approach to optimize labeling conditions, often fails to resolve proteins from these more complex sample matrices (Snelling and Wallace 2017). Two-dimensional SDS-PAGE gives discrete spots for highly abundant proteins, but still exhibits strikingly poor resolution of these more complex samples. Thus, there is a clear need for a high-throughput way to optimize labeling conditions for such samples; for instance, surface immobilized chemical probes may enable more stringent wash methods to remove contaminating content. It is important to note that these contaminating substances carry through sample preparation (enrichment, washing, and protein digestion) and are present to some degree in the final sample. They are also prone to precipitating during LC, potentially plugging columns and emitter tips during proteomics analyses. Because of this, liquid chromatography–mass spectrometry (LC-MS) analysis of samples from complex matrices requires additional strategies to protect the instrument and produce quality quantitative data.

3 Challenges of Microbiome Proteomics and How ABPP Can Facilitate

Compared to the exponential growth in studies using HTS for metagenomics and metatranscriptomics, the use of mass spectrometry-based metaproteomic analyses of microbial communities has been substantially more limited. The reasons for this have been discussed in depth elsewhere (Arnold et al. 2016; Heyer et al. 2017; Hettich et al. 2012; Lee et al. 2017). However, there are two major ABPP-related hurdles that are important to highlight here. The first hurdle is sample complexity, which ABPP could help address. While sample complexity is a concern for all mass spectrometry-based proteomics studies, it is especially important to consider in metaproteomes from microbial communities. The wide dynamic range of peptide abundances, co-elution limiting detection of low abundance peptides, and difficulties developing sequence databases are even more pronounced in these highly complex systems where hundreds to thousands of different taxa may be present. To address this problem enrichment strategies, sample fractionation, and two-dimensional chromatography have all been employed (Biteen et al. 2016; Moon et al. 2018; Mayers et al. 2017; Gilbert et al. 2016, 2018; Xiao et al. 2017; Jansson and Hofmockel 2018; Jansson and Baker 2016). However, these steps can dramatically limit the number of samples that can be analyzed. ABPP could serve as an excellent tool to address this problem. Using ABPP, it is possible to reduce sample complexity while retaining information about relevant proteins. It is important to note that the diverse nature of microbial communities and their environments mean that each sample type will require to optimization of labeling and enrichment conditions which may be difficult for low biomass samples. To circumvent this, similar but more easily obtained matrices can be employed for protocol development.

The other major hurdle facing proteomic analysis of microbial communities is the development of a suitable sequence database. Ideally, a matched metagenome is available for that specific sample. This is rarely possible due to cost and limited available biomass. As such, a database will often need to be constructed from available genomes. There are three specific problems that must be considered: sequence availability, database size, and functional prediction.

The first is the lack of complete, closed genome sequences for many environmental organisms. While genomes for phylogenetically similar organisms may be available, it is difficult to know how well matched the available sequence and the actual sequence are without doing metagenomic sequencing. For some frequently studied microbial communities (e.g., human gut microbiota), multiple metagenomes are available to the scientific community. This raises the second problem: for many less studied microbial communities, these resources are unavailable. In these cases, one approach is to use a large database of all available protein sequences. This comes with its own difficulties, including redundant protein sequences and failure of traditional statistical methods to work with such large databases. Several approaches to address this problem have been developed over the past few years (ComPIL, MetaLab, etc.) (Chatterjee et al. 2016; Zhang et al. 2016; Cheng et al. 2017). Such tools will likely play a key role in successfully drawing conclusions out of massive proteomics datasets.

Regardless of how the sequence database is constructed, the identified proteins need to have some predicted function associated with them in order to use these data to begin to tease out the biological mechanisms. This presents the third challenge: the current state of functional annotation and prediction. While the use of homology-based functional prediction is widespread and can be incredibly powerful, misannotations and lack of annotation for genes in many environmental or unculturable organisms introduce a major caveat to the analysis of proteomics data from complex microbial communities. This is also another challenge that ABPP could help to address. Probes for specific activities such as serine-catalyzed hydrolysis or ATP hydrolysis have been used by our group and others to predict function for previously unannotated proteins in *Mycobacterium tuberculosis* (Ortega et al. 2016). Such a strategy could be readily extended to complex microbial communities to identify proteins of interest for further study.

Additionally, as mentioned in Sect. 4.1, probe-assisted cell sorting can provide a way to enhance database quality for particular functional guilds within a microbiome and provide much improved proteomics analyses.

4 Biological Applications for ABPP

The use of ABPP in microbial communities has thus far been very limited, due in part to the technical challenges discussed above. However, there have been some excellent studies employing this strategy or a similar approach to characterize the active members within communities. Here, we will highlight the use of ABPP and other strategies employing activity-based probes to study function within both host-associated and environmental microbial communities.

4.1 Host-Associated Microbial Communities

Host-associated microbiome research has grown exponentially in the past decade, with the human gut microbiome being the major focus. The microorganisms that inhabit the human gut play a role in an ever-expanding list of human biological processes ranging from production of neurotransmitters to the development of the immune system [reviewed in Rooks and Garrett (2016), Mayer et al. (2015)]. If the mechanisms underlying these community interactions could be better understood, this could represent a new point of intervention for a number of diseases. This will require answering the two major questions of microbial communities: 'who's there?' and 'what are they doing?'. The development of HTS has helped to answer the former. The latter question has been more difficult to answer due to the epistemological limits discussed above; metagenomics and metatranscriptomics can describe potential function but cannot themselves demonstrate that an activity is truly present. Here, we highlight two particular areas of host-associated microbiome research where ABPP could be a powerful tool: xenobiotic metabolism and bile acid metabolism and signaling. However, the potential utility of ABPP in gut microbiome research is far ranging, and it will be of interest to see its role in this field moving forward.

4.1.1 Xenobiotic Metabolism

Broadly speaking, host metabolism of xenobiotics (compounds foreign to the mammalian body including dietary compounds, environmental pollutants, and drugs) occurs in two steps called phase I and phase II metabolism. Phase I metabolism typically consists of a redox reaction that increases compound polarity. This is most often a cytochrome P450- or flavin-containing monooxygenase-catalyzed oxidation to generate or uncover a reactive moiety on the xenobiotic. In phase II metabolism, a transferase typically conjugates a hydrophilic group such as a sulfate or glucuronate to the previously generated reactive moiety to increase solubility and aid in excretion (Koppel et al. 2017). Importantly, gut microbiota can impact metabolism of xenobiotics through multiple routes (i.e., deconjugation, reduction, or modification of gene expression). Modulation of xenobiotic metabolism by gut microbiota has been shown to impact liver function, immune development, and drug efficacy and may serve as a major system through which host–microbiota interaction occurs (Sousa et al. 2008; Wallace et al. 2010).

While our understanding of the underlying molecular mechanisms is in its infancy, there are an appreciable number of pre-HTS studies focused on the ability of gut microbes to metabolize xenobiotics (Sousa et al. 2008; Scheline 1968; Savage 2001). Much of the early work describing these reactions was done comparing metabolism in colonized and germ-free animals. If a particular transformation was observed in the colonized animals but not in the germ-free animals, the microbiota was presumed to be responsible. Alternatively, microbiota isolated from a host was incubated in vitro with the xenobiotic of interest, and if biotransformation was observed then the microbiota could potentially be responsible for the same reaction in vivo. While such approaches were sufficient to implicate microbiota in xenobiotic metabolism, there is a major limitation to such studies; these approaches cannot identify the (potentially multiple) enzymes or taxa responsible for that biotransformation in vivo. Such information would be useful in the design of probiotics, inhibitors, or screening tools depending on the impact of the biotransformation on the host. ABPP could serve as a powerful tool to aid in these identifications.

As discussed in Sect. 2, one of the principle strategies for probe design in ABPP is to use known mechanism-based, irreversible inhibitors for the enzyme class of interest as the basis for the probe binding and reactive groups of the probe. Unfortunately, very little knowledge is available regarding the protein structure or exact reaction mechanism of many of the enzymes produced by gut microbiota. Chemical transformations that are mediated by the microbiome have been characterized, which may be sufficient in some cases for probe design. These include hydrolytic reactions, lyase reactions, reductive transformations, polysaccharide degradation and utilization, functional group transfers, conjugations, and radical catalyzed reactions (Koppel et al. 2017; Spanogiannopoulos et al. 2016; Koppel and Balskus 2016). For some of these reactions, proteolysis and polysaccharide degradation in particular, probe designs already exist and have been employed eukaryotic and prokaryotic systems. For other reactions such as reductions and conjugations novel probes will need to be developed and validated.

Reduction of alkenes, azo-groups, nitro-groups, and sulfoxide groups by gut microbiota has been well documented (Sousa et al. 2008). In fact, one of the best examples of connecting a metabolic reaction to a particular organism is the reduction of digoxin to dihydrodigoxin by the gut microbe *Eggerthella lenta* (Haiser et al. 2014; Haiser et al. 2013). While general reductase reaction mechanisms are understood, they often lack a covalent substrate–enzyme complex that a probe can be designed to capture. A reductase probe strategy could be inspired by probes for their enzymatic opposites, oxidases. Probes for cytochrome P450 enzymes and monoamine oxidases involve the generation of a reactive intermediate (a ketene or iminium-containing Michael acceptor, respectively), which can react with nearby residues in the enzyme (Wright et al. 2009; Krysiak et al. 2012). Depending on the reactivity of this intermediate, however, the probe can diffuse and label nearby enzymes other than the target. An alternative would be photoreactive reversible inhibitors, although these can result in off-target labeling as well.

One of the important functions of gut microbiota is the degradation of polysaccharides via glycoside hydrolases (GHs) and polysaccharide lyases. Importantly, microbiota possesses not only enzymes capable of degrading dietary polysaccharides, but also GHs capable of removing glucuronic acids from xenobiotics. One of the major reactions in phase II metabolism is the conjugation of a glucuronic acid to xenobiotics to form a glucuronide. Glucuronides can then be transported to the gut where microbial GHs (specifically β-glucuronidases) can deconjugate the glucuronic acid from the xenobiotic, regenerating the metabolite. This can lead to altered pharmacodynamics and potentially severe side effects. The chemotherapeutic irinotecan can interact with microbial glucuronidases in this manner, leading to dose-limiting severe diarrhea. In seminal work, Redinbo and colleagues showed that this process could be inhibited by co-administration of bacterial-specific β-glucuronidases inhibitors (Wallace et al. 2010; Wallace et al. 2015). Nonetheless, the specific taxa responsible for this activity remain unidentified. Recently, our group has developed a FACS-based platform to isolate and identify cells responsible for deglucuronidation in vivo (manuscript pending). Additionally, Overkleeft and colleagues recently described synthesis and application of irreversible inhibitors for retaining β-glucuronidases (Wu et al. 2017). While initially demonstrated in human systems, such probes can be applied to label and identify microbial β-glucuronidases as well (unpublished observations).

4.1.2 Bile Acid Metabolism and Signaling

In addition to xenobiotics, gut microbiota can also act on endogenous metabolites. Some of the most abundant metabolites present in the gut are bile acids and salts. Primary bile acids are amphipathic, cholesterol-derived steroids that are produced in the liver and aid in the emulsification of lipids during digestion (Joyce and Gahan 2016; Li and Chiang 2014). When a bile acid is conjugated to an amino acid (taurine and glycine in humans), the more water-soluble bile salt is produced. Eventually, the bile salt enters the gut where it can be altered by bacterial enzymes. There are two subsequent modifications that can occur. The first is deconjugation of the amino acid by bile salt hydrolases (BSHs), regenerating the bile acid. The newly regenerated bile acid can then undergo the second modification, which is an alteration of the sterol core via redox reaction or isomerization. The resulting compound is called a secondary bile acid.

In addition to their role in emulsification bile acids are also potent signaling molecules, activating host nuclear receptors such as the farnesoid X receptor

(FXR) and G-protein coupled receptors such as TGR5. These signaling events influence energy metabolism and dysregulation that may play a role in the development of liver disease and metabolic disorders such as diabetes and obesity (Li and Chiang 2014). Secondary bile acids, which result from gut microbiota transformation of primary bile acids, are structurally distinct from their parent compounds. This can lead to altered receptor binding and thus signaling activity (Joyce and Gahan 2016).

Recently, Lei and colleagues developed bile acid-derived photoreactive probes (Zhuang et al. 2017). Using a competition-based approach, they identified known and novel bile acid-interacting proteins in HeLa cells including carnitine palmitoyl transferase 1A (CPT1A) and ADP-dependent glucokinase (ADPGK). Such probes could be useful in microbial systems as well and could help identify the enzymes and microbes that are responsible for primary bile acid transformation. Furthermore, bile acids are known to trigger germination of some microbial spores, indicating that there are likely other important bile acid binding proteins to be identified (Francis et al. 2013).

The opportunity to utilize ABPP for characterization and quantification bile acid regulation and signaling in the gut microbiota and host tissues is significant. The development of function-dependent approaches in this research realm has been hindered by the lack of technologies capable of identifying mechanistic relationships between bile salt hydrolase activity, the specific enzymes, and taxa within the microbiome that are directly responsible for metabolism, and the host pathways that respond to bile acids. ABPP will have a role in identifying the levels of functionally active bile salt hydrolases in the gut microbiome in healthy and diseased individuals, determining the specific microbes capable of transporting and metabolizing specific bile acids, and revealing interactions between specific bile acids and host nuclear receptors. ABPP can be used to dissect bile acid metabolism and signaling to yield an improved understanding of the mechanisms at play and the role of bile acids in metabolic disease and response to various exposures.

4.2 Environmental Microbial Communities

Like host-associated microbial communities, interest in environmental microbial communities (such as those found in soil or in marine environments) has increased recently, in large part due to the ability to use HTS as an approach to study communities in a culture-free manner. Also, like host-associated communities, environmental communities have seen limited application of ABPP. ABPP has been used to study axenic cultures of environmental microbes such as nitrifying bacteria, extremophilic archaea, and cellulolytic bacteria (Chauvigne-Hines et al. 2012; Bennett et al. 2016; Zweerink et al. 2017; Sadler et al. 2014; Ansong et al. 2014). Because ABPP does not require the ability to genetically manipulate organisms or the use of antibodies, it may prove especially useful in the study of microbial communities where those tools are often not available.

4.2.1 Metabolic Activity and Translation

Heterogeneity between microbial cells has is increasingly being understood to play a key role in community function and survival. This is especially relevant when considering structurally complex communities such as biofilms, where concentration gradients of nutrients and terminal electron acceptors impact cell physiology (Jansson and Hofmockel 2018; Deveau et al. 2018). Where HTS-based analysis of community composition cannot easily distinguish between active and inactive cells, ABPP can. The capacity to switch the probes' reporter groups from affinity reagents (e.g., biotin) to fluorophores enables imaging as well. This provides structural information about where the active cells and proteins are localized that would otherwise require genetic manipulation—often not possible in environmental isolates—or the use of immunoreagents such as antibodies, which are not often available commercially and must be generated and validated.

While not targeted at a specific class of enzymes such as proteases or glycoside hydrolases, bioorthogonal noncanonical amino acid targeting (BONCAT) has been employed to detect newly translated proteins and translationally active cells within complex microbial communities. BONCAT has been successfully utilized in the study of planktonic marine and slow-growing archaeal communities to identify actively translating microbes (Leizeaga et al. 2017; Hatzenpichler et al. 2016). Unlike axenic systems, however, BONCAT in microbial communities has not yet been extended to identification of the newly translated proteins themselves. In part, this is due to the complications mentioned above.

General measures of metabolic activity, such as active respiration, DNA/RNA synthesis, and membrane integrity have all been used to identify active members within microbial communities (Mou et al. 2008; Maurice et al. 2013). As with BONCAT, these tools have also been coupled to FACS to allow isolation of the active population as well. While these approaches may help to identify live/active cells, they are not designed to directly identify which organisms and enzymes are responsible for a given activity, which is often the more interesting question. Tools capable of labeling and isolating microbes and enzymes based on a more specific function, such as ABPs, will be needed to fill these knowledge gaps.

4.2.2 Vitamin Metabolism

Vitamins and other micronutrients can influence microbial communities in multiple ways primarily by acting as cofactors or precursors to cofactors. As these compounds are often energetically 'expensive' to produce, microbial communities often consist of both organisms capable of producing the vitamin themselves (auxotrophs) and those that must uptake the vitamin from the environment (heterotrophs). Thus, the exchange of vitamins between cells within an environment may play a key role in community dynamics (Konopka et al. 2015). The dynamic microbial interactions involving vitamins also means that they are critical to the formation and fitness of microbiomes in general (Degnan et al. 2014).

As vitamins are involved in a variety of biochemical reactions, a single reaction mechanism-based strategy would likely be insufficient to capture vitamin utilizing enzymes. Additionally, vitamins often affect biological change via non-covalent binding events and characterizing the transporters involved in their cellular incorporation is also important. In order to capture vitamin-interacting proteins, ABPP focused on vitamins has utilized photoreactive tags. By incorporating diazirine and alkyne-containing amino acids into the parent molecules, our group has applied B_1 (thiamine), B_2 (riboflavin), and B_7 (biotin)-based ABPs to the photosynthetic thermophile *Chloroflexus aurantiacus*, which can be found in microbial mats (Anderson et al. 2016). This work successfully identified a number of proteins including transporters, kinases, and fatty acid biosynthesis machinery. Importantly, we were also able to enrich a number of proteins that were undetectable using a global proteomics approach, demonstrating that ABPP can be used to enrich otherwise undetectable targets in microbial systems.

One B vitamin group, B_{12} (cobalamin), has attracted particular attention in regard to microbial communities. B_{12} is exclusively synthesized by bacteria and archaea, and its de novo synthesis is energetically and genetically costly, requiring nearly 30 enzymes (Martens et al. 2002; Roth et al. 1993). As such, most organisms will uptake B_{12} from the environment rather than produce it for themselves (auxotrophy). B_{12} availability has been postulated to influence microbial communities in the gut and marine ecosystems, but exact mechanisms remain poorly understood. Our group has also synthesized a B_{12} mimic that contains a diazirine for photocrosslinking to nearby proteins and an alkyne to enable CuAAC (Romine et al. 2017). Using this probe, we were able to identify proteins involved in folate, methionine, and ubiquinone metabolism as well as a light-sensing transcription factor in the environmental microbe *Halomonas* sp. HL-48 (Romine et al. 2017). We determined that vitamin B_{12} acts as a novel regulator of various microbial functions and intercellular interactions. Most recently, we have shown that our B_{12} probe mimics the natural vitamin almost perfectly, such that microbes can be cultured with the probe as the sole source of B_{12} (Rosnow et al. 2018). This allows for the direct identification of macromolecule-B_{12} interactions including proteins, DNA, and RNA throughout the growth phase of a microbe or microbiome. We anticipate that future studies will incorporate vitamin or other metabolite-based affinity probes to understand their roles and interactions in complex communities.

4.2.3 Community Signaling

The ability of individual cells to interact with each other via signaling systems is a key feature of microbial communities. One of the best examples of such signaling is quorum sensing (QS), in which a microbe is able to sense the presence and abundance of nearby organisms. While this phenomenon is best understood in proteobacteria, analogous systems in microbes form all domains of life (Atkinson and Williams 2009). Fundamentally, QS systems consist of two components: a small, diffusible signal and a response regulator that controls the physiological response. The diffusible signals are chemically diverse, ranging from short peptides to isoprenoids to *N*-acyl homoserine lactones. As QS systems play a critical role in community-level behaviors such as biofilm formation, the ability to identify organisms and enzymes involved in QS signal transduction will likely be key to understanding how microbial communities develop and function. Photoaffinity probes based on QS molecules have been designed and applied to axenic *Pseudomonas aeruginosa* and mammalian cells (Dubinsky et al. 2009; Garner et al. 2011; Baker et al. 2017). Application of such probes could be extended to microbial communities as well and may help to better establish the role QS systems play in interspecies interaction.

In addition to QS systems, certain metabolites such as peptidoglycan, vitamin metabolites, and indoles can act as signaling molecules. Peptidoglycan makes up the cell wall of most bacteria, and peptidoglycan synthesis is required for cell growth and division. Thus, the availability of free peptidoglycan may serve as an indicator of favorable growth conditions to nearby bacterial cells (Dworkin 2014). Indeed, metabolic activity, cell growth, and cell division are regulated by peptidoglycan-sensing serine/threonine kinases in taxa such as *Bacillus subtilis*, *Mycobacterium tuberculosis*, and *Streptococcus pneumoniae* (Shah et al. 2008; Mir et al. 2011; Beilharz et al. 2012). Peptidoglycan metabolites can also influence eukaryotic microbes as well. In *Candida albicans*, a gut-associated fungal taxa, N-acetylglucosamine signaling was shown to influence physiological changes necessary for mating (Huang et al. 2010). Peptidoglycan is also an important pathogen-/microbial- associated molecular pattern (PAMP or MAMP respectively) sensed by the mammalian immune system and plays a key role in the induction and management of the inflammatory response (Dworkin 2014). This is also true of folate, tryptophan, and retinoid metabolites, which are known to influence immune development and activity at mucosal sites (van de Pavert et al. 2014; Kjer-Nielsen et al. 2012; Schiering et al. 2017).

Probes based on these metabolites would be unique and powerful tools to define these interactions in actual community systems. However, the design of such probes is challenging. The binding modes between these signals and their receptors are non-covalent, necessitating the use of a photoaffinity tag. Depending on the size of the metabolite, this tag may introduce a relatively small change in size (such as vitamin B_{12}) or a large change (such as indole derivatives). The placement of such a tag is critical to whether or not the probe will exhibit the same binding behavior as the native metabolite. If available, protein crystal structures can be helpful to identify potential modification sites that may have limited impact on binding. Given the limited number of protein structures for many environmental microbes, structure prediction tools such as Phyre2 are particularly useful in this area (Kelley et al. 2015). For smaller metabolites probe design can be especially challenging, as most of the synthetically accessible sites (amines, carboxylates, alcohols, etc.) are often essential for protein binding. As such, it may be necessary to employ a subtractive approach where a general reactive group is competed with the native substrate in order to successfully profiling small metabolite–protein interactions.

5 Current Challenges and Outlook

The application of ABPP to study microbial communities remains promising, though critical challenges in sample preparation, probe design, and database availability will need to be addressed (Fig. 2). The presence of a complex, often undefined sample matrix complicates peptide analysis, click chemistry, and potentially even probe labeling. All ABPP strategies fundamentally rely on a chemical 'warhead,' a group that is prone to making covalent bonds with an enzyme when that enzyme participates in a reaction. The presence of both biotic and abiotic molecules such as carbohydrates, salts, and surfactants can all interfere with this reaction by quenching the reactive group or inhibiting the enzyme. While this can be somewhat controlled for in laboratory culture, 'real-world' samples consist of such chemically diverse substituents that removing them all from the cells is often impossible. Strategies that incorporate a non-denaturing purification step, such as size-exclusion chromatography prior to assaying for activity, may be useful to clean up samples before labeling, at the cost of destroying the native cellular environment (Beller et al. 2018). Alternatively, strategies such as an enzymatic digest to break up a defined matrix can be employed. Post-probe labeling, SDS-PAGE or 2D-DIGE to cut out fluorescence protein bands can also be effective in some sample types.

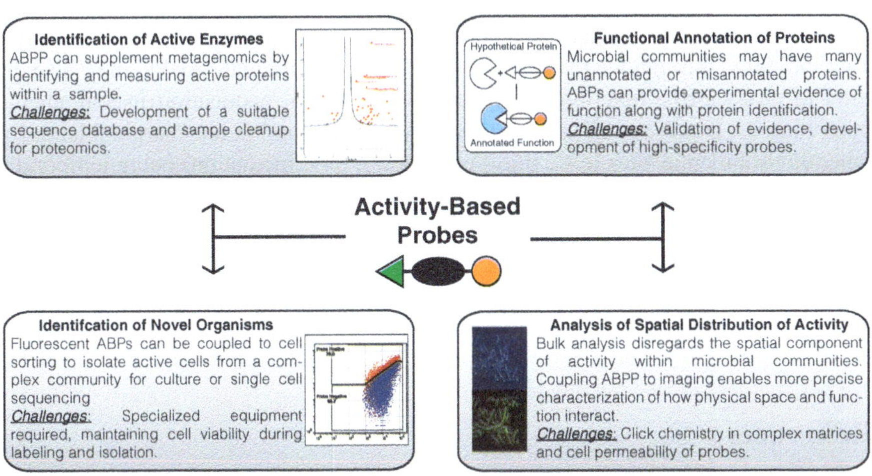

Fig. 2 Potential applications and challenges for ABPs in microbial communities. Once designed, probes can be used coupled to an enrichable moiety or surface. The probes can then be used for ABPP to identify and quantify active enzymes (top left) or used to functionally annotate proteins of unknown function (top right). Alternatively, the probe can be coupled to a fluorophore and then applied to the sample. This allows the use of FACS to isolate microbes possessing that function for cultivation or sequencing (bottom left) as well as imaging to study the spatial distribution of activity (bottom right)

Probe development for microbial communities also remains a challenge. Most activity-based probes take inspiration from known irreversible inhibitors of an enzyme or enzyme class. Historically, the development and characterization of these compounds have mainly been to study or act as a therapeutic for a mammalian system. The same enzyme class or metabolic pathway may not be present in the phylogenetically distant fungi and prokaryotes. While there are some irreversible inhibitors of microbial systems (penicillin, artemisinin, difluoromethylornithine, etc.), they are mostly antimicrobial and cannot be used studies where the organisms need to be viable. Alternatively, probes can be designed with some knowledge of the reaction mechanism. For enzymes where homologs are present in commonly studied microbes, such as *E. coli* or *C. albicans*, mechanisms and structures may already be described. Given that such probes often need to be optimized and validated, assumptions based on homology may not be optimal for preliminary studies. In the near term, it will likely be more beneficial to utilize either a subtractive approach or photoaffinity probes based on the substrate of interest.

Despite these challenges, activity-based approaches are powerful tools for the study of complex microbial communities. At present, the functional capacity of microbiomes is largely inferred from genomic and transcriptomic data. Determining the contribution of individual microbial species and enzymes for a particular function is largely outside the capability of existing technologies. ABPP has a real potential to make specific measurements to report upon the functional 'health' of a microbiome. Within medicine, ABPP provides a platform to evaluate individual variability and susceptibility to diseases, to understand consequences associated with exposures in adults and developing children, and track microbe functions longitudinally throughout the development of an individual. ABPP may also be used to characterize functions associated with nutrient acquisition in soil/aquatic microbiomes. If successful, one could imagine agricultural fields, marginal lands, or other environmental sites (e.g., those undergoing bioremediation) being temporally tested and the data used for 'prescriptive' amendments to alter plant productivity, test for other features of soil health, or to improve ecosystem health. Such insights offered by ABPP constitute a major advancement in our understanding of complex microbial communities and will lay a foundation for new strategies to improve agricultural productivity, bioremediation efforts, and human health.

References

Agard NJ, Baskin JM, Prescher JA, Lo A, Bertozzi CR (2006) A comparative study of bioorthogonal reactions with azides. ACS Chem Biol 1(10):644–648

Anderson LN, Koech PK, Plymale AE, Landorf EV, Konopka A, Collart FR, Lipton MS, Romine MF, Wright AT (2016) Live cell discovery of microbial vitamin transport and enzyme-cofactor interactions. ACS Chem Biol 11(2):345–354. https://doi.org/10.1021/acschembio.5b00918

Ansong C, Sadler NC, Hill EA, Lewis MP, Zink EM, Smith RD, Beliaev AS, Konopka AE, Wright AT (2014) Characterization of protein redox dynamics induced during light-to-dark transitions and nutrient limitation in cyanobacteria. Front Microbiol 5:325. https://doi.org/10.3389/fmicb.2014.00325

Arnold JW, Roach J, Azcarate-Peril MA (2016) Emerging technologies for gut microbiome research. Trends Microbiol 24(11):887–901. https://doi.org/10.1016/j.tim.2016.06.008

Atkinson S, Williams P (2009) Quorum sensing and social networking in the microbial world. J R Soc Interface 6(40):959–978. https://doi.org/10.1098/rsif.2009.0203

Baker YR, Hodgkinson JT, Florea BI, Alza E, Galloway W, Grimm L, Geddis SM, Overkleeft HS, Welch M, Spring DR (2017) Identification of new quorum sensing autoinducer binding partners in *Pseudomonas aeruginosa* using photoaffinity probes. Chem Sci 8(11):7403–7411. https://doi.org/10.1039/c7sc01270e

Beilharz K, Novakova L, Fadda D, Branny P, Massidda O, Veening JW (2012) Control of cell division in *Streptococcus pneumoniae* by the conserved Ser/Thr protein kinase StkP. Proc Natl Acad Sci U S A 109(15):E905–E913. https://doi.org/10.1073/pnas.1119172109

Beller HR, Rodrigues AV, Zargar K, Wu YW, Saini AK, Saville RM, Pereira JH, Adams PD, Tringe SG, Petzold CJ, Keasling JD (2018) Discovery of enzymes for toluene synthesis from anoxic microbial communities. Nat Chem Biol. https://doi.org/10.1038/s41589-018-0017-4

Bennett K, Sadler NC, Wright AT, Yeager C, Hyman MR (2016) Activity-based protein profiling of ammonia monooxygenase in *Nitrosomonas europaea*. Appl Environ Microbiol 82(8):2270–2279. https://doi.org/10.1128/AEM.03556-15

Biteen JS, Blainey PC, Cardon ZG, Chun M, Church GM, Dorrestein PC, Fraser SE, Gilbert JA, Jansson JK, Knight R, Miller JF, Ozcan A, Prather KA, Quake SR, Ruby EG, Silver PA, Taha S, van den Engh G, Weiss PS, Wong GC, Wright AT, Young TD (2016) Tools for the microbiome: nano and beyond. ACS Nano 10(1):6–37. https://doi.org/10.1021/acsnano.5b07826

Chatterjee S, Stupp GS, Park SK, Ducom JC, Yates JR 3rd, Su AI, Wolan DW (2016) A comprehensive and scalable database search system for metaproteomics. BMC Genom 17(1):642. https://doi.org/10.1186/s12864-016-2855-3

Chauvigne-Hines LM, Anderson LN, Weaver HM, Brown JN, Koech PK, Nicora CD, Hofstad BA, Smith RD, Wilkins MJ, Callister SJ, Wright AT (2012) Suite of activity-based probes for cellulose-degrading enzymes. J Am Chem Soc 134(50):20521–20532. https://doi.org/10.1021/ja309790w

Cheng K, Ning Z, Zhang X, Li L, Liao B, Mayne J, Stintzi A, Figeys D (2017) MetaLab: an automated pipeline for metaproteomic data analysis. Microbiome 5(1):157. https://doi.org/10.1186/s40168-017-0375-2

Cravatt BF, Wright AT, Kozarich JW (2008) Activity-based protein profiling: from enzyme chemistry to proteomic chemistry. Annu Rev Biochem 77:383–414. https://doi.org/10.1146/annurev.biochem.75.101304.124125

Degnan PH, Taga ME, Goodman AL (2014) Vitamin B12 as a modulator of gut microbial ecology. Cell Metab 20(5):769–778. https://doi.org/10.1016/j.cmet.2014.10.002

Deveau A, Bonito G, Uehling J, Paoletti M, Becker M, Bindschedler S, Hacquard S, Herve V, Labbe J, Lastovetsky OA, Mieszkin S, Millet LJ, Vajna B, Junier P, Bonfante P, Krom BP, Olsson S, van Elsas JD, Wick LY (2018) Bacterial-fungal interactions: ecology, mechanisms and challenges. FEMS Microbiol Rev 42(3):335–352. https://doi.org/10.1093/femsre/fuy008

Dubinsky L, Jarosz LM, Amara N, Krief P, Kravchenko VV, Krom BP, Meijler MM (2009) Synthesis and validation of a probe to identify quorum sensing receptors. Chem Commun (Camb) 47:7378–7380. https://doi.org/10.1039/b917507e

Dubinsky L, Krom BP, Meijler MM (2012) Diazirine based photoaffinity labeling. Bioorg Med Chem 20(2):554–570. https://doi.org/10.1016/j.bmc.2011.06.066

Dworkin J (2014) The medium is the message: interspecies and interkingdom signaling by peptidoglycan and related bacterial glycans. Annu Rev Microbiol 68:137–154. https://doi.org/10.1146/annurev-micro-091213-112844

Fierer N (2017) Embracing the unknown: disentangling the complexities of the soil microbiome. Nat Rev Microbiol 15(10):579–590. https://doi.org/10.1038/nrmicro.2017.87
Francis MB, Allen CA, Shrestha R, Sorg JA (2013) Bile acid recognition by the Clostridium difficile germinant receptor, CspC, is important for establishing infection. PLoS Pathog 9(5): e1003356. https://doi.org/10.1371/journal.ppat.1003356
Garner AL, Yu J, Struss AK, Lowery CA, Zhu J, Kim SK, Park J, Mayorov AV, Kaufmann GF, Kravchenko VV, Janda KD (2011) Synthesis of 'clickable' acylhomoserine lactone quorum sensing probes: unanticipated effects on mammalian cell activation. Bioorg Med Chem Lett 21 (9):2702–2705. https://doi.org/10.1016/j.bmcl.2010.11.122
Gilbert JA, Blaser MJ, Caporaso JG, Jansson JK, Lynch SV, Knight R (2018) Current understanding of the human microbiome. Nat Med 24(4):392–400. https://doi.org/10.1038/nm.4517
Gilbert JA, Jansson JK, Knight R (2014) The Earth Microbiome project: successes and aspirations. BMC Biol 12:69. https://doi.org/10.1186/s12915-014-0069-1
Gilbert JA, Quinn RA, Debelius J, Xu ZZ, Morton J, Garg N, Jansson JK, Dorrestein PC, Knight R (2016) Microbiome-wide association studies link dynamic microbial consortia to disease. Nature 535(7610):94–103. https://doi.org/10.1038/nature18850
Haiser HJ, Gootenberg DB, Chatman K, Sirasani G, Balskus EP, Turnbaugh PJ (2013) Predicting and manipulating cardiac drug inactivation by the human gut bacterium Eggerthella lenta. Science 341(6143):295–298. https://doi.org/10.1126/science.1235872
Haiser HJ, Seim KL, Balskus EP, Turnbaugh PJ (2014) Mechanistic insight into digoxin inactivation by Eggerthella lenta augments our understanding of its pharmacokinetics. Gut Microbes 5(2):233–238. https://doi.org/10.4161/gmic.27915
Hatzenpichler R, Connon SA, Goudeau D, Malmstrom RR, Woyke T, Orphan VJ (2016) Visualizing in situ translational activity for identifying and sorting slow-growing archaeal-bacterial consortia. Proc Natl Acad Sci U S A 113(28):E4069–4078. https://doi.org/10.1073/pnas.1603757113
Hettich RL, Sharma R, Chourey K, Giannone RJ (2012) Microbial metaproteomics: identifying the repertoire of proteins that microorganisms use to compete and cooperate in complex environmental communities. Curr Opin Microbiol 15(3):373–380. https://doi.org/10.1016/j.mib.2012.04.008
Heyer R, Schallert K, Zoun R, Becher B, Saake G, Benndorf D (2017) Challenges and perspectives of metaproteomic data analysis. J Biotechnol 261:24–36. https://doi.org/10.1016/j.jbiotec.2017.06.1201
Huang G, Yi S, Sahni N, Daniels KJ, Srikantha T, Soll DR (2010) N-acetylglucosamine induces white to opaque switching, a mating prerequisite in Candida albicans. PLoS Pathog 6(3): e1000806. https://doi.org/10.1371/journal.ppat.1000806
Jansson JK, Baker ES (2016) A multi-omic future for microbiome studies. Nat Microbiol 1:16049. https://doi.org/10.1038/nmicrobiol.2016.49
Jansson JK, Hofmockel KS (2018) The soil microbiome-from metagenomics to metaphenomics. Curr Opin Microbiol 43:162–168. https://doi.org/10.1016/j.mib.2018.01.013
Joyce SA, Gahan CG (2016) Bile acid modifications at the microbe-host interface: potential for nutraceutical and pharmaceutical interventions in host health. Annu Rev Food Sci Technol 7:313–333. https://doi.org/10.1146/annurev-food-041715-033159
Kelley LA, Mezulis S, Yates CM, Wass MN, Sternberg MJ (2015) The Phyre2 web portal for protein modeling, prediction and analysis. Nat Protoc 10(6):845–858. https://doi.org/10.1038/nprot.2015.053
Kim D, Hofstaedter CE, Zhao C, Mattei L, Tanes C, Clarke E, Lauder A, Sherrill-Mix S, Chehoud C, Kelsen J, Conrad M, Collman RG, Baldassano R, Bushman FD, Bittinger K (2017) Optimizing methods and dodging pitfalls in microbiome research. Microbiome 5(1):52. https://doi.org/10.1186/s40168-017-0267-5
Kjer-Nielsen L, Patel O, Corbett AJ, Le Nours J, Meehan B, Liu L, Bhati M, Chen Z, Kostenko L, Reantragoon R, Williamson NA, Purcell AW, Dudek NL, McConville MJ, O'Hair RA, Khairallah GN, Godfrey DI, Fairlie DP, Rossjohn J, McCluskey J (2012) MR1 presents

microbial vitamin B metabolites to MAIT cells. Nature 491(7426):717–723. https://doi.org/10.1038/nature11605

Kolb HC, Finn MG, Sharpless KB (2001) Click chemistry: diverse chemical function from a few good reactions. Angew Chem Int Ed Engl 40(11):2004–2021

Konopka A, Lindemann S, Fredrickson J (2015) Dynamics in microbial communities: unraveling mechanisms to identify principles. ISME J 9(7):1488–1495. https://doi.org/10.1038/ismej.2014.251

Koppel N, Balskus EP (2016) Exploring and understanding the biochemical diversity of the human microbiota. Cell Chem Biol 23(1):18–30. https://doi.org/10.1016/j.chembiol.2015.12.008

Koppel N, Maini Rekdal V, Balskus EP (2017) Chemical transformation of xenobiotics by the human gut microbiota. Science 356(6344). https://doi.org/10.1126/science.aag2770

Krysiak JM, Kreuzer J, Macheroux P, Hermetter A, Sieber SA, Breinbauer R (2012) Activity-based probes for studying the activity of flavin-dependent oxidases and for the protein target profiling of monoamine oxidase inhibitors. Angew Chem Int Ed Engl 51 (28):7035–7040. https://doi.org/10.1002/anie.201201955

Lee PY, Chin SF, Neoh HM, Jamal R (2017) Metaproteomic analysis of human gut microbiota: where are we heading? J Biomed Sci 24(1):36. https://doi.org/10.1186/s12929-017-0342-z

Leizeaga A, Estrany M, Forn I, Sebastian M (2017) Using click-chemistry for visualizing in situ changes of translational activity in planktonic marine bacteria. Front Microbiol 8:2360. https://doi.org/10.3389/fmicb.2017.02360

Li T, Chiang JY (2014) Bile acid signaling in metabolic disease and drug therapy. Pharmacol Rev 66(4):948–983. https://doi.org/10.1124/pr.113.008201

Martens JH, Barg H, Warren MJ, Jahn D (2002) Microbial production of vitamin B12. Appl Microbiol Biotechnol 58(3):275–285. https://doi.org/10.1007/s00253-001-0902-7

Maurice CF, Haiser HJ, Turnbaugh PJ (2013) Xenobiotics shape the physiology and gene expression of the active human gut microbiome. Cell 152(1–2):39–50. https://doi.org/10.1016/j.cell.2012.10.052

Mayer EA, Tillisch K, Gupta A (2015) Gut/brain axis and the microbiota. J Clin Invest 125 (3):926–938. https://doi.org/10.1172/JCI76304

Mayers MD, Moon C, Stupp GS, Su AI, Wolan DW (2017) Quantitative metaproteomics and activity-based probe enrichment reveals significant alterations in protein expression from a mouse model of inflammatory bowel disease. J Proteome Res 16(2):1014–1026. https://doi.org/10.1021/acs.jproteome.6b00938

Mir M, Asong J, Li X, Cardot J, Boons GJ, Husson RN (2011) The extracytoplasmic domain of the *Mycobacterium tuberculosis* Ser/Thr kinase PknB binds specific muropeptides and is required for PknB localization. PLoS Pathog 7(7):e1002182. https://doi.org/10.1371/journal.ppat.1002182

Moon C, Stupp GS, Su AI, Wolan DW (2018) Metaproteomics of colonic microbiota unveils discrete protein functions among colitic mice and control groups. Proteomics 18(3–4). https://doi.org/10.1002/pmic.201700391

Mou X, Sun S, Edwards RA, Hodson RE, Moran MA (2008) Bacterial carbon processing by generalist species in the coastal ocean. Nature 451(7179):708–711. https://doi.org/10.1038/nature06513

Nair RN, Rosnow JJ, Murphree TA, Bowden ME, Lindemann SR, Wright AT (2017) De novo synthesis of alkyne substituted tryptophans as chemical probes for protein profiling studies. Org Chem Front 4(4):495–499. https://doi.org/10.1039/C6QO00819D

Ortega C, Anderson LN, Frando A, Sadler NC, Brown RW, Smith RD, Wright AT, Grundner C (2016) Systematic survey of serine hydrolase activity in mycobacterium tuberculosis defines changes associated with persistence. Cell Chem Biol 23(2):290–298. https://doi.org/10.1016/j.chembiol.2016.01.003

van de Pavert SA, Ferreira M, Domingues RG, Ribeiro H, Molenaar R, Moreira-Santos L, Almeida FF, Ibiza S, Barbosa I, Goverse G, Labao-Almeida C, Godinho-Silva C, Konijn T, Schooneman D, O'Toole T, Mizee MR, Habani Y, Haak E, Santori FR, Littman DR, Schulte-Merker S, Dzierzak E, Simas JP, Mebius RE, Veiga-Fernandes H (2014) Maternal retinoids control type 3 innate lymphoid cells and set the offspring immunity. Nature 508 (7494):123–127. https://doi.org/10.1038/nature13158

Pollock J, Glendinning L, Wisedchanwet T, Watson M (2018) The madness of microbiome: attempting to find consensus "best practice" for 16S microbiome studies. Appl Environ Microbiol 84(7). https://doi.org/10.1128/aem.02627-17

Romine MF, Rodionov DA, Maezato Y, Anderson LN, Nandhikonda P, Rodionova IA, Carre A, Li X, Xu C, Clauss TR, Kim YM, Metz TO, Wright AT (2017) Elucidation of roles for vitamin B12 in regulation of folate, ubiquinone, and methionine metabolism. Proc Natl Acad Sci U S A 114(7):E1205–E1214. https://doi.org/10.1073/pnas.1612360114

Rooks MG, Garrett WS (2016) Gut microbiota, metabolites and host immunity. Nat Rev Immunol 16(6):341–352. https://doi.org/10.1038/nri.2016.42

Rosnow JJ, Hwang S, Killinger BJ, Kim YM, Moore RJ, Lindemann SR, Maupin-Furlow JA, Wright AT (2018) Cobalamin activity-based probe enables microbial cell growth and finds new cobalamin-protein interactions across domains. Appl Environ Microbiol. https://doi.org/10.1128/aem.00955-18

Roth JR, Lawrence JG, Rubenfield M, Kieffer-Higgins S, Church GM (1993) Characterization of the cobalamin (vitamin B12) biosynthetic genes of Salmonella typhimurium. J Bacteriol 175 (11):3303–3316

Sadaghiani AM, Verhelst SH, Bogyo M (2007) Tagging and detection strategies for activity-based proteomics. Curr Opin Chem Biol 11(1):20–28. https://doi.org/10.1016/j.cbpa.2006.11.030

Sadler NC, Melnicki MR, Serres MH, Merkley ED, Chrisler WB, Hill EA, Romine MF, Kim S, Zink EM, Datta S, Smith RD, Beliaev AS, Konopka A, Wright AT (2014) Live cell chemical profiling of temporal redox dynamics in a photoautotrophic cyanobacterium. ACS Chem Biol 9 (1):291–300. https://doi.org/10.1021/cb400769v

Sadler NC, Wright AT (2015) Activity-based protein profiling of microbes. Curr Opin Chem Biol 24:139–144. https://doi.org/10.1016/j.cbpa.2014.10.022

Savage DC (2001) Microbial biota of the human intestine: a tribute to some pioneering scientists. Curr Issues Intest Microbiol 2(1):1–15

Scheline RR (1968) The metabolism of drugs and other organic compounds by the intestinal microflora. Acta Pharmacol Toxicol (Copenh) 26(4):332–342

Schiering C, Wincent E, Metidji A, Iseppon A, Li Y, Potocnik AJ, Omenetti S, Henderson CJ, Wolf CR, Nebert DW, Stockinger B (2017) Feedback control of AHR signalling regulates intestinal immunity. Nature 542(7640):242–245. https://doi.org/10.1038/nature21080

Shah IM, Laaberki MH, Popham DL, Dworkin J (2008) A eukaryotic-like Ser/Thr kinase signals bacteria to exit dormancy in response to peptidoglycan fragments. Cell 135(3):486–496. https://doi.org/10.1016/j.cell.2008.08.039

Snelling TJ, Wallace RJ (2017) The rumen microbial metaproteome as revealed by SDS-PAGE. BMC Microbiol 17(1):9. https://doi.org/10.1186/s12866-016-0917-y

Sousa T, Paterson R, Moore V, Carlsson A, Abrahamsson B, Basit AW (2008) The gastrointestinal microbiota as a site for the biotransformation of drugs. Int J Pharm 363(1–2):1–25. https://doi.org/10.1016/j.ijpharm.2008.07.009

Spanogiannopoulos P, Bess EN, Carmody RN, Turnbaugh PJ (2016) The microbial pharmacists within us: a metagenomic view of xenobiotic metabolism. Nat Rev Microbiol 14(5):273–287. https://doi.org/10.1038/nrmicro.2016.17

Speers AE, Adam GC, Cravatt BF (2003) Activity-based protein profiling in vivo using a copper (i)-catalyzed azide-alkyne [3 + 2] cycloaddition. J Am Chem Soc 125(16):4686–4687. https://doi.org/10.1021/ja034490h

Wallace BD, Roberts AB, Pollet RM, Ingle JD, Biernat KA, Pellock SJ, Venkatesh MK, Guthrie L, O'Neal SK, Robinson SJ, Dollinger M, Figueroa E, McShane SR, Cohen RD, Jin J, Frye SV, Zamboni WC, Pepe-Ranney C, Mani S, Kelly L, Redinbo MR (2015) Structure and inhibition of microbiome beta-glucuronidases essential to the alleviation of cancer drug toxicity. Chem Biol 22(9):1238–1249. https://doi.org/10.1016/j.chembiol.2015.08.005

Wallace BD, Wang H, Lane KT, Scott JE, Orans J, Koo JS, Venkatesh M, Jobin C, Yeh LA, Mani S, Redinbo MR (2010) Alleviating cancer drug toxicity by inhibiting a bacterial enzyme. Science 330(6005):831–835. https://doi.org/10.1126/science.1191175

Weiss S, Amir A, Hyde ER, Metcalf JL, Song SJ, Knight R (2014) Tracking down the sources of experimental contamination in microbiome studies. Genome Biol 15(12):564. https://doi.org/10.1186/s13059-014-0564-2

Wright AT, Song JD, Cravatt BF (2009) A suite of activity-based probes for human cytochrome P450 enzymes. J Am Chem Soc 131(30):10692–10700. https://doi.org/10.1021/ja9037609

Wu L, Jiang J, Jin Y, Kallemeijn WW, Kuo CL, Artola M, Dai W, van Elk C, van Eijk M, van der Marel GA, Codee JDC, Florea BI, Aerts J, Overkleeft HS, Davies GJ (2017) Activity-based probes for functional interrogation of retaining beta-glucuronidases. Nat Chem Biol 13(8):867–873. https://doi.org/10.1038/nchembio.2395

Xiao M, Yang J, Feng Y, Zhu Y, Chai X, Wang Y (2017) Metaproteomic strategies and applications for gut microbial research. Appl Microbiol Biotechnol 101(8):3077–3088. https://doi.org/10.1007/s00253-017-8215-7

Zarraonaindia I, Smith DP, Gilbert JA (2013) Beyond the genome: community-level analysis of the microbial world. Biol Philos 28(2):261–282. https://doi.org/10.1007/s10539-012-9357-8

Zhang X, Ning Z, Mayne J, Moore JI, Li J, Butcher J, Deeke SA, Chen R, Chiang CK, Wen M, Mack D, Stintzi A, Figeys D (2016) MetaPro-IQ: a universal metaproteomic approach to studying human and mouse gut microbiota. Microbiome 4(1):31. https://doi.org/10.1186/s40168-016-0176-z

Zhuang S, Li Q, Cai L, Wang C, Lei X (2017) Chemoproteomic profiling of bile acid interacting proteins. ACS Cent Sci 3(5):501–509. https://doi.org/10.1021/acscentsci.7b00134

Zweerink S, Kallnik V, Ninck S, Nickel S, Verheyen J, Blum M, Wagner A, Feldmann I, Sickmann A, Albers SV, Brasen C, Kaschani F, Siebers B, Kaiser M (2017) Activity-based protein profiling as a robust method for enzyme identification and screening in extremophilic Archaea. Nat Commun 8:15352. https://doi.org/10.1038/ncomms15352

Activity-Based Protein Profiling Methods to Study Bacteria: The Power of Small-Molecule Electrophiles

Shabnam Sharifzadeh, Joshua D. Shirley and Erin E. Carlson

Contents

S. Sharifzadeh · E. E. Carlson (✉)
Department of Chemistry, University of Minnesota-Twin Cities,
Minneapolis, MN 55455, USA
e-mail: carlsone@umn.edu

J. D. Shirley · E. E. Carlson
Department of Medicinal Chemistry, University of Minnesota-Twin Cities,
Minneapolis, MN 55455, USA

E. E. Carlson
Department of Biochemistry, Molecular Biology and Biophysics,
University of Minnesota-Twin Cities, Minneapolis, MN 55455, USA

Current Topics in Microbiology and Immunology (2019) 420:23–48
DOI 10.1007/82_2018_135

Published Online: 20 September 2018

Abstract ABPP methods have been utilized for the last two decades as a means to investigate complex proteomes in all three domains of life. Extensive use in eukaryotes has provided a more fundamental understanding of the biological processes involved in numerous diseases and has driven drug discovery and treatment campaigns. However, the use of ABPP in prokaryotes has been less common, although it has gained more attention over the last decade. The urgent need for understanding bacteriophysiology and bacterial pathogenicity at a foundational level has never been more apparent, as the rise in antibiotic resistance has resulted in the inadequate and ineffective treatment of infections. This is not only a result of resistance to clinically used antibiotics, but also a lack of new drugs and equally as important, new drug targets. ABPP provides a means for which new, clinically relevant drug targets may be identified through gaining insight into biological processes. In this chapter, we place particular focus on the discussion of ABPP strategies that have been applied to study different classes of bacterial enzymes.

1 Introduction

Advancement of genomic and proteomic methods has enabled rapid characterization and functional assignment of poorly annotated genes and proteins. While proteomics has tremendously accelerated such assignments through the development of analytical methods for large mixtures of proteins, cellular proteins are constantly generated, modified, and degraded. Many enzymes are produced as inactive zymogens requiring specific post-translational modifications (PTMs) for activation. Therefore, comparative profiling techniques which enable assessment of these dynamic changes in proteome abundance, modification, intracellular location, and functional state are employed (Vogel and Marcotte 2012). Activity-based protein profiling (ABPP) has emerged as a powerful complimentary approach to monitor the activity of proteins in complex proteomes. Taking advantage of active-site-directed chemical probes, called activity-based probes (ABPs), this strategy enables visualization of the active form of enzymes. ABPs are mechanism-based irreversible inhibitors, consisting of an *active warhead*, which reacts with components of the active site, a binding or *spacer group* that provides spacing between the active warhead and the *reporter*, while also modulating the reactivity and/or specificity of the probe and a reporter tag for identification and purification purposes (Fig. 1a). The active warhead is typically an electrophile, which reacts with an active-site nucleophile and is retained on-site due to the formation of a covalent bond. Since only catalytically competent proteins can undergo this reaction, active isoforms are visualized in this way. The permanent covalent nature of the ABPs' bond with their targets has made them amenable for diverse applications including live-cell imaging, target enrichment, and direct biochemical analysis of the targeted proteins (Fig. 1b). Moreover, assessment of the selectivity of enzyme inhibitors is another potential application for activity-based proteomics. In bacterial systems, ABPs have been applied for various applications

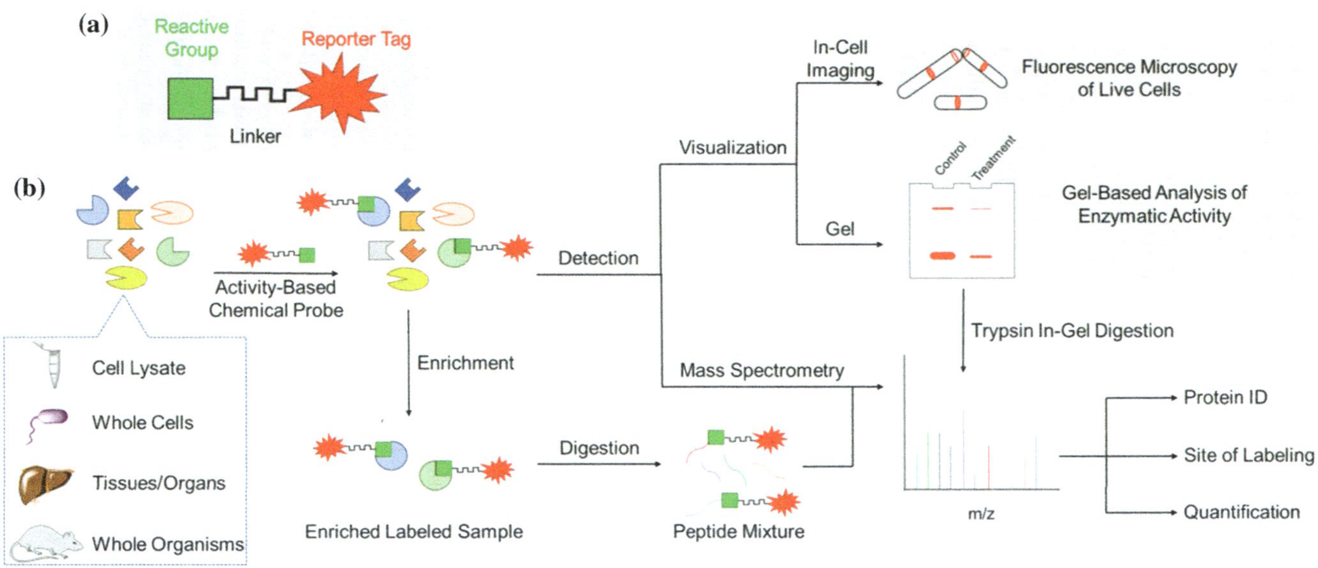

Fig. 1 **a** General structure of an ABP. Reactive warhead, usually an electrophilic group, is linked to a reporter tag such as a fluorophore or biotin via an inert linker. **b** General workflow of ABPP. A sample is treated with an ABP, followed by the detection of the interaction between enzyme and ABP. Additional steps may be carried out in between, including washes or enrichment. Labeled proteome can be detected through either visualization or mass spectrometry methods

such as target discovery, the study of microbial pathogenesis and host-pathogen interactions, live-cell imaging, and inhibitor discovery (Sadler and Wright 2015; Bottcher and Sieber 2008b; Heal and Tate 2012; Willems et al. 2014; Sharifzadeh et al. 2017).

1.1 Activity-Based Probes: What They Are and How They Work

Since ABPP has been applied to a broad range of probes with various modes of action, one must be aware of the technical differences. By definition, ABPs represent any active-site-directed chemical probe that reacts with the enzyme in a mechanism-based manner, covalently attaches to a catalytic residue and is retained in the active site. Other groups of covalent protein modifiers, which could potentially be mistaken for ABPs, include substrate-based probes and affinity probes. Substrate-based probes are fluorogenic molecules that depend on the reactivity of the enzyme to become fluorescent. This group of probes consists of a recognition element that binds within the active site, usually designed based on the enzyme's natural substrate, and a reporter group that generates a signal once processed by the enzyme. The enzymatic cleavage leads to either increase or quenching of fluorescence (Edgington et al. 2011).

Proteases are perhaps the best example for which both classes of probes have been established and applied. Protease enzymes constitute an extensive family of enzymes in both prokaryotes and eukaryotes, comprising almost 2% of the human genome. Their primary function to cleave peptide bonds is critical for maintaining normal physiology of cells and any dysregulation of their expression levels and activity could lead to serious pathologies such as inflammatory diseases and cancer. Accordingly, direct assessment of the proteolytic activity and regulation under normal and pathologic conditions is critical. Based on their mechanism of action, proteases are classified into seven subfamilies, three of which, cysteine, serine, and threonine proteases, use nucleophilic amino acids to perform catalysis. Protease enzymes are often highly selective and only perform limited cleavage of specific sites. The primary amino acid sequence surrounding the scissile amide bond is one of the important factors that determines the selectivity of proteases. Accordingly, substrate- and activity-based probes have been devised based on the structural preferences of individual enzymes. In the former group of probes, the proteolytic function of the enzyme leads to a shift in fluorescent properties of the probe. In the latter group, as mentioned earlier, covalent modification of the enzyme occurs in an activity-dependent manner. While substrate-based probes are processed by the enzyme, ABPs act as an irreversible covalent inhibitor of the enzyme. Moreover, ABPs can be modified to carry different groups such as fluorophores, biotin, and clickable functionalities. Due to irreversible modification of targets by ABPs, labeled proteins can be identified, enriched, and quantified if needed. The activity of proteases has been visualized by fluorescent ABPs, as a non-invasive imaging technique under various pathologic conditions such as inflammatory diseases and cancer (Withana et al. 2016).

Finally, affinity-based probes are compounds that are recognized by enzymes, but do not rely explicitly on enzymatic activity. Photoactivatable groups such as benzophenone or diazirine are often installed on affinity probes in order to covalently tag their targets. Similarly, chemically reactive groups such as alkyl halides and succinimidyl esters can be incorporated to enable covalent tagging of nucleophilic residues in the proximity of the binding site (Weerapana et al. 2008). Although this latter strategy has been exploited in the same applications as ABPs, such as target discovery and binding site identification, it must be noted that this class of probes often functions regardless of the activity state of their target (Eirich et al. 2011; Wright and Sieber 2016).

2 Probe Design Strategies

2.1 Reactive Warhead and Recognition Elements

Ideal warheads must possess proper reactivity with proteins as well as biocompatibility. Several small-molecule electrophilic probes have been developed that show class-specific reactivity toward nucleophilic active-site residues of certain enzyme classes. Such electrophiles are designed to exhibit mechanism-based reactivity toward enzyme family members, while remaining inert to physiologically abundant nucleophiles such as hydroxyls, thiols, and amines. Therefore, these small-molecule probes could be exploited to irreversibly label enzymes within their native environment. Available design strategies take advantage of the unique intrinsic reactivity of each enzyme class, as well as the structural features required for substrate recognition. Selectivity of reactive residues toward specific electrophiles could be explained by the Hard and Soft, Acids and Bases (HSAB) theory (Pearson 1990). According to HSAB theory, reacting species are classified into "hard" or "soft" based on the polarizability, i.e., the inherent property of an atom or functional group that describes how easily the electron density can be delocalized to form covalent bonds, and electrophiles react with nucleophiles of similar hardness or softness. The sulfhydryl group of cysteine residues is highly polarizable and "soft," due to the large atomic radius of the sulfur atom. In contrast, the hydroxyl group of serine residue is considered to be "hard." Therefore, the serine hydrolase family of enzymes is targeted by hard electrophiles such as fluorophosphonates, while the cysteine hydrolase family is targeted by softer electrophiles such as α,β-unsaturated ketones (Weerapana et al. 2008). However, some electrophiles such as sulfonyl fluoride show promiscuity toward diverse nucleophiles, demonstrating the complexities associated with probe design (Narayanan and Jones 2015).

The active warhead of several successful activity-based probes has been designed based on known covalent, mechanism-based inhibitors of that class of enzymes. In this regard, serine and cysteine hydrolases were the first enzyme classes that were successfully targeted upon the introduction of the ABPP strategy. In the absence of such cognate inhibitors, de novo probes have been designed that were inspired by structural features found in natural products. In the event of broad applications, such as profiling the activity of a class of enzymes rather than a single isoform, fine-tuning the selectivity of ABPs is of utmost importance. In those instances, off-target labeling can be an issue, and appropriate controls and statistical measures must be devised in order to limit false discoveries. While concentration-dependent selectivity can be achieved through adjusting the amount of probe applied in some cases, additional controls such as competitive inhibition and no probe added samples are essential in most cases.

2.2 Linker

The linker group, which connects the active warhead to the reporter group or the click handle, is a crucial component and affects target identification. Ideally, the linker must be inert and does not form any interactions with the target, while distancing the reporter group from the active site. Hydrophobic linkers, such as alkyl chains, have been reported. However, the use of hydrophobic linkers presents the challenge of non-specific binding of proteins through hydrophobic interactions. This non-specificity, along with solubility issues, presents unfavorable aspects of using hydrophobic linkers. A hydrophilic linker group, such as polyethyleneglycol, PEG, is the linker of choice in most cases.

Cleavable linkers, such as a Tobacco Etch Virus (TEV) protease-cleavable linker, allow for selective peptide elution upon enrichment and avoid background signals resulting from non-specific binding. TEV-cleavable linkers have been incorporated in isotopic Tandem Orthogonal Proteolysis (isoTOP)-MS analysis strategies for the characterization and quantification of ABP labeling (see Detection Methods). Moreover, cleavable linkers enable determination of the site of labeling using MS-MS techniques, since the reactive portion of the probe remains attached to the target protein after cleavage. Additional cleavable linkers that have been employed in ABPP are trypsin-cleavable linkers, diazobenzenes that undergo cleavage with sodium dithionite, vicinal diols cleaved with sodium periodate, and light-sensitive linkers that are cleaved with UV-light (Yang et al. 2013; Leriche et al. 2012; Rudolf et al. 2013; Kim et al. 2009; Orth and Sieber 2009). An intrinsic drawback of these types of linkers, however, is incomplete cleavage, which can result in low signal.

2.3 Reporter

Upon enzyme binding, ABPs must possess a feature that facilitates detection of labeling. The reporter tag of ABPs enables this to happen through direct visualization of the probe (gel-based or imaging) or through enrichment and protein identification (mass spectrometry) (Sadaghiani et al. 2007). Initial ABPP studies took advantage of direct radiochemical detection, through incorporation of ^{125}I (Bogyo et al. 2000). Radiochemical detection is not applicable in large-scale analysis, due to limitations imposed by radioactive material. Instead, fluorescent and affinity tags were developed and have been utilized to detect enzymatic labeling of ABPs for several decades (Patricelli et al. 2001). Biotinylated ABPs have also been developed to enable target enrichment, a more time-consuming detection strategy. Fluorescent ABPs are the most direct method for in-gel detection of activity with high sensitivity and throughput.

Tags can either be directly conjugated to probes, in a one-step labeling process, or can be attached post-labeling, in a two-step labeling strategy. The latter carries a significant advantage in allowing the probe to retain more native properties, as a bulky tag can disrupt protein–probe interactions and/or decreases probe cell permeability. Installation of a bioorthogonal handle on the probe can replace the fluorescent or affinity tag used in the one-step process. Following an incubation period and binding to the enzyme, a bioorthogonal reaction can be utilized to add on a tag, in a two-step labeling process. This method enables the probe to interact with the protein targets without the potential interference of the bulky read-out tag.

2.3.1 Quenched Activity-Based Probes

A major limitation of ABPs bearing fluorophore reporters is the inherent fluorescence both when bound to a target and free in solution. Quenched activity-based probes (qABPs) are a subclass of ABPs, which are quenched or in a dark state per se and become fluorescent only upon mechanism-based reaction with their target and elimination of the quencher group. Since the fluorescence requires enzymatic activity and covalent modification of the target, background signal is insignificant. Therefore, qABPs are excellent candidates for non-invasive imaging and dynamic monitoring of proteins in vivo. Examples of qABPs have been developed to study a variety of cysteine proteases in mammalian cells (Blum et al. 2005; Bender et al. 2015; Withana et al. 2016).

2.3.2 Bioorthogonal Ligation Reactions

In order to preform two-step labeling, conjugation strategies must be employed in which reactions are fast, highly selective, and can be carried out under physiological conditions (McKay and Finn 2014). The key component of the two-step labeling strategy is the incorporation of a bioorthogonal handle onto the probe. Bioorthogonal handles are unique functional groups that are unreactive to native biological functionalities, such as water and other abundant nucleophiles. This property, along with their relatively small size, allows bioorthogonal handles to be attached to activity-based probes without disrupting the desired interactions between probe and protein (Hang et al. 2003; Sletten and Bertozzi 2011). A number of different strategies have been developed in the field of bioorthogonal chemistry over the last two decades (Lang and Chin 2014). The copper-catalyzed Hüisgen [3 + 2] azide-alkyne cycloaddition (CuAAC), an example of click chemistry optimized for biological applications by Sharpless and coworkers, has been used widely to ligate reporter tags and probes (Kolb et al. 2001; Kolb and Sharpless 2003; Rostovtsev et al. 2002; Speers and Cravatt 2004). The CuAAC reaction has superiority over the Staudinger ligation, which utilizes phosphines and azides as highly selective reactants, but has poor reaction kinetics. However, the CuAAC strategy also has a key limitation in the use of copper, which is cytotoxic. The strain-promoted [3 + 2] cycloaddition between an azide and a strained alkyne eliminates the need for copper, while retaining superior reaction kinetics (Agard et al. 2004, 2006; Lang and Chin 2014) (Fig. 2). However, it has also been noted that strained alkynes can undergo reactions with nucleophilic groups (Beatty et al. 2010; Chang et al. 2010; Ekkebus et al. 2013; McKay and Finn 2014).

An alternative strategy that reigns superior when considering reaction kinetics is the inverse electron-demand Diels–Alder reaction between strained alkenes or alkynes and tetrazines. These reactions are highly selective, in addition to incredibly fast, and release nitrogen gas as a byproduct and dihydropyridazines or pyridazines as products (Lang and Chin 2014). Significant advancements in the use of this

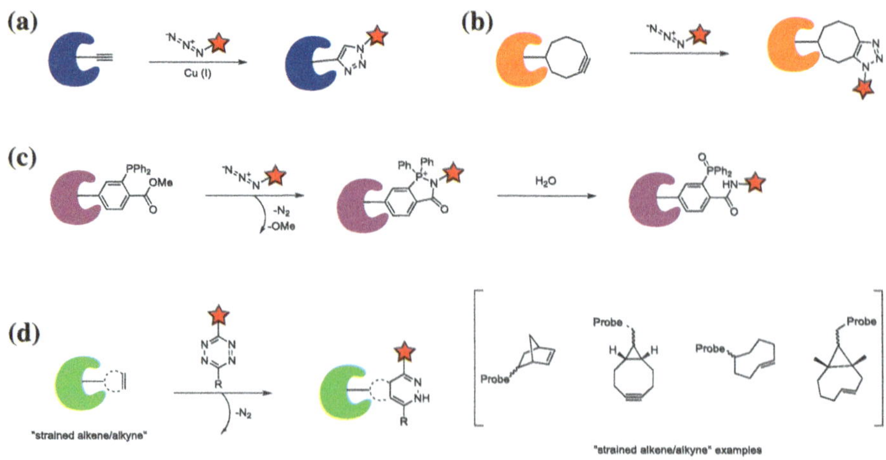

Fig. 2 Common bioorthogonal reactions. Colored semi-circles represent probe-labeled enzymes, and red stars indicate fluorescent or affinity tags. **a** Copper(I)-catalyzed alkyne-azide cycloaddition (CuAAc), **b** strain-promoted alkyne-azide cycloaddition, **c** Staudinger ligation, **d** inverse electron-demand Diels–Alder reaction

reaction have been developed in the last few years, and the use of unnatural amino acids containing these functional groups has been reported for improving the specificity of labeling. Additionally, development of quenched tetrazine–fluorophore conjugate probes that upon reaction with a bioorthogonal partner yield high fluorescence that does not require a washing step to remove unreacted probe has been reported. This has provided a useful tool for in-cell and in-animal imaging (Wu and Devaraj 2018).

3 Detection Methods

3.1 Gel-Based Detection

One of the first detection strategies employed for the analysis of ABPP was gel electrophoresis. Following a labeling step with the ABPs, proteomes can be separated and analyzed by either one-dimensional (1D) or two-dimensional (2D) gel electrophoresis. 1D-gel electrophoresis, most often sodium dodecyl sulfate polyacrylamide gel electrophoresis (SDS-PAGE), is utilized to separate proteins based on their molecular weight. 2D-gel electrophoresis separates proteins based on isoelectric point (first dimension), followed by separation based on size (second dimension).

These strategies can be applied for ABPs possessing both affinity and fluorescent tags. Upon separation of the proteome, labeled proteins can be visualized via fluorescence gel-imaging or blotting strategies. Gel-based detection is heavily relied upon, as it is robust, simple, high-throughout, and requires relatively little material. Upward of hundreds of proteome samples can be analyzed by gel electrophoresis per day; however, this strategy has its limitations. The number of proteins that can be resolved is often a limitation of gel-based strategies, as well as the inability to determine the molecular targeting of a probe (i.e., the identity of most labeled proteins and which amino acid is being labeled). As a result of these limitations, strategies were developed to increase the resolving power of ABPP and to determine specific interactions between a probe and its targets.

3.2 Live-Cell Imaging

Gel-based fluorescence detection has been the most utilized application of fluorescent probes; however, cell imaging strategies have also been employed. For example, the Carlson group has utilized both strategies to study the penicillin-binding proteins (PBPs) in bacteria. In an initial effort to study individual PBPs in *Bacillus subtilis* and *Streptococcus pneumoniae*, fluorophore-containing cephalosporin derivatives were synthesized and assessed for their ability to bind

PBPs. In vivo labeling of the PBPs was observed through fluorescence microscopy and gel-based analyses, and it was determined that these cephalosporin-analogues are more selective than Bocillin-FL, a commercially available analogue of penicillin-V that labels all PBPs (Fig. 5a). Furthermore, dual labeling experiments suggested that subpopulations of PBPs are localized separately during cell growth and division (Kocaoglu et al. 2012). β-lactone-based probes, developed to extend the PBP-targeting toolbox, enabled selective tagging of PBP2x and 2b in *S. pneumoniae*. Super-resolution fluorescence microscopy of pneumococcal cells treated with two lactone analogs, labeling PBPs 2x and 2b, respectively, demonstrated that PBP2b activity is restricted to a ring around the division site. PBP2x activity, on the other hand, was detected in both the surrounding ring and the septum at the division center (Sharifzadeh et al. 2017). This was the first time that individual PBP activity was detected with high spatial resolution, demonstrating the prowess of ABPs.

3.3 Mass Spectrometry

Gel-based strategies enable visualization of tagged proteins; however, they do not provide insight about the identity of the labeled proteins. In order to address this issue, liquid chromatography-mass spectrometry (LC-MS) strategies have been developed in which labeled proteins can be enriched from complex mixtures, using proper affinity tags, and subsequently characterized. A typical target identification experiment includes treatment of cells with ABPs, followed by enrichment of target–probe complexes via affinity purification. A biotin tag enables facile enrichment of probe–protein complexes by avidin beads and hence, is the most common reporter used in ABPP target identification studies. Given that biotin affects cell permeability, a click group is often appended to the probe, which can subsequently be ligated to a biotin or fluorescent reporter. Following enrichment of labeled proteins, proteomics can be carried out using LC-MS to characterize the proteome. Additionally, the sensitivity of MS instruments enables detection of very low-abundant proteins, which is not possible using gel-based methodologies.

Quantitative methods for chemical proteomics have been developed to more accurately quantify probe targets. It should be noted that these methods generally rely on relative quantification and do not provide absolute quantification. Activity-based protein profiling and multidimensional protein identification technology (ABPP-MudPIT) is a quantitative method that enables comparing enzymatic activities between different proteome samples (Jessani et al. 2005; Speers and Cravatt 2009). However, ABPP-MudPIT does not provide site-specific labeling information, in addition to being low throughput, which are key limitations of this platform. This issue was overcome with the development of tandem orthogonal proteolysis (TOP). TOP introduces a TEV-biotin tag onto probe-labeled proteins in complex mixtures. Upon streptavidin enrichment, proteins are first cleaved with trypsin and the resulting peptides eluted. Next, the remaining resin-bound peptides are incubated with TEV protease to selectively release probe-labeled species, which are analyzed via LC-MS/MS. Site-specific labeling of the probe can be elucidated via algorithms such as SEQUEST (Weerapana et al. 2007). SEQUEST is one example of peptide analysis software, but there are numerous programs that can be utilized (Verheggen et al. 2017). Further advancement of this methodology has led to isotopic-TOP (isoTOP)-ABPP, in which an isotopically tagged probe can be utilized to differentiate the extent of protein labeling in multiple proteomes (Weerapana et al. 2010).

LC-MS/MS, often the use of "label free" methods where spectral counting is used to quantify relative amounts of labeled peptides, is commonly utilized in ABPP. This is due to the fact that enrichment steps utilizing an affinity tag, such as biotin, remove a large amount of excess proteins from the proteome that have not been labeled by an ABP. On the other hand, "labeled" methods have been used less often. In this case, isotopically labeled amino acids are utilized in ABP-treated and ABP-untreated samples to enable quantification of the peptides from each sample. Such methods include SILAC, SILAM, iTRAQ, and iCAT. These methods are used more often in systems where the entire proteome is to be analyzed, as opposed to an enriched subpopulation of the proteome. However, this is not always the case and both "labeled" methods and enrichment have been used. An in-depth discussion of these methods is outside the scope of this chapter, and readers should refer to more comprehensive reviews (Cardoza et al. 2012; Chen et al. 2017; Wang et al. 2018; Wright and Sieber 2016).

4 ABPP in Bacteria

4.1 Serine-Reactive Probes

Well-established ABPP methods to assess the functional state of serine hydrolases (SHs) exist due to the broad distribution and importance of SHs in living organisms. SHs contain a conserved serine residue in their active site, which participates in the hydrolysis of amide, ester, and thioester substrates of the enzyme. The catalytic mechanism of SHs involves the formation of an acyl–enzyme intermediate, which has enabled the design and development of ABPs that target the active serine residue. Fluorophosphonates (FPs) have been used to profile the activity of SHs in complex proteomes. A common example of such a reactive group is shown in Fig. 3. Different modifications on the FP scaffold, such as inclusion of various click functionalities, fluorophore groups, or affinity tags, have been implemented in order to accommodate multiple applications. FP-based molecules show very broad reactivity toward SHs, enabling simultaneous investigation of a vast number of enzymes. For example, comparative analysis of SH activity in cancer cells has provided useful insight on dysregulated enzymes in eukaryotic systems (Nomura et al. 2010).

SHs are known to perform crucial functions in bacterial physiology and virulence, as well. Ortega et al. monitored the activity of SHs in *Mycobacterium tuberculosis* (*Mtb*), which is a clinically significant pathogen causing tuberculosis (TB), under replicating and non-replicating conditions. Unlike most infections that clear within a few days of antibiotic treatment, TB requires very long chemotherapy, which is attributed to the microorganism's exceptional ability to persist in host tissues. A quiescent or non-replicating population of *Mtb* bacilli causes this phenotypic drug tolerance state, rendering many TB therapies ineffective. This long treatment period, further complicated by patient compliance, has given rise to the emergence of drug-resistant mutants. Indeed, multidrug resistant (MDR) and extensively drug-resistant (XDR) *Mtb* strains are currently among the most challenging pathogens to treat. Understanding the physiological adaptations that are responsible for transition to the persistent state is essential in order to efficiently combat *Mtb*. The enzymes that remain active in the persistent state could be manipulated as potential diagnostic biomarkers or antimycobacterial targets.

SHs play central roles during all stages of *Mtb* growth and metabolism, including persistence. Assessment of the global activity of the SHs in replicating and persistent *Mtb*, using a general FP-based probe (**FP-ABP**; Fig. 3), revealed substantial differences between the two states. A significant number of SHs that remained active during persistence, including the essential protease ClpP, represent potential new drug targets in non-replicating *Mtb* (Ortega et al. 2016). Another study investigated the activity of *Mtb* esterases using ABPs (**ActivX TAMRA-FP**;

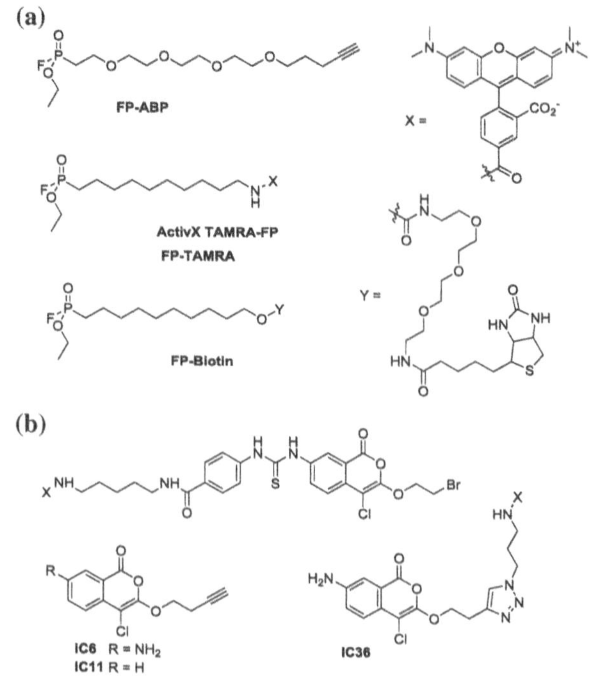

Fig. 3 Serine-reactive ABPs. **a** Fluorophosphonate probes; **b** isocoumarin probes

Fig. 3) in conjunction with substrate-based probes in active, dormant, and reactivating culture conditions. Reactivation of *Mtb*, which is transition from dormancy to the active state, is a crucial pathophysiological phenomenon, which remains poorly understood. As a result of this study, three esterase enzymes that become functional during early reactivation phase were determined for the first time (Tallman et al. 2016). In a recent study, selective substrate- and activity-based probes were developed to target *Mtb* "hydrolase important for pathogenesis 1" (*Mtb* Hip1) as a tool for monitoring mycobacterial infection and response to antibiotic therapy. Hip1 is a serine protease showing minimal homology with mammalian serine proteases, making it an ideal target for this purpose. A fluorogenic substrate assay was used to screen Hip1 against a library of serine-reactive small molecules, among which a series of 7-amino-4-chloro-3-(2-bromoethoxy)isocoumarin compounds were identified as hits and were used to develop an ABP (Fig. 3) (Lentz et al. 2016).

In order to shed light on host–pathogen interactions, FP-based probes (**FP-TAMRA** and **FP-Biotin**; Fig. 3) were applied to animal models infected with *Vibrio cholera* and secreted bacterial and active host SHs were identified. The results of this study revealed four *V. cholera* proteases that were active in infected rabbits, and one that was found to be active in human choleric stool (Hatzios et al. 2016).

Due to the high reactivity of the electrophilic group in FP-based probes, very complex activity profiles are generated in most cases, making it impractical to track the activity of low-abundant enzymes. To address this challenge, a series of diphenylphosphonate (DPP)-based ABPs was developed to target trypsin-like serine proteases selectively. This subgroup of SHs consists of several enzymes with unique physiological regulatory roles in humans, such as thrombin, factor VIIa, factor Xa and tryptase, have been exploited as drug targets in cardiovascular and inflammatory diseases (Pan et al. 2006). Initially generated as potent, irreversible serine protease inhibitors (Powers et al. 2002), DPP-derived probes label trypsin-like serine protease enzymes as pure proteins or in-cell lysates (Pan et al. 2006). However, the bulky diphenylphosphonate group might hinder the covalent reaction between the probe and the active-site serine residue, as was the case in DPP-based probes for Hip1 enzyme and rhomboid proteases in *Mtb* and *Escherichia coli*, respectively (Lentz et al. 2016; Vosyka et al. 2013). Rhomboid proteases are intramembrane serine proteases involved in critical biochemical processes such as signaling, making them potentially important drug targets. Screening of a library of different electrophiles against AarA and GlpP revealed several isocoumarin-containing compounds as potent inhibitors of either or both enzymes. Isocoumarin compounds conjugated with click and fluorophore groups, **IC6**, **11,** and **36**, tagged the serine residue in the active site of GlpG in *E. coli* cell lysates and whole cells (Vosyka et al. 2013) (Fig. 3).

4.2 Cysteine-Reactive Probes

Enzymes containing catalytic cysteine residues are ubiquitous and crucial in all domains of life, highlighting the importance of developing specific tools for studying them. The papain family of cysteine proteases comprises a group of closely related enzymes, with essential roles in many important pathological conditions such as cancer, rheumatoid arthritis, asthma, and osteoporosis. The enzymatic mechanism used by the papain family is highly conserved and hence, electrophilic inhibitors reacting in this context could be exploited in the generation of chemical probes with broad utility. The first group of cysteine-reactive ABPs was generated by Greenbaum et al. based on the natural product cysteine protease inhibitor, E64 (Greenbaum et al. 2000). E64 contains an epoxide group linked to a peptide moiety in which the leucine and agmatine groups mimic the natural substrate. JPM-565 was synthesized based on E64, which was further modified to develop probe DCG-04 (Fig. 4). Today, a wide range of small-molecule electrophiles have been developed as mechanism-based inhibitors of the cysteine proteases (Siklos et al. 2015; Powers et al. 2002; Shannon and Weerapana 2015).

In prokaryotes, cysteine proteases are among the most abundant hydrolytic enzymes, playing crucial roles in bacterial viability, metabolism, and environmental interactions. Acyloxymethyl ketone (AOMK)-based irreversible inhibitors for bacterial cysteine proteases have been reported, which target diverse families of cysteine proteases. These compounds show high selectivity for this class of enzymes, while maintaining low reactivity toward weakly nucleophilic groups (Powers et al. 2002). A common strategy to target these proteases is inclusion of an AOMK group in a peptide scaffold that is recognized by the enzyme and facilitates the irreversible binding of the reactive group (Kato et al. 2005). Several substrate-based probes and a number of ABPs have been developed using this strategy (Roncase et al. 2017).

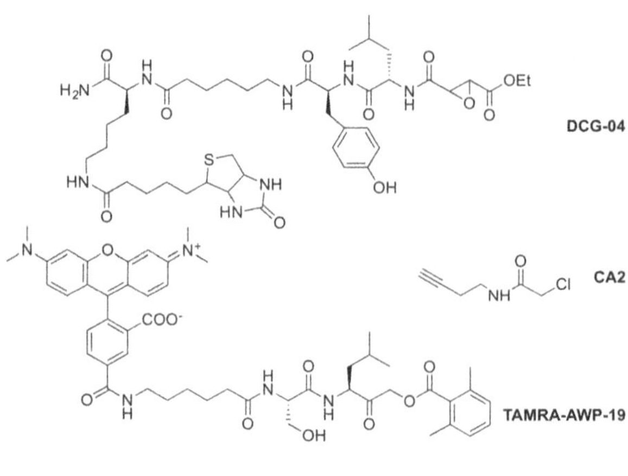

Fig. 4 Cysteine-reactive probes which incorporate electrophiles such as epoxide, α-chloroacetamide, and AOMK

AOMK-containing probes have been used to study the cysteine-protease domain (CPD) of *Clostridium difficile* toxins A and B (AOMK-Leu-Ser-TAMRA; **TAMRA-AWP-19**; Fig. 4). This compound covalently labels the active site in CPD and enables the use of an activity-based probe as a tool for investigating potential covalent binders of the active site. Bender et al. utilized TAMRA-AWP-19 as a competitor to investigate the ability of a library of FDA-approved drugs to inhibit CPD and found that ebselen, a clinical trial candidate for multiple indications, was a potent inhibitor of these proteins (Bender et al. 2015).

Bacteria use quorum sensing (QS), an inter-cellular communication mechanism that depends on cell density, to communicate with other bacteria and their environment. *Pseudomonas aeruginosa* is an opportunistic human pathogen that causes severe chronic infections in immunocompromised and cystic fibrosis patients. Given the high clinical significance of *P. aeruginosa*, it has become one of the main model organisms in which to study QS systems (Lee and Zhang 2015). One of the best characterized systems in this microorganism, the PQS system, regulates the production of a variety of virulence factors, using 2-alkyl-4-quinolones (AQs) as signal molecules (Prothiwa et al. 2016). In order to identify the exact roles of different AQs, cysteine-reactive chemical probes were synthesized to monitor the activity of the PqsD enzyme, which is involved in the biosynthesis of AQs. Among the tested electrophiles, α-chloroacetamide probes (**CA2**, Fig. 4) selectively labeled the active-site cysteine, which were further used in fluorescence-based PqsD inhibition assays. Several compounds were found to inhibit probe binding, which could be further used to expand the toolkit to study PqsD enzyme (Prothiwa et al. 2016).

4.3 Antibiotic-Derived Probes

β-lactams comprise the largest class of antibiotics in clinical use and have shown great success in combating Gram-positive and Gram-negative bacteria. β-lactams prevent cell growth and division through inhibition of a class of bacterial enzymes known as the penicillin-binding proteins (PBPs). PBPs catalyze the polymerization (transglycosylation) and cross-linking (transpeptidation) of bacterial cell wall, also known as the peptidoglycan (PG). β-lactams structurally resemble the terminal D-Ala-D-Ala group of PG stem peptides, and hence, the transpeptidase (TP) domain of PBPs recognize them as their natural substrate. Although PBPs were discovered as targets of the β-lactam antibiotics multiple decades ago, the specific roles of individual PBP homologs remain enigmatic.

During the transpeptidation step, a conserved serine residue from the active-site attacks the β-lactam ring in a mechanism-based manner, leading to formation of an inert acyl–enzyme complex that inhibits biosynthesis of the cell wall. Given the chemical stability of this acyl–PBP intermediate, β-lactams have long been used as chemical probes to study the PBPs. While radioactive penicillins were commonly

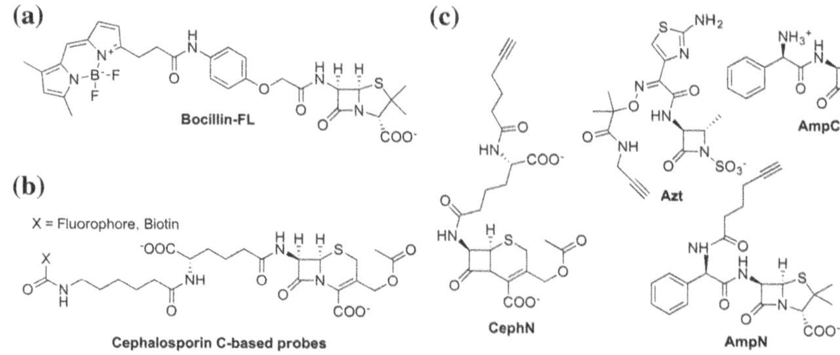

Fig. 5 Antibiotic-based probes containing an electrophilic lactam ring

applied in this regard for several decades, fluorescent analogues were more recently produced and quickly replaced their radioactive counterparts for profiling the activity of PBPs. Bocillin-FL (Boc-FL), comprised of penicillin-V and BODIPY FL as a reporter group, is commercially available. Showing global affinity for the entire PBP content in a given microorganism, Boc-FL has been exploited to monitor TP activity in pure protein samples, as well as cell lysates and whole cells in multiple microbes (Fig. 5a). In order to obtain selectivity for individual or a subset of PBPs, cephalosporin C molecules tagged with different fluorophores were developed which targeted a subset of PBPs in *B. subtilis* and *S. pneumoniae* (Fig. 5b) (Kocaoglu et al. 2012). Staub and Sieber produced ABPs using natural product-based antibiotics, ampicillin, cephalosporin C, or aztreonam, that were functionalized with an alkyne handle to target penicillin-binding proteins (PBPs; Fig. 5c) (Staub and Sieber 2008). These antibiotic-based probes were tested in multiple organisms including *Pseudomonas putida*, *Listeria welshimeri,* and *Bacillus licheniformis*, and were shown to target PBPs specifically. In the same study, synthetic β-lactams were also developed, which were shown to target a diverse set of resistance- and virulence-associated proteins in addition to the PBPs. Among the targeted proteins, caseinolytic protein protease (ClpP) stood out, which is a highly conserved serine protease with established crucial roles in the virulence of pathogenic bacteria such as *Staphylococcus aureus* and MRSA strains (Staub and Sieber 2008).

4.4 Carbohydrate-Based Probes

Glycoside hydrolases (GHs), also known as glycosidases, are a large class of enzymes with central roles in both prokaryotic and eukaryotic systems. Their enzymatic function leads to either inversion or retention of anomeric configuration upon hydrolysis. The latter mechanism employs two carboxylate residues; one that acts as a nucleophile and one that acts as a general acid/base catalyst (Zechel and Withers 2000). This catalytic mechanism proceeds through the formation of a covalent glycosyl–enzyme complex, and hence, chemical probes could be devised to trap this intermediate (Fig. 6a). It must be noted that GH enzymes, especially *exo*-glycosidases which cleave only the terminal ends of glycoconjugates, have a pocket-shaped active site, enabling extensive enzyme–substrate interactions, yet hindering the development of chemical probes due to limitation in accommodating bulky groups (Vocadlo and Bertozzi 2004). To characterize glycosidases that do not proceed through formation of a covalent glycosyl–enzyme intermediate, affinity-based probes containing photoreactive groups such as diazirines, benzophenones, or aryl azides are employed (Gandy et al. 2011), which is beyond the scope of this review. 2-Deoxy-2-fluoroglycosides have long been known as mechanism-based glycosidase inhibitors (Withers et al. 1987). In fact, inclusion of a fluorine group at the C2 or C5 of the sugar moiety would destabilize the transition state for deglycosylation of glycosyl–enzyme intermediate (Wicki et al. 2002). Placing a good leaving group at the anomeric position is very important and would kinetically facilitate formation of the covalent intermediate. Another key consideration is the reporter group, which must meet strict size requirements, due to the aforementioned limitations posed by the active site of this enzyme family. Considering all of these factors, fluorosugar-based probes have been developed and applied in ABPP of GHs. In most cases, click-compatible groups, such as alkynes or azides, are exploited to reduce probe size and enable ligation with reporters and affinity tags. The Bertozzi group pioneered the field by modifying 2-deoxy-2-fluoro-β-D-galactosyl fluoride to include a 6-azido group, which was later used to attach a FLAG tag through the Staudinger ligation. This probe, called **6Az2FGalF**, successfully labeled *E. coli* LacZ both as the pure enzyme and in cell lysates, demonstrating the power of ABPs for rapid detection of GHs (Vocadlo and Bertozzi 2004). Stubbs et al. developed a 2-azidoacetyl-2-deoxy-5-fluoro-β-D-glucopyranosyl fluoride (**2AA5FGF**) probe and utilized it to profile the activity of the NagZ enzyme in *V. cholera*, a key enzyme involved in peptidoglycan recycling pathways in Gram-negative bacteria. Using the Staudinger ligation or Sharpless–Meldal click reaction, reporters and affinity groups were added (Stubbs et al. 2008) (Fig. 6b).

ABPP methods have been employed to increase our understanding of microbial metabolism. Among the numerous microbial metabolic pathways, those that are related to biofuel conversion have received significant attention due to the world's growing demand for renewable and sustainable energy sources. In this regard, ABPP would enable functional characterization of unknown enzymes, measurement of changes in activity under different conditions, and the discovery and manipulation of new biofuel pathways. A recent review summarizes the application of ABPP in bioenergy conversion (Liu et al. 2015). Hydrolysis of the glycosidic bond, catalyzed by microbial GHs, is the rate-limiting step in the biofuel conversion process. An excellent study sought to target *Clostridium thermocellum*, an anaerobic bacterium capable of utilizing cellulose to efficiently produce ethanol through a multienzyme complex known as the cellulosome. A diverse suite of activity- and

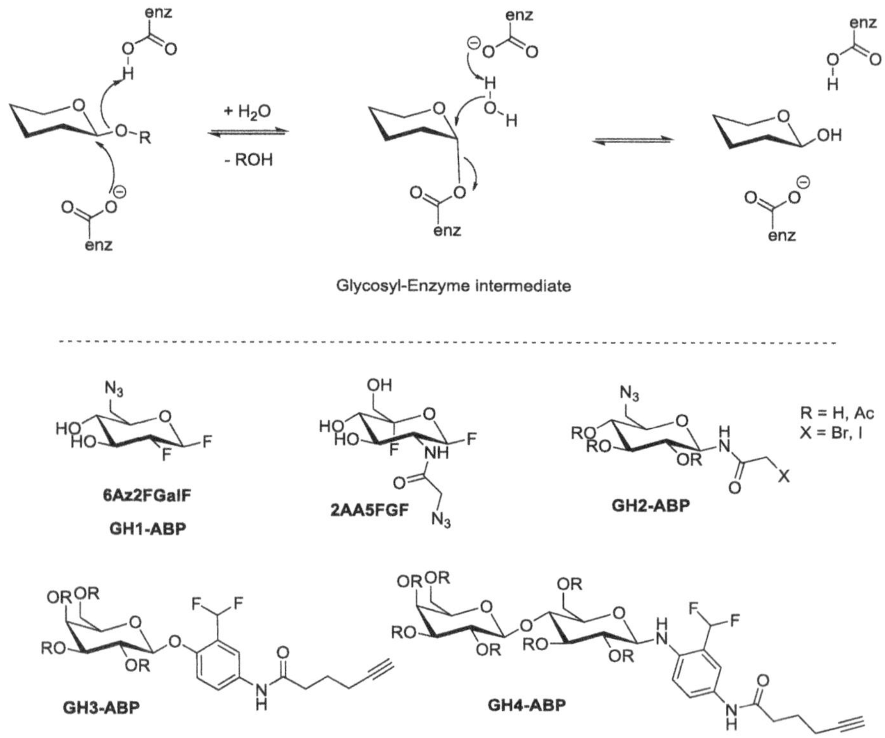

Fig. 6 **a** Mechanism of hydrolysis of glycosidic bond by β-glucosidases. **b** Structures of carbohydrate-based ABPs

affinity-based probes were generated and applied to the *C. thermocellum* cellulosome, revealing various probes with selectivity for GH enzymes. Different reactive groups, structural variation of the carbohydrate backbone, e.g. mono- or dissaccharide, as well as modifications such as *O*-acetylation of sugar units, were employed to tune the protein labeling profile (**GH-ABPs**, Fig. 6b). Derivatization of ABPs led to significant variability in GH labeling profiles, broadening the application of such GH-ABPs for study of cellulose-degrading enzymes in both aerobic and anaerobic systems (Chauvigne-Hines et al. 2012).

4.5 Probes Inspired by Natural Products

Activity-based profiling of enzyme families for which irreversible covalent inhibitors are not known can be challenging. Moreover, selective targeting of individual enzymes within a family may not be possible by merely using the known inhibitors and novel chemical tools must be devised. In cases that nucleophilic residues are involved in the catalytic process, probes could be designed using the variety of electrophiles available for incorporation into ABPs.

β-lactone (2-oxetanone) scaffolds, inspired by several natural products (Fig. 7a), have been used in the design of ABPs to target various enzymes and enzyme classes. β-lactones represent a group of compounds with promising biological

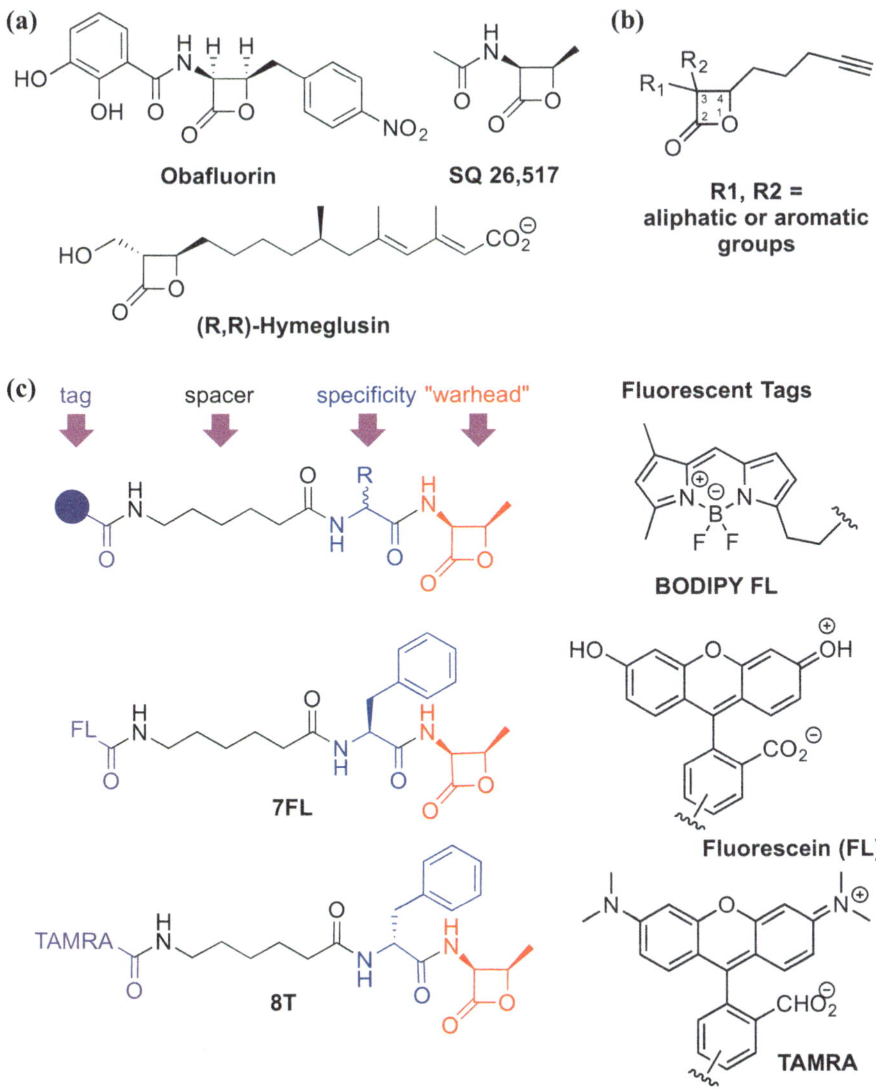

Fig. 7 Probes inspired by natural products. **a** Structure of natural products containing a β-lactone group. **b** General structure of β-lactone probe library developed by Bottcher et al. **c** General structure of PBP-selective β-lactone probes, as well as those used for imaging the PBPs; 7FL and 8T

activity and electrophilic potential, facilitating covalent addition of active-site residues. Böttcher and Sieber applied β-lactones functionalized with an alkyne group to prokaryotes in order to identify their target enzymes in *P. putida*, *L. welshimeri*, *B. licheniformis*, *B. subtilis*, and *E. coli*, as well as mouse liver cytosol as a eukaryotic control system. Different aliphatic and aromatic groups, inspired by the ones found in natural products, were inserted on the C2-position of the β-lactone ring to introduce diversity and increase targeting (Bottcher and Sieber 2008a) (Fig. 7b). Several enzymes involved in bacterial metabolism, antibiotic resistance, and virulence, such as ClpP, were identified (Bottcher and Sieber 2008a). Application of these molecules to *S. aureus* enabled specific targeting of ClpP in this pathogen, leading to development of ClpP inhibitors as antivirulence agents for the first time (Bottcher and Sieber 2008a, b; Weinandy et al. 2014).

To label the PBPs, especially those that are least targeted by the β-lactams, Sharifzadeh et al. developed a library of probes containing a β-lactone electrophilic group, various amino acids to induce selectivity, a spacer group, and a fluorophore tag as a reporter (Fig. 7c). The lactone scaffold, including the configuration of functional groups located on the ring, was designed to mimic the natural D-Ala-D-Ala portion of PG, i.e., the natural substrate of the PBPs (See Sect. 3.2). Interestingly, the *2S, 3R* configuration of the amine and methyl groups, which mimics the substrate geometry, was proven essential for PBP binding as the *2R, 3S* isomer rendered the probes inactive (Sharifzadeh et al. 2017). Given that most β-lactone natural products are active in the *trans* configuration, this study demonstrates the importance of structural considerations as part of probe design. The PBP labeling profile of these β-lactone probes was assessed in *S. pneumoniae*. Subsequently, two of the devised probes were used to visualize the activity of PBPs 2x and 2b in pneumococcal cells for the first time using dual color super-resolution fluorescence microscopy, highlighting the potential of ABPs in solving biological problems (Fig. 7c).

5 Application of ABPP in Natural Product Target Discovery

Natural products have evolved to encompass various bioactivity profiles and have found medicinal applications for many centuries. Due to their biosynthetic origin and often complex chemical structures, natural compounds interact with a vast number of biological macromolecules, usually in a highly selective manner. Some natural products react with their targets to form a covalent complex. This covalent reactivity, along with the high selectivity of these compounds, makes this group of natural products highly relevant for ABPP studies and target discovery. Two such cases pertaining to antimicrobial natural products are discussed here, and the avid reader is referred to other existing reviews for more examples (Drahl et al. 2005; Carlson 2010; Wright and Sieber 2016; Krysiak and Breinbauer 2012; Chen et al. 2017). However, it must be noted that generation of ABPs requires modifying the parent molecule to include a reporter tag, which might affect the bioactivity of the molecule and may be synthetically challenging.

4-Hydroxyderricin (**4-HD**) is a chalcone derived from the *Angelica keiskei* plant that has received much attention for its antimicrobial activity against crucial human pathogens such as *S. aureus*. To identify its macromolecular target in *S. aureus*, **4-HD** was functionalized with an alkyne handle and ABPP studies were undertaken. This study suggested that seryl tRNA synthetases (STS) are targeted by **4-HD** via the formation of covalent bonds with essential cysteine residues from the active site (Fig. 8) (Battenberg et al. 2013).

Showdomycin is an antibacterial nucleoside analogue isolated from *Streptomyces showdoensis*. A closer look at the structure of showdomycin reveals close resemblance with uridine, in which the 1′-pyrimidine group is replaced with an electrophilic maleimide moiety (Fig. 8). In order to uncover its biological target, a showdomycin-based ABP was synthesized through conjugation of an alkyne moiety at the 5′-O position (Bottcher and Sieber 2010). This investigation indicated that the antibacterial activity of showdomycin could be attributed to interactions with various enzymes, among which MurA1 and MurA2 stood out as essential enzymes involved with bacterial cell wall biosynthesis. Moreover, significant differences in the activities of a number of virulence-associated enzymes were elucidated between pathogenic and non-pathogenic bacteria (Bottcher and Sieber 2010).

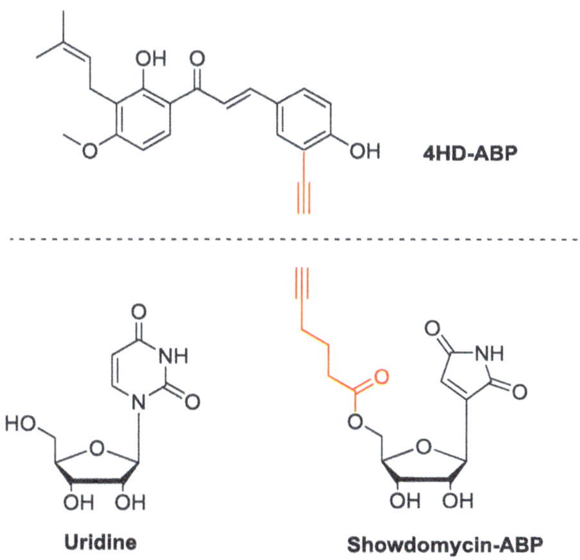

Fig. 8 Natural product ABPs used in target discovery. The natural product is modified with addition of a click handle (shown in red) or a reporter group to enable detection, enrichment, and characterization of targeted proteins

6 Conclusion

ABPP has significantly advanced the fields of proteomics, enzymology, and probe and inhibitor development, enabling the precise study of enzymes in an activity-dependent manner. Although ABPP has been applied extensively in eukaryotic systems, use in prokaryotes has lagged. Here, we have described ABP design strategies and examples of the use of ABPP in bacterial systems. The need for elucidating new drug targets in bacterial organisms is becoming dire, as antimicrobial resistance involving clinically targeted enzymes has led to pan-resistant organisms that can no longer be effectively treated. ABPP presents a platform from which new drug targets can be discovered and exploited. As described with the Sieber and Carlson groups' work with β-lactones, the ability to rationally design enzyme-specific probes provides a powerful and invaluable tool to better understand the mechanisms by which bacteria function, including pathogenicity. Additionally, use of natural products as a basis to create selective scaffolds has been described and should continue to be exploited. Future work must strive to build and design scaffolds that are capable of selectively targeting individual enzymes or enzyme classes in prokaryotes. Although it has only been applied in eukaryotes, the use of nucleophilic hydrazine-based probes to target cellular electrophiles in reversed-polarity (RP)-ABPP represents a new direction in the field of ABPP, extending the range of targetable proteins (Matthews et al. 2017). Moreover, this methodology could aid in understanding the understudied field of post-translational modifications in bacteria. Continuing research into the advancement of MS-based detection methods will also increase the utility of ABPP. Lastly, multiple research areas such as microbial pathogenesis, renewable energy, and biotechnology would benefit from advancements in this powerful analytical platform.

References

Agard NJ, Prescher JA, Bertozzi CR (2004) A strain-promoted [3 + 2] azide-alkyne cycloaddition for covalent modification of biomolecules in living systems. J Am Chem Soc 126:15046–15047

Agard NJ, Baskin JM, Prescher JA, Lo A, Bertozzi CR (2006) A comparative study of bioorthogonal reactions with azides. ACS Chem Biol 1:644–648

Battenberg OA, Yang Y, Verhelst SH, Sieber SA (2013) Target profiling of 4-hydroxyderricin in *S. aureus* reveals seryl-tRNA synthetase binding and inhibition by covalent modification. Mol BioSyst 9:343–351

Beatty KE, Fisk JD, Smart BP, Lu YY, Szychowski J, Hangauer MJ, Baskin JM, Bertozzi CR, Tirrell DA (2010) Live-cell imaging of cellular proteins by a strain-promoted azide-alkyne cycloaddition. ChemBioChem 11:2092–2095

Bender KO, Ofori L, Van Der Linden WA, Mock ED, Datta GK, Chowdhury S, Li H, Segal E, Lopez MS, Ellman JA, Figdor CG, Bogyo M, Verdoes M (2015) Design of a highly selective quenched activity-based probe and its application in dual color imaging studies of cathepsin S activity localization. J Am Chem Soc 137:4771–4777

Blum G, Mullins SR, Keren K, Fonovic M, Jedeszko C, Rice MJ, Sloane BF, Bogyo M (2005) Dynamic imaging of protease activity with fluorescently quenched activity-based probes. Nat Chem Biol 1:203–209

Bogyo M, Verhelst S, Bellingard-Dubouchaud V, Toba S, Greenbaum D (2000) Selective targeting of lysosomal cysteine proteases with radiolabeled electrophilic substrate analogs. Chem Biol 7:27–38

Bottcher T, Sieber SA (2008a) Beta-lactones as privileged structures for the active-site labeling of versatile bacterial enzyme classes. Angew Chem Int Ed Engl 47:4600–4603

Bottcher T, Sieber SA (2008b) Beta-lactones as specific inhibitors of ClpP attenuate the production of extracellular virulence factors of *Staphylococcus aureus*. J Am Chem Soc 130:14400–14401

Bottcher T, Sieber SA (2010) Showdomycin as a versatile chemical tool for the detection of pathogenesis-associated enzymes in bacteria. J Am Chem Soc 132:6964–6972

Cardoza JD, Parikh JR, Ficarro SB, Marto JA (2012) Mass spectrometry-based proteomics: qualitative identification to activity-based protein profiling. Wiley Interdiscip Rev Syst Biol Med 4:141–162

Carlson EE (2010) Natural products as chemical probes. ACS Chem Biol 5:639–653

Chang PV, Prescher JA, Sletten EM, Baskin JM, Miller IA, Agard NJ, Lo A, Bertozzi CR (2010) Copper-free click chemistry in living animals. Proc Natl Acad Sci U S A 107:1821–1826

Chauvigne-Hines LM, Anderson LN, Weaver HM, Brown JN, Koech PK, Nicora CD, Hofstad BA, Smith RD, Wilkins MJ, Callister SJ, Wright AT (2012) Suite of activity-based probes for cellulose-degrading enzymes. J Am Chem Soc 134:20521–20532

Chen X, Wong YK, Wang J, Zhang J, Lee YM, Shen HM, Lin Q, Hua ZC (2017) Target identification with quantitative activity based protein profiling (ABPP). Proteomics 17

Drahl C, Cravatt BF, Sorensen EJ (2005) Protein-reactive natural products. Angew Chem Int Ed Engl 44:5788–5809

Edgington LE, Verdoes M, Bogyo M (2011) Functional imaging of proteases: recent advances in the design and application of substrate-based and activity-based probes. Curr Opin Chem Biol 15:798–805

Eirich J, Orth R, Sieber SA (2011) Unraveling the protein targets of vancomycin in living *S. aureus* and *E. faecalis* cells. J Am Chem Soc 133:12144–12153

Ekkebus R, Van Kasteren SI, Kulathu Y, Scholten A, Berlin I, Geurink PP, De Jong A, Goerdayal S, Neefjes J, Heck AJ, Komander D, Ovaa H (2013) On terminal alkynes that can react with active-site cysteine nucleophiles in proteases. J Am Chem Soc 135:2867–2870

Gandy MN, Debowski AW, Stubbs KA (2011) A general method for affinity-based proteomic profiling of exo-alpha-glycosidases. Chem Commun (Camb) 47:5037–5039

Greenbaum D, Medzihradszky KF, Burlingame A, Bogyo M (2000) Epoxide electrophiles as activity-dependent cysteine protease profiling and discovery tools. Chem Biol 7:569–581

Hang HC, Yu C, Kato DL, Bertozzi CR (2003) A metabolic labeling approach toward proteomic analysis of mucin-type O-linked glycosylation. Proc Natl Acad Sci U S A 100:14846–14851

Hatzios SK, Abel S, Martell J, Hubbard T, Sasabe J, Munera D, Clark L, Bachovchin DA, Qadri F, Ryan ET, Davis BM, Weerapana E, Waldor MK (2016) Chemoproteomic profiling of host and pathogen enzymes active in cholera. Nat Chem Biol 12:268–274

Heal WP, Tate EW (2012) Application of activity-based protein profiling to the study of microbial pathogenesis. Top Curr Chem 324:115–135

Jessani N, Niessen S, Wei BQ, Nicolau M, Humphrey M, Ji Y, Han W, Noh DY, Yates JR 3rd, Jeffrey SS, Cravatt BF (2005) A streamlined platform for high-content functional proteomics of primary human specimens. Nat Methods 2:691–697

Kato D, Boatright KM, Berger AB, Nazif T, Blum G, Ryan C, Chehade KA, Salvesen GS, Bogyo M (2005) Activity-based probes that target diverse cysteine protease families. Nat Chem Biol 1:33–38

Kim HY, Tallman KA, Liebler DC, Porter NA (2009) An azido-biotin reagent for use in the isolation of protein adducts of lipid-derived electrophiles by streptavidin catch and photorelease. Mol Cell Proteomics 8:2080–2089

Kocaoglu O, Calvo RA, Sham LT, Cozy LM, Lanning BR, Francis S, Winkler ME, Kearns DB, Carlson EE (2012) Selective penicillin-binding protein imaging probes reveal substructure in bacterial cell division. ACS Chem Biol 7:1746–1753

Kolb HC, Sharpless KB (2003) The growing impact of click chemistry on drug discovery. Drug Discov Today 8:1128–1137

Kolb HC, Finn MG, Sharpless KB (2001) Click chemistry: diverse chemical function from a few good reactions. Angew Chem Int Ed Engl 40:2004–2021

Krysiak J, Breinbauer R (2012) Activity-based protein profiling for natural product target discovery. Top Curr Chem 324:43–84

Lang K, Chin JW (2014) Bioorthogonal reactions for labeling proteins. ACS Chem Biol 9:16–20

Lee J, Zhang L (2015) The hierarchy quorum sensing network in *Pseudomonas aeruginosa*. Protein Cell 6:26–41

Lentz CS, Ordonez AA, Kasperkiewicz P, La Greca F, O'donoghue AJ, Schulze CJ, Powers JC, Craik CS, Drag M, Jain SK, Bogyo M (2016) Design of selective substrates and activity-based probes for Hydrolase Important for Pathogenesis 1 (HIP1) from *Mycobacterium tuberculosis*. ACS Infect Dis 2:807–815

Leriche G, Chisholm L, Wagner A (2012) Cleavable linkers in chemical biology. Bioorg Med Chem 20:571–582

Liu Y, Fredrickson JK, Sadler NC, Nandhikonda P, Smith RD, Wright AT (2015) Advancing understanding of microbial bioenergy conversion processes by activity-based protein profiling. Biotechnol Biofuels 8:156

Matthews ML, He L, Horning BD, Olson EJ, Correia BE, Yates JR 3rd, Dawson PE, Cravatt BF (2017) Chemoproteomic profiling and discovery of protein electrophiles in human cells. Nat Chem 9:234–243

Mckay CS, Finn MG (2014) Click chemistry in complex mixtures: bioorthogonal bioconjugation. Chem Biol 21:1075–1101

Narayanan A, Jones LH (2015) Sulfonyl fluorides as privileged warheads in chemical biology. Chem Sci 6:2650–2659

Nomura DK, Dix MM, Cravatt BF (2010) Activity-based protein profiling for biochemical pathway discovery in cancer. Nat Rev Cancer 10:630–638

Ortega C, Anderson LN, Frando A, Sadler NC, Brown RW, Smith RD, Wright AT, Grundner C (2016) Systematic survey of serine hydrolase activity in *Mycobacterium tuberculosis* defines changes associated with persistence. Cell Chem Biol 23:290–298

Orth R, Sieber SA (2009) A photolabile linker for the mild and selective cleavage of enriched biomolecules from solid support. J Org Chem 74:8476–8479

Pan Z, Jeffery DA, Chehade K, Beltman J, Clark JM, Grothaus P, Bogyo M, Baruch A (2006) Development of activity-based probes for trypsin-family serine proteases. Bioorg Med Chem Lett 16:2882–2885

Patricelli MP, Giang DK, Stamp LM, Burbaum JJ (2001) Direct visualization of serine hydrolase activities in complex proteomes using fluorescent active site-directed probes. Proteomics 1:1067–1071

Pearson RG (1990) Hard and soft acids and bases—the evolution of a chemical concept. Coord Chem Rev 100:403–425

Powers JC, Asgian JL, Ekici OD, James KE (2002) Irreversible inhibitors of serine, cysteine, and threonine proteases. Chem Rev 102:4639–4750

Prothiwa M, Szamosvari D, Glasmacher S, Bottcher T (2016) Chemical probes for competitive profiling of the quorum sensing signal synthase PqsD of *Pseudomonas aeruginosa*. Beilstein J Org Chem 12:2784–2792

Roncase EJ, Moon C, Chatterjee S, Gonzalez-Paez GE, Craik CS, O'donoghue AJ, Wolan DW (2017) Substrate profiling and high resolution co-complex crystal structure of a secreted C11 protease conserved across commensal bacteria. ACS Chem Biol 12:1556–1565

Rostovtsev VV, Green LG, Fokin VV, Sharpless KB (2002) A stepwise huisgen cycloaddition process: copper(I)-catalyzed regioselective "ligation" of azides and terminal alkynes. Angew Chem Int Ed Engl 41:2596–2599

Rudolf GC, Heydenreuter W, Sieber SA (2013) Chemical proteomics: ligation and cleavage of protein modifications. Curr Opin Chem Biol 17:110–117

Sadaghiani AM, Verhelst SH, Bogyo M (2007) Tagging and detection strategies for activity-based proteomics. Curr Opin Chem Biol 11:20–28

Sadler NC, Wright AT (2015) Activity-based protein profiling of microbes. Curr Opin Chem Biol 24:139–144

Shannon DA, Weerapana E (2015) Covalent protein modification: the current landscape of residue-specific electrophiles. Curr Opin Chem Biol 24:18–26

Sharifzadeh S, Boersma MJ, Kocaoglu O, Shokri A, Brown CL, Shirley JD, Winkler ME, Carlson EE (2017) Novel electrophilic scaffold for imaging of essential penicillin-binding proteins in *Streptococcus pneumoniae*. ACS Chem Biol 12:2849–2857

Siklos M, Benaissa M, Thatcher GR (2015) Cysteine proteases as therapeutic targets: does selectivity matter? A systematic review of calpain and cathepsin inhibitors. Acta Pharm Sin B 5:506–519

Sletten EM, Bertozzi CR (2011) From mechanism to mouse: a tale of two bioorthogonal reactions. Acc Chem Res 44:666–676

Speers AE, Cravatt BF (2004) Profiling enzyme activities in vivo using click chemistry methods. Chem Biol 11:535–546

Speers AE, Cravatt BF (2009) Activity-Based Protein Profiling (ABPP) and Click Chemistry (CC)-ABPP by MudPIT mass spectrometry. Curr Protoc Chem Biol 1:29–41

Staub I, Sieber SA (2008) Beta-lactams as selective chemical probes for the in vivo labeling of bacterial enzymes involved in cell wall biosynthesis, antibiotic resistance, and virulence. J Am Chem Soc 130:13400–13409

Stubbs KA, Scaffidi A, Debowski AW, Mark BL, Stick RV, Vocadlo DJ (2008) Synthesis and use of mechanism-based protein-profiling probes for retaining beta-D-glucosaminidases facilitate identification of *Pseudomonas aeruginosa* NagZ. J Am Chem Soc 130:327–335

Tallman KR, Levine SR, Beatty KE (2016) Small-molecule probes reveal esterases with persistent activity in dormant and reactivating *Mycobacterium tuberculosis*. ACS Infect Dis 2:936–944

Verheggen K, Raeder H, Berven FS, Martens L, Barsnes H, Vaudel M (2017) Anatomy and evolution of database search engines-a central component of mass spectrometry based proteomic workflows. Mass Spectrom Rev

Vocadlo DJ, Bertozzi CR (2004) A strategy for functional proteomic analysis of glycosidase activity from cell lysates. Angew Chem Int Ed Engl 43:5338–5342

Vogel C, Marcotte EM (2012) Insights into the regulation of protein abundance from proteomic and transcriptomic analyses. Nat Rev Genet 13:227–232

Vosyka O, Vinothkumar KR, Wolf EV, Brouwer AJ, Liskamp RM, Verhelst SH (2013) Activity-based probes for rhomboid proteases discovered in a mass spectrometry-based assay. Proc Natl Acad Sci U S A 110:2472–2477

Wang S, Tian Y, Wang M, Wang M, Sun GB, Sun XB (2018) Advanced activity-based protein profiling application strategies for drug development. Front Pharmacol 9:353

Weerapana E, Speers AE, Cravatt BF (2007) Tandem orthogonal proteolysis-activity-based protein profiling (TOP-ABPP)—a general method for mapping sites of probe modification in proteomes. Nat Protoc 2:1414–1425

Weerapana E, Simon GM, Cravatt BF (2008) Disparate proteome reactivity profiles of carbon electrophiles. Nat Chem Biol 4:405–407

Weerapana E, Wang C, Simon GM, Richter F, Khare S, Dillon MB, Bachovchin DA, Mowen K, Baker D, Cravatt BF (2010) Quantitative reactivity profiling predicts functional cysteines in proteomes. Nature 468:790–795

Weinandy F, Lorenz-Baath K, Korotkov VS, Bottcher T, Sethi S, Chakraborty T, Sieber SA (2014) A beta-lactone-based antivirulence drug ameliorates *Staphylococcus aureus* skin infections in mice. ChemMedChem 9:710–713

Wicki J, Rose DR, Withers SG (2002) Trapping covalent intermediates on beta-glycosidases. Methods Enzymol 354:84–105

Willems LI, Overkleeft HS, Van Kasteren SI (2014) Current developments in activity-based protein profiling. Bioconjug Chem 25:1181–1191

Withana NP, Garland M, Verdoes M, Ofori LO, Segal E, Bogyo M (2016) Labeling of active proteases in fresh-frozen tissues by topical application of quenched activity-based probes. Nat Protoc 11:184–191

Withers SG, Street IP, Bird P, Dolphin DH (1987) 2-Deoxy-2-Fluoroglucosides—a novel class of mechanism-based glucosidase inhibitors. J Am Chem Soc 109:7530–7531

Wright MH, Sieber SA (2016) Chemical proteomics approaches for identifying the cellular targets of natural products. Nat Prod Rep 33:681–708

Wu H, Devaraj NK (2018) Advances in tetrazine bioorthogonal chemistry driven by the synthesis of novel tetrazines and dienophiles. Acc Chem Res 51:1249–1259

Yang Y, Hahne H, Kuster B, Verhelst SH (2013) A simple and effective cleavable linker for chemical proteomics applications. Mol Cell Proteomics 12:237–244

Zechel DL, Withers SG (2000) Glycosidase mechanisms: anatomy of a finely tuned catalyst. Acc Chem Res 33:11–18

Opportunities and Challenges in Activity-Based Protein Profiling of Mycobacteria

Hiren V. Patel, Michael Li and Jessica C. Seeliger

Contents

Abstract Mycobacteria, from saprophytic to pathogenic species, encounter diverse environments that demand metabolic versatility and rapid adaptation from these bacteria for their survival. The human pathogen *Mycobacterium tuberculosis*, for example, can enter a reversible state of dormancy in which it is metabolically active, but does not increase in number, and which is believed to enable its survival in the human host for years, with attendant risk for reactivation to active tuberculosis. Driven by the need to combat mycobacterial diseases like tuberculosis,

H. V. Patel
Department of Molecular Genetics and Microbiology, Stony Brook University, 11794 Stony Brook, NY, USA

M. Li · J. C. Seeliger (✉)
Department of Pharmacological Sciences, Stony Brook University, 11794 Stony Brook, NY, USA
e-mail: jessica.seeliger@stonybrook.edu

Current Topics in Microbiology and Immunology (2019) 420:49–72
DOI 10.1007/82_2018_125

Published Online: 04 September 2018

efforts to understand such adaptations have benefitted in recent years from application of activity-based probes. These studies have been inspired by the potential of these chemical tools to uncover protein function for previously unannotated proteins, track shifts in protein activity as a function of environment, and provide a streamlined method for screening and developing inhibitors. Here we seek to contextualize progress thus far with achieving these goals and highlight the unique challenges and opportunities for activity-based probes to further our understanding of protein function and regulation, bacterial physiology, and antibiotic development.

1 Introduction

The past six years have seen the burgeoning application of activity-based probes (ABPs) to mycobacteria, from protein profiling with well-established probes to the development of probes for particular enzymes to the identification of inhibitor targets. Mycobacteria include numerous pathogenic species, including *Mycobacterium tuberculosis* (Mtb), the causative agent of tuberculosis in humans and the focus of most studies to date that use ABPs. Mtb is a major human pathogen, causes millions of deaths annually, and infects an estimated one-third of the world's population. Most infected individuals are asymptomatic, a clinical state known as latency, but are at risk of reactivation to active disease by mechanisms that are as yet poorly understood (Ernst 2012). The complexity of Mtb infection and disease and the host–pathogen interactions that underlie different disease states remain poorly understood. Efforts to unravel the mechanisms by which Mtb can survive and thrive within humans often use in vitro culture models that mimic the host environment. In particular, various environmental factors induce a state known as non-replicating persistence, or dormancy, in which the bacteria remain metabolically active and in fact grow and divide, but do not "replicate" in the sense of increasing in number (Rittershaus et al. 2013). This is a phenotypic state that is reversible: Mtb recovers and grows logarithmically upon removal of the relevant factor. Mtb in this state displays phenotypic drug tolerance compared to actively replicating Mtb. In human disease, phenotypically drug-tolerant subpopulations of Mtb are believed to contribute to the long antibiotic treatment times necessary to achieve clinical cure (Gold and Nathan 2017). Thus, a major goal in mycobacterial research is to understand mechanisms that enable Mtb to enter into and reactivate from dormancy. One of the best-studied models is hypoxia, which has been used in many of the activity-based protein-profiling (ABPP) studies in Mtb discussed in this review (Rustad et al. 2009).

The Mtb genome was first sequenced in 1998 and revealed that Mtb encodes many proteins with little or no homology to proteins of known function (Cole et al. 1998). To date, a significant proportion of the proteome still bears the annotation "conserved hypothetical protein." ABPP thus has obvious appeal as a method to investigate protein activity in response to conditions that mimic the infection context and to aid the functional annotation of the proteome.

This review assumes a general knowledge of ABPs, which we define as chemical tools that covalently modify proteins and enable subsequent detection of the resulting adducts. For more detailed coverage of ABPs and ABPP methodologies, we refer the reader to several reviews (Barglow and Cravatt 2007; Cravatt et al. 2008; Simon and Cravatt 2010). We focus below on the handful of studies on mycobacteria, and for studies performed with similar probes, we present additional comparative analyses for the proteins that were identified and discuss remaining opportunities.

2 Activity-Based Protein Profiling in *Mycobacterium tuberculosis*

ABPP studies in Mtb have been pursued for two well-characterized classes of activity-based probes: a lysine-reactive ATP analogue for ATP-binding proteins and a serine-modifying fluorophosphonate for serine hydrolases. In this section, we review and compare the resulting profiles, especially with respect to novel protein annotation and comparison between culture conditions.

2.1 Profiling of ATP-Binding Proteins

Two studies in *M. tuberculosis* were published nearly concurrently in 2013 (Ansong et al. 2013; Wolfe et al. 2013). Both used acylphosphate ATP analogues that covalently modify lysine side chain(s) proximal to an ATP-binding site (Table 1). The primary differences between the two studies were (1) the nature of the acyl modification on the probe, (2) the method used to enrich and detect modified proteins, and (3) whether a comparative condition was also profiled. We first summarize the outcomes of each study, highlighting important methodological distinctions, and then compare the results.

Ansong et al. used a custom "label-free" diacyl ATP analogue with a terminal alkyne on each of the acyl chains (Ansong et al. 2013). Treatment of whole-cell lysates of Mtb with this probe was followed by copper-mediated azide-alkyne cycloaddition with commercially available biotin-azide. The intact biotinylated proteins were enriched by streptavidin affinity chromatography and detected by accurate mass and time (AMT) tag mass spectrometry (Zimmer et al. 2006), which in this context the authors term AMT-ABPP. The primary advantage of the AMT approach is to enable higher sample throughput by comparing accurate mass and retention time information from experimental samples to a rigorously validated AMT peptide database based on MS/MS assignments of peptides from reference

Table 1 Activity-based probes applied to or developed for mycobacteria

Application	Target Protein/Class	Name	Structure	References
Profiling	ATP-binding	ATP-ABP	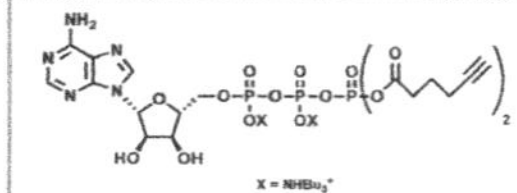	Ansong et al. (2013)
Profiling	ATP-binding	Desthiobiotin-ATP	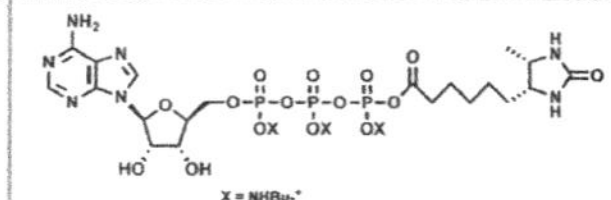	Wolfe et al. (2013)
Profiling	Serine hydrolases	FP-ABP	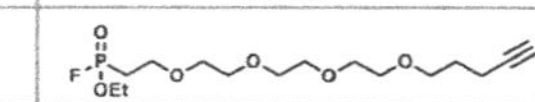	Ortega et al. (2016)
Profiling	Serine hydrolases	Desthiobiotin-FP	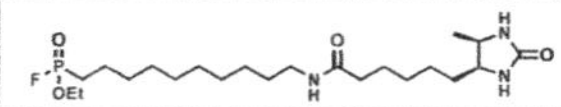	Tallman et al. (2016)
Target ID	MbtA/Adenylating enzymes	Sal-AMS	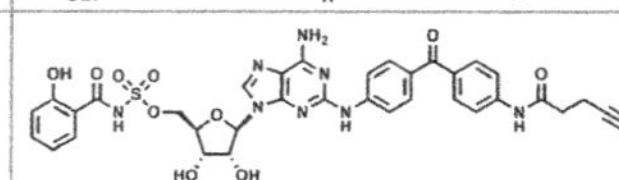	Duckworth et al. (2012)
Profiling	Hip1/Serine hydrolases	Compound 4	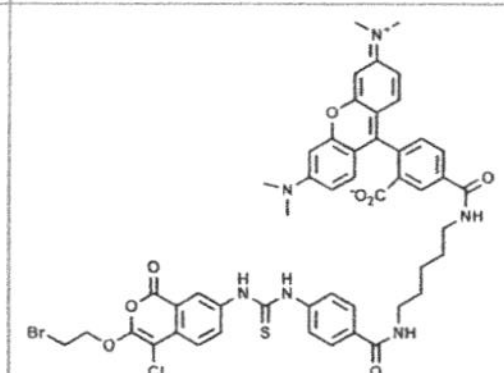	Lentz et al. (2016)
Target ID	Serine hydrolases	THL-alk1	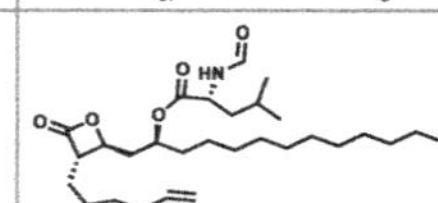	Ravindran et al. (2014)
Target ID	Serine hydrolases	EZ120P	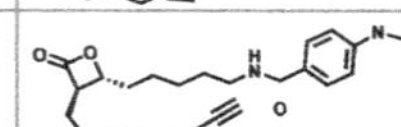	Lehmann et al. (2017)

samples. By assigning peptides in experimental samples by comparison to the database, this approach largely obviates the need to do full MS/MS analysis (although very high run-to-run LC-MS reproducibility is a must) and thus makes acquiring data for multiple replicates and conditions more efficient. As such, Ansong et al. analyzed Mtb lysates treated with dimethyl sulfoxide (DMSO) to control for non-specific binding to streptavidin and lysates pretreated with ATPγS to control for non-specific labeling of their probe. Importantly, the conditions for the ATPγS treatment were validated by dose-dependent competition experiments using fluorescence detection to determine the ATPγS concentration range that competes successfully with their probe (approximately $\geq$ 1 mM). For the MS analyses, their rigorous cutoff criteria for hit proteins included detection of $\geq$ 2 unique peptides, a statistically significant difference across all replicates of probe-treated and the two negative controls, and an intensity ratio of $\geq$5 for probe versus DMSO treatment, which is higher than the more common twofold literature precedent. Nevertheless, they detected 317 proteins (77% of the predicted proteome) that met these criteria and thereby validated the current functional annotation for 245 members of various ATP-binding protein families. These results indicate surprisingly widespread adenosine nucleotide binding within Mtb, as noted by the title of the study.

Their results validated known and predicted ATP-binding or ATP-processing proteins such as kinases and nucleotide-binding proteins and also proteins that bind to or react with acyl-CoA molecules. Interestingly, only 5 of the 11 predicted Ser/Thr kinases (PknA, B, D, G, F) were detected, highlighting the potential of ABPP to validate expression and activity (or integrity) at the protein level under a given growth condition. With respect to the potential of ABPP to yield novel protein annotation, 72 of the 317 proteins were annotated as hypothetical and therefore constitute potential novel ATP-binding proteins. In addition, 9 proteins had annotation but no known nucleotide binding or phosphohydrolase activity. Although these proteins were considered the result of non-selective "false labeling," the authors acknowledge that their methods, which isolated proteins under native conditions, allowed for the potential enrichment and detection of proteins that *interact* with ATP-binding proteins.

Ansong et al. went on to complete a more detailed bioinformatics analysis of the 73 hypothetical proteins, which yielded an additional 36 proteins with some homology consistent with nucleotide binding and/or ATPase activity. The two bioinformatics methods used for these predictions were not in complete agreement with each other, however. The authors note that this discrepancy highlights the power of experimental probe-labeling data to validate computational approaches.

Most importantly, the authors validated ATP binding for a selection of the remaining 36 proteins with no sequence similarity to known nucleotide-binding domains. They recombinantly expressed and demonstrated labeling with ATP-ABP and competition of labeling with ATP or ATPγS for 5 proteins, confirmed the modified lysine by mutation for 2 proteins, and identified the modified peptide for 1 protein. They also identified the modified peptide for two Ser/Thr protein kinases, PknB and PknD, and confirmed that the identity of the modified lysine is consistent with the expected position of the ATP-binding site. As such, the study provides ample confirmation for their method as well as strong evidence for a group of probable novel nucleotide-binding domains that provide initial functional information for proteins of otherwise unknown function and likely represent novel nucleotide-binding motifs and folds.

On the heels of the Ansong et al. study, Wolfe et al. published an analogous investigation using the commercially available desthiobiotin-ATP monoacylphosphate probe (Table 1) (Wolfe et al. 2013). They performed profiling on lysates from Mtb grown not only under standard normoxic culture conditions for 7 days, but also under hypoxia upon gradual oxygen depletion over 14 days. In contrast to the other study, they trypsinized the probe-treated lysates *prior* to streptavidin affinity enrichment and detection by traditional shotgun proteomics methods. As such, their most rigorous cutoff for hit identification is detection of the lysine-modified peptide. This approach guarantees bona fide reactivity of the protein with the probe, so non-reactive but interacting proteins are not a consideration as they were for Ansong et al., although non-specific reactivity is still a possibility as noted further below. At the same time, detection of the modified peptide is the greatest potential limitation, as it restricts the sensitivity of their approach based on the ion yields of modified peptides for LC-MS/MS detection. Thus, it is not surprising that Wolfe et al. report 139 proteins that met their criteria of ≥ 1 peptide detected in ≥ 2 biological replicates with $\geq 90\%$ confidence under either condition (multiple normoxic and hypoxic biological replicates). Importantly, the original publication reports 122 proteins, but this number erroneously does not include 17 additional high-confidence identifications that are reported in Table S1, although they are included in the analysis of differentially abundant proteins under normoxia versus hypoxia (Wolfe and Dobos, personal communication). A DMSO-treated control was reported qualitatively and revealed only a handful of proteins, most of which were deemed by the authors to be probable promiscuous binders due to their abundance in the proteome. As in Ansong et al., only a subset of Ser/Thr kinases was detected although one additional kinase, PknH, was also identified.

Importantly, as noted by the authors, the desthiobiotin-labeled probe can also modify biotin-binding or metabolizing enzymes, and indeed, the 139 proteins include the biotin-dependent enzymes AccA1, A2, A3, and AccD5. These were not identified by Ansong et al., consistent with their use of a "label-free" alkyne probe, and this outcome highlights the chemical nature of the probe and the use of appropriate controls as key considerations in activity-based profiling studies.

Otherwise, Wolfe et al. focused primarily on comparative analyses between treatment conditions rather than mining their results for novel ATP-binding proteins. Under hypoxia, they detected several changes consistent with known shifts in RNA levels, including the upregulation of proteins such as HspX, Acg, TB31.7, Rv2624c, and Rv1738, whose expression is controlled by the hypoxic-responsive transcriptional regulator DosR (Galagan et al. 2013). Similarly, the upregulation of isocitrate lyase (Icl, Rv0467) and malate synthase G (GlcB), as also confirmed by Western blot, is consistent with a hypoxia-induced metabolic shift to alternative carbon metabolism pathways such as the glyoxylate shunt. Perhaps most obviously novel is the failure to detect most protein kinases (with the exception of PknA) under hypoxia. Although no interpretation is offered, this suggests a drastic shift in signaling and raises questions about how processes influenced by these kinases, including cell growth and division and cell wall synthesis, are regulated under hypoxia, especially given that PknG and PknB have important roles in virulence and survival (Cowley et al. 2004; Fernandez et al. 2006) and Mtb remains metabolically active and continues to grow and divide even during dormancy.

The authors also compare probe labeling under normoxia and hypoxia with or without ATPγS pre-treatment, although they consider this condition not as a negative control condition as Ansong et al. did, but as a proxy for ATP-competitive inhibitors. It is not clear that the non-covalent binding of ATPγS would be fully competitive with their probe under the conditions used (0.5 mM ATPγS, 5 mM probe), and as such, it is not surprising that they detected many known ATP-binding proteins that were labeled by the probe even in the presence of ATPγS. More surprising is the difference for several proteins in apparent ATPγS competition under normoxia and hypoxia. For example, in the presence of ATPγS, the chaperone DnaK is labeled to a greater degree by ATP-ABP under hypoxia versus normoxia, while overall probe labeling in the absence of ATPγS remains largely unchanged. The authors propose that these effects reflect dynamic binding constants for ATP based on the differential availability of co-factors during different growth states, an intriguing idea that remains to be further tested.

The two studies employing ATP-based activity-based probes were performed under different growth conditions using different labeling, enrichment and detection strategies and perhaps as a consequence yielded labeled proteomes with limited

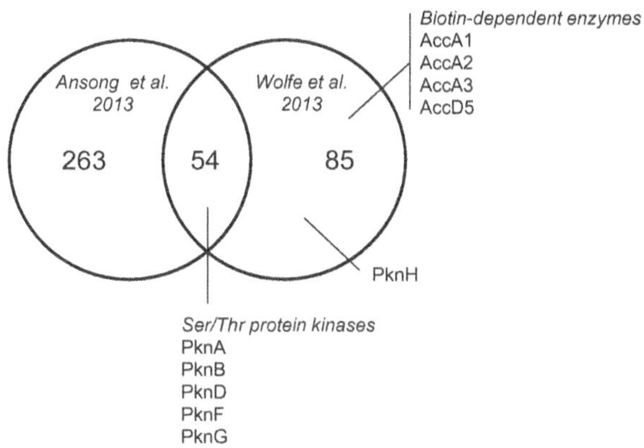

Fig. 1 Distribution of observed ATP probe-labeled proteins in *M. tuberculosis*.Diagram showing the distribution of proteins detected under any conditions reported by Ansong et al. and Wolfe et al. Five Ser/Thr protein kinases were detected by both; PknH was uniquely identified by Wolfe et al. Use of desthiobiotin-ATP may have influenced the labeling and detection of biotin-dependent enzymes via protein binding to desthiobiotin rather than ATP by Wolfe et al. 2013

Table 2 Functional categories of *M. tuberculosis* proteins identified by ATP probes

	H37Rv proteome[a]		Ansong et al. (2013)		Wolfe et al. (2013)	
Category	Number	Percent	Number	Percent	Number	Percent
Intermediary metabolism and respiration	877	22	80	25	47	34
Information pathways (transcription/ translation)	206	5	62	19	23	17
Conserved hypotheticals	907	23	49	15	13	9
Lipid Metabolism	225	6	34	11	23	17
Unknown	653	16	32	10	0	0
Cell wall and cell processes	513	13	26	8	11	8
Regulatory proteins	188	5	23	7	8	6
Virulence, detoxification and adaptation	90	2	12	4	14	10
PE/PPE	165	4	1	0	0	0
Insertion sequences and phages	136	3	0	0	0	0
Total	3960		319[b]		139	

[a]As reported by Ansong et al. (2013)

[b]This total is 2 more than the reported total number of detected proteins (317), but is consistent with the functional category distribution in Table 1 of Ansong et al. (2013)

overlap (Fig. 1). Only 54 proteins were detected in common, although their distribution across the various functional categories was largely similar, suggesting that they are both broadly sampling the ATP-binding proteome (Table 2). Compared to the overall predicted proteome, both detected proteomes had overrepresentation in the functional categories of information pathways (called transcription and translation by Ansong et al.) and lipid metabolism (both by 2- to 3-fold), but underrepresentation in the cell wall and cell processes ($\sim$2-fold). Both studies detected all 5 kinases found by Ansong et al. The failure to detect PknE, I, K, and L by either study is at first surprising, especially as a Δ*pknE* mutant has growth and virulence defects, implying that PknE is expressed (Kumar et al. 2013). On the other hand, a post-translational regulatory mechanism may keep these kinases in an ATP-binding incompetent state under the growth conditions used, or technical limitations such as low protein abundance or the absence of an appropriate lysine for labeling prevented their detection. The last point highlights a potential limitation of the probe that is not explicitly addressed by either study, but should be kept in mind when considering potential false negatives.

Overall several key implications of ATP-ABP profiling in Mtb are (1) the large group of putative ATP-binding proteins that bear no homology to known nucleotide-binding domains and (2) the apparent wholesale downregulation of protein kinases under hypoxia. Unfortunately, there have been no further studies following up on either finding.

2.2 *Profiling of Serine Hydrolase Enzymes*

As with the ATP probe studies, two studies profiling serine hydrolases in *M. tuberculosis* were published close together in 2016 (Ortega et al. 2016; Tallman et al. 2016). Ortega et al. (2016) came from the same team of Christoph Grundner (CIDR) and Aaron Wright (PNNL) that gave rise to the Ansong et al. study; Tallman et al. came from the laboratory of Kimberly Beatty (OHSU), who has specialized in fluorogenic probes for general esterase activity. While both compared serine hydrolase labeling under replicating and non-replicating conditions, Tallman et al. analyzed non-replicating cells induced by nutrient starvation in addition to hypoxia. Otherwise, the studies again differ primarily in the nature of the label on the fluorophosphonate, the methods used to detect and identify the modified proteins, and the precise growth conditions compared. We review first the overall serine hydrolase profiles detected by each of the two studies and then compare the results from both for cells cultured under hypoxia.

Ortega et al. (2016) once again employed a "label-free" alkyne probe, this time connected via a PEG linker to a fluorophosphonate group, which covalently and selectively modifies the active site serine residue of serine hydrolases. Their analysis included comparisons among multiple replicates not only of probe-treated and vehicle-treated controls, but also of probe-treated samples from cells growth grown in normoxic versus hypoxic conditions. Using similar cutoff criteria for hit identification as in their ATP-ABP study, the authors detected 208 proteins. Interestingly, the majority (130 proteins) had annotations other than hydrolase, while the remaining 78 were annotated as either hydrolases or hypothetical proteins. Ortega et al. (2016) focused on the 78 proteins for further analysis. Notably, these hits validated previous annotation for 51 serine hydrolases (as esterases, proteases, synthases, hydratases, amidases, or other hydrolases). The category "other hydrolases" included enzymes that are not hydrolytic, such as the putative peroxidase/oxidoreductase BpoC (Rv0554). Ortega et al. (2016) note that the Rv0554 crystal structure is, however, consistent with a serine hydrolase, with an alpha/beta hydrolase fold and a putative catalytic triad (Johnston et al. 2010). Rv0554 is thus an important example of a likely mis-annotation revealed by ABPP.

As for novel annotation, 27 hypothetical proteins were identified as serine hydrolases and bioinformatic analysis supported hydrolase assignment for 17 of these. ABPP provides important validation of these computational predictions, as it did in the ATP-ABP profiling, but in terms of the potential of ABPP to supply truly unique information, the set of 130 "non-hydrolases" is arguably the more interesting group to consider. Ortega et al. (2016) note that the serine hydrolase activity that they detect for these proteins may be due to an additional function or another domain in these enzymes. While the results of this ABP study raise intriguing questions about functional annotation, further biochemical studies are clearly required not only to validate assignments as hydrolases, but to determine physiological functions. Ortega et al. (2016) did follow up by expressing and purifying six of the hypothetical proteins (Rv0525, Rv0613c, Rv1192, Rv1794, Rv3311, and Rv3528c) and providing evidence for proteolytic activity for four of them, including two (Rv0613c, Rv3311) that had no bioinformatic support for hydrolase assignment. Unlike for the ATP-ABP study, however, they did not confirm activity by ABP labeling or serine modification by mutation for any of these candidates. We have confirmed Rv1192 as a serine hydrolase by these assays, but were unable to replicate the protease activity (Patel, Seeliger et al., unpublished data). Our results underscore the challenge of going beyond hydrolase assignment to characterizing the cellular functions for these enzymes, given the broad range of activities encompassed by the serine hydrolase superfamily.

Tallman et al. used a commercially available desthiobiotin-labeled probe, which also differed from the ABP used by Ortega et al. (2016) in having a saturated alkyl chain rather than a PEG linker to the fluorophosphonate-reactive group (Tallman et al. 2016). They did not compare probe-treated samples to vehicle-treated controls; the absence of this key comparative control may be one reason they limited their analysis to annotated serine hydrolases. Of the 52 proteins that met their cutoff and annotation criteria, they compared only annotated lipases and esterases to results from other studies, revealing that 14 were also detected by Ortega et al. (2016). We performed a more comprehensive comparison of the serine hydrolases and all detected proteins from both studies and found greater overlap between the two data sets (Fig. 2). Overall, 97 hydrolases (as defined by either study) were detected by at least one study with 34 detected by both (35% overlap). Interestingly, among these are the potentially mis-annotated Rv0445c (previously labeled as the peroxidase BpoC) and 6 conserved hypothetical proteins, thus corroborating the assignments by Ortega et al. (2016).

Tallman et al. noted that they detected additional annotated lipases and esterases compared to Ortega et al., perhaps due to the more lipophilic nature of their probe. Our analysis of all detected proteins showed that the serine hydrolases that they uniquely identified include additional lipid-processing enzymes, including the polyketide synthases Pks2, Pks5, and Pks6 (Quadri 2014) and the probable mycolic acid epoxide hydrolase EphD (Madacki et al. 2018). Finally, 45 "non-hydrolases" were detected by both studies. This overlap supports the hypothesis by Ortega et al. (2016) that this group represents proteins that have an additional hydrolase domain or hydrolase function. On the other hand, some may be due to non-specific detection of abundant proteins (e.g., ribosomal proteins), although this should be unlikely for Ortega et al. (2016) given their stringent comparison to vehicle-treated controls. Further, these and other proteins may have been identified due to their *interaction* with serine hydrolases. This possibility was raised by Ansong et al. for the ATP-ABP profiling.

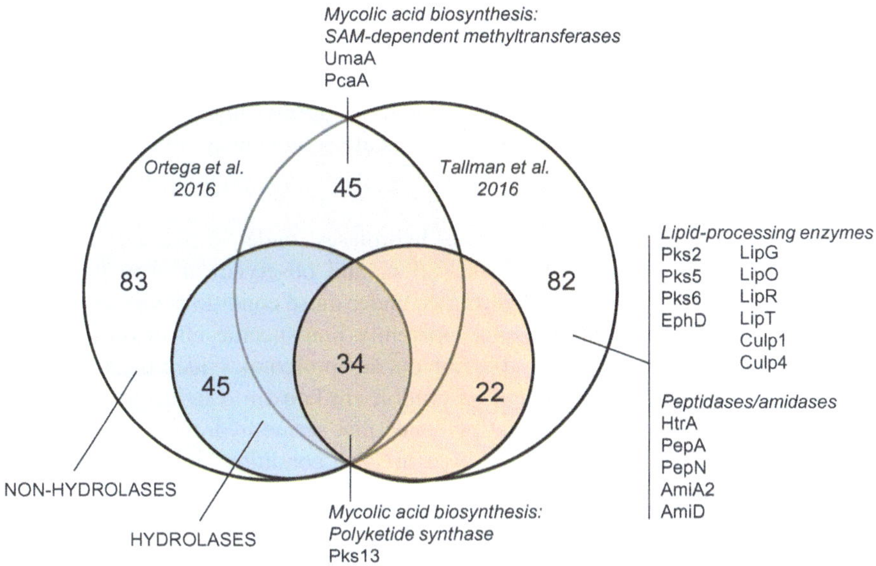

Fig. 2 Distribution of observed fluorophosphonate probe-labeled proteins in *M. tuberculosis*. Diagram showing the distribution of proteins detected under any conditions reported by Ansong et al. and Wolfe et al. For each study, the smaller shaded circle represents detected hydrolases as defined by either study. Several mycolic acid biosynthesis enzymes were detected in common, although UmaA and PcaA are unlikely serine hydrolases and may have been enriched via their interaction with Pks13. Tallman et al. uniquely identified many lipid-processing enzymes, suggesting that their probe is biased toward lipid-binding proteins. They also uniquely detected several putative peptidases and amidases

In comparing serine hydrolase labeling for Mtb cultured under normoxia versus hypoxia, Ortega et al. (2016) limited their analysis to the 78 hydrolases and hypothetical proteins that they detected under both conditions. Strikingly, the majority appear to be regulated in an on/off manner, with 41 detected only under aerated/normoxic conditions. Even for those detected under both conditions, most are downregulated (2- to 25-fold), indicating a general reduction in activity. Tallman et al. corroborated these results qualitatively by both gel-based analysis and proteomics. Both studies found few hydrolases that were active exclusively under hypoxia (1 by Tallman et al. in their limited analysis of annotated serine hydrolases; 3 by Ortega et al. (2016) among their list of serine hydrolases and hypothetical proteins). Interestingly, Tallman et al. also profiled the cells returned to normoxic conditions for a limited 4.5 h to identify serine hydrolases that change rapidly in activity during this so-called reactivation period. Perhaps due to the very short adaptation period ($\sim 1/5$ of the doubling time for Mtb under normoxia), few changes in annotated serine hydrolases were detected: Only three did not overlap with those detected under hypoxia. Importantly, Ortega et al. (2016) also performed quantitative proteomics to obtain relative protein abundances and showed that changes in protein abundance correlated poorly with changes in activity detected by

ABPP between normoxic and hypoxic cultures. Thus, they provide experimental evidence for the oft-cited unique ability of ABPP to report on enzyme function rather than simply protein expression and suggest that the lack of correlation indicates extensive post-translational regulation during hypoxia. Both studies highlight the detection of serine hydrolases that are active under both or exclusively hypoxic conditions as a unique outcome that points to potential drug targets or biomarkers.

Tallman et al. also profiled serine hydrolases under a second model of non-replicating persistence, carbon starvation, and observed an even more substantial downregulation of serine hydrolases under those conditions versus hypoxia. Unsurprisingly, given this broad and apparently non-specific effect on the entire serine hydrolase proteome, they did not pursue proteomics identification under these conditions. While hypoxia and starvation are known to induce distinct transcriptional responses, this is the first evidence that serine hydrolase activities also experience distinct shifts under these two in vitro conditions that are commonly used to induce persistence.

The remainder of the Tallman et al. study was devoted to comparing their ABP results to applications of fluorogenic esterase substrates developed in their laboratory. Notably, they identified at least one esterase, Rv3036c, that was not labeled by the fluorophosphonate probe. While we do not cover their work with fluorogenic substrates here since these are not activity-based probes, their results with these substrate analogues and their comparisons to the Ortega et al. (2016) results underscore the substrate selectivity that reactive probes exhibit in general, even if the reactive moiety is identical. Especially as such tools become used more widely, this is an important reminder that ABPP is not necessarily comprehensive for the population the probe reports on.

Limitations are also related to detection methods. The Tallman study briefly addresses at least one example of known serine hydrolases that is not detected by ABPP, LipY. They attribute lack of detection to a methodological limitation, in that the tryptic cut sites in LipY are only in domains that are not part of the mature lipase. Other examples include the lipid acyltransferases Chp1 and Chp2. Neither was detected by serine hydrolase ABPP, even though they were identified as serine hydrolases in part by using a fluorophosphonate ABP and both are expressed and active under replicating conditions based on genetic knockout and chemical inhibition experiments (Belardinelli et al. 2014; Seeliger et al. 2012; Touchette et al. 2015). While Ortega showed that changes in protein abundance and enzyme activity are not well correlated, both contribute to the probability of detection by ABPP. Protein abundance and, presumably, enzyme activity span orders of magnitude, which presents a major challenge for comprehensive detection by ABPP. Also, as proteomics is biased toward the most abundant, and for ABPP, the most active enzymes, this method may miss the potentially most vulnerable drug targets, assuming that a target must be inhibited to a significant degree to compromise function and bacterial survival. Factors that affect a protein's desirability as a drug target are of course myriad and diverse and go beyond abundance and activity, but it is important to keep in mind what may be missed by ABPP. This is particularly important when using ABPP to identify targets of inhibitors and elucidate inhibitor mechanism of action, as detailed further below.

Many of the challenges of applying ABPP and interpreting the results as discussed above are not specific to mycobacteria. However, comparisons across in vitro culture models of key phenotypic states such as dormancy highlight how the results may be influenced by particular growth conditions, such as the means and kinetics by which hypoxia is achieved. The metabolic versatility of Mtb including its ability to survive under multiple stressors that induce dormancy and to metabolize diverse carbon sources means that minor changes in culture conditions may induce significant regulatory changes, including of enzyme activity, that are detected by ABPP. Thus, we should be wary of comparisons and generalizations across studies and cautious of interpretations that may be specific to the experimental setup, rather than reporting on biologically significant phenomena. However, ABPP in mycobacteria is still in its infancy and its expanded application for profiling and inhibitor studies will reinforce these initial results.

3 Development of Activity-Based Probes for Applications to Mycobacteria

In addition to the general profiling studies reviewed above, there are two examples of activity-based probes developed specifically for applications to Mtb (Duckworth et al. 2012; Lentz et al. 2016). Both were based on inhibitors developed against specific target proteins within Mtb, and their application thus far has focused on in vitro characterization with their targets. Thus, we review them here as potential tools and separately from inhibitor-based probes used to identify inhibitor targets in the next section. Below we discuss the motivation for applying these probes in Mtb and note some of the challenges to development and application that are common to creating new ABPs.

3.1 Activity-Based Probe for the Adenylating Enzyme MbtA

The first probe, developed by Duckworth et al., is a photo-crosslinking inhibitor of MbtA. MbtA is an adenylating enzyme in the biosynthesis of the iron-chelating siderophore mycobactin, which is required for Mtb survival under iron-limiting conditions (Duckworth et al. 2012). The ABP is based on the bisubstrate inhibitor 5′-O-[N-(salicyl)sulfamoyl]adenosine (Sal-AMS) (Table 1), which not only inhibits MtbA in vitro, but exhibits a low micromolar minimum inhibitory concentration against Mtb under iron-deficient conditions. However, Sal-AMS also has activity under conditions in which mycobactin biosynthesis is not stringently required, suggesting that it has secondary activity due to off-target effects. Thus, the authors’ goal was to develop an ABP to identify the targets of Sal-AMS. Since Sal-AMS is a non-covalent inhibitor, it was necessary to add a reactive group to covalently link the inhibitor to its binding partners. Their strategy was to extend Sal-AMS at the C-2 position of the adenosine ring, where previous SAR indicated that MtbA would be tolerant to modifications. They noted that the Sal-AMS derivative, 2-phenylamino-Sal-AMS, was even more potent against MbtA in vitro and against Mtb and extended at this position further with a UV-reactive benzophenone group. To enable detection of inhibitor–protein adducts, they also added an alkyne group connected via an alkyl linker to the benzophenone. Importantly for their goal of eventually identifying inhibitor targets, they confirmed that their Sal-AMS ABP had comparable and selective activity against Mtb under iron-replete versus iron-limiting conditions.

The bulk of their study concerned the validation of Sal-AMS ABP in a series of careful assays with purified MbtA enzyme and in crude bacterial lysates. In summary, they demonstrated UV-dependent reactivity, selectivity for MbtA over bovine serum albumin as a control protein, dose-dependent labeling, competition with Sal-AMS in labeling purified MbtA, and labeling of MtbA overexpressed in *E. coli* and natively expressed in *M. smegmatis*, a non-pathogenic relative of Mtb. Interestingly, they showed that Sal-AMS ABP also labeled the *E. coli* adenylating enzyme EntE, demonstrating that their ABP has potential as a general probe for adenylating enzymes or as a scaffold for developing adenylating enzyme-specific probes. Unfortunately, despite these promising initial data, there have been no further reports regarding applications of this adenylation ABP.

3.2 Activity-Based Probe for the Serine Protease Hip1

The second ABP was developed by Lentz et al. against the virulence-associated serine protease Hip1 (Rv2224c) with the ultimate goal of tracking Hip1 activity in cells, especially in the context of host cell infection, since Hip1 is implicated in anti-inflammatory host–pathogen interactions (Lentz et al. 2016). Their broad-ranging study pursued multiple strategies to develop reporters for Hip1 activity, including screening for optimized substrates and covalent inhibitors as scaffolds for fluorogenic and activity-based probes. Their first strategy was to turn an optimized peptidomimetic substrate into a covalent inhibitor by appending a serine protease-specific reactive group, diphenyl phosphonate (DPP). However, further characterization revealed that this inhibitor was in fact reversible, which they hypothesized was due to incompatibility between the Hip1 active site and the bulky DPP group.

Their next approach was to screen a library of ~500 small molecules containing serine-reactive electrophiles. They identified a series of 7-amino-4-chloro-3-(2-bromoethoxy)isocoumarins as hits and created an ABP from the most potent compound by attaching a rhodamine dye (Table 1). Unfortunately, this probe labeled numerous enzymes in crude lysates of mouse-derived macrophage-like cells, but only weakly labeled purified Hip1 that was spiked into the sample. This result suggested that the probe would not be useful as a specific Hip1 reporter in an infection context, although the authors did demonstrate Hip1-dependent labeling of Mtb within macrophage-like cells using fluorogenic peptidomimetic substrates. Lentz et al. then applied their newfound knowledge of Hip1 substrate specificity to create a lysine-modified isocoumarin derivative that more selectively inhibited Hip1 in macrophage lysates. They point out that this modified inhibitor has the potential to become an ABP via the attachment of a tag at one of several available sites, but no further work in this direction has been reported.

Both Duckworth et al. and Lentz et al. made significant progress toward their goals, but have not yet applied their tools in the cellular context. The adenylation probe is poised to be used in Mtb but the probe may have become less useful as a tool to elucidate the targets of a potential drug candidate since subsequent development of the inhibitor scaffold has introduced additional modifications to improve its pharmacological properties. However, Sal-AMS ABP remains intriguing as a probe for adenylating enzymes, pending validation studies to examine its generality across this enzyme class. Duckworth et al. chose to emphasize its potential as a starting point for enzyme-specific probes. This application arguably faces the same challenges illustrated by the development of the serine protease probe by Lentz et al. As in antibiotic development in general, target-based approaches are laborious undertakings that are poised to yield sensitive but not necessarily selective probes. Moreover, Lentz et al. focused on testing with Hip1 added to crude mammalian cell lysates; the goal of imaging in the infection context will depend further on the ability of the probe to enter both mammalian and Mtb cells.

4 Identification of Inhibitor Targets Using ABPP in Mycobacteria

In contrast to the previous section, we here review probes that, while also derived from inhibitors, are not motivated by or designed for specific targets. Thus, the studies discussed below validated their probes for retention of activity against Mtb, but otherwise proceeded immediately to identify their targets in mycobacteria (Lehmann et al. 2017; Ravindran et al. 2014). The two inhibitor-based probes are similar: Both are designed to mimic lipids or fatty acids and are based on a β-lactone-reactive group that covalently modifies serine hydrolase active sites.

Since ABPP uniquely provides biochemical evidence for target inhibition, target profiles have the potential to guide follow-up studies to identify inhibitor mechanism of action and to provide a method with advantages over traditional genetic approaches using spontaneous resistant mutants. Using the two studies on β-lactones as well as one example identifying the targets of the non-covalent kinase inhibitor staurosporine (Hatzios et al. 2013), we discuss the exciting potential and inherent challenges in exploiting ABPP for mechanism of action studies in mycobacteria.

4.1 Targets of Tetrahydrolipstatin

The first report sought to identify the targets of the β-lactone tetrahydrolipstatin (THL), which is clinically approved for the treatment of severe obesity (as Orlistat), but also has moderate in vitro activity as an antibiotic (Ravindran et al. 2014). Ravindran et al. were motivated by earlier studies characterizing THL activity against both replicating and hypoxia-induced non-replicating mycobacteria to examine THL targets as a function of growth condition. THL has several alkyl chains that provide convenient positions for adding an alkyne to allow labeling and detection of covalent inhibitor–enzyme adducts formed upon inhibitor treatment of whole Mtb cells. Ravindran et al. synthesized three possible derivatives and chose THL-alk1 (Table 1) based on retention of activity against *M. bovis* BCG, an avirulent vaccine strain, and efficient protein target labeling that was largely competed by an excess of THL.

Their approach to identifying the targets emphasized reproducibility across multiple labels (biotin-azide or biotin-rhodamine-azide) and protein isolation strategies, including from individual gel bands or from entire gel lanes following avidin enrichment. In all replicates, positive identification was made by comparison to a DMSO vehicle-treated control. Since detection and isolation of proteins for identification was gel-based, the analysis was qualitative and likely uncovered the most abundant and active proteins modified by THL-alk1. After manually removing proteins detected in the DMSO-treated controls, the definitive criterion for positive protein identification was detection in at least two of the three labeling replicates from log-phase conditions and led to a short list of 14 proteins. Unsurprisingly, given the structure of THL-alk1, this group was dominated by proteins associated with lipid processing, including putative lipases (Lip D, G, H, I, M, O, N, V, W) and lipid biosynthetic enzymes (thioesterase TesA and the lipid acyltransferase Ag85C) (Table 2).

More interestingly, Ravindran et al. tracked the activity of abundant THL-alk1-labeled proteins over several growth conditions: log-phase culture, 18 days under hypoxia by gradual oxygen depletion in sealed flasks, and 3 days of regrowth upon re-introduction of oxygen. Similar to the serine hydrolase profiles observed by Ortega et al. (2016) and Tallman et al., targeted hydrolases were overall downregulated during hypoxia, although unlike Tallman et al. the authors observed recovery of activity upon re-aeration, likely because the incubation period was significantly longer (3 days vs. 4.5 h). Importantly, two major bands in fluorescent dye-labeled samples retained activity during hypoxia and based on MS identification included LipM, Ag85C, and TesA. This result is consistent with the serine hydrolase profiling studies, which detected these three proteins under both aerating and hypoxic conditions (Tallman et al. detected Ag85C only after re-aeration). Ravindran et al. further confirmed that the observed changes in THL-alk1 reactivity did not correlate with corresponding transcript levels, suggesting that strong regulation occurs on the protein level. The authors went on to confirm that overexpression of LipH and TesA resulted in reduced susceptibility to

THL. Despite these data, they do not comment on the potential contribution of these or other hydrolases to the mechanism of action for THL, but rather emphasize the utility of THL-alk1 as a functional probe for the observed targets and in particular as a diagnostic tool for monitoring lipase activity.

4.2 *Targets of the β-Lactone EZ210*

The second study by Lehmann et al. identified a β-lactone with a low micromolar minimum inhibitory concentration (MIC) against Mtb from a small library of related compounds originally developed as inhibitors of the protease ClpP in *S. aureus* (Lehmann et al. 2017). Interestingly, their hit compound EZ120 was selective for mycobacteria versus other bacteria and had desirable drug-like properties including a low micromolar minimal bactericidal concentration and toxicity to mouse macrophages only at >200-fold the MIC. They introduced an alkyne at the equivalent position to Ravindran et al. to create a probe (EZ120P) for identifying the targets of EZ120 (Table 1). To maximize protein identification, they analyzed both soluble and membrane fractions of lysates from probe-treated cells. Unfortunately, due to experimental limitations, they were able to analyze only the soluble fraction from Mtb and otherwise analyzed identically treated *M. smegmatis* to identify proteins in the insoluble fraction. The quantitative MS analysis comparing EZ120P- and DMSO-treated samples yielded a limited number of hits that met their statistical and fold-change criteria: 5 Mtb and 14 *M. smegmatis* proteins from the soluble fraction and 9 *M. smegmatis* proteins from the insoluble fraction. Our analysis of these lists showed that among the soluble proteins detected in *M. smegmatis* as targets of EZ120P, 12 have Mtb orthologues, but none of these were detected in the corresponding experiment in Mtb (Table 3). This raises questions about the validity of extrapolating results between different mycobacterial species. Notably, both Mtb and *M. smegmatis* were treated with the same concentration of probe for the same amount of time despite their very different growth characteristics (doubling time of 24 vs. 3 h) and susceptibilities to EZ120 (MIC of 1.6 vs. 50 μM).

Several proteins labeled by EZ120P overlap with those labeled by THL-alk1, namely LipH (also called LipI), LipM, and TesA. In addition, both probes labeled one or two isoforms of the Ag85 complex, a set of homologous lipid acyltransferases. The overlap in target profiles is not surprising given the similarities between the inhibitors. Differences in analysis methods and choice of mycobacterial species make it difficult to correlate differences in the target lists with structural differences between the two probes, although THL-alk1 targets included more lipases and were overall more restricted to lipid-processing enzymes.

For further functional studies on targets of EZ120P, Lehmann et al. considered only hits that were predicted by genetic studies to be essential to Mtb survival in vitro, presumably under the assumption that these were most likely to contribute to the mechanism of action for EZ120. Specifically, they focused on Pks13 and Ag85A, both proteins involved in the synthesis of cell wall lipids containing essential fatty acids known as mycolic acids. In addition to validating biochemically that EZ120P modifies both proteins, they showed that EZ120 treatment reduces mycolic acid levels in a dose-dependent manner, at least in *M. smegmatis*. However, the effect was significantly weaker than that of isoniazid, a frontline TB drug and known inhibitor of mycolic acid biosynthesis. Thus, it is not clear whether inhibition of mycolic acids contributes to the mechanism of action for EZ120.

4.3 Targets of Staurosporine

In the final example of inhibitor target identification in mycobacteria, Ortega et al. (2014) sought to investigate the role of kinases in Mtb reactivation upon re-aeration from hypoxia by identifying the targets of the promiscuous kinase inhibitor staurosporine (Ortega et al. 2014). In contrast to the above studies, Ortega et al. (2014) did not create a probe from staurosporine, but instead performed competitive ABPP with their ATP-ABP probe (Table 1). In this format, targets are identified by competition with the inhibitor of interest and detected by loss rather than gain of signal. The advantage is that the inhibitor is not modified to enable subsequent tagging and therefore does not need to be validated for activity as a distinct chemical entity, e.g., as in the case of THL-alk1 and EZ120P versus respective parent compounds. The disadvantage is that the inhibitor and ABP must be well matched; that is, the assumption is made that the inhibitor does not exert its biological effects via binding to targets other than those detected by ABPP. This is valid for staurosporine, an extensively characterized ATP-competitive inhibitor. However, staurosporine is a non-covalent inhibitor, which necessitated the use of sub-saturating concentrations of ATP-ABP to avoid competing staurosporine off its targets. This strategy identified PknF, PknD, and PknB as the most strongly inhibited proteins, although another 13 proteins were inhibited $\geq$twofold. The authors went on to demonstrate using null mutants that of the kinases inhibited by staurosporine, only loss of function in PknB compromised Mtb recovery from hypoxia.

4.4 Using Activity-Based Protein Profiling to Investigate Inhibitor Mechanism of Action

Overall, although both Lehmann et al. and Ravindran et al. aimed to identify inhibitor targets, neither ultimately attempted to correlate their ABPP results with the mechanism(s) by which the inhibitors exert their effects on Mtb. These interpretations are appropriately conservative: Although ABPP reveals targeted proteins, it does not reveal which targets are biologically important. It is likely a fallacy to assume that any inhibitor will target only one or two proteins. More likely, the

Table 3 Additional annotation of proteins targeted by EZ120P in *M. smegmatis* and *M. tuberculosis* (adapted from Lehmann et al. 2017)

Fraction	Organism	UnitProt	Msm mc^2155 Locus	Mtb H37Rv Locus	Label	Mtb Predicted Essentiality	Annotation[a]
S	Msm	A0R347	MSMEG_5341				Dipeptidyl aminopeptidase/acylaminoacyl peptidase
S	Msm	A0QYJ6	MSMEG_3682				Conserved hypothetical protein
S	Msm	A0QQY9	MSMEG_0924	Rv0480c		–	Possible amidohydrolase
S	Msm	A0QS51	MSMEG_1352	Rv0646c	LipG	–	Probable lipase/esterase
S	Msm	A0R425	MSMEG_5830	Rv0798c	Cfp29	–	29 kDa antigen
S	Msm	A0QY24	MSMEG_3508	Rv1191		–	Conserved protein
S	Msm	A0R1Z7	MSMEG_4934	Rv1314c		–	Conserved protein
S	Msm	A0QZ36	MSMEG_3793	Rv1641	InfC	+	Probable initiation factor IF-3
S	Msm	A0QYI2	MSMEG_3667	Rv2045c	LipT	–	Carboxylesterase
S	Msm	A0R070	MSMEG_4280	Rv2214c	EphD	–	Possible short-chain dehydrogenase
S	Msm	A0R0N6	MSMEG_4450	Rv2307c		–	Conserved hypothetical protein
S	Msm	A0QVP3	MSMEG_2644	Rv2800		–	Possible hydrolase
S	Msm	A0QVM9	MSMEG_2630	Rv2837c		+	Hypothetical protein
S	Msm	A0QVC0	MSMEG_2521	Rv2888c	AmiC	–	Probable amidase/aminohydrolase
S	Mtb	I6XD69; A0A089QNB9	MSMEG_0408	Rv2048c	Pks12[b]	+	Polyketide synthase
S	Mtb	O53512; I6X3I6	MSMEG_4244	Rv2178c	**AroG**	+	3-deoxy-D-arabino-heptulosonate 7-phosphate synthase
S	Mtb	Q50681	MSMEG_0302	Rv2284	LipM	–	Probable esterase
S	Mtb	P9WQD5; I6X5R8	MSMEG_4514	Rv2928	TesA	–	Probable thioesterase
S	Mtb	I6X8D2	MSMEG_6392	Rv3800c	**Pks13**	+	Polyketide synthase
I	Msm	A0R404	MSMEG_5659	Rv0194		–	Probable transmembrane

(continued)

Table 3 (continued)

Fraction	Organism	UnitProt	Msm mc^{2}155 Locus	Mtb H37Rv Locus	Label	Mtb Predicted Essentiality	Annotation[a]
							multidrug efflux pump
I	Msm	A0R2D2	MSMEG_5070	Rv1223	**HtrA/ DegP**	+	Probable serine protease
I	Msm	A0QWT7	MSMEG_3059	Rv1400c	LipH	–	Probable lipase
I	Msm	A0QU51	MSMEG_2078	Rv1886c	**Ag85B**[b]	–	Mycolyltransferase
I	Msm	A0R198	MSMEG_4673	Rv2460c	**ClpP2**	+	Probable ATP-dependent CLP protease proteolytic subunit 2
I	Msm	A0R1G6	MSMEG_4745	Rv2518c	LptB[b]	+	Probable LD-transpeptidase
I	Msm	A0QWJ2	MSMEG_2961	Rv2587c	SecD[b]	+	Probable protein-export membrane protein

[a]Annotation from Tuberculist or Smegmalist

[b]Modified assignment of predicted essentially based on FLUTE (http://orca2.tamu.edu/U19/)

S = soluble; *I* = insoluble; *Msm* = *M. smegmatis*; *Mtb* = *M. tuberculosis*

Bold = originally assigned by Lehmann et al. (2017) as predicted essential majority of the observed targets are not biologically important. Alternatively, inhibitors may exert their effects via the inhibition of multiple targets. The first hypothesis is supported by Ortega et al. (2014), in which the inhibition of PknB was responsible for staurosporine-sensitive reactivation of Mtb while the other major targets PknF and PknD proved dispensable. However, the different possibilities raised by ABPP-based target identification have not otherwise been thoroughly explored for novel inhibitors.

ABPP has the potential to accelerate mechanism of action studies compared to genetic methods, especially in slow-growing pathogens like Mtb, if results could be more effectively leveraged to prioritize observed targets. However, there are several key issues to consider that the studies above exemplify in their methods, but do not discuss. First, the inhibitor treatment conditions used to measure biological activity differ significantly from those used to identify targets: While minimal inhibitory concentrations were determined over a period of days with cultures starting at a low density of Mtb, target identification was performed after treating highly concentrated Mtb cell suspensions for a few hours. In the case of Ortega et al. (2014), competitive ABPP was performed on lysates.

In considering this discrepancy in experimental timescales, we note that the kinetics of ABPP in live cells has, to our knowledge, never been explored. Although ABPs (i.e., the inhibitors from which they are derived) are considered irreversible modifiers, we do not know what the stabilities of the acyl intermediates are, nor how protein turnover influences the effective concentration of active enzymes in the cell. Since observed inhibitor activity is the integrated result of all these factors, the kinetics of target modification is a crucial and as yet unexplored aspect of ABPP that could extend the usefulness of ABPP for investigating inhibitor mechanism of action. One hypothesis is that some targets may be abundant and/or promiscuous, but are not inhibited on a timescale consistent with biological effects that would be observed in, for example, a pulse-labeling/washout experiment that examines the persistent suppression of growth after the removal of antibiotic, also known as the post-antibiotic effect. Those few targets, ideally, that are targeted with kinetics similar to observed post-antibiotic effects would be strong candidates for contributing to the mechanism of action.

ABPP as a function of time is potentially a resource-heavy endeavor requiring significant investment in proteomics; on the other hand, a well-resolved set of targets could more easily be followed using gel-based fluorescence detection, as was done by Ravindran et al. However, fluorescence detection raises a second issue: the specter of dynamic range. As noted earlier, methods used for isolation and detection in ABPP, particularly gel-based but also affinity enrichment methods followed by proteomic identification, risk bias toward the most abundant and most active proteins. Proteins expressed at low levels or inhibited by only a few fold may contribute significantly to inhibitor activity, but may not be detected or prioritized using current approaches and cutoff criteria. Indeed, inducible proteolysis has revealed that the depletion of different drug targets has diverse effects on mycobacterial growth (Wei et al. 2011). While protein degradation is not directly equivalent to chemical inhibition, these results support the idea that the degree of biochemical inhibition as detected by ABPP does not necessarily translate into biological activity.

5 Summary and Outlook

In summary, ABPP studies in mycobacteria have provided key information about different subsets of the proteome and their activities upon adaptation to different stresses, primarily hypoxia. The results have indicated many tantalizing possibilities and future directions, including the use of probes as diagnostic tools for tracking the activity of particular enzyme subsets and, in certain cases, following activity within particular metabolic pathways.

Mtb encounters and adapts to unique extracellular and intracellular environments during infection that likely require concerted changes in regulatory pathways that sense the environment and allow the bacterium to adapt appropriately. In particular, the studies discussed above shed light on the importance of protein regulation, likely independent of genetic or transcriptional regulation, as a mechanism for adapting to stressors. ABPP is poised to provide unique insights into how bacteria adapt to various environments via post-translational regulation of protein activity. Furthermore, applying ABPP to other pathogenic mycobacterial species such as *M. leprae* and *M. abscessus* remains an unexplored area that may provide important information about bacterial physiology and virulence in these clinically important but understudied pathogens.

Inhibitor development directed by ABPP is currently a nascent effort in mycobacteria. The studies discussed here demonstrated the initial potential to influence drug development via the identification of protein targets, but also illustrated key challenges related to selectivity and potential target redundancy. Many understudied and overlooked aspects of drug development have influenced our understanding of inhibitor kinetics and stability. Studies in mammalian and protozoal systems have demonstrated that well-focused and multidimensional approaches that are coupled with ABPP can yield potentially exciting inhibitors for drug development and interesting tools for understanding the physiology of these organisms (Arastu-Kapur et al. 2008; Greenbaum et al. 2002; Viader et al. 2016). The challenges of pursuing ABPP in mycobacteria should be viewed as opportunities to expand our understanding of what targets are druggable in mycobacteria and how they contribute to bacterial survival.

Acknowledgements We thank Dr. Benjamin Cravatt, Dr. Peter Tonge, and members of the Seeliger laboratory for helpful discussions on topics and concepts discussed in this review.

References

Ansong C et al (2013) Identification of widespread adenosine nucleotide binding in *Mycobacterium tuberculosis*. Chem Biol 20:123–133

Arastu-Kapur S et al (2008) Identification of proteases that regulate erythrocyte rupture by the malaria parasite *Plasmodium falciparum*. Nat Chem Biol 4:203–213. https://doi.org/10.1038/nchembio.70

Barglow KT, Cravatt BF (2007) Activity-based protein profiling for the functional annotation of enzymes. Nat Methods 4:822–827. https://doi.org/10.1038/nmeth1092

Belardinelli JM, Larrouy-Maumus G, Jones V, Sorio de Carvalho LP, McNeil MR, Jackson M (2014) Biosynthesis and translocation of unsulfated acyltrehaloses in *Mycobacterium tuberculosis*. J Biol Chem 289:27952–27965. https://doi.org/10.1074/jbc.M114.581199

Cole ST et al (1998) Deciphering the biology of *Mycobacterium tuberculosis* from the complete genome sequence. Nature 393:537–544

Cowley S et al (2004) The *Mycobacterium tuberculosis* protein serine/threonine kinase PknG is linked to cellular glutamate/glutamine levels and is important for growth in vivo. Mol Microbiol 52:1691–1702. https://doi.org/10.1111/j.1365-2958.2004.04085.x

Cravatt BF, Wright AT, Kozarich JW (2008) Activity-based protein profiling: from enzyme chemistry to proteomic chemistry. Annu Rev Biochem 77:383–414. https://doi.org/10.1146/annurev.biochem.75.101304.124125

Duckworth BP, Wilson DJ, Nelson KM, Boshoff HI, Barry CE 3rd, Aldrich CC (2012) Development of a selective activity-based probe for adenylating enzymes: profiling MbtA Involved in siderophore biosynthesis from *Mycobacterium tuberculosis*. ACS Chem Biol 7:1653–1658. https://doi.org/10.1021/cb300112x

Ernst JD (2012) The immunological life cycle of tuberculosis. Nat Rev Immunol 12:581–591. https://doi.org/10.1038/nri3259

Fernandez P, Saint-Joanis B, Barilone N, Jackson M, Gicquel B, Cole ST, Alzari PM (2006) The Ser/Thr protein kinase PknB is essential for sustaining mycobacterial growth. J Bacteriol 188:7778–7784. https://doi.org/10.1128/JB.00963-06

Galagan JE et al (2013) The *Mycobacterium tuberculosis* regulatory network and hypoxia. Nature 499:178–183. https://doi.org/10.1038/nature12337

Glickman MS, Cox JS, Jacobs WR (2000) A novel mycolic acid cyclopropane synthetase is required for cording persistence, and virulence of *Mycobacterium tuberculosis*. Mol Cell 5:717–727

Gold B, Nathan C (2017) Targeting phenotypically tolerant *Mycobacterium tuberculosis*. Microbiol Spectr 5 https://doi.org/10.1128/microbiolspec.tbtb2-0031-2016

Greenbaum DC et al. (2002) A role for the protease falcipain 1 in host cell invasion by the human malaria parasite. Science 298:2002–2006 https://doi.org/10.1126/science.1077426

Hatzios SK et al. (2013) Osmosensory signaling in *Mycobacterium tuberculosis* mediated by a eukaryotic-like Ser/Thr protein kinase. Proc Natl Acad Sci USA 110:E5069–5077. https://doi.org/10.1073/pnas.1321205110

Johnston JM, Jiang M, Guo Z, Baker EN (2010) Structural and functional analysis of Rv0554 from *Mycobacterium tuberculosis*: testing a putative role in menaquinone biosynthesis. Acta Crystallogr D Biol Crystallogr 66:909–917. https://doi.org/10.1107/S0907444910025771

Kumar D, Palaniyandi K, Challu VK, Kumar P, Narayanan S (2013) PknE, a serine/threonine protein kinase from *Mycobacterium tuberculosis* has a role in adaptive responses. Arch Microbiol 195:75–80. https://doi.org/10.1007/s00203-012-0848-4

Lehmann J et al (2017) An antibacterial beta-lactone kills *Mycobacterium tuberculosis* by infiltrating mycolic acid biosynthesis. Angew Chem Int Ed Engl. https://doi.org/10.1002/anie.201709365

Lentz CS et al (2016) Design of selective substrates and activity-based probes for hydrolase important for pathogenesis 1 (HIP1) from *Mycobacterium tuberculosis*. ACS Infect Dis https://doi.org/10.1021/acsinfecdis.6b00092

Madacki J et al (2018) Impact of the epoxide hydrolase EphD on the metabolism of mycolic acids in mycobacteria. J Biol Chem 293:5172–5184. https://doi.org/10.1074/jbc.RA117.000246

Meena LS, Chopra P, Vishwakarma RA, Singh Y (2013) Biochemical characterization of an S-adenosyl-l-methionine-dependent methyltransferase (Rv0469) of *Mycobacterium tuberculosis*. Biol Chem 394:871–877. https://doi.org/10.1515/hsz-2013-0126

Ortega C et al (2016) Systematic survey of serine hydrolase activity in *Mycobacterium tuberculosis* defines changes associated with persistence cell. Chem Biol 23:290–298. https://doi.org/10.1016/j.chembiol.2016.01.003

Ortega C et al (2014) *Mycobacterium tuberculosis* Ser/Thr protein kinase B mediates an oxygen-dependent replication switch. PLoS Biol 12:e1001746. https://doi.org/10.1371/journal.pbio.1001746

Portevin D, De Sousa-D'Auria C, Houssin C, Grimaldi C, Chami M, Daffe M, Guilhot C (2004) A polyketide synthase catalyzes the last condensation step of mycolic acid biosynthesis in mycobacteria and related organisms. Proc Natl Acad Sci U S A 101:314–319. https://doi.org/10.1073/pnas.0305439101

Quadri LE (2014) Biosynthesis of mycobacterial lipids by polyketide synthases and beyond. Crit Rev Biochem Mol Biol 49:179–211. https://doi.org/10.3109/10409238.2014.896859

Ravindran MS, Rao SP, Cheng X, Shukla A, Cazenave-Gassiot A, Yao SQ, Wenk MR (2014) Targeting lipid esterases in mycobacteria grown under different physiological conditions using activity-based profiling with Tetrahydrolipstatin (THL). Mol Cell Proteomics 13:435–448 https://doi.org/10.1074/mcp.m113.029942

Rittershaus ES, Baek SH, Sassetti CM (2013) The normalcy of dormancy: common themes in microbial quiescence. Cell Host Microbe 13:643–651. https://doi.org/10.1016/j.chom.2013.05.012

Rustad TR, Sherrid AM, Minch KJ, Sherman DR (2009) Hypoxia: a window into *Mycobacterium tuberculosis* latency. Cell Microbiol 11:1151–1159. https://doi.org/10.1111/j.1462-5822.2009.01325.x

Seeliger JC et al (2012) Elucidation and chemical modulation of sulfolipid-1 biosynthesis in *Mycobacterium tuberculosis*. J Biol Chem 287:7990–8000. https://doi.org/10.1074/jbc.M111.315473

Simon GM, Cravatt BF (2010) Activity-based proteomics of enzyme superfamilies: serine hydrolases as a case study. J Biol Chem 285:11051–11055. https://doi.org/10.1074/jbc.R109.097600

Tallman KR, Levine SR, Beatty KE (2016) Small-molecule probes Reveal Esterases with persistent activity in dormant and reactivating *Mycobacterium tuberculosis* ACS. Infect Dis 2:936–944. https://doi.org/10.1021/acsinfecdis.6b00135

Touchette MH, Holsclaw CM, Previti ML, Solomon VC, Leary JA, Bertozzi CR, Seeliger JC (2015) The rv1184c locus encodes Chp2, an acyltransferase in *Mycobacterium tuberculosis* polyacyltrehalose lipid biosynthesis. J Bacteriol 197:201–210. https://doi.org/10.1128/JB.02015-14

Viader A et al. (2016) A chemical proteomic atlas of brain serine hydrolases identifies cell type-specific pathways regulating neuroinflammation Elife 5 https://doi.org/10.7554/elife.12345

Wei J-R et al (2011) Depletion of antibiotic targets has widely varying effects on growth. Proc Natl Acad Sci U S A 108:4176–4181. https://doi.org/10.1073/pnas.1018301108

Wolfe LM et al (2013) A chemical proteomics approach to profiling the ATP-binding proteome of *Mycobacterium tuberculosis*. Mol Cell Proteomics 12:1644–1660. https://doi.org/10.1074/mcp.M112.025635

Zimmer JS, Monroe ME, Qian WJ, Smith RD (2006) Advances in proteomics data analysis and display using an accurate mass and time tag approach. Mass Spectrom Rev 25:450–482. https://doi.org/10.1002/mas.20071

Activity-Based Protein Profiling at the Host–Pathogen Interface

Yekaterina Kovalyova and Stavroula K. Hatzios

Contents

Abstract Activity-based protein profiling (ABPP) is a technique for selectively detecting reactive amino acids in complex proteomes with the aid of chemical probes. Using probes that target catalytically active enzymes, ABPP can rapidly define the functional proteome of a biological system. In recent years, this approach has been increasingly applied to globally profile enzymes active at the host–pathogen interface of microbial infections. From in vitro co-culture systems to

Y. Kovalyova · S. K. Hatzios (✉)
Microbial Sciences Institute, Yale University, West Haven, CT 06516, USA
e-mail: stavroula.hatzios@yale.edu

S. K. Hatzios
Department of Molecular, Cellular, and Developmental Biology,
Yale University, New Haven, CT 06511, USA

Y. Kovalyova · S. K. Hatzios
Department of Chemistry, Yale University, New Haven, CT 06511, USA

Current Topics in Microbiology and Immunology (2019) 420:73–92
DOI 10.1007/82_2018_129

Published Online: 11 September 2018

animal models of infection, these studies have revealed enzyme-mediated mechanisms of microbial pathogenicity, host immunity, and metabolic adaptation that dynamically shape pathogen interactions with the host.

1 Introduction

Enzymes regulate the metabolic response to an infection, can directly influence host–pathogen interactions, and are excellent drug targets. However, studying active enzymes in complex infection models using classic transcriptional or proteomic profiling methods presents several challenges: post-translational modifications that alter protein function are often overlooked, enzymes expressed at low levels in infected tissues may remain undetected, and dynamic changes in enzyme function (e.g., in response to changing pH or protein–protein interactions) can be difficult to measure (Heal and Tate 2012). Activity-based protein profiling (ABPP) addresses these problems by using chemical probes to facilitate the selective detection and enrichment of enzymes that are not only expressed, but active during an infection (Cravatt et al. 2008). Activity-based probes (ABPs) form selective, irreversible covalent linkages to defined classes of reactive amino acids, thereby appending a fluorescent dye or affinity tag onto labeled enzymes. In this manner, ABPs can reveal changes in the functional proteome of host and microbial cells through fluorescence imaging or mass spectrometry (Sadler and Wright 2015).

Activity-based approaches offer several key advantages for studies of enzymes functional at the host–pathogen interface. While genetically encoded reporters are routinely used to monitor the localization and expression of low-abundance proteins (Kanca et al. 2017), gene-fusion strategies are only possible in genetically tractable organisms. In contrast, ABPs can be used to elucidate these biochemical properties in microbial pathogens and more complex animal hosts that are not amenable to routine genetic manipulation. Furthermore, ABPP can identify secreted enzymes that are active during infection, but whose potential contributions to microbial survival or virulence may be missed by high-throughput, transposon-insertion sequencing techniques. Transposon libraries are frequently used to identify genes required for pathogen growth in infected cells or animals; however, the fitness defect of a transposon mutant lacking a secreted enzyme may be suppressed by *trans*-complementation (i.e., supplementation of the missing factor by other members of the transposon library) (Carey et al. 2018). In addition, ABPP can accelerate the discovery and development of drugs targeting enzymes that shape host–pathogen interactions. Broad-spectrum ABPs can provide a template for the design of more specific inhibitors or be used to identify novel inhibitors from a compound library via functional screens (Moellering and Cravatt 2012). Finally, ABPP can reveal enzymes that are only active during infection and could potentially serve as diagnostic markers. Thus, activity-based approaches provide fundamental insights into enzymes functional at the host–pathogen interface that can be harnessed for translational applications.

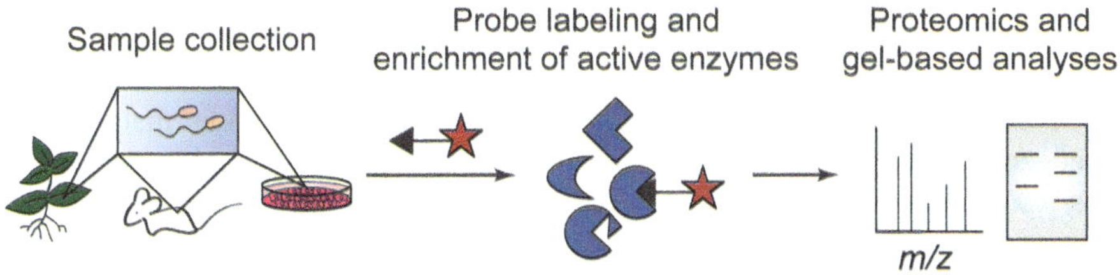

Fig. 1 Identifying enzymes active at the host–pathogen interface with ABPP. The proteomes of infected plants, animals, and cells can be labeled with ABPs targeting the active site of a particular enzyme class to facilitate detection of catalytically active enzymes by gel-based or mass spectrometry analyses

Here, we describe how ABPP has been used to study the functional proteome of microbial infections. Though activity-based approaches have also been applied to study specific mechanisms of microbial pathogenesis, including antibiotic resistance, virulence-associated enzyme activities, and virus–host interactions, a number of recently penned review articles offer excellent coverage of these topics (Heal and Tate 2012; Puri and Bogyo 2013; Sadler and Wright 2015; Shahiduzzaman and Coombs 2012; Strmiskova et al. 2016). In this review, we focus on applications of ABPP to studies of pathogen (principally bacterial) interactions with a variety of hosts, including mammalian cells, plants, and animal models of infection (Fig. 1). We discuss how ABPs can facilitate global profiling of enzymes active at the host–pathogen interface, uncover enzyme-mediated mechanisms of microbial virulence, and identify novel inhibitors of enzymes that modulate host–pathogen interactions. Finally, we describe how future applications of this technology are poised to deepen our understanding of the biochemical interactions that shape microbial infections.

2 Applications of ABPP to In Vitro Models of Microbial Infection

Cell-culture models offer a natural starting point for studies of host–pathogen interactions. By distilling the infection interface to a single host-cell type, these systems facilitate detailed investigations of biochemical factors that shape host–pathogen dynamics at the molecular level. Furthermore, relative to infected animals, in vitro infection models provide superior flexibility and economy for evaluating a wide range of infection conditions (Duell et al. 2011). For these reasons, mammalian cells, in addition to blood serum and plasma, have been particularly useful for uncovering activity-based mechanisms of microbial virulence and host immunity during infection.

2.1 Detecting Host Enzymes Active in Infected Cells

ABPP is a valuable technique for studying biochemical processes that promote pathogen clearance from the host. The gastrointestinal pathogen *Salmonella enterica* serovar Typhimurium has been widely used to investigate enzyme-mediated immune responses to bacterial infection (Behnsen et al. 2015; Broz et al. 2012). In this section, we describe how ABPP has been applied to uncover changes in protein degradation, cell death, and metabolism that underlie immune cell interactions with *S.* Typhimurium, illustrating the breadth of cellular functions that can be evaluated in infected cells using activity-based approaches.

Ubiquitination is a post-translational modification that induces protein degradation in eukaryotic cells (Swatek and Komander 2016). Deubiquitinases (DUBs) regulate this process by cleaving ubiquitin modifications from proteins, thus inhibiting their degradation (Mevissen and Komander 2017). Manipulation of DUB activity by bacterial pathogens can alter the abundance of proteins that mediate host responses to infection (Ashida et al. 2014). Kummari et al. applied ABPP to determine how *S.* Typhimurium impacts DUB activity in infected avian macrophages (Kummari et al. 2015). Using a ubiquitin-based ABP, the authors identified a DUB, UCH-L5, with enhanced activity in *S.* Typhimurium-infected cells. Overexpression of UCH-L5 in avian macrophages correlated with increased production of the pyroptosis-associated factors caspase-1 and IL-1β; in addition, production of these factors was suppressed by a chemical inhibitor of UCH-L5 activity (Kummari et al. 2015). These findings suggest a possible role for UCH-L5 in modulating pyroptosis. Though additional studies are needed to determine how this enzyme influences *S.* Typhimurium survival in infected macrophages, this work illustrates how ABPP can be used to uncover pathogen-induced proteolytic pathways in the host.

Activity-based approaches can also be used to study host proteases that mediate immune responses to infection. The caspase family of cysteine proteases plays an important role in programmed cell death pathways (Creagh 2014). Activation of caspase-1 initiates cell death by pyroptosis (Place and Kanneganti 2018), whereas caspase-3 and caspase-7 promote apoptosis (Walsh et al. 2008). To elucidate cell death mechanisms triggered by *S.* Typhimurium, a fluorescent acyloxymethyl ketone, AWP28, was used to monitor relative levels of caspase-1, 3, and 7 activity in bone marrow-derived macrophages infected with *S.* Typhimurium (Puri et al. 2012). Though caspase-1-mediated pyroptosis was the principal form of cell death observed in infected macrophages, knocking down the *Casp1* gene led to an increase in caspase-3 and 7 activities, accompanied by enhanced apoptosis. Thus, crosstalk between these caspases appears to mediate the fate of host cells during *S.* Typhimurium infection.

Host responses to infection can also be influenced by pathogen-induced changes in cellular metabolism. Several host lipids, including the endogenous cannabinoids anandamide (AEA) and 2-arachidonoylglycerol (2-AG), are believed to play immunomodulatory roles in infected cells (Hernandez-Cervantes et al. 2017).

The serine hydrolase-specific ABP fluorophosphonate-biotin (FP-biotin) was used by Lee et al. to monitor the activity of cannabinoid-catabolizing enzymes in avian macrophages infected with *S.* Typhimurium (Lee et al. 2018). Two of these enzymes, fatty acid amide hydrolase (FAAH) and α/β-hydrolase domain 6 (ABHD6), were less active in infected macrophages. Furthermore, inhibition of FAAH and ABHD6 activity increased the intracellular concentration of 2-AG, which in turn enhanced the phagocytic activity of avian macrophages toward dye-conjugated bioparticles (Lee et al. 2018). Together, these studies suggest host metabolism of endogenous cannabinoids is modulated during *S.* Typhimurium infection to promote pathogen uptake by immune cells.

Activity-based analyses of *S.* Typhimurium-infected macrophages have uncovered shifts in protein degradation, cell death, and lipid metabolism that are triggered by pathogen interaction with host cells in vitro. Similar techniques could be applied to determine which of these pathways are stimulated by other pathogens at the post-translational level. Recently developed lysosome-specific ABPs (Wiedner et al. 2014) promise to enhance the detection of host enzymes that directly interact with microbes following phagocytosis and could provide further insight into enzymatic processes that promote pathogen clearance from the host.

2.2 *Profiling Pathogen Adaptation to the Host Environment*

ABPs can be used to understand how the functional proteome of a pathogen is altered in response to external cues, such as nutrient availability and environmental stress. Multiplexing approaches, in which multiple ABPs are used to simultaneously profile distinct functional classes of enzymes, can facilitate the rapid identification of proteins that may promote pathogen adaptation to the host environment. Wiedner et al. used two ABPs to monitor the activities of cysteine-reactive proteins and serine hydrolases produced by the fungal pathogen *Aspergillus fumigatus* in human serum (Wiedner et al. 2012). These studies revealed growth-dependent shifts in the primary metabolism, oxidative stress response, and nutrient-scavenging pathways of *A. fumigatus* over time that may facilitate fungal survival in the human lung. Moreover, this work illustrates how multiple ABPs can be used to broadly profile mechanisms of pathogen adaptation to complex environments.

ABPP has also been used to study the localization of pathogen enzymes that are active at the host–microbe interface. The parasitic flatworm *Schistosoma mansoni* expresses several surface-exposed proteins that are believed to promote pathogen survival in blood vessels by modulating host immune and hemostatic responses (Wang et al. 2017). Previous studies have identified membrane-associated calpains on the schistosome cell surface (Karcz et al. 1991), but the roles of these calcium-dependent cysteine proteases in *S. mansoni* infection are not well understood. Using an ABP comprised of the cell-impermeable calpain inhibitor E64c conjugated to biotin, Wang et al. detected two active calpains, SmCalp1 and

SmCalp2, on the parasite cell surface (Wang et al. 2017). When incubated with murine plasma, *S. mansoni* was found to induce the calpain-dependent cleavage of fibronectin, a blood-clotting factor, suggesting SmCalp1 and SmCalp2 may prevent coagulation to suppress the host inflammatory response (Esmon et al. 2011) and enhance pathogen mobility through the vascular system. Though additional studies are needed to clarify the specific contributions of these proteases to infection, this work illustrates how ABPP can be used to identify pathogen enzymes in direct contact with the host environment that could serve as potential therapeutic targets.

ABPP can also be used to identify the targets of microbial enzymatic products within the host. The human gut microbiome encodes a number of nonribosomal peptide synthetases (NRPS) that produce metabolites of unknown function. Guo and coworkers characterized several of these compounds by cloning diverse NRPS clusters into multiple bacterial strains and analyzing their peptide products by mass spectrometry (Guo et al. 2017). To identify host enzymes targeted by these metabolites, the cysteine-reactive ABP iodoacetamide-alkyne was used to analyze membrane preparations of human monocytes incubated with a representative dipeptide product (Phe-Phe-H) versus a vehicle control. These experiments identified the cysteine protease cathepsin L as the primary target of Phe-Phe-H, suggesting cathepsin inhibition by this microbial peptide may attenuate host immune responses to promote intestinal colonization by the microbiota (Guo et al. 2017).

Changes in enzyme activity over the course of an infection can provide clues regarding biochemical processes that accompany different stages of a pathogen's life cycle. ABPs labeled with unique, isobaric tags (Gygi et al. 1999) could facilitate quantitative proteomic analyses of pathogen enzymes active under different infection conditions. Combining ABPP with other powerful techniques such as bio-orthogonal noncanonical amino acid tagging (BONCAT) (Stone et al. 2017) or stable isotope labeling with amino acids in cell culture (SILAC) (Surmann et al. 2015) could also enhance the identification of low-abundance microbial enzymes active in infected cells. Collectively, these techniques are poised to expand the depth of future activity-based analyses at the host–pathogen interface.

3 Applications of ABPP to Infected Plants

The early adoption of ABPP by the plant biology community has resulted in the broad application of this technique to studies of host–microbe interactions spanning multiple plant and pathogen species. ABPP was first applied by van der Hoorn et al. to survey cysteine proteases active in plant tissue extracts using DCG-04, a biotinylated analog of the papain-like cysteine protease inhibitor E-64 (van der Hoorn et al. 2004). Mass spectrometry analysis of DCG-04-enriched proteins identified three previously uncharacterized proteases produced by the well-studied model for flowering plants, *Arabidopsis thaliana*. This study also established DCG-04 as a valuable tool for monitoring the inhibition of plant protease activity, which has since played a pivotal role in the discovery and characterization of many pathogen-secreted protease inhibitors (Rooney et al. 2005; Shindo et al. 2016; Song et al. 2009; Tian et al. 2007; van Esse et al. 2008). From global-profiling strategies to mechanistic studies of targeted enzymes, ABPP has been used to uncover a wide range of plant–pathogen interactions (Kaschani et al. 2012). Indeed, an impressive number of ABPP studies has been performed in infected plants, perhaps more so than in any other multicellular model of microbial infection.

3.1 Discovery of Pathogen-Secreted Protease Inhibitors

Pathogen-secreted effectors often subvert plant responses to infection by modulating host enzyme activity (Toruno et al. 2016). Enzyme inhibition, degradation, and activation are just a few of the mechanisms by which pathogen factors can alter biochemical processes in the host. For example, protease inhibitors are commonly deployed by plant pathogens to thwart host immune responses (Jashni et al. 2015). Many of these inhibitors have been discovered using ABPP, which can be used to rapidly assess changes in host protease activity during infection.

Early work by Rooney and coworkers established a direct link between Avr2, a protein secreted by the fungal pathogen *Cladosporium fulvum*, and Rcr3, an extracellular cysteine protease present in the tomato apoplast (Rooney et al. 2005). Purified Avr2 was shown to selectively inhibit DCG-04-labeling of Rcr3 at low pH, suggesting this interaction may be functional in the acidic environment of the apoplast. The authors subsequently demonstrated binding of Avr2 to Rcr3 in co-immunoprecipitation assays and inhibition of native Rcr3 protease activity in tomato apoplastic fluid by purified Avr2 (Rooney et al. 2005). In addition, co-administration of purified Rcr3 and Avr2 to tomato leaves was found to induce programmed cell death consistent with the plant defense mechanism known as the hypersensitive response. The hypersensitive response can be triggered by the tomato protein Cf-2, which promotes resistance to *C. fulvum* infection (Jashni et al. 2015). Notably, Cf-2-dependent cell death was observed in tomato leaves infiltrated with both Avr2 and Rcr3, but not in leaves infiltrated with either protein alone, or in combination with the protease inhibitor E-64 (Rooney et al. 2005). These findings suggest that Cf-2-mediated host defense is induced by the Rcr3-Avr2 complex. In addition, activation of Cf-2 by a host protease bound to a pathogen effector demonstrated a key provision of the guard hypothesis, which posits that Cf proteins indirectly sense Avr proteins through their host targets (van der Hoorn and Kamoun 2008) (Fig. 2). Using DCG-04, the authors also demonstrated that other tomato proteases are inhibited by Avr2, indicating that pathogen-secreted protease inhibitors have pleiotropic roles on host protease activity that can be revealed using ABPP (Rooney et al. 2005).

To identify other host proteases targeted by Avr2, van Esse and colleagues compared DCG-04-enriched proteins from wild-type and Avr2-expressing *A. thaliana* extracts by mass spectrometry (van Esse et al. 2008). Expression of the *Avr2* transgene correlated with a decrease in the relative abundance of three

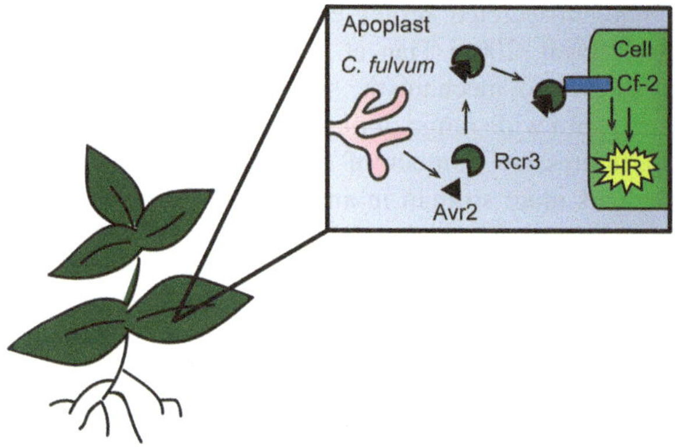

Fig. 2 Proposed model of the guard hypothesis revealed by ABPP. The *C. fulvum*-secreted protease inhibitor Avr2 binds to the protease Rcr3 in the tomato apoplast. The receptor protein Cf-2 senses the Avr2-Rcr3 complex and triggers the hypersensitive response (HR)

Arabidopsis proteases (CPR1, XCP1, and XCP2). Interestingly, pre-treatment of wild-type extracts with purified Avr2 prior to DCG-04 labeling had a more pronounced effect on host protease activity: of the seven proteases detected in untreated wild-type extracts, only one (cathepsin B) remained active following incubation with Avr2 (van Esse et al. 2008). Similar results were obtained using tomato apoplastic fluids: fluids incubated with purified Avr2 contained fewer active cysteine proteases than fluids from plants engineered to express the pathogen protein. These findings suggest Avr2 binds to host proteases with variable affinity and that the relative inhibition of these proteases by Avr2 can be resolved through activity-based analyses.

ABPP has also been used to characterize new protease inhibitors at the plant–pathogen interface. Using bioinformatic tools, Tian and colleagues identified a family of putative protease inhibitors secreted by the oomycete pathogen *Phytophthora infestans* during infection of the tomato plant (Tian et al. 2007). One of these proteins, EPIC2B, was found to bind to a newly identified papain-like extracellular cysteine protease, PIP1, in tomato apoplastic fluids. To determine whether EPIC2B could inhibit the activity of PIP1 or other cysteine proteases in the apoplast, the authors treated tomato fluid extracts with purified EPIC2B prior to DCG-04 labeling and observed inhibition of several host proteases (Tian et al. 2007). In addition, purified EPIC2B was found to directly inhibit PIP1 activity in transgenic *Nicotiana benthamiana* plants expressing epitope-tagged PIP1. However, time-course analyses of PIP1 labeling by DCG-04 in the presence of EPIC2B revealed an increase in PIP1 activity over time, suggesting that EPIC2B is a reversible PIP1 inhibitor. Using a similar assay, Song et al. demonstrated that the closely related tomato cysteine protease Rcr3, a previously established target of the *C. fulvum* effector Avr2, is also reversibly inhibited by EPIC2B (Song et al. 2009). Consistent with this finding, EPIC2B was unable to trigger the hypersensitive response in tomato plants expressing Cf-2 and Rcr3. Together, these results suggest that plant pathogens may have evolved weak protease inhibitors to transiently disarm host enzyme activity while subverting innate immune recognition by Cf proteins.

Like EPIC2B, other protease inhibitors have been discovered by using ABPP to measure the inhibition of native plant proteases in situ. Shindo and coworkers mined the genome of *Pseudomonas syringae* PtoDC3000, a Gram-negative bacterial pathogen, for genes encoding small, secreted proteins of unknown function (Shindo et al. 2016). They purified and assayed 43 of these proteins for the ability to inhibit the activity of C14, an extracellular immune protease produced by the tomato plant. C14-expressing leaf extracts were pre-incubated with each putative effector, then labeled with MV201, a fluorescent ABP modeled after E-64. Only one protein, dubbed C14-inhibiting protein (Cip1), was found to inhibit C14 in this assay (Shindo et al. 2016). Genetic deletion of the *cip1* gene from *P. syringae* attenuated bacterial growth and virulence in the tomato plant. Like EPIC2B, Cip1 was also shown to weakly inhibit Rcr3 and subvert the Cf-2-dependent hypersensitive response (Shindo et al. 2016). In sum, these studies provide an elegant example of how ABPP can be used to link the biochemical activity of a novel pathogen effector to virulence phenotypes in vivo.

ABPP has also been applied to determine how the secretion and activity of a pathogen protease influences plant immune responses. Lu et al. designed a specific ABP to study the post-translational processing of AvrPphB, a papain-like cysteine protease secreted by the Type III secretion system of *P. syringae* (Lu et al. 2013). A truncated AvrPphB mutant lacking the enzyme's prodomain remained active, but could no longer be secreted. In contrast, when mutated to prevent prodomain cleavage, both AvrPphB secretion and activity were retained, but the protein was incapable of inducing the hypersensitive response in *A. thaliana* (Lu et al. 2013). These findings indicate that protein secretion precedes prodomain cleavage of AvrPphB and that prodomain removal is required for the activation of innate immune responses by this protease. Selective ABPs can thus provide valuable insights regarding the molecular regulation of protease-activated surveillance pathways in plants.

Over the past several years, applications of ABPP to plant–pathogen biology have evolved from studies of specific effector–protease interactions to large-scale inhibitor discovery campaigns. These efforts have established a rich history of ABPP at the plant–pathogen interface and have set the standard for applications of this technology to studies of microbial interactions with multicellular hosts. As the field continues to grow, global-profiling efforts coupled with more sophisticated bioinformatic tools are likely to reveal additional mechanisms of biochemical antagonism enacted by pathogen-secreted enzyme inhibitors.

3.2 Global Profiling of Active Enzymes

ABPP can rapidly uncover tissue-specific, enzyme-mediated responses to infection. Different ABPs can be applied to catalogue the repertoire of enzymes active under infection conditions as a starting point for target-based assays. Dynamic trends in enzyme activity can also provide insight into the physiology of an infection.

For example, ABPP analysis of *A. thaliana* infected with the fungal pathogen *Botrytis cinerea* revealed a decrease in the activities of several plant serine hydrolases over time, suggesting pathogen interference with host enzyme activities (Kaschani et al. 2009).

Combined with other global-profiling methods, ABPP can also be used to relate dynamic changes in transcript or protein abundance during infection to changes in the functional proteome. Grosse-Holz et al. took a multi-omics approach to identify proteases whose activity is modulated by agroinfiltration, the use of attenuated *Agrobacterium tumefaciens* as a vector for the transient production of antibodies, vaccines, and other commercially important biomolecules in plants (Grosse-Holz et al. 2018). Proteolysis is believed to be a confounding factor in agroinfiltration that limits the yield of transgenically expressed proteins. To identify extracellular proteases produced in response to agroinfiltration, the authors analyzed agroinfiltrated *N. benthamiana* leaves with probes targeting serine hydrolases and papain-like cysteine proteases (Grosse-Holz et al. 2018). In parallel, these samples were analyzed by RNA-seq. Comparison of these datasets revealed that a large number of proteases are transcribed during agroinfiltration, but only few are active extracellularly, suggesting these enzymes are under post-transcriptional and/or post-translational control (Grosse-Holz et al. 2018). The subset of active proteases produced by agroinfiltrated leaves may represent promising targets for improving the efficiency of recombinant protein expression.

3.3 Mechanistic Studies of Proteasome Activity and Inhibition

Another strength of ABPP is its utility for studying enzyme complexes encoded by essential genes that cannot easily be interrogated using genetic approaches. The plant proteasome is a striking example of a multi-subunit molecular machine that has been challenging to study due to the pleiotropic effects induced by subunit mutations, post-translational regulation of proteasome activity, and the potential for dynamic proteasome remodeling with paralogous subunits (Misas-Villamil et al. 2017). ABPs have been instrumental in characterizing the differential activation of proteasome subunits during bacterial infection (Kolodziejek et al. 2011; Misas-Villamil et al. 2017). Using subunit-selective probes with distinct fluorescent dyes, Misas-Villamil et al. detected three active proteasome subunits in *A. thaliana* leaf extracts: β1, β2, and β5 (Misas-Villamil et al. 2017). Interestingly, when applied to *N. benthamiana* extracts, these probes revealed the presence of an additional β1 subunit. Consistent with these data, mass spectrometry analyses using the pan-specific subunit probe MVB072 detected unique peptides from two distinct β1 subunits. In addition, ABPP of *N. benthamiana* leaf extracts infected with *P. syringae* revealed significant phenotypic variation in subunit activation across multiple experiments, suggesting that subunit activities can be independently modulated during infection (Misas-Villamil et al. 2017). Applications of these probes to study the subunit architecture and activation of other plant proteasomes during infection could reveal a new mode of bacterial manipulation of an essential protein complex and potentially uncover the assembly of inducible, infection-dependent proteasomes that shape host immunity.

4 Applications of ABPP to Infected Animals

Animal models of infection that recapitulate clinical symptoms of human disease are powerful systems for elucidating the effects of pathogen factors on host physiology. However, identifying biochemical interactions that underlie host responses to infection in the context of an infected animal can be technically challenging. Microbial enzymes are often present at low concentrations in host tissues, limiting their detection by unenriched proteomic analyses (Mayers et al. 2017). Furthermore, the intrinsic complexity of tissue homogenates, which typically include cells distal to the host–microbe interface, can inhibit the detection of pathogen enzymes and the host proteins they interact with during infection. ABPP addresses these challenges by enabling the enrichment of a targeted enzyme class prior to mass spectrometry analysis. By facilitating the discovery of bacterial and host enzymes active in infected animals, ABPP can uncover biochemical processes that shape host responses to infection.

Tissue inflammation is a common response to microbial infections that is often triggered by enzymatic processes (Karin et al. 2006). For example, proteolytic activation of certain cellular receptors, such as proteinase-activated receptor 2 (PAR2), can induce intestinal inflammation (Cenac et al. 2002). *Citrobacter rodentium*, an enteric pathogen that causes murine colitis, stimulates trypsin-like serine protease activity in the colonic fluid of infected mice that promotes PAR2 activation (Hansen et al. 2005). To identify proteases active during *C. rodentium* infection, Hansen and coworkers incubated colonic luminal fluid from infected animals with agarose beads conjugated to soybean trypsin inhibitor (STI) (Hansen et al. 2005). Mass spectrometry analysis of the STI-enriched enzymes identified three host serine proteases: trypsinogen 16, granzyme A, and kallikrein B. Granzyme A was also detected in the colonic fluid of *C. rodentium*-infected mice by Western blot, and a protein of similar molecular weight was detected using the ABP Biotin-Pro-Lys-diphenylphosphonate. Furthermore, granzyme A and *C. rodentium*-infected fluid were both shown to cleave a peptide containing the PAR2 activation sequence in vitro (Hansen et al. 2005). Taken together, these findings suggest granzyme A is produced by host tissues during *C. rodentium* infection and promotes intestinal inflammation via PAR2 cleavage. More broadly, this work provides an early example of how activity-based approaches can be applied to survey infection-associated changes in host enzyme activity.

While host enzyme activity can play an important role in modulating responses to infection, pathogen enzymes that are active in an infected animal can also provide important clues regarding the virulence mechanisms and metabolic pathways that shape microbial interactions with the host. We recently applied ABPP to identify secreted host and pathogen serine hydrolases active in cholera, a severe diarrheal disease caused by the enteric pathogen *Vibrio cholerae* (Hatzios et al. 2016). Using the serine hydrolase-selective probe FP-biotin, we identified 233 and 71 probe-enriched proteins in the cecal fluid of *V. cholerae*-infected rabbits and in human choleric stool, respectively. Notably, ABPP detected several pathogen-secreted proteases that were not identified by mass spectrometry analyses of these samples in the absence of ABP enrichment. One of these proteases, IvaP, was previously shown to be transcribed at similar levels in exponential-phase cultures and in rabbit cecal fluid (Mandlik et al. 2011), yet gel-based ABPP and Western blot analyses revealed the enzyme is neither expressed nor active under conditions of exponential growth in vitro (Hatzios et al. 2016). Further experiments established that IvaP undergoes significant post-translational processing to yield the functional protease that is active in cecal fluid. In addition, comparative ABPP analyses of cecal fluid from rabbits infected with either wild-type *V. cholerae* or a mutant strain expressing catalytically deficient IvaP suggest the activity of several host and pathogen serine hydrolases is influenced by this secreted protease (Hatzios et al. 2016). Finally, genetic deletion of IvaP, along with three other pathogen-secreted proteases active in rabbit cecal fluid, increased the abundance of a host lectin that binds to *V. cholerae* in the intestine, suggesting a broader role for these enzymes in modulating gut-microbe interactions. In sum, these findings demonstrate how ABPP can be applied to draw activity-based relationships between bacterial and host enzymes in complex infection models.

In addition to studies of infection biology, activity-based approaches can be used to explore the biochemistry of endogenous microbial communities in the host. Recent work by Mayers et al. applied ABPP to study the functional proteome of gut microbes in a mouse model of inflammatory bowel disease (IBD) (Mayers et al. 2017). IBD mice and isogenic controls were fed ^{14}N- or ^{15}N-labeled spirulina chow, respectively, to enable quantitative comparisons of their fecal proteomes by mass spectrometry. Using BioGlyCMK, a novel ABP targeting reactive cysteines, roughly 1000 microbial proteins were enriched from the fecal pellets of IBD mice, enabling the detection of enzymes with active-site cysteines that were overlooked by unenriched proteomic analyses of these samples (Mayers et al. 2017). In addition, comparison of the BioGlyCMK-enriched proteins in IBD versus control mice revealed a significant increase in the abundance of microbial cysteine-type peptidases and hydrolases, suggesting a possible role for these enzymes in IBD pathology. This study represents an important advance in metaproteomic analyses of the fecal microbiome, which could potentially be used to correlate the dysregulation of key microbial proteins with intestinal inflammation.

ABPP is emerging as a valuable tool for studies of host–microbe interactions in infected animals. Future challenges will include enriching pathogen enzymes from more complex tissue homogenates, which are more difficult to extract and process than fecal or intestinal fluid samples, and adapting ABPP to resolve the proteomic contributions of distinct microbial species in the host. Improvements in mass spectrometry technology, enrichment strategies, and methods that couple metaproteomic and metagenomic analyses will likely accelerate these efforts and their application to additional disease models.

5 Applications of ABPP to Inhibitor Discovery and Characterization

Enzymes dynamically regulate biochemical processes that influence host–pathogen interactions. ABPP provides a versatile platform for identifying inhibitors that can disrupt these processes and potentially alter infection outcomes (Moellering and Cravatt 2012). ABPs are themselves irreversible inhibitors designed to facilitate the detection of active enzymes via a synthetically incorporated dye or affinity tag (Heal et al. 2011). Proteins or small molecules that competitively bind to an enzyme's active site can diminish probe labeling and subsequent detection of a targeted enzyme. For example, pathogen proteins that inhibit host enzyme activity, such as the pathogen-secreted protease inhibitors discussed in Sect. 3.1, can be characterized through competition assays with ABPs (Rooney et al. 2005; Shindo et al. 2016; Song et al. 2009). Here, we focus on activity-based approaches that facilitate the discovery, design, and target identification of small-molecule inhibitors for disrupting enzyme-mediated mechanisms of microbial pathogenesis.

5.1 Identifying the Targets of Bioactive Inhibitors

High-throughput chemical genetic screens can be used to identify compounds that alter infection phenotypes. When a compound library is designed to target a defined enzyme class, bioactive molecules can be competed against broad-spectrum ABPs to facilitate identification of inhibited enzymes (Moellering and Cravatt 2012) (Fig. 3). Candidate inhibitors can also be fashioned into ABPs for direct labeling and detection of targeted proteins. Together, these activity-based approaches can provide fundamental insight into the molecular mechanisms triggered by chemical modulators of infection.

Using a chemical library comprised of protease and hydrolase inhibitors, Child et al. discovered a group of chloroisocoumarins that enhance host-cell invasion by *Toxoplasma gondii*, an obligate intracellular parasite (Child et al. 2013). To identify proteins targeted by these 'enhancer' compounds, FP-rhodamine, a serine hydrolase-specific ABP, was used to detect enzymes active in enhancer-treated *T. gondii* lysates versus lysates treated with a structurally related, but biologically inactive compound. Probe labeling of a 35-kDa species was selectively inhibited by all enhancers, suggesting these compounds target a common serine hydrolase.

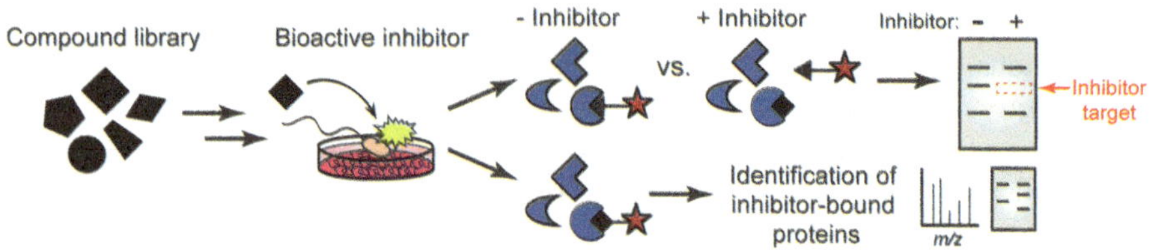

Fig. 3 ABPP strategies for identifying inhibitor targets. Bioactive inhibitors identified via chemical screens of infected cells can be competed against broad-spectrum ABPs (top) or adapted into custom ABPs (bottom) to identify inhibitor-bound proteins using gel-based or proteomic analyses

To identify the targeted enzyme, *T. gondii* lysates pretreated with or without a functional enhancer were labeled with FP-biotin or an alkyne-containing enhancer analog designed to enrich labeled enzymes via click chemistry-mediated attachment of an affinity tag (Child et al. 2013). Mass spectrometry analysis of these probe-labeled proteins revealed a putative thioesterase, TgPPT1, with a predicted molecular weight of ~35 kDa that was substantially less abundant in enhancer-treated lysates. Activity-based analyses demonstrated that recombinant TgPPT1 is a serine hydrolase selectively inhibited by enhancer addition. Furthermore, deletion of the *Tgppt1* gene from *T. gondii* suppressed enhancer-induced effects on host-cell invasion, confirming TgPPT1 mediates the enhancer phenotype (Child et al. 2013). In sum, these studies illustrate how the molecular basis of a pharmacologically driven, infection-associated phenotype can be uncovered through creative applications of ABPP.

ABPP can also be used to validate the targets of bioactive inhibitors discovered through biochemical assays. Using a serine protease inhibitor library, Gloeckl and coworkers identified two phosphonate-containing peptides, JCP83 and JO146, that inhibit the in vitro activity of CtHtrA, a serine protease produced by the sexually transmitted bacterial pathogen *Chlamydia trachomatis* (Gloeckl et al. 2013). Both compounds were found to significantly decrease *C. trachomatis* viability at early stages of its developmental cycle in infected cells. To identify proteins targeted by JO146 during *C. trachomatis* infection, the authors developed fluorescently labeled and biotinylated analogs of JO146 for ABPP. Both gel-based and proteomic analyses confirmed JO146 binding to CtHtrA in *C. trachomatis*-infected cells, demonstrating how activity-based approaches can be used to connect the targets of small-molecule inhibitors to virulence phenotypes during infection (Gloeckl et al. 2013). These strategies should be broadly useful for mapping the interacting partners of other bioactive inhibitors at the host–pathogen interface.

5.2 *High-Throughput Screens for Selective Inhibitors*

Though gel-based ABPP assays can be used to identify small molecules that inhibit probe labeling of a targeted enzyme, these approaches are fairly low-throughput.

Using fluorescent ABPs, Bachovchin and coworkers developed a competitive, activity-based platform that is compatible with high-throughput screening (HTS) (Bachovchin et al. 2009). This technology, dubbed fluopol-ABPP, identifies compounds that decrease the rate of ABP binding to an enzyme's active site by monitoring changes in the fluorescence polarization signal of the probe-labeling reaction. Because enzyme-bound ABPs emit more highly polarized light than unbound probes in solution, small molecules that inhibit ABP binding will result in a lower fluorescence polarization signal than inactive compounds. Unlike most HTS assays, fluopol-ABPP offers a way to identify selective enzyme inhibitors in the absence of a substrate, making it a versatile tool for studies of uncharacterized enzymes.

Bender and coworkers recently applied this strategy to identify inhibitors of the cysteine protease domain (CPD) of TcdB, a secreted toxin produced by the colitis-inducing pathogen *Clostridium difficile* (Bender et al. 2015). A fluopol-ABPP screen using TAMRA-AWP-19, a CPD-reactive ABP, revealed several potent CPD inhibitors, including ebselen, a selenium-containing compound previously shown to be safe for human use. Ebselen inhibited cell rounding of TcdB-treated cells in vitro and enhanced the survival of TcdB-treated mice. In addition, compound administration reduced the tissue pathology of mice infected with *C. difficile* without diminishing the stool-associated colony-forming units of these bacteria, demonstrating that ebselen reduces disease symptoms by targeting toxin function, as opposed to bacterial survival, in the host (Bender et al. 2015). In sum, these studies provide a striking example of how high-throughput ABPP methods can be used to identify antivirulence compounds for bacterial infections. Additional applications of this technology could advance future efforts to discover inhibitors of other virulence-promoting pathogen factors and improve treatment options for antibiotic-resistant bacterial infections.

5.3 Rational Design and Optimization of Inhibitors

The rational design of selective inhibitors is often guided by the structural features of native enzyme substrates. Substrate-profiling techniques and targeted screens of substrate-based inhibitor libraries can serve as a starting point for the development of more potent inhibitors. Competition assays using broad-spectrum ABPs can also be used to evaluate and optimize inhibitor selectivity. These approaches were recently applied by Lentz et al. to develop a selective inhibitor for Hip1, a serine protease with immunomodulatory properties produced by the tuberculosis pathogen *Mycobacterium tuberculosis* (Lentz et al. 2016). Multiple combinatorial peptide libraries were used to generate substrate-specificity data for Hip1, which informed the design of several fluorogenic protease substrates. The authors screened a small library of serine-reactive electrophiles against a fluorogenic substrate and identified a series of irreversible, chloroisocoumarin-based Hip1 inhibitors (Lentz et al. 2016). To assess inhibitor selectivity, lysates of RAW 264.7 cells were pre-incubated with the lead chloroisocoumarin compound, then labeled with the broad-spectrum ABP

FP-rhodamine. Because inhibitor addition diminished probe labeling of several host serine hydrolases, the authors used the substrate-specificity data for Hip1 to further optimize inhibitor potency and selectivity. Together, these approaches illustrate how ABPs can be used in tandem with substrate-profiling methods to guide the design of novel inhibitors. In turn, these inhibitors can provide a scaffold for the synthesis of novel ABPs that could facilitate future studies of enzyme function and trafficking in infected cells, or the early detection of bacterial infections in clinical isolates.

6 Concluding Remarks

ABPP is a powerful tool for identifying catalytically active enzymes that promote biochemical crosstalk at the host–pathogen interface. Probes that expand the functional and temporal scope of this technology promise to enhance the depth of future profiling efforts. Multiplexing approaches that sample diverse functional classes of enzymes will provide a more holistic view of the biochemical processes that underlie microbial infections. In addition, caged probes that allow precise monitoring of enzyme activity over rapid timescales and within specific cellular compartments should enhance dynamic and spatial studies of protein function within infected cells. ABPP is also poised to improve our understanding of the biochemical crosstalk between complex microbial communities and infected hosts. Innovative computational and barcoding tools will be needed to map specific enzymes to their producing strains within a population of closely related bacteria. Focused screens for microbe-secreted protease inhibitors, akin to those functional at the plant–pathogen interface, could also be pursued to identify bacterial factors that promote immune adaptation to pathogenic or commensal microbes in the gut. Such activity-based approaches will continue to accelerate the discovery of enzymes that dynamically shape pathogen interactions with the host.

Acknowledgements We gratefully acknowledge Pamela Chang, Joshua Gendron, Yannick Jacob, and Lindsay Triplett for providing helpful feedback on this manuscript. Y. K. was supported by an NIH Predoctoral Training Grant (T32GM067543).

References

Ashida H, Kim M, Sasakawa C (2014) Exploitation of the host ubiquitin system by human bacterial pathogens. Nat Rev Microbiol 12:399–413

Bachovchin DA, Brown SJ, Rosen H, Cravatt BF (2009) Identification of selective inhibitors of uncharacterized enzymes by high-throughput screening with fluorescent activity-based probes. Nat Biotechnol 27:387–394

Behnsen J, Perez-Lopez A, Nuccio SP, Raffatellu M (2015) Exploiting host immunity: the *Salmonella* paradigm. Trends Immunol 36:112–120

Bender KO, Garland M, Ferreyra JA, Hryckowian AJ, Child MA, Puri AW, Solow-Cordero DE, Higginbottom SK, Segal E, Banaei N et al (2015) A small-molecule antivirulence agent for treating *Clostridium difficile* infection. Sci Transl Med 7:306ra148

Broz P, Ohlson MB, Monack DM (2012) Innate immune response to *Salmonella typhimurium*, a model enteric pathogen. Gut Microbes 3:62–70

Carey AF, Rock JM, Krieger IV, Chase MR, Fernandez-Suarez M, Gagneux S, Sacchettini JC, Ioerger TR, Fortune SM (2018) TnSeq of *Mycobacterium tuberculosis* clinical isolates reveals strain-specific antibiotic liabilities. PLoS Pathog 14:e1006939

Cenac N, Coelho AM, Nguyen C, Compton S, Andrade-Gordon P, MacNaughton WK, Wallace JL, Hollenberg MD, Bunnett NW, Garcia-Villar R et al (2002) Induction of intestinal inflammation in mouse by activation of proteinase-activated receptor-2. Am J Pathol 161:1903–1915

Child MA, Hall CI, Beck JR, Ofori LO, Albrow VE, Garland M, Bowyer PW, Bradley PJ, Powers JC, Boothroyd JC et al (2013) Small-molecule inhibition of a depalmitoylase enhances *Toxoplasma* host-cell invasion. Nat Chem Biol 9:651–656

Cravatt BF, Wright AT, Kozarich JW (2008) Activity-based protein profiling: from enzyme chemistry to proteomic chemistry. Annu Rev Biochem 77:383–414

Creagh EM (2014) Caspase crosstalk: integration of apoptotic and innate immune signalling pathways. Trends Immunol 35:631–640

Duell BL, Cripps AW, Schembri MA, Ulett GC (2011) Epithelial cell coculture models for studying infectious diseases: benefits and limitations. J Biomed Biotechnol 2011:852419

Esmon CT, Xu J, Lupu F (2011) Innate immunity and coagulation. J Thromb Haemost 9(Suppl 1):182–188

Gloeckl S, Ong VA, Patel P, Tyndall JD, Timms P, Beagley KW, Allan JA, Armitage CW, Turnbull L, Whitchurch CB et al (2013) Identification of a serine protease inhibitor which causes inclusion vacuole reduction and is lethal to *Chlamydia trachomatis*. Mol Microbiol 89:676–689

Grosse-Holz F, Kelly S, Blaskowski S, Kaschani F, Kaiser M, van der Hoorn RAL (2018) The transcriptome, extracellular proteome and active secretome of agroinfiltrated *Nicotiana benthamiana* uncover a large, diverse protease repertoire. Plant Biotechnol J 16:1068–1084

Guo CJ, Chang FY, Wyche TP, Backus KM, Acker TM, Funabashi M, Taketani M, Donia MS, Nayfach S, Pollard KS et al (2017) Discovery of reactive microbiota-derived metabolites that inhibit host proteases. Cell 168:517–526

Gygi SP, Rist B, Gerber SA, Turecek F, Gelb MH, Aebersold R (1999) Quantitative analysis of complex protein mixtures using isotope-coded affinity tags. Nat Biotechnol 17:994–999

Hansen KK, Sherman PM, Cellars L, Andrade-Gordon P, Pan Z, Baruch A, Wallace JL, Hollenberg MD, Vergnolle N (2005) A major role for proteolytic activity and proteinase-activated receptor-2 in the pathogenesis of infectious colitis. Proc Natl Acad Sci U S A 102:8363–8368

Hatzios SK, Abel S, Martell J, Hubbard T, Sasabe J, Munera D, Clark L, Bachovchin DA, Qadri F, Ryan ET et al (2016) Chemoproteomic profiling of host and pathogen enzymes active in cholera. Nat Chem Biol 12:268–274

Heal WP, Dang TH, Tate EW (2011) Activity-based probes: discovering new biology and new drug targets. Chem Soc Rev 40:246–257

Heal WP, Tate EW (2012) Application of activity-based protein profiling to the study of microbial pathogenesis. Top Curr Chem 324:115–135

Hernandez-Cervantes R, Mendez-Diaz M, Prospero-Garcia O, Morales-Montor J (2017) Immunoregulatory role of cannabinoids during infectious disease. NeuroImmunoModulation 24:183–199

Jashni MK, Mehrabi R, Collemare J, Mesarich CH, de Wit PJ (2015) The battle in the apoplast: further insights into the roles of proteases and their inhibitors in plant-pathogen interactions. Front Plant Sci 6:584

Kanca O, Bellen HJ, Schnorrer F (2017) Gene tagging strategies to assess protein expression, localization, and function in *Drosophila*. Genetics 207:389–412

Karcz SR, Podesta RB, Siddiqui AA, Dekaban GA, Strejan GH, Clarke MW (1991) Molecular cloning and sequence analysis of a calcium-activated neutral protease (calpain) from *Schistosoma mansoni*. Mol Biochem Parasitol 49:333–336

Karin M, Lawrence T, Nizet V (2006) Innate immunity gone awry: linking microbial infections to chronic inflammation and cancer. Cell 124:823–835

Kaschani F, Gu C, Niessen S, Hoover H, Cravatt BF, van der Hoorn RA (2009) Diversity of serine hydrolase activities of unchallenged and *Botrytis*-infected *Arabidopsis thaliana*. Mol Cell Proteomics 8:1082–1093

Kaschani F, Gu C, van der Hoorn RA (2012) Activity-based protein profiling of infected plants. Methods Mol Biol 835:47–59

Kolodziejek I, Misas-Villamil JC, Kaschani F, Clerc J, Gu C, Krahn D, Niessen S, Verdoes M, Willems LI, Overkleeft HS et al (2011) Proteasome activity imaging and profiling characterizes bacterial effector syringolin A. Plant Physiol 155:477–489

Kummari E, Alugubelly N, Hsu CY, Dong B, Nanduri B, Edelmann MJ (2015) Activity-based proteomic profiling of deubiquitinating enzymes in *Salmonella*-infected macrophages leads to identification of putative function of UCH-L5 in inflammasome regulation. PLoS ONE 10: e0135531

Lee JH, Hou X, Kummari E, Borazjani A, Edelmann MJ, Ross MK (2018) Endocannabinoid hydrolases in avian HD11 macrophages identified by chemoproteomics: inactivation by small-molecule inhibitors and pathogen-induced downregulation of their activity. Mol Cell Biochem 444:125–141

Lentz CS, Ordonez AA, Kasperkiewicz P, La Greca F, O'Donoghue AJ, Schulze CJ, Powers JC, Craik CS, Drag M, Jain SK et al (2016) Design of selective substrates and activity-based probes for Hydrolase Important for Pathogenesis 1 (HIP1) from *Mycobacterium tuberculosis*. ACS Infect Dis 2:807–815

Lu H, Wang Z, Shabab M, Oeljeklaus J, Verhelst SH, Kaschani F, Kaiser M, Bogyo M, van der Hoorn RA (2013) A substrate-inspired probe monitors translocation, activation, and subcellular targeting of bacterial type III effector protease AvrPphB. Chem Biol 20:168–176

Mandlik A, Livny J, Robins WP, Ritchie JM, Mekalanos JJ, Waldor MK (2011) RNA-Seq-based monitoring of infection-linked changes in *Vibrio cholerae* gene expression. Cell Host Microbe 10:165–174

Mayers MD, Moon C, Stupp GS, Su AI, Wolan DW (2017) Quantitative metaproteomics and activity-based probe enrichment reveals significant alterations in protein expression from a mouse model of inflammatory bowel disease. J Proteome Res 16:1014–1026

Mevissen TET, Komander D (2017) Mechanisms of deubiquitinase specificity and regulation. Annu Rev Biochem 86:159–192

Misas-Villamil JC, van der Burgh AM, Grosse-Holz F, Bach-Pages M, Kovacs J, Kaschani F, Schilasky S, Emon AE, Ruben M, Kaiser M et al (2017) Subunit-selective proteasome activity profiling uncovers uncoupled proteasome subunit activities during bacterial infections. Plant J 90:418–430

Moellering RE, Cravatt BF (2012) How chemoproteomics can enable drug discovery and development. Chem Biol 19:11–22

Place DE, Kanneganti TD (2018) Recent advances in inflammasome biology. Curr Opin Immunol 50:32–38

Puri AW, Bogyo M (2013) Applications of small molecule probes in dissecting mechanisms of bacterial virulence and host responses. Biochemistry 52:5985–5996

Puri AW, Broz P, Shen A, Monack DM, Bogyo M (2012) Caspase-1 activity is required to bypass macrophage apoptosis upon *Salmonella* infection. Nat Chem Biol 8:745–747

Rooney HC, Van't Klooster JW, van der Hoorn RA, Joosten MH, Jones JD, de Wit PJ (2005) *Cladosporium* Avr2 inhibits tomato Rcr3 protease required for Cf-2-dependent disease resistance. Science 308:1783–1786

Sadler NC, Wright AT (2015) Activity-based protein profiling of microbes. Curr Opin Chem Biol 24:139–144

Shahiduzzaman M, Coombs KM (2012) Activity based protein profiling to detect serine hydrolase alterations in virus infected cells. Front Microbiol 3:308

Shindo T, Kaschani F, Yang F, Kovacs J, Tian F, Kourelis J, Hong TN, Colby T, Shabab M, Chawla R et al (2016) Screen of non-annotated small secreted proteins of *Pseudomonas syringae* reveals a virulence factor that inhibits tomato immune proteases. PLoS Pathog 12: e1005874

Song J, Win J, Tian M, Schornack S, Kaschani F, Ilyas M, van der Hoorn RA, Kamoun S (2009) Apoplastic effectors secreted by two unrelated eukaryotic plant pathogens target the tomato defense protease Rcr3. Proc Natl Acad Sci U S A 106:1654–1659

Stone SE, Glenn WS, Hamblin GD, Tirrell DA (2017) Cell-selective proteomics for biological discovery. Curr Opin Chem Biol 36:50–57

Strmiskova M, Desrochers GF, Shaw TA, Powdrill MH, Lafreniere MA, Pezacki JP (2016) Chemical methods for probing virus-host proteomic interactions. ACS Infect Dis 2:773–786

Surmann K, Simon M, Hildebrandt P, Pfortner H, Michalik S, Stentzel S, Steil L, Dhople VM, Bernhardt J, Schluter R et al (2015) A proteomic perspective of the interplay of *Staphylococcus aureus* and human alveolar epithelial cells during infection. J Proteomics 128:203–217

Swatek KN, Komander D (2016) Ubiquitin modifications. Cell Res 26:399–422

Tian M, Win J, Song J, van der Hoorn R, van der Knaap E, Kamoun S (2007) A *Phytophthora infestans* cystatin-like protein targets a novel tomato papain-like apoplastic protease. Plant Physiol 143:364–377

Toruno TY, Stergiopoulos I, Coaker G (2016) Plant-pathogen effectors: cellular probes interfering with plant defenses in spatial and temporal manners. Annu Rev Phytopathol 54:419–441

van der Hoorn RA, Kamoun S (2008) From guard to decoy: a new model for perception of plant pathogen effectors. Plant Cell 20:2009–2017

van der Hoorn RA, Leeuwenburgh MA, Bogyo M, Joosten MH, Peck SC (2004) Activity profiling of papain-like cysteine proteases in plants. Plant Physiol 135:1170–1178

van Esse HP, Van't Klooster JW, Bolton MD, Yadeta KA, van Baarlen P, Boeren S, Vervoort J, de Wit PJ, Thomma BP (2008) The *Cladosporium fulvum* virulence protein Avr2 inhibits host proteases required for basal defense. Plant Cell 20:1948–1963

Walsh JG, Cullen SP, Sheridan C, Luthi AU, Gerner C, Martin SJ (2008) Executioner caspase-3 and caspase-7 are functionally distinct proteases. Proc Natl Acad Sci U S A 105:12815–12819

Wang Q, Da'dara AA, Skelly PJ (2017) The human blood parasite *Schistosoma mansoni* expresses extracellular tegumental calpains that cleave the blood clotting protein fibronectin. Sci Rep 7:12912

Wiedner SD, Anderson LN, Sadler NC, Chrisler WB, Kodali VK, Smith RD, Wright AT (2014) Organelle-specific activity-based protein profiling in living cells. Angew Chem Int Ed Engl 53:2919–2922

Wiedner SD, Burnum KE, Pederson LM, Anderson LN, Fortuin S, Chauvigne- Hines LM, Shukla AK, Ansong C, Panisko EA, Smith RD et al (2012) Multiplexed activity-based protein profiling of the human pathogen *Aspergillus fumigatus* reveals large functional changes upon exposure to human serum. J Biol Chem 287:33447–33459

Chemical Proteomic Profiling of Protein Fatty-Acylation in Microbial Pathogens

Tao Peng and Howard C. Hang

Contents

Abstract Protein fatty-acylation describes the covalent modification of protein with fatty acids during or after translation. Chemical proteomic profiling methods have provided new opportunities to explore protein fatty-acylation in microbial pathogens. Recent studies suggest that protein fatty-acylation is essential to survival and pathogenesis of eukaryotic pathogens such as parasites and fungi. Moreover,

T. Peng (✉)
State Key Laboratory of Chemical Oncogenomics, School of Chemical Biology and Biotechnology, Peking University Shenzhen Graduate School, Shenzhen 518055, China
e-mail: pengtao@pkusz.edu.cn

H. C. Hang (✉)
Laboratory of Chemical Biology and Microbial Pathogenesis, The Rockefeller University, New York, NY 10065, USA
e-mail: hhang@mail.rockefeller.edu

Current Topics in Microbiology and Immunology (2019) 420:93–110
DOI 10.1007/82_2018_126

Published Online: 21 August 2018

fatty-acylation in host cells can be exploited or manipulated by pathogenic bacteria. Herein, we first review the prevalent classes of fatty-acylation in microbial pathogens and the chemical proteomic profiling methods for their global analysis. We then summarize recent fatty-acylation profiling studies performed in eukaryotic pathogens and during bacterial infections, highlighting how they contribute to functional characterization of fatty-acylation under these contexts.

1 Introduction

Protein fatty-acylation refers to the covalent modification of protein with fatty-acyl groups of 8–20 saturated or unsaturated carbon chain length after or during protein translation (Resh 2016). The modification occurs on a variety of amino acid residues and therefore encompasses different types, such as *N*-myristoylation (Fig. 1a), *S*-palmitoylation (Fig. 1b), N^{ε}-Lys-fatty-acylation (Fig. 1c), and *O*-fatty-acylation (Hang and Linder 2011; Peng et al. 2016; Resh 2016; Jiang et al. 2018). The attachment of a hydrophobic fatty-acyl group to a protein substrate can have dramatic effects on its function by targeting to specific membrane compartments, mediating protein–protein interactions, and controlling protein stability (Schultz et al. 1988; Resh 1999, 2016; Jiang et al. 2018).

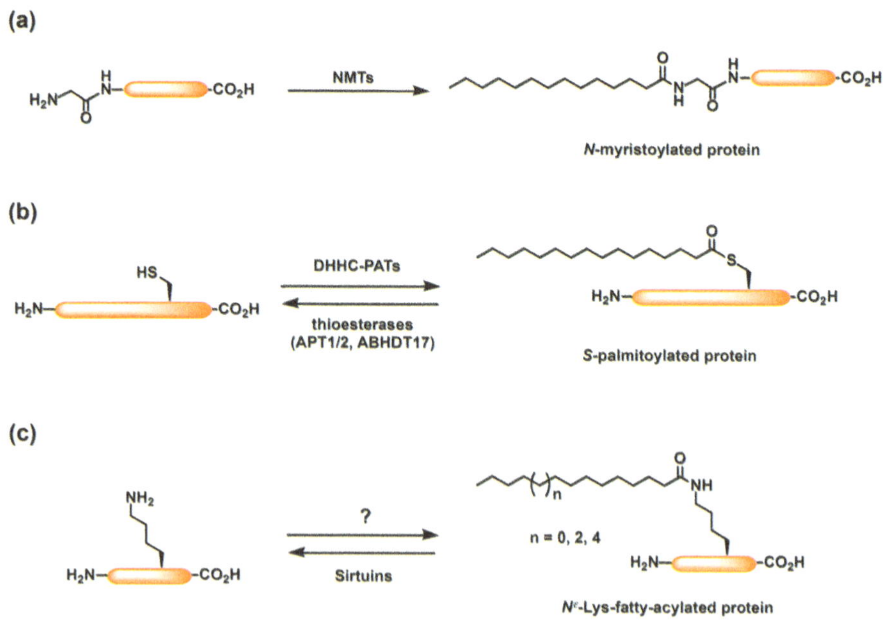

Fig. 1 Different classes of protein fatty-acylation. **a** *N*-myristoylation. **b** *S*-palmitoylation. **c** N^{ε}-Lys-fatty-acylation

Historically, protein fatty-acylation had been difficult to study, largely due to limit in detection techniques and unavailability of identification methods. In the last decade, the advent of chemical proteomic profiling methods for fatty-acylation (Hang and Linder 2011; Peng et al. 2016; Tate et al. 2015), along with the rapid advances of mass spectrometry, has greatly improved our ability to identify fatty-acylated proteins in a proteome-wide scale under various biological contexts. As a result, the discovery of numerous fatty-acylated proteins by recent proteomic studies (Blanc et al. 2015) suggests that the modification is ubiquitous and plays much broader roles in regulating eukaryotic biology than previously appreciated. While most of these fundamental studies were performed in mammalian cell lines and the model species such as yeast *Saccharomyces cerevisiae*, the importance of fatty-acylation in microbial pathogens has now been attracting greater attention. As demonstrated in recent reports, fatty-acylation is likely to be extensively used in microbial pathogens such as protozoan parasites, fungi, and bacteria to regulate key aspects of their biology, including those essential for pathogenesis (Maurer-Stroh and Eisenhaber 2004; Blanc et al. 2013; Ritzefeld et al. 2017). Providing a global understanding of the modification prevalence, chemical proteomic profiling of protein fatty-acylation in microbial pathogens has become fascinating to help unravel its regulatory mechanisms for microbial virulence.

In this review, we briefly introduce different classes of fatty-acylation and the chemical proteomic profiling methods for this modification. We then summarize the large-scale proteomic profiling studies of fatty-acylation performed in microbial pathogens, with a focus on pathogenic parasites and fungi. Finally, we highlight two proteomic studies investigating how bacterial effector proteins alter fatty-acylation landscape of host cells for pathogenesis.

2 Protein Fatty-Acylation

Protein fatty-acylation is subcategorized into several classes according to the modified amino acid residue, carbon chain length, and saturation of the fatty-acyl group, which is excellently summarized in recent reviews (Resh 2016; Jiang et al. 2018). Only the most prevalent and relevant classes are briefly introduced here.

2.1 N-myristoylation

N-myristoylation describes the covalent attachment of a myristoyl group, i.e., the 14-carbon fatty-acyl (C14:0) group, to the amino group of the *N*-terminal glycine residue of a protein through an amide bond (Fig. 1a). While the majority of *N*-myristoylated proteins are modified during translation, post-translational *N*-myristoylation also occurs during apoptosis following caspase cleavage of the protein to expose an *N*-terminal glycine residue (Wright et al. 2010).

N-myristoylation is catalyzed by *N*-myristoyltransferase (NMT) using myristoyl-CoA as the cofactor (Fig. 1a) and is conserved across eukaryotic species. There are two genes in vertebrates encoding NMT enzymes, NMT1 and NMT2, whereas the enzyme is expressed from a single gene in lower eukaryotes. *N*-myristoylation had been thought to be irreversible, until a recent report showing that *Shigella* type III secretion system (T3SS) effector invasion plasmid antigen J (IpaJ) cleaves the peptide bond between *N*-myristoylated glycine and the following residue of protein (Burnaevskiy et al. 2013). *N*-myristoylation is involved in a variety of cellular processes in higher eukaryotes (Wright et al. 2010; Udenwobele et al. 2017). NMT has also been identified to be essential for viability in eukaryotic pathogens, such as parasites and fungi, and presents a potential drug target for treating parasitic and fungal infections (Tate et al. 2013; Georgopapadakou 2002; Fang et al. 2015; Schlott et al. 2018).

2.2 *S*-palmitoylation

Originally discovered on viral glycoproteins (Veit 2012; Schmidt and Schlesinger 1979), *S*-palmitoylation (or *S*-fatty-acylation) is the covalent modification of cysteine residues of a protein with palmitoyl groups, i.e., the 16-carbon fatty-acyl (C16:0) groups, through thioester bonds (Fig. 1b). *S*-palmitoylation is catalyzed by a family of evolutionarily conserved Asp-His-His-Cys (DHHC)-containing protein acyltransferases (PATs); 23 PATs are known in humans, 5–7 in yeast, and 12 in parasites (Resh 2016; Hodson et al. 2015). Different from *N*-myristoylation, the thioester bond of *S*-palmitoylation is enzymatically reversible. Two acyl-protein thioesterases (APTs), as well as the α/β-hydrolase domain 17 (ABHD17) protein family (Lin and Conibear 2015), have been identified to cleave palmitoyl groups from *S*-palmitoylated proteins (Fig. 1b). *S*-palmitoylation is ubiquitous and estimated to occur on as much as 10% of the human proteome (Blanc et al. 2015), including receptors, ion channels, signaling proteins, and so on. Therefore, *S*-palmitoylation is extensively used as a regulatory mechanism in eukaryotes (Linder and Deschenes 2007; Yount et al. 2013; Zhang and Hang 2017) and specifically plays central roles in the pathogenesis of eukaryotic pathogens, such as parasites and fungi, as well as in bacterial and viral infections (Blanc et al. 2013; Sobocińska et al. 2018).

2.3 N^{ε}-Lys-Fatty-Acylation

Covalent attachment of a fatty-acyl group to the epsilon amino group of a lysine residue via amide bond linkage is referred to as N^{ε}-Lys-fatty-acylation (Fig. 1c), with exact nature of the fatty-acyl group undetermined. Although N^{ε}-Lys-fatty-acylation was first reported more than 20 years ago (Stevenson et al. 1992), it has been understudied until recent studies uncovering sirtuins can remove fatty-acyl groups from modified lysines (Fig. 1c) (Jiang et al. 2013; Liu et al. 2015). Many fundamental questions about N^{ε}-Lys-fatty-acylation remain to be answered. For example, the enzymes responsible for the modification have not yet been identified; its abundance and biological functions in eukaryotes are still unknown. Nevertheless, it has been reported that some bacterial effectors exhibit the N^{ε}-fatty-acyltransferase activity during infection to alter the landscape of host fatty-acylation (Zhou et al. 2017; Liu et al. 2018), which is reviewed below.

3 Chemical Proteomic Methods for Profiling Fatty-Acylation

3.1 Chemical Reporter Strategy for Studying Fatty-Acylation

The emergence of bioorthogonal chemical reactions (Prescher and Bertozzi 2005; Raghavan and Hang 2009) has inspired the development of various chemical reporters for studying protein post-translational modifications (PTMs) (Grammel and Hang 2013). Currently, fatty-acylation chemical reporters have been broadly used for detection and proteome-wide profiling of protein fatty-acylation when coupled with bioorthogonal reactions and quantitative proteomic analysis (Fig. 2a) (Charron et al. 2009; Hang et al. 2011; Thinon and Hang 2015; Tate et al. 2015). Fatty-acylation chemical reporters are fatty acid analogues functionalized with unique bioorthogonal handles, such as alkynyl and azido groups (Fig. 2b). These analogues can be utilized by native fatty-acylation machinery in cells to metabolically label fatty-acylated proteins. Then the labeled proteins are further derivatized via bioorthogonal reactions with either fluorophores for rapid detection or affinity tags (e.g., biotin) for selective enrichment and large-scale proteomic identification. Currently, the most widely used fatty-acylation chemical reporters are the alkynyl derivatives of myristic acid and palmitic acid, i.e., alk-12 (or YnMyr) and alk-16 (or 17-ODYA) (Fig. 2b), which preferentially label *N*-myristoylated and *S*-palmitoylated proteins (Peng et al. 2016; Thinon and Hang 2015), respectively. Other palmitic acid analogues, such as YnPal (Wright et al. 2016) and x-alk-16 (Peng and Hang 2015), with one carbon less than alk-16 and an additional diazirine group (Fig. 2b), respectively, have also been used for labeling *S*-palmitoylated proteins. It is worth mentioning that both alk-14 and alk-16 have been used for studying N^{ε}-Lys-fatty-acylation (Jiang et al. 2013; Liu et al. 2015).

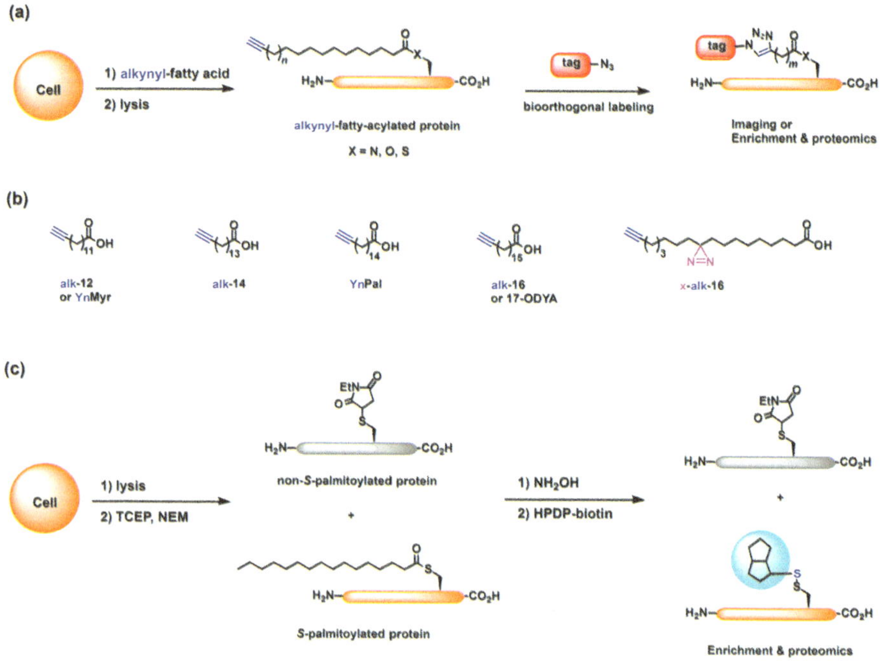

Fig. 2 Chemical proteomic profiling methods for protein fatty-acylation. **a** Bioorthogonal detection and enrichment of fatty-acylated proteins with chemical reporter strategy. **b** Representative fatty-acylation chemical reporters. **c** Chemical enrichment of *S*-palmitoylated proteins with ABE strategy

3.2 Acyl-Biotin Exchange (ABE) Strategy for Studying S-palmitoylation

The characteristic hydroxylamine (NH_2OH) sensitivity of thioester linkage in *S*-palmitoylation has been exploited to develop the ABE strategy for selective enrichment of *S*-palmitoylated proteins (Drisdel and Green 2004; Roth et al. 2006; Kang et al. 2008). In this strategy, the disulfide bonds of proteins are first reduced and the resulting free cysteines capped with *N*-ethyl maleimide (NEM). Then NH_2OH is used to selectively cleave thioester bonds, followed by capture of newly liberated cysteines with *N*-(6-(biotinamido)hexyl)-3′-(2′-pyridyldithio)propionamide (biotin–HPDP) or thiosepharose resin (Forrester et al. 2011) for enrichment and subsequent proteomic analysis (Fig. 2c).

Compared to the chemical reported strategy, ABE does not require metabolic labeling of cells with exogenous fatty acid analogues and therefore is ideal for studying *S*-palmitoylation in animal tissues. However, ABE cannot distinguish *S*-palmitoylated proteins from other thioester-modified proteins, such as intermediates in ubiquitin conjugation system (Jones et al. 2012). In contrast, the chemical reporter strategy enables investigation of *S*-palmitoylation dynamics by classic pulse-chase labeling, but may not differentiate *S*-palmitoylation from other classes of fatty-acylation (Wright et al. 2016). Overall, these two strategies are largely complementary and both provide unique coverages of *S*-palmitoylated proteins, as demonstrated in a recent comparative study (Jones et al. 2012).

4 Proteomic Profiling of Fatty-Acylation in Eukaryotic Pathogens

4.1 Proteomic Profiling of N-myristoylation in Parasites

Protozoan parasites, such as *Plasmodia*, *Toxoplasma*, *Trypanosoma*, and *Leishmania* that cause malaria, toxoplasmosis, trypanosomiasis, and leishmaniases, respectively, are still among the most prevalent eukaryotic pathogens around the world, posing a significant health threat to billions of people worldwide, especially those living in tropical areas of developing countries. The co-translational *N*-myristoylation is essential for the growth and survival of many protozoan parasites (Tate et al. 2013). Therefore, inhibition of this modification has been proposed to be a potential therapeutic strategy for treatment of parasitic infections. However, relatively little was known about the specific roles of *N*-myristoylation in parasite biology, largely due to limited availability of global *N*-myristoylation profiles in various parasites. Over the past several years, a series of proteome-wide profiling studies for protein *N*-myristoylation in a variety of parasites enabled by the chemical reporter strategy have greatly expanded our appreciation of the significance of this modification in parasitic pathogens.

N*-myristoylation in *Plasmodium falciparum

In 2014, Wright et al. applied the well-established *N*-myristoylation chemical reporter alk-12 and label-free quantitative proteomic analysis to globally identify *N*-myristoylated proteins in asexual stage *Plasmodium falciparum* (Wright et al. 2014), the parasite responsible for malaria. Taking advantage of a cleavable azido-biotin reagent (Broncel et al. 2015), they uncovered more than 30 reliable *N*-myristoylated proteins that are associated with a wide range of functions such as motility, protein transport, development, and protein phosphorylation. Specifically, among these identified proteins are the small inner-membrane complex (IMC) proteins ISP1 and ISP3 that are essential for the formation of IMC and parasitic invasion of red blood cells. More interestingly, the authors evaluated several NMT inhibitors and further demonstrated that these compounds blocked *N*-myristoylation effectively at EC50 values that correlate with parasite growth inhibition, thus linking parasite death in vitro and in vivo directly to loss of protein *N*-myristoylation (Wright et al. 2014). The seminal study not only provided a list of *N*-myristoylated proteins in *P. falciparum*, but also highlighted the potential of NMT as an antimalarial drug target.

N-myristoylation in *Trypanosoma brucei* and *Trypanosoma cruzi*

Recently, the chemical reporter strategy, in combination with proteomic analysis, has also been applied to globally profile protein *N*-myristoylation in the sleeping sickness parasite *Trypanosoma brucei* (Wright et al. 2016). An initial label-free quantitative proteomic attempt using *N*-myristoylation chemical reporter alk-12 identified over 100 robustly enriched proteins in *T. brucei*, with only half possessing the putative *N*-terminal glycine for *N*-myristoylation and others being known GPI-anchored or *S*-palmitoylated proteins and potentially N^{ε}-Lys-fatty-acylated proteins (Wright et al. 2016). These results are not surprising and consistent with previous observation that alk-12 can also label other types of fatty-acylated proteins in addition to *N*-myristoylated proteins, due to high promiscuity of the fatty-acylation machineries to fatty acid chain length. To differentiate *N*-myristoylated proteins from other fatty-acylated proteins and simplify data interpretation, well-characterized NMT inhibitors were then employed to selectively reduce alk-12 incorporation into *N*-myristoylated, but not other fatty-acylated proteins in *T. brucei*. Following enrichment and label-free quantitative proteomics, ~50 proteins were identified to have reduced abundance in response to NMT inhibitors and considered to be highly confident *N*-myristoylated proteins, with many of them bearing the *N*-terminal glycine motif (Wright et al. 2016). In a similar study, Roberts and Fairlamb applied the azido analogue of myristic acid and quantitative proteomics to identify *N*-myristoylated proteins in *Trypanosoma cruzi*, the causative agent of Chagas' disease (Roberts and Fairlamb 2016). The authors identified 50 high-confidence *N*-myristoylated proteins using a combination of label-free and stable isotope labeling of cells in culture (SILAC) (Ong et al. 2002) quantitative proteomic analyses in the absence or presence of a well-characterized NMT inhibitor. Interestingly, most of these proteins are of unknown function, suggesting that more mechanistic studies are required to understand *N*-myristoylation in this parasite.

N-myristoylation in *Leishmania donovani*

Leishmania donovani is the major causative agent of leishmaniases, a spectrum of tropical parasitic diseases endangering millions of people around the world. Until recently, a proteome-wide profile of *N*-myristoylation in *L. donovani* was still lacking. Taking a similar approach to other parasites as introduced above, Wright et al. integrated metabolic labeling with *N*-myristoylation chemical reporter alk-12, chemical knockdown of NMT with selective inhibitors, and label-free quantitative proteomics to globally profile *N*-myristoylated proteins in *L. donovani* (Wright et al. 2015). This analysis allowed them to identify ~30 candidates to be high-confidence *N*-myristoylated proteins, with some of them involved in protein phosphorylation, transport, and degradation, highlighting diverse roles of *N*-myristoylation in modulating multiple pathways in this parasite.

4.2 Proteomic Profiling of S-palmitoylation in Parasites

As in higher eukaryotes, *S*-palmitoylation has also been used extensively to regulate key aspects of biology in parasites (Ritzefeld et al. 2017). While NMT inhibitors are readily available, there are no specific chemical inhibitors for PATs in parasites. Although the broadly reactive *S*-palmitoylation inhibitor 2-bromopalmitate (2-BP) has been widely used, its lack of specificity on lipid metabolic pathways (Zheng et al. 2013) is of great concern, which may complicate global analyses of *S*-palmitoylation. Nevertheless, a series of chemical proteomic profiling studies in various parasites have revealed much more *S*-palmitoylated proteins than previously estimated and provided important insights into the regulatory mechanisms of *S*-palmitoylation for parasitic pathogenesis.

S-palmitoylation in *Plasmodium falciparum*

In a landmark study, Jones et al. reported the first global profile of protein *S*-palmitoylation in the asexual stage of *P. falciparum* (Jones et al. 2012). The authors employed two powerful chemical proteomic methods, i.e., ABE and the chemical reporter alk-16, coupled with SILAC quantitative proteomic analysis to confidently identify *S*-palmitoylated proteins. This comprehensive study identified more than 400 putative *S*-palmitoylated proteins, including those essential for drug resistance, schizont development, parasitic invasion, and other virulence-associated processes (Jones et al. 2012). Comparison of ABE with chemical reporter strategies suggested that they were largely complementary and both methods provided unique coverage of *S*-palmitoylated proteins. The authors further combined ABE and SILAC quantitative proteomics with 2-BP treatment to examine *S*-palmitoylation dynamics and demonstrated that some *S*-palmitoylated proteins were stably modified while others were dynamically regulated (Jones et al. 2012). Overall, this study has provided a wealth of data on *S*-palmitoylated proteins in *P. falciparum*, as well as an exemplary application of the chemical proteomic methods for profiling *S*-palmitoylation.

S-palmitoylation in *Toxoplasma gondii*

S-palmitoylation has been implicated to play an important role in host invasion of *Toxoplasma gondii*. For instance, inhibition of *S*-palmitoylation with 2-BP can inhibit host cell invasion and disrupt parasite morphology (Alonso et al. 2012). In addition, small-molecule inhibition of *T. gondii* palmitoyl protein thioesterase-1 (TgPPT1) has been shown to trigger secretion of invasion-associated organelles and markedly enhance infectivity of this parasite (Child et al. 2013). In order to understand the roles of *S*-palmitoylation in parasitic invasion, Foe et al. performed a global analysis of *S*-palmitoylated proteins using *S*-palmitoylation chemical reporter alk-16 and label-free quantitative proteomics (Foe et al. 2015). Notably, NH_2OH treatment of the proteome sample was included for comparative proteomic analysis to differentiate *S*-palmitoylation from other types of fatty-acylation, as NH_2OH selectively cleaves *S*-palmitoylation while leaving other fatty-acylation intact. This study revealed a total of 282 high-confidence *S*-palmitoylated proteins, some of which were validated to be involved in motility, cell morphology, and host cell invasion. Follow-up studies on a newly identified *S*-palmitoylated protein AMA1 indicated that blocking its *S*-palmitoylation enhanced the secretion of AMA1 and other invasion-related proteins (Foe et al. 2015). In a complementary study, the ABE method coupled with semi-quantitative proteomic analysis was applied to *T. gondii*, identifying more than 400 *S*-palmitoylated protein candidates (Caballero et al. 2016), with half of them overlapping with the high-confidence *S*-palmitoylated protein list reported by Foe et al. Together, these studies suggest that *S*-palmitoylation is ubiquitous in *T. gondii* and plays an essential role in the biology of this pathogen.

S-palmitoylation in *Trypanosoma brucei*

Global analyses of *S*-palmitoylation in *T. brucei* have also been performed with both ABE and chemical reporter strategies. An early proteome-wide profiling study in 2011 using ABE and semi-quantitative proteomic analysis identified a total of 124 putative *S*-palmitoylated proteins in *T. brucei* (Emmer et al. 2011). Interestingly, the study showed that the broad *S*-palmitoylation inhibitor 2-BP caused serious growth inhibition of the parasite while no growth effect was observed by the selective inhibition of each of the 12 individual PATs, suggesting functional redundancy of PATs or essential *S*-palmitoylated proteins. The *S*-palmitoylation chemical reporter YnPal, coupled with label-free quantitative proteomics, has also been applied to *T. brucei*, enabling the identification of about 100 putative *S*-palmitoylated proteins (Wright et al. 2016). It is worth mentioning that around 70 proteins identified with the chemical reporter strategy were not found by the ABE method, again highlighting the complementarity of these two strategies for *S*-palmitoylation analysis.

4.3 Proteomic Profiling of N-myristoylation and S-palmitoylation in Fungal Pathogens

Opportunistic fungal pathogens, such as *Aspergillus fumigatus*, *Candida albicans*, and *Cryptococcus neoformans* that cause disease conditions Aspergillosis, Candidiasis, and Cryptococcosis, respectively, are endangering billions of people every year around the world with more than 50% mortality rates, resulting in over 1.5 million deaths each year (Brown et al. 2012). However, research on the pathophysiology of human fungal pathogens is largely lagging behind that of other microorganism pathogens. Like parasitic pathogens and other eukaryotes, major fungal pathogens such as *A. fumigatus*, *C. albicans*, and *C. neoformans* all decode for their own NMTs (Santiago-Tirado and Doering 2016), the activity of which is essential for their viability. Therefore, NMT presents a potential target for development of antifungals. Although several fungal proteins, for example, the ADP-ribosylation factor (ARF), have been shown to be *N*-myristoylated, the global picture of protein *N*-myristoylation in these pathogens is still lacking and represents an attractive direction for future efforts to understand the significance of this modification in fungal pathogens.

Fungal pathogens also express their own PATs, and inhibition of these enzymes affects their growth and survival (Santiago-Tirado and Doering 2016). As in mammalian cells, the best-studied *S*-palmitoylated protein in fungal pathogens is the signaling protein Ras, whose modification is required for proper membrane localization of the protein and fungal pathogenesis. While chemical proteomic profiling of *S*-palmitoylation has been performed in model yeast *S. cerevisiae* (Roth et al. 2006) and *Schizosaccharomyces pombe* (Zhang et al. 2013), the global understanding of this modification in fungal pathogens has been very limited.

Recently, we reported a proteomic analysis to globally identify the specific substrates of an important PAT pfa4 in *C. neoformans* (Santiago-Tirado et al. 2015). An initial screening for regulators of *C. neoformans* internalization identified pfa4 to be essential for fungal virulence in vitro and in vivo. In order to mechanistically explain the functions of pfa4, we metabolically labeled the wild-type and pfa4 knockout strains with *S*-palmitoylation chemical reporter alk-16, and then employed label-free quantitative proteomic analysis to determine the pfa4-specific substrates by comparison of the *S*-palmitoylated protein abundance in these two strains (Fig. 3). This analysis allowed us to uncover the pfa4-specific *S*-palmitoylome, which included proteins involved in cell wall synthesis, membrane trafficking, signal transduction, and protein transport. Follow-up studies further suggested that defects on *S*-palmitoylation of a specific subset of proteins combined to result in avirulence of the pfa4 knockout strain (Santiago-Tirado et al. 2015). Overall, this study presents the first global *S*-palmitoylation analysis in fungal pathogens and provides important insights into development of new antifungals targeting pfa4.

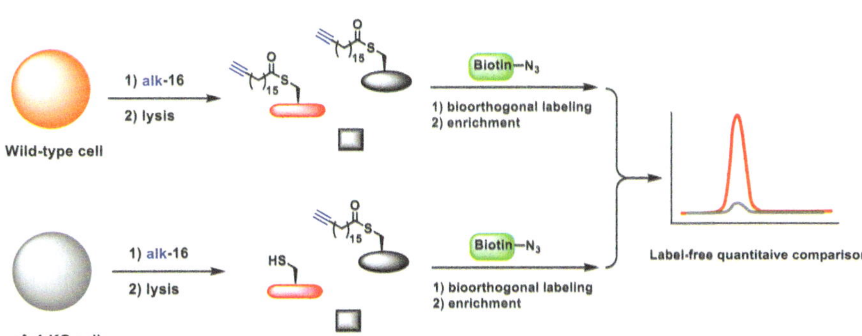

Fig. 3 Schematic for proteomic identification of pfa4-specific protein substrates in *C. neoformans* using the chemical reporter strategy

5 Protein Fatty-Acylation in Bacterial Virulence

5.1 Host-Mediated Fatty-Acylation of Bacterial Effectors

Bacterial pathogens have evolved sophisticated protein secretion systems to transport effector proteins into the cytosol of infected host cells. These effectors normally possess unique biochemical activities to modulate a variety of host cell processes, such as membrane trafficking, transcription, and signal transduction, for bacterial survival and replication in host cells. Recent findings have suggested that bacterial effectors can be post-translationally modified by host enzymatic machinery inside infected cells. The main function of these modifications seems to target the effectors into precise subcellular compartments to exert their functions in a spatially coordinated manner (Hicks and Galán 2013; Ivanov and Roy 2014; Popa et al. 2016).

Among these modifications is fatty-acylation that increases the hydrophobicity of proteins and dramatically affects protein localization by tethering to intracellular or plasma membranes. While *N*-myristoylation has been largely described in many effectors from plant bacterial pathogens, for example, the *Pseudomona syringae* effectors AvrRpm1 and AvrB (Nimchuk et al. 2000; Maurer-Stroh and Eisenhaber 2004), *S*-palmitoylation is more commonly reported in effectors from human bacterial pathogens. For example, three *Salmonella* effector proteins have been reported to be *S*-palmitoylated, i.e., SifA, SspH2, and SseI. SifA is prenylated at its *C*-terminus and *S*-palmitoylated at an adjacent cysteine residue (Reinicke et al. 2005). In contrast, SspH2 and SseI are *S*-palmitoylated at a conserved cysteine residue within their *N*-terminal domains, resulting in targeting of these effectors to the host cell plasma membrane (Fig. 4a) (Hicks et al. 2011). In addition, *Legionella* effectors GobX and LpdA were also shown to be *S*-palmitoylated recently (Lin et al. 2015; Schroeder et al. 2015). Nevertheless, a global analysis of host-mediated fatty-acylation of bacterial effectors has not yet been reported and requires further exploration in future.

5.2 Modulation of Host Fatty-Acylation by Bacterial Effectors

In addition to taking advantage of host PTM machinery for their own modification, bacterial effector proteins may also manipulate PTMs on host proteins to facilitate bacterial survival and infection. These effector proteins are normally enzymes that directly catalyze the formation or removal of PTMs on host proteins by imitating the activities of host PTM machinery or by using unique activities (Cui and Shao 2011; Salomon and Orth 2013; Reddick and Alto 2014). The manipulated host

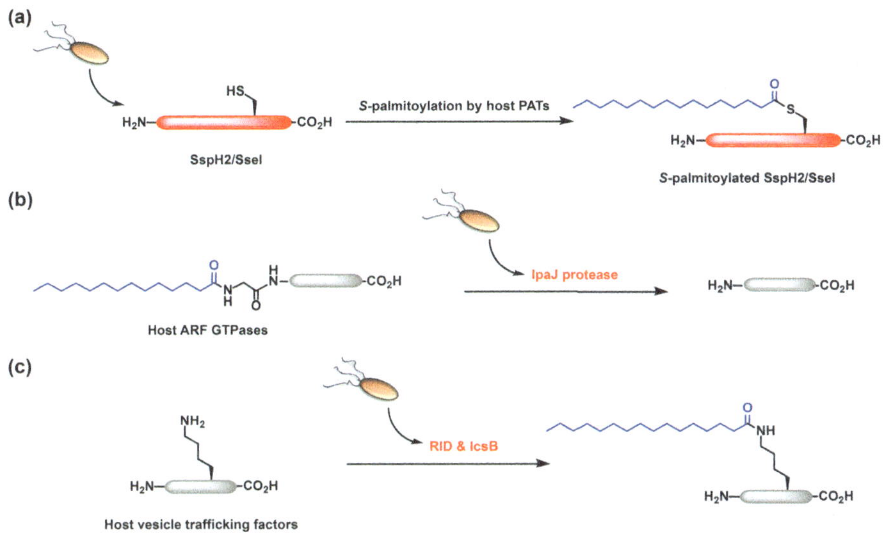

Fig. 4 Protein fatty-acylation in bacterial virulence. **a** SspH2 and SseI are *S*-palmitoylated by host PATs. **b** IpaJ cleaves *N*-myristoylation of host GTPases. **c** RID and IcsB are N^{ε}-lysine-fatty-acyltransferases of bacterial origin

targets are generally those required for host immune responses, such as the mitogen-activated protein kinase (MAPK) pathway, nuclear factor-κB (NF-κB) signaling, and the ubiquitin system (Cui and Shao 2011). Apart from these, recent intriguing studies have shown that bacterial pathogens have evolved effector proteins with unprecedented enzymatic activities to modulate host fatty-acylation for survival or escape from clearance.

In 2013, Burnaevskiy et al. reported an unprecedented enzymatic activity of *Shigella flexneri* T3SS effector IpaJ that acts as a cysteine protease to cleave *N*-myristoylated glycines of small GTPases for inhibition of host secretion and trafficking systems during infection (Fig. 4b) (Burnaevskiy et al. 2013). To determine the protein substrates of IpaJ and the mechanism of substrate selection, we performed a global *N*-myristoylation profiling of host cells treated with IpaJ under different physiological contexts (Burnaevskiy et al. 2015). In an initial in vitro study, we treated the alk-12-labeled cell lysates with recombinant IpaJ and applied label-free quantitative proteomics to determine the *N*-myristoylation reduction in the presence of IpaJ. This analysis showed that IpaJ eliminates modification of the majority of *N*-myristoylated proteins, implicating no substrate preference and selection of this effector in vitro. On the contrary, a subsequent in vivo *N*-myristoylation profiling on alk-12-labeled cells that were infected with wild-type or IpaJ knockout *Shigella* strain demonstrated that IpaJ is remarkably specific to cleave Golgi-associated ARF/ARL family GTPases during infection (Burnaevskiy et al. 2015). This finding then led us to propose that binding of IpaJ with GTPase domain is the prerequisite for substrate selection. Overall, the chemical proteomic profiling greatly expanded our understanding of *N*-myristoylation in the context of host–bacteria interaction.

Very recently, the Rho guanosine triphosphatase (GTPase) inactivation domain (RID), a conserved effector domain of multifunctional autoprocessing repeats-in-toxin (MARTX) toxins from *Vibrio cholera*, has been reported to possess a lysine N^{ε}-fatty-acyltransferase activity that covalently modifies lysine residues of host RhoGTPases with palmitoyl groups (Fig. 4c) (Zhou et al. 2017). This modification results in disruption of RhoGTPase-mediated signaling in host cell, and thus potentially facilitates bacterial evasion from host defense. During the progression of this study, we were focusing on the RID homologue in *S. flexneri*, IcsB, which is a T3SS effector essential for bacterial evasion of host autophagy. Not surprisingly, we discovered that IcsB is also an N^{ε}-fatty-acyltransferase that modifies RhoGTPases like RID (Fig. 4c) (Liu et al. 2018). To determine the substrate scope and biological consequence of this enzymatic activity, a chemical proteomic profiling involving alk-16 metabolic labeling of host cells infected with wild-type or IcsB knockout *Shigella* strain and SILAC quantitative proteomics was performed. This unbiased global analysis enabled the identification of about 60 IcsB substrates that are associated with a variety of cellular processes, such as endosomal recycling and membrane trafficking (Liu et al. 2018), thus providing us important insights into substrate recognition mechanism and biological function of this effector.

6 Conclusions

Protein fatty-acylation is ubiquitous and regulates almost every aspect of cellular biology. Specifically in the interface of host–pathogen interaction, fatty-acylation, e.g., *N*-myristoylation and *S*-palmitoylation, is essential for the viability and pathogenesis of eukaryotic pathogens such as parasites and fungi. Pathogenic bacteria have also evolved the surprising capacity to hijack host fatty-acylation machinery for modifying their own effector proteins or manipulate host fatty-acylation levels to facilitate survival and invasion during infection. The application of chemical proteomic profiling strategies to microbial pathogens for global analysis of fatty-acylation has significantly expanded our understanding of the importance of this modification in microbial pathogenesis. Unraveling the regulatory roles of fatty-acylation under this context will ultimately pave a new avenue for developing anti-infective strategies. Future efforts in this area should be focused on discovery of specific inhibitors for fatty-acylation and characterization of its biological consequence during microbial infections.

References

Alonso AM, Coceres VM, De Napoli MG, Nieto Guil AF, Angel SO, Corvi MM (2012) Protein palmitoylation inhibition by 2-bromopalmitate alters gliding, host cell invasion and parasite morphology in *Toxoplasma gondii*. Mol Biochem Parasitol 184(1):39–43. https://doi.org/10.1016/j.molbiopara.2012.03.006

Blanc M, Blaskovic S, van der Goot FG (2013) Palmitoylation, pathogens and their host. Biochem Soc Trans 41(1):84–88. https://doi.org/10.1042/bst20120337

Blanc M, David F, Abrami L, Migliozzi D, Armand F, Bürgi J, van der Goot F (2015) SwissPalm: protein palmitoylation database [version 1; referees: 3 approved], F1000Research 4:261. https://doi.org/10.12688/f1000research.6464.1

Broncel M, Serwa RA, Ciepla P, Krause E, Dallman MJ, Magee AI, Tate EW (2015) Multifunctional reagents for quantitative proteome-wide analysis of protein modification in human cells and dynamic profiling of protein lipidation during vertebrate development. Angew Chem Int Ed 54(20):5948–5951. https://doi.org/10.1002/anie.201500342

Brown GD, Denning DW, Gow NAR, Levitz SM, Netea MG, White TC (2012) Hidden killers: human fungal infections. Sci Transl Med 4(165):165rv113. https://doi.org/10.1126/scitranslmed.3004404

Burnaevskiy N, Fox TG, Plymire DA, Ertelt JM, Weigele BA, Selyunin AS, Way SS, Patrie SM, Alto NM (2013) Proteolytic elimination of N-myristoyl modifications by the Shigella virulence factor IpaJ. Nature 496:106–109. https://doi.org/10.1038/nature12004

Burnaevskiy N, Peng T, Reddick LE, Hang Howard C, Alto Neal M (2015) Myristoylome profiling reveals a concerted mechanism of ARF GTPase deacylation by the bacterial protease IpaJ. Mol Cell 58(1):110–122. https://doi.org/10.1016/j.molcel.2015.01.040

Caballero MC, Alonso AM, Deng B, Attias M, de Souza W, Corvi MM (2016) Identification of new palmitoylated proteins in *Toxoplasma gondii*. Biochim Biophys Acta 1864(4):400–408. https://doi.org/10.1016/j.bbapap.2016.01.010

Charron G, Wilson J, Hang HC (2009) Chemical tools for understanding protein lipidation in eukaryotes. Curr Opin Chem Biol 13(4):382–391. https://doi.org/10.1016/j.cbpa.2009.07.010

Child MA, Hall CI, Beck JR, Ofori LO, Albrow VE, Garland M, Bowyer PW, Bradley PJ, Powers JC, Boothroyd JC, Weerapana E, Bogyo M (2013) Small-molecule inhibition of a depalmitoylase enhances Toxoplasma host-cell invasion. Nat Chem Biol 9:651–656. https://doi.org/10.1038/nchembio.1315

Cui J, Shao F (2011) Biochemistry and cell signaling taught by bacterial effectors. Trends Biochem Sci 36(10):532–540. https://doi.org/10.1016/j.tibs.2011.07.003

Drisdel RC, Green WN (2004) Labeling and quantifying sites of protein palmitoylation. Biotechniques 36(2):276–285. https://doi.org/10.2144/04362RR02

Emmer BT, Nakayasu ES, Souther C, Choi H, Sobreira TJP, Epting CL, Nesvizhskii AI, Almeida IC, Engman DM (2011) Global analysis of protein palmitoylation in African trypanosomes. Eukaryot Cell 10(3):455–463. https://doi.org/10.1128/ec.00248-10

Fang W, Robinson DA, Raimi OG, Blair DE, Harrison JR, Lockhart DEA, Torrie LS, Ruda GF, Wyatt PG, Gilbert IH, van Aalten DMF (2015) N-myristoyltransferase is a cell wall target in *Aspergillus fumigatus*. ACS Chem Biol 10(6):1425–1434. https://doi.org/10.1021/cb5008647

Foe IT, Child MA, Majmudar JD, Krishnamurthy S, van der Linden WA, Ward GE, Martin BR, Bogyo M (2015) Global analysis of palmitoylated proteins in *Toxoplasma gondii*. Cell Host Microbe 18(4):501–511. https://doi.org/10.1016/j.chom.2015.09.006

Forrester MT, Hess DT, Thompson JW, Hultman R, Moseley MA, Stamler JS, Casey PJ (2011) Site-specific analysis of protein S-acylation by resin-assisted capture. J Lipid Res 52(2):393–398. https://doi.org/10.1194/jlr.D011106

Georgopapadakou NH (2002) Antifungals targeted to protein modification: focus on protein N-myristoyltransferase. Expert Opin Investig Drugs 11(8):1117–1125. https://doi.org/10.1517/13543784.11.8.1117

Grammel M, Hang HC (2013) Chemical reporters for biological discovery. Nat Chem Biol 9 (8):475–484. https://doi.org/10.1038/nchembio.1296
Hang HC, Linder ME (2011) Exploring protein lipidation with chemical biology. Chem Rev 111 (10):6341–6358. https://doi.org/10.1021/cr2001977
Hang HC, Wilson JP, Charron G (2011) Bioorthogonal chemical reporters for analyzing protein lipidation and lipid trafficking. Acc Chem Res 44(9):699–708. https://doi.org/10.1021/ar200063v
Hicks SW, Galán JE (2013) Exploitation of eukaryotic subcellular targeting mechanisms by bacterial effectors. Nat Rev Microbiol 11:316–326. https://doi.org/10.1038/nrmicro3009
Hicks SW, Charron G, Hang HC, Galán JE (2011) Subcellular targeting of *Salmonella* virulence proteins by host-mediated S-palmitoylation. Cell Host Microbe 10(1):9–20. https://doi.org/10.1016/j.chom.2011.06.003
Hodson N, Invergo B, Rayner Julian C, Choudhary Jyoti S (2015) Palmitoylation and palmitoyl-transferases in *Plasmodium* parasites. Biochem Soc Trans 43(2):240–245. https://doi.org/10.1042/bst20140289
Ivanov SS, Roy C (2014) Host lipidation: a mechanism for spatial regulation of legionella effectors. In: Hilbi H (ed) Molecular mechanisms in legionella pathogenesis. Springer Berlin Heidelberg, Berlin, Heidelberg, pp 135–154. https://doi.org/10.1007/82_2013_344
Jiang H, Khan S, Wang Y, Charron G, He B, Sebastian C, Du J, Kim R, Ge E, Mostoslavsky R, Hang HC, Hao Q, Lin H (2013) SIRT6 regulates TNF-alpha secretion through hydrolysis of long-chain fatty acyl lysine. Nature 496(7443):110–113. https://doi.org/10.1038/nature12038
Jiang H, Zhang X, Chen X, Aramsangtienchai P, Tong Z, Lin H (2018) Protein lipidation: occurrence, mechanisms, biological functions, and enabling technologies. Chem Rev 118 (3):919–988. https://doi.org/10.1021/acs.chemrev.6b00750
Jones ML, Collins MO, Goulding D, Choudhary JS, Rayner JC (2012) Analysis of protein palmitoylation reveals a pervasive role in *Plasmodium* development and pathogenesis. Cell Host Microbe 12(2):246–258. https://doi.org/10.1016/j.chom.2012.06.005
Lin DTS, Conibear E (2015) ABHD17 proteins are novel protein depalmitoylases that regulate N-Ras palmitate turnover and subcellular localization. eLife 4:e11306. https://doi.org/10.7554/elife.11306
Lin Y-H, Doms AG, Cheng E, Kim B, Evans TR, Machner MP (2015) Host cell-catalyzed S-palmitoylation mediates golgi targeting of the legionella ubiquitin ligase GobX. J Biol Chem 290(42):25766–25781. https://doi.org/10.1074/jbc.M115.637397
Linder ME, Deschenes RJ (2007) Palmitoylation: policing protein stability and traffic. Nat Rev Mol Cell Biol 8:74–84. https://doi.org/10.1038/nrm2084
Liu Z, Yang T, Li X, Peng T, Hang HC, Li XD (2015) Integrative chemical biology approaches for identification and characterization of "erasers" for fatty-acid-acylated lysine residues within proteins. Angew Chem Int Ed 54(4):1149–1152. https://doi.org/10.1002/anie.201408763
Liu W, Zhou Y, Peng T, Zhou P, Ding X, Li Z, Zhong H, Xu Y, Chen S, Hang HC, Shao F (2018) N^{ε}-fatty acylation of multiple membrane-associated proteins by Shigella IcsB effector to modulate host function. Nature Microbiology in press. https://doi.org/10.1038/s41564-018-0215-6
Maurer-Stroh S, Eisenhaber F (2004) Myristoylation of viral and bacterial proteins. Trends Microbiol 12(4):178–185. https://doi.org/10.1016/j.tim.2004.02.006
Nimchuk Z, Marois E, Kjemtrup S, Leister RT, Katagiri F, Dangl JL (2000) Eukaryotic fatty acylation drives plasma membrane targeting and enhances function of several type iii effector proteins from *Pseudomonas syringae*. Cell 101(4):353–363. https://doi.org/10.1016/s0092-8674(00)80846-6
Ong S-E, Blagoev B, Kratchmarova I, Kristensen DB, Steen H, Pandey A, Mann M (2002) Stable isotope labeling by amino acids in cell culture, SILAC, as a simple and accurate approach to expression proteomics. Mol Cell Proteomics 1(5):376–386. https://doi.org/10.1074/mcp.M200025-MCP200

Peng T, Hang HC (2015) Bifunctional fatty acid chemical reporter for analyzing S-palmitoylated membrane protein-protein interactions in mammalian cells. J Am Chem Soc 137(2):556–559. https://doi.org/10.1021/ja502109n

Peng T, Thinon E, Hang HC (2016) Proteomic analysis of fatty-acylated proteins. Curr Opin Chem Biol 30:77–86. https://doi.org/10.1016/j.cbpa.2015.11.008

Popa CM, Tabuchi M, Valls M (2016) Modification of bacterial effector proteins inside eukaryotic host cells. Front Cell Infect Microbiol 6:73. https://doi.org/10.3389/fcimb.2016.00073

Prescher JA, Bertozzi CR (2005) Chemistry in living systems. Nat Chem Biol 1:13–21. https://doi.org/10.1038/nchembio0605-13

Raghavan AS, Hang HC (2009) Seeing small molecules in action with bioorthogonal chemistry. Drug Discovery Today 14(3):178–184. https://doi.org/10.1016/j.drudis.2008.09.014

Reddick LE, Alto NM (2014) Bacteria fighting back: how pathogens target and subvert the host innate immune system. Mol Cell 54(2):321–328. https://doi.org/10.1016/j.molcel.2014.03.010

Reinicke AT, Hutchinson JL, Magee AI, Mastroeni P, Trowsdale J, Kelly AP (2005) A *Salmonella typhimurium* effector protein SifA is modified by host cell prenylation and S-acylation machinery. J Biol Chem 280(15):14620–14627. https://doi.org/10.1074/jbc.M500076200

Resh MD (1999) Fatty acylation of proteins: new insights into membrane targeting of myristoylated and palmitoylated proteins. Biochim Biophys Acta 1451(1):1–16. https://doi.org/10.1016/S0167-4889(99)00075-0

Resh MD (2016) Fatty acylation of proteins: the long and the short of it. Prog Lipid Res 63:120–131. https://doi.org/10.1016/j.plipres.2016.05.002

Ritzefeld M, Wright MH, Tate EW (2017) New developments in probing and targeting protein acylation in malaria, leishmaniasis and African sleeping sickness. Parasitology 145(2):157–174. https://doi.org/10.1017/S0031182017000282

Roberts AJ, Fairlamb AH (2016) The N-myristoylome of *Trypanosoma cruzi*. Sci Rep 6:31078. https://doi.org/10.1038/srep31078

Roth AF, Wan J, Bailey AO, Sun B, Kuchar JA, Green WN, Phinney BS, Yates JR III, Davis NG (2006) Global analysis of protein palmitoylation in yeast. Cell 125(5):1003–1013. https://doi.org/10.1016/j.cell.2006.03.042

Salomon D, Orth K (2013) What pathogens have taught us about posttranslational modifications. Cell Host Microbe 14(3):269–279. https://doi.org/10.1016/j.chom.2013.07.008

Santiago-Tirado FH, Doering TL (2016) All about that fat: lipid modification of proteins in *Cryptococcus neoformans*. J Microbiol 54(3):212–222. https://doi.org/10.1007/s12275-016-5626-6

Santiago-Tirado FH, Peng T, Yang M, Hang HC, Doering TL (2015) A single protein S-acyl transferase acts through diverse substrates to determine cryptococcal morphology, stress tolerance, and pathogenic outcome. PLoS Path 11(5):e1004908. https://doi.org/10.1371/journal.ppat.1004908

Schlott AC, Holder AA, Tate EW (2018) N-myristoylation as a drug target in malaria: exploring the role of N-myristoyltransferase substrates in the inhibitor mode of action. ACS Infect Dis 4(4):449–457. https://doi.org/10.1021/acsinfecdis.7b00203

Schmidt MFG, Schlesinger MJ (1979) Fatty acid binding to vesicular stomatitis virus glycoprotein: a new type of post-translational modification of the viral glycoprotein. Cell 17(4):813–819. https://doi.org/10.1016/0092-8674(79)90321-0

Schroeder GN, Aurass P, Oates CV, Tate EW, Hartland EL, Flieger A, Frankel G (2015) Legionella pneumophila effector LpdA is a palmitoylated phospholipase D virulence factor. Infect Immun 83(10):3989–4002. https://doi.org/10.1128/iai.00785-15

Schultz AM, Henderson LE, Oroszlan S (1988) Fatty acylation of proteins. Annu Rev Cell Biol 4(1):611–647. https://doi.org/10.1146/annurev.cb.04.110188.003143

Sobocińska J, Roszczenko-Jasińska P, Ciesielska A, Kwiatkowska K (2018) Protein palmitoylation and its role in bacterial and viral infections. Front Immunol 8:2003. https://doi.org/10.3389/fimmu.2017.02003

Stevenson FT, Bursten SL, Locksley RM, Lovett DH (1992) Myristyl acylation of the tumor necrosis factor alpha precursor on specific lysine residues. J Exp Med 176(4):1053–1062. https://doi.org/10.1084/jem.176.4.1053

Tate EW, Bell AS, Rackham MD, Wright MH (2013) N-myristoyltransferase as a potential drug target in malaria and leishmaniasis. Parasitology 141(1):37–49. https://doi.org/10.1017/S0031182013000450

Tate EW, Kalesh KA, Lanyon-Hogg T, Storck EM, Thinon E (2015) Global profiling of protein lipidation using chemical proteomic technologies. Curr Opin Chem Biol 24:48–57. https://doi.org/10.1016/j.cbpa.2014.10.016

Thinon E, Hang HC (2015) Chemical reporters for exploring protein acylation. Biochem Soc Trans 43(2):253–261. https://doi.org/10.1042/bst20150004

Udenwobele DI, Su R-C, Good SV, Ball TB, Varma Shrivastav S, Shrivastav A (2017) Myristoylation: an important protein modification in the immune response. Front Immunol 8:751. https://doi.org/10.3389/fimmu.2017.00751

Veit M (2012) Palmitoylation of virus proteins. Biol Cell 104(9):493–515. https://doi.org/10.1111/boc.201200006

R, Wan J, Arstikaitis P, Takahashi H, Huang K, Bailey AO, Thompson JX, Roth AF, Drisdel RC, Mastro R, Green WN, Yates Iii JR, Davis NG, El-Husseini A (2008) Neural palmitoyl-proteomics reveals dynamic synaptic palmitoylation. Nature 456:904–909. https://doi.org/10.1038/nature07605

Wright MH, Heal WP, Mann DJ, Tate EW (2010) Protein myristoylation in health and disease. J Chem Biol 3(1):19–35. https://doi.org/10.1007/s12154-009-0032-8

Wright MH, Clough B, Rackham MD, Rangachari K, Brannigan JA, Grainger M, Moss DK, Bottrill AR, Heal WP, Broncel M, Serwa RA, Brady D, Mann DJ, Leatherbarrow RJ, Tewari R, Wilkinson AJ, Holder AA, Tate EW (2014) Validation of N-myristoyltransferase as an antimalarial drug target using an integrated chemical biology approach. Nat Chem 6:112–121. https://doi.org/10.1038/nchem.1830

Wright MH, Paape D, Storck EM, Serwa RA, Deborah DF, Tate EW (2015) Global analysis of protein *N*-myristoylation and exploration of *N*-myristoyltransferase as a drug target in the neglected human pathogen *Leishmania donovani*. Chem Biol 22(3):342–354. https://doi.org/10.1016/j.chembiol.2015.01.003

Wright MH, Paape D, Price HP, Smith DF, Tate EW (2016) Global profiling and inhibition of protein lipidation in vector and host stages of the sleeping sickness parasite *Trypanosoma brucei*. ACS Infect Dis 2(6):427–441. https://doi.org/10.1021/acsinfecdis.6b00034

Yount JS, Zhang MM, Hang HC (2013) Emerging roles for protein S-palmitoylation in immunity from chemical proteomics. Curr Opin Chem Biol 17(1):27–33. https://doi.org/10.1016/j.cbpa.2012.11.008

Zhang MM, Hang HC (2017) Protein S-palmitoylation in cellular differentiation. Biochem Soc Trans 45(1):275–285. https://doi.org/10.1042/bst20160236

Zhang MM, Wu P-YJ, Kelly FD, Nurse P, Hang HC (2013) Quantitative control of protein S-palmitoylation regulates meiotic entry in fission yeast. PLoS Biol 11(7):e1001597. https://doi.org/10.1371/journal.pbio.1001597

Zheng B, DeRan M, Li X, Liao X, Fukata M, Wu X (2013) 2-Bromopalmitate analogues as activity-based probes to explore palmitoyl acyltransferases. J Am Chem Soc 135(19):7082–7085. https://doi.org/10.1021/ja311416v

Zhou Y, Huang C, Yin L, Wan M, Wang X, Li L, Liu Y, Wang Z, Fu P, Zhang N, Chen S, Liu X, Shao F, Zhu Y (2017) N^{ε}-Fatty acylation of Rho GTPases by a MARTX toxin effector. Science 358(6362):528–531. https://doi.org/10.1126/science.aam8659

How to Target Viral and Bacterial Effector Proteins Interfering with Ubiquitin Signaling

Gerbrand J. van der Heden van Noort and Huib Ovaa

Contents

Abstract Ubiquitination is a frequently occurring and very diverse posttranslational modification influencing a wide scope of cellular processes. Ubiquitin (Ub) has the unique ability to form eight different lysine-linked polymeric chains, mixed chains and engages with ubiquitin-like (Ubl) molecules. The distinct signals evoked by specific enzymes play a crucial role in, for instance, proteasome-mediated protein degradation, cell cycle regulation, and DNA damage responses. Due to the large variety of cellular functions that this posttranslational modification influences, the enzymes that construct such Ub modifications, and subsequently controle and degrade these signals, is enormous. In this chapter, we will discuss the current

G. J. van der Heden van Noort (✉) · H. Ovaa
Department of Cell and Chemical Biology, Chemical Immunology, Oncode Institute, Leiden University Medical Centre, 2333 ZC Leiden, The Netherlands
e-mail: gvanderheden@lumc.nl

H. Ovaa
e-mail: h.ovaa@lumc.nl

Current Topics in Microbiology and Immunology (2019) 420:111–130
DOI 10.1007/82_2018_134

Published Online: 04 September 2018

state-of-the-art of activity-based probes, reporter substrates, and other relevant tools based on Ub as recognition element, to study the enzymes involved in the complex system of ubiquitination.

1 Introduction

Protein activity is regulated by the attachment and removal of posttranslational modifications (PTMs). Nucleophilic side chain functionalities in the protein can be decorated with such PTMs aided by a large class of enzymes that can monitor the substrate and the need to activate or deactivate its activity. Some of those PTMs are intensively studied such as, for instance, phosphorylation, but other PTMs are less well understood due to their high complexity, such as glycosylation and ubiquitination. In the latter case, a 76 amino acid protein ubiquitin (Ub) is transferred mostly to the ε-amine of a lysine residue in a target protein. Ubiquitination is involved in almost all aspects of eukaryotic biology including the regulation of immune responses, cell cycle progression, and protein degradation. Ub is a stable protein with a denaturation temperature of over 80 °C that is highly conserved from yeast to man with only three respective changes. Another interesting feature of Ub is its ability to form polymeric Ub chains in which the ε-amine functionality of any of the seven internal lysine residues or the N-terminal amine of Ub can be linked to the C-terminal carboxylic acid of a subsequent Ub. Eight different topologies can thus be formed all exerting different signaling outcomes. Lys-48-linked poly-Ub, for instance, is involved in proteasome-mediated protein degradation, whereas the Lys-63-linked poly-Ub is implicated to play a role in DNA damage repair. Most linkage types have been connected to one or more specific functions but new discoveries keep expanding the repertoire of involvement of Ub in cell signaling. On top of this already complex system with eight different homotypic poly-Ub chains, the possibility to form mixed heterotypic chains and branched chains complicates the Ub system even further. Posttranslational modifications such as phosphorylation, acetylation and newly discovered ADP-ribosylation of Ub and hybrid chains with ubiquitin-like modifiers such as SUMO and NEDD8 are also found to play distinct roles. The numerical amount of combinations one could think of our vast, but are all biologically relevant?

The general mechanism of Ub attachment to a target protein or a predecessor Ub (forming a poly-Ub chain) depends on a series of so-called E1-, E2-, and E3-enzymes that work together to activate and ligate Ub to its target using ATP as energy source. This process, known as the canonical Ub cascade, can undergo repetitive cycles and lead to poly-ubiquitinated substrates. One of two E1 activating enzymes, known to date, adenylates the C-terminal carboxylic acid of Ub and subsequently forms a reactive thioester intermediate. One of ~40 E2 conjugating enzymes takes over the Ub cargo by means of a trans-thiolation reaction (Stewart et al. 2016). One of over 600 E3-enzymes encoded in the human genome either mediates the transfer of Ub from the E2 to the substrate or takes over the Ub cargo itself prior to transferring it to the target, depending on whether the E3 belongs to the RING, HECT or RBR class of conjugating enzymes (Buetow and Huang 2016). The specific combination of E2/E3 not only ensures the correct target proteins are ubiquitinated but also dictates the type of poly-Ub linkage that is installed; actively controlling the signaling outcome of the ubiquitination event.

Counteracting the buildup of (poly-)ubiquitinated proteins is a group of deubiquitinating proteases (DUBs) that break down the ubiquitin modification, liberating the substrate protein, recycling Ub, and ending the Ub invoked signaling. Nearly a hundred of such deubiquitinating proteases have been identified in human cells that can be classified into seven distinct families. The subfamilies of ubiquitin-specific proteases (USPs), ubiquitin C-terminal hydrolases (UCHs), ovarian tumor domain proteases (OTUs), Machado–Joseph disease proteases (MJD), motif interacting with Ub-containing novel DUB family (MINDYs), and zinc finger with UFM1-specific peptidase domain protein (ZUFSPs) are cysteine proteases, whereas JAB1/MPN/MOV34 proteases (JAMMs) are zinc-dependent metallo-proteases.

To understand the complex signaling network brought about by the writers, readers, and erasers of the ubiquitin code, tools to study them in detail and on molecular level are of great interest. Below we describe the current state of the molecular toolbox available and three examples of their use in unraveling some of the secrets in viral and bacterial interference with ubiquitin biology.

2 The Current Toolbox

A diversity of tools has been developed to study the activities of enzymes involved in constructing and deconstructing ubiquitin chains and ubiquitinated proteins. The probes based on the Ub core as recognition element will be discussed below classified on their ability to target different enzymatic functions. Both chemical and semi-synthetic techniques are used to construct such probes, each having distinct advantages and drawbacks that, however, will not be discussed in great detail in this chapter.

2.1 DUB Targeting Covalent Probes

The first activity-based probe based on Ub forming a covalent complex between probe and target protease was based on replacing Gly75-Gly76 with 4-aminobutyraldehyde (Pickart and Rose 1986). The ubiquitin aldehyde (Ubal) reagent was found to form a complex and inhibit the activity of the UCH protease studied. Modifications on this reactive C-terminal element with, for instance, a nitrile moiety (Lam et al. 1997) and glycine vinyl sulfone (VS) or glycine vinyl methyl ester amine (VME) (Borodovsky et al. 2001, 2002) led to the development of a bigger panel of activity-based probes able to capture the active site cysteines of deubiquitinating enzymes. These electron poor vinyl motifs act as Michael acceptor element and allow the sulfur nucleophile of the active site cysteine to be trapped by forming a covalent intermediate in an irreversible reaction. Later on, the total chemical synthesis of Ub (mutants) using solid phase peptide chemistry (El Oualid et al. 2010; Kumar et al. 2009; Pasunooti et al. 2009) opened the way to prepare Ub probes carrying fluorescent labels or affinity handles on large scale (de Jong et al. 2012). Exploiting a similar mode of action as Ub-VME is Ub-propargyl (Prg), where the terminal alkyne of the propargyl moiety unexpectedly is able to react with the active site cysteine leading to a covalent adduct in the form of a vinyl thioether linkage (see Fig. 1a) (Ekkebus et al. 2013). Major advantage of this last variant, Ub-Prg, is its unreactiveness toward other cysteine proteases, making it really a DUB-specific probe. In general, specificity of such probes for DUBs is based on the Ub element of the probe to be recognized by the protease on a binding interface or so-called S1-pocket preceding the active site placing the reactive

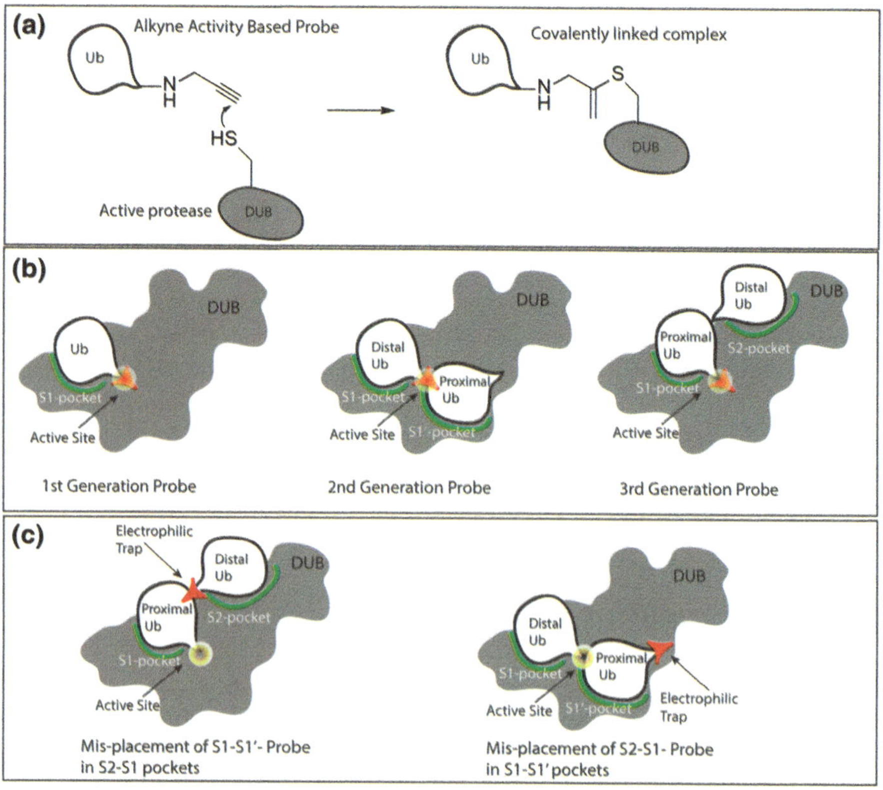

Fig. 1 Activity-based probes to target DUB activity. **a** Reaction details of cysteine protease reaction with Ub-Prg, **b** different probes targeting S1, S1–S1′, or S1–S2 interactions, **c** misplacement of probes on DUBS allowing the study of different binding pockets

element directly over the active site cysteine. In this situation, the two reacting partners are optimally aligned and lead to the formation of a covalent adduct (see Fig. 1a).

Some DUBs have been shown to have a preference to specifically cleave certain poly-Ub chains. In order to investigate such possible preferences, in a classical experiment all native diUb molecules are incubated with a purified recombinant DUB and cleavage of the diUb is monitored over time. Biggest drawback of this method is that it is not compatible with complex biological settings (such as cell lysate), and it is limited to isolated DUBs. To overcome such issues, a second generation of probes to investigate linkage-specific proteolysis of DUBs has emerged more recently. These probes generally consist of two ubiquitin moieties carrying a Michael acceptor element in the isopeptide linkage region in-between the two Ub moieties. Initial reports show the two Ub regions to be linked together using non-native connections such as a triazole (McGouran Joanna et al. 2013) and thiol ether linkage (Li et al. 2014). Two type of probes presenting either a dehydroalanine (Dha) (Haj-Yahya et al. 2014) or VME-like electrophilic trap (Mulder et al. 2014) mimic the native lysine–glycine linkage the closest using amide linkages. A panel of all seven isopeptide-linked diUb probes can be constructed and used to covalently capture the active site cysteine of the DUB showing its reactivity and preference toward certain linkage types. Using such probes the N-terminal or distal Ub molecule will be positioned before the active site in the so-called S1-pocket, and the C-terminal or proximal Ub molecule will be positioned after the active site cysteine in the so-called S1′-pocket. Due to the geometrical differences between al Lys-linked diUb probes, the DUB will only be able to position the probes mimicking its natural substrates in such a way that the active site cysteine is able to react with the reactive element (see Fig. 1b).

Some DUBs are able to recognize Ub chain topologies using other binding surfaces further away from the active site, such as, for instance, an S2-site preceding the S1-site. A third generation of probes targeting such S2 binding sites has been developed where a diUb molecule is equipped with a reactive element at the proximal C-terminus (Flierman et al. 2016). These probes are only able to react with DUBs that contain a S2-site that plays a determining role in positioning the diUb molecule in the S2- and S1-sites thereby placing the alkyne directly over the active site cysteine (see Fig. 1b). Noteworthy is that the isopeptide linkage between proximal and distal Ub has been replaced by a protease stable triazole linkage, prohibiting the protease of interest to degrade the probe during assays.

If a DUB recognizes such a third generation diUb probe using its S1- and S1′-sites, the alkyne will not be in the vicinity of the active site cysteine and no covalent adduct will be formed. Conversely, if a second generation probe will be reacted with a DUB recognizing the diUb moiety using it's S2- and S1-sites, no reaction will occur since the reactive element will not be aligned with the reactive cysteine (see Fig. 1c). Having access to both second and third generation probes offers an exciting combination to investigate the binding interfaces that play a role in determining binding preferences of DUBs and cast a light on their molecular mechanism of action.

More recently a new probe able to reversibly capture cysteine DUBs followed by the release of the still active proteases has been published, making use of subtle chemistry to form, and disrupt disulfide bridges between the active site cysteine and Ub-based thiol-containing probe (de Jong et al. 2017). Although only the proof of principal studies has been performed, this novel technology holds great promise for the future capture, release, and follow-up investigations of native active cysteine DUBs.

Not all DUBs are cysteine proteases, however, and the metallo-DUB family currently is understudied due to the lack of probes. Selective probes targeting such proteases are needed and will no doubtably be developed in the future.

The covalent capture of active DUBs with Ub-based activity-based probes allows for purification of the formed complexes and subsequent crystallization efforts to study the interactions between protease and Ub in detail. A large number of crystal structures has been solved using such Ub-based ABPs as is reviewed by van Tilburg et al. (2016). Synthetic procedures to prepare above-mentioned probes and reagents on large scale are in place and are finding their way into the related fields of Ub-like proteins such as, for instance, SUMO biology (Mulder et al. 2018).

2.2 E1–E2–E3 Targeting Covalent Probes

Due to the diversity and complexity in the conjugating and ligation machinery, development of probes targeting the constructing of ubiquitinated substrates has only emerged more recently than probes targeting the deconstructing proteases, for the simple reason that targeting a sequential enzymatic cascade is more difficult than targeting a single proteolytic step.

The first step in the cascade is activation of the C-terminal carboxylate of Ub by and E1-activating enzyme. The E1 firstly adenylates the carboxylic acid group of glycine 76 at the expensive of ATP in one region of the E1 and subsequently the catalytic cysteine of the E1 in another region of the enzyme takes over the Ub cargo by forming a thioester and expelling AMP. Initial probes based on Ub_{71}-CGG-vinyl sulfonamides show reactivity toward the cysteine of yeast E1 Uba1, although the C-terminal Arg-Leu-Arg-Gly-Gly region of Ub is replaced by the smaller Cys-Gly-Gly vinyl adenyl sulfonamide, rendering a shorter Ub mutant with a RL deletion (see Fig. 2a) (Lu et al. 2010). These Ub probes are based on small molecule adenyl sulfonamide analogues that can also be used to monitor intracellular E1 activity and specific inhibitory potential toward different E1's (An and Statsyuk 2013; Misra et al. 2017). Other probes based on Ub_{77}-dehydroalanine adenyl amides also show reactivity toward a. o. Uba1 although these probes are longer than the native Ub_{76} substrate (An and Statsyuk 2016).

Combined this data implies that unlike the active sites of DUBs the E1's are more tolerant toward their substrates or the flexibility of the C-terminal tail of Ub corrects for possible misalignment during the Ub transfer from the ATP binding

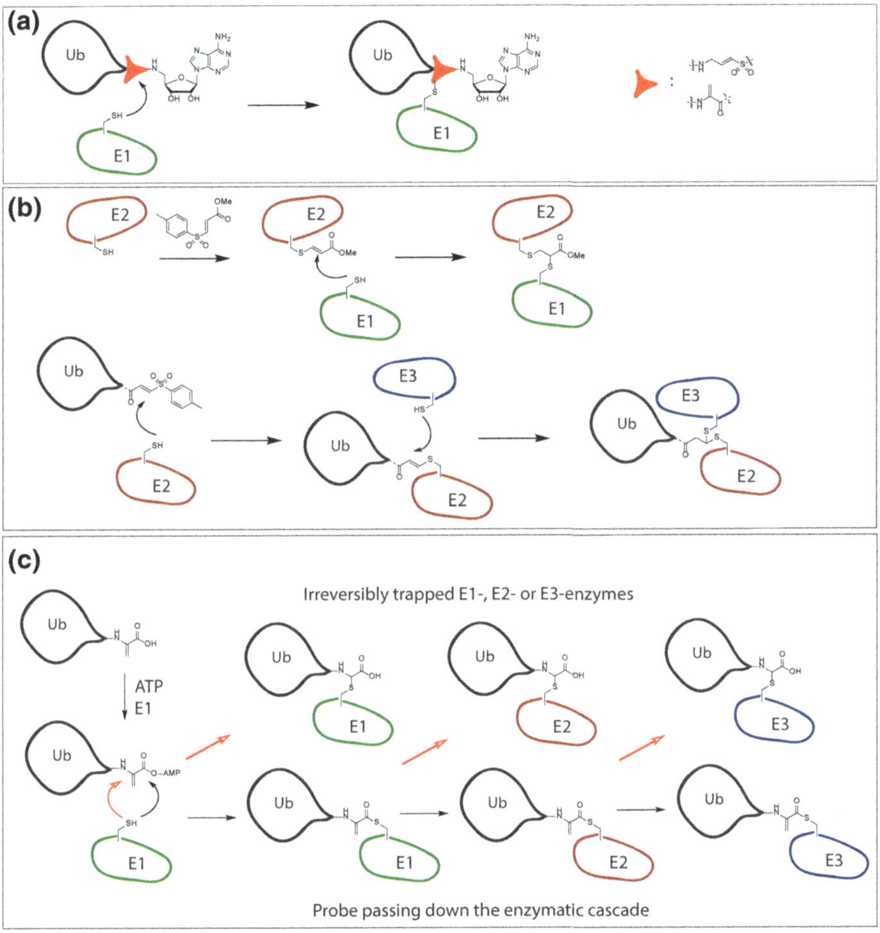

Fig. 2 Activity-based probes to capture **a** E1-enzymes, **b** E1–E2 interactions or Ub–E2–E3 interactions, **c** Ub–E1, Ub–E2, and Ub–E3 interactions

domain to the catalytic cysteine domain of the E1. These adenylate mimic ABPs are important to study E1's but in view of the bigger picture are quite restricted. Due to the stabilized adenylate (either amide or sulfonamide) linkage, these probes lack the ability to be transferred down the enzymatic cascade, being stuck in the first stage.

The second step in the cascade is transfer of the Ub cargo from E1 to E2, a processes that can be trapped and studied using a E2 based–ABP (Stanley et al. 2015). Recombinant expression of an E2 and modification with an tosyl-substituted double-activated ene-reagent (TDAE) forms an electron poor activated vinyl sulfide that upon juxta-positioning of the E1's cysteine is able to form a stable bis thioether E1–E2 complex (see Fig. 2b). In an analogues approach, the third step in the cascade, Ub transfer from E2 to E3 can be interrogated. A ubiquitin-charged E2 carrying a TDAE element allows for trapping the trans-thiolation event toward an E3, forming a stable complex between Ub, E2, and E3 (see Fig. 2b) (Pao et al. 2016). Of note is that in this last TDAE derived probe, the C-terminal RGG motif of Ub is replaced by the reactive TDAE element, which might limit the generality of such probes as it is implicated that R74 and the diGly motif can play an important role of recognition of the appropriate Ubl.

All approaches mentioned above are useful in specifically studying one step in the cascade of Ub-ligation. To amend the need to be able to pass down the probe through the complete cascade and target E1's, E2's, and E3's; simultaneously, a 'cascading'-probe based on a C-terminal dehydroalanine (Dha) moiety was developed (Mulder et al. 2016). This reactive element has been employed earlier in diUb probes targeting cysteine DUBs (Haj-Yahya et al. 2014), but was found in other studies to show low reactivity. This low reactivity can be harnessed in trapping the E1–E2–E3 enzymatic cascade consecutively (see Fig. 2c). The Ub_{75}Dha can be activated by an E1 to form a thioester intermediate and transferred to an E2 and subsequent RBR- or HECT-E3 via a trans-thiolation reaction. In each step, only a small portion of the active cysteines will be trapped by the Dha-moiety, while the rest of the cysteine will continue to function normal in the trans-thiolation step. By doing so, the probe will capture all components in the cascade, which has been shown to indeed be detectable using proteomic approaches in a proof of principle study.

The above-mentioned arsenal of probes has emerged very recently, and most likely the near future holds exciting insights in the ubiquitination cascade through implementation of these probes. Complicating factor in the investigation of E3-ligases is that of the three major classes of E3's only the HECT and RBR classes have an active cysteine residue that takes over the Ub cargo from an E2 and transfer it to the final substrate protein. RING E3s do not possess such an active site cysteine and merely serve as platforms to bring Ub charged E2's and substrates together, thereby making them unsuited for direct probing using ABPs.

2.3 *Assay Reagents*

Other type of probes or assay reagents based on a Ub scaffold have also been developed allowing the real-time monitoring of DUB or ligase activity. In contrast to the probes described above, these reagents do not have a Michael acceptor element and thus do not form a covalent complex with the targeted enzyme, but mostly rely on a fluorescent signal that is altered. Advantage of the probes described below is that one enzyme can perform multiple catalytic cycles, and hence, signal amplification occurs allowing a more accurate readout of the enzymes native activity and/or specificity.

An important class of Ub-based assay reagents is the fluorogenic Ub reagents, where a quenched fluorophore is conjugated at the C-terminus of Ub. DUB activity will cleave the amide bond at position 76, and the fluorophore will be released from the Ub and simultaneously start to fluoresce (see Fig. 3a). Hence, the increase in fluorescence is a direct measure of DUB activity. Fluorogenic reporters used in such type of reagents are aminomethylcoumarin (AMC) (Dang et al. 1998) or substituted rhodamine-110 (RHO) (Hassiepen et al. 2007) scaffolds that show favorable fluorescent properties. In a similar setup, DUB mediated aminoluciferin release can be assayed in a bioluminescence approach using a luciferase assay (Orcutt et al. 2012). All of these reporters are conjugated to a mono-Ub recognition element, and hence, the preference of a DUB to specific poly-Ub topologies cannot be assessed. To allow monitoring chain-specific proteolysis mediated by S1–S2 interactions on the DUB, diUb-AMC substrates were generated (Flierman et al. 2016). In analogy to the diUb-Prg covalent probes, these diUb-AMC substrates are linked via a non-hydrolysable triazole linker preventing proteolysis of the diUb entity during the assay.

One important note is that the fluorogenic substrates do not contain an isopeptide linkage at the side where the DUB would perform its proteolytic action, whereas the natural substrates for most DUBs would. To mimic this isopeptide link more closely, fluorescent polarization (FP) reagents were developed where Ub is conjugated via a native isopeptide linkage to a fluorophore carrying substrate derived peptide (Tirat et al. 2005; Geurink et al. 2012). Rationale behind these reagents is that once the fluorophore-containing peptide is linked to Ub, the large construct tumbles slowly in solution and light remains polarized, whereas after proteolysis the small fluorophore-containing peptide tumbles faster and hence the polarization of light decreases (see Fig. 3b). Such tools are not only reported for Ub but also for other Ubl's such as the three SUMO's, Nedd8, and ISG15.

Another type of reagent makes use of the fact that a fluorophore and quenching moiety can be placed in close proximity of each other on a diUb or Ub-substrate peptide and upon DUB action both entities are separated and hence fluorescence is restored (see Fig. 3c). Probes based on such a time-resolved fluorescence resonance energy transfer (FRET) principle consisting of Ub-peptide (Ohayon et al. 2012) and diUb moieties (Geurink et al. 2016; Ye et al. 2012) have been described. All seven lysine-linked diUb FRET pairs are available which form a platform to measure DUB preference and kinetic parameters in real time. Using similar technology, the E2 mediated constructing of poly-Ub chains equipped with a FRET acceptor onto a FRET donor-containing Ub results in a FRET signal (Madiraju et al. 2012). In this particular study, inhibitors for the E2 can be identified when a decrease in FRET signal is observed in the assay.

Although it was long assumed that in order to ubiquitinate the target substrate the complete cascade, so E1, E2, and E3 functions are needed, recent reports show that Ub-thioesters can be used in in vitro studies to monitor HECT- or RBR-E3 ligase activity. In these assays, the E1- and E2-functions are bypassed and

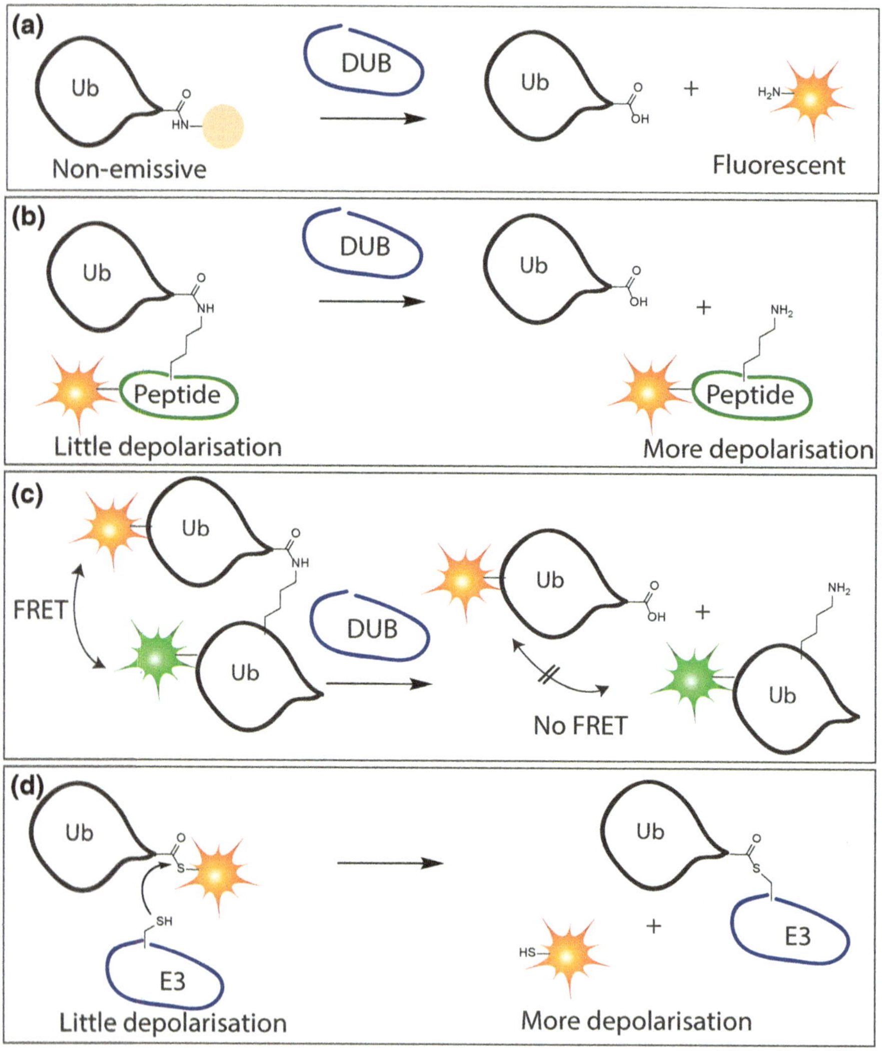

Fig. 3 Assay reagents based on Ub. **a** Fluorogenic substrates, **b** fluorescent polarization reagents to monitor DUB activity, **c** fluorescent energy transfer reagents, **d** fluorescent polarization reagents to monitor E3-ligase activity

Ub-thioesters [Ub-MES (Park et al. 2015) or fluorescent analogue Ub-FLUOR (Krist et al. 2016)] are directly trans-thiolated by the active cysteines in several E3's. The fluorescent thioester allows real-time quantitative monitoring of E3-ligase activity using a fluorescence polarization set up (Park et al. 2017) (see Fig. 3d).

2.4 (Poly)Ub Chains

All probes mentioned above are based on mono-ubiquitinated peptides, mono-Ub or diUb carrying a reactive or fluorescent entity. Moving forward in our understanding of the Ub system, it would be of great value to have access to well defined larger Ub chains and probes made thereof. Synthetic, semi-synthetic and biochemical methods are in place to produce poly-Ub chains, carrying native isopeptide linkages or artificial (non-hydrolysable) linkages.

Biochemical methodology to obtain all except K27-poly-Ub chains, are reported and rely on the use of linkage-specific E2/E3 enzyme combinations as has been reviewed elsewhere (Faggiano et al. 2016) (see Fig. 4a). Although the majority of poly-Ub chains can be produced, several drawbacks come with such methods. Crucial in biochemical production of Ub chains is that in order to have a defined polymer length, the mixture of simultaneously produced poly-Ub's needs to be separated, most frequently done using cation exchange chromatography. Inherent in producing polymers is that the isolated yield of one specific poly-Ub species is low and scalability of biochemical methodology hence can be an issue. Some of the E2/E3 combinations are not absolute topology specific and additional treatment using DUBs might be necessary to guarantee a homogenous preparation, further complicating procedures. All in all six out of seven lysine-linked poly-Ub chains can be prepared biochemically although preparation of high quantities of pure material still is a laborious task.

On the other hand, the total chemical synthesis of poly-Ub chains has been undertaken. Several methods using auxiliaries, different thiol modified amino acids and ligation/desulfurization strategies are in place as reviewed elsewhere (Spasser and Brik 2012), allowing for the sequential construction of Ub chains with the impressive highlight of K48-linked tetra-Ub (Kumar et al. 2011). Although for such time-consuming approaches a high level of expertise is required, it does allow the introduction of fluorophores, affinity handles or conjugation to substrate peptides or proteins. Of note is that using an isoUb native chemical ligation approach also K27 tetra-Ub chains and mixed chains have been prepared and employed in structural studies (Tang et al. 2017). In contrast to sequential strategies, chemical polymerization approaches using a bifunctional thiolysine/thioester-Ub mutants (van der Heden van Noort et al. 2017; Moyal et al. 2012) (see Fig. 4b) leading to native isopeptide-linked Ub chains, bifunctional thiol/ene Ub mutants leading to thioether-linked Ub chains (Trang et al. 2012) (see Fig. 4c) or azide/alkyne-Ub mutants leading to triazole-linked Ub chains (Schneider et al. 2014) (see Fig. 4d) give rise to larger poly-Ub analogues including the biochemical unavailable K27-linked Ub chains. Such stable poly-Ub chains can be employed to study the interacting proteins or 'readers' of the Ub code by performing pull-down experiments and subsequent proteomic analysis using mass spectrometry approaches (Zhao et al. 2017; Zhang et al. 2017).

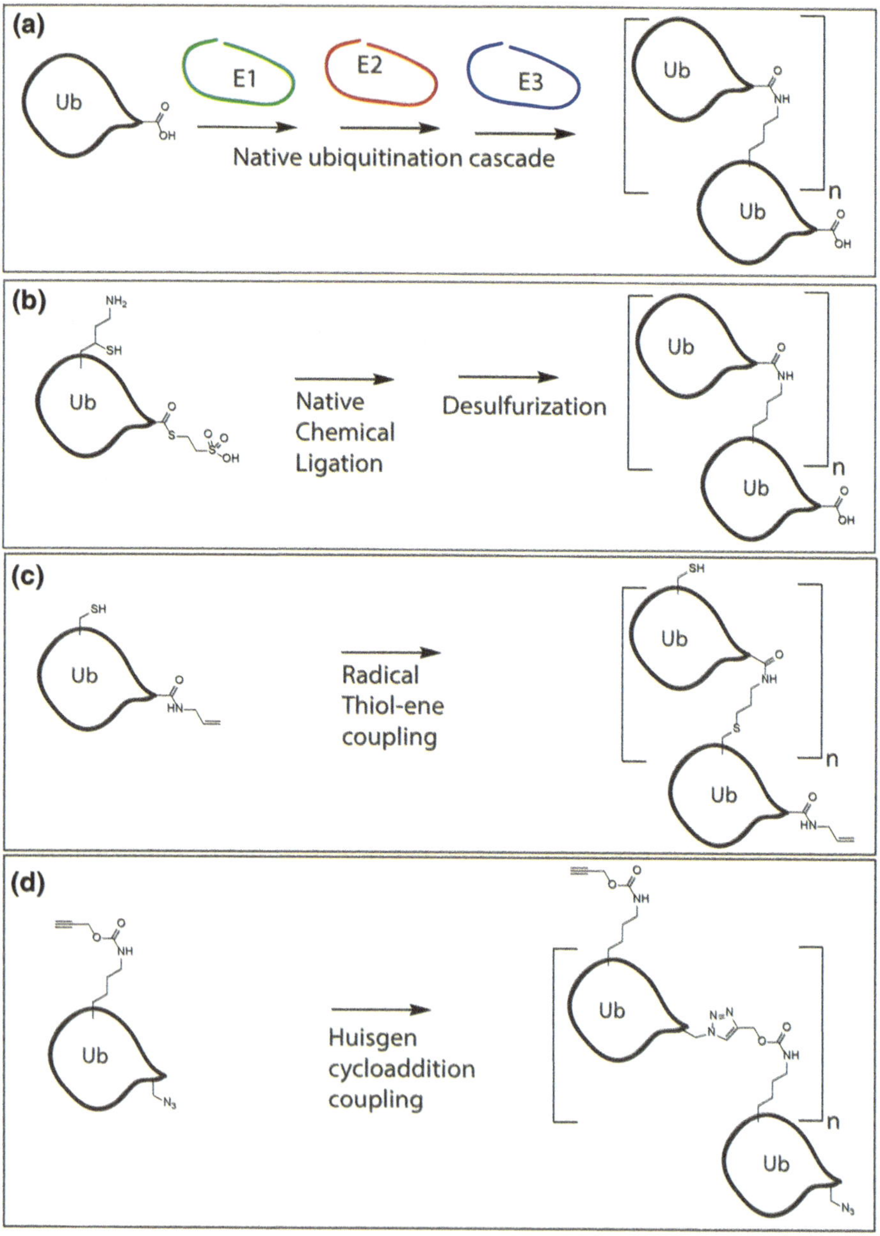

Fig. 4 Poly-Ub generation using **a** the native E1–E2–E3 cascade, **b** a native chemical ligation approach, **c** a thiol–ene radical coupling approach, and **d** a copper-catalyzed Huisgen cycloaddition approach

3 Applications

The above-described arsenal of ubiquitin-based probes, substrates, chains, and analogues are each designed to target-specific enzymatic functions or give insight into unanswered question regarding the ubiquitin–proteasome system. Apart from the proof of principle studies conducted during the development of these probes, it is important to note that these tools have made the translational step from the drawing board to actual biochemical and structural studies on viral and bacterial enzymes playing a role in the manipulation of (human) host pathways.

3.1 Applications in Virology

Viruses have defense mechanism in place to either utilize the host's ubiquitin system in their own advantage or to combat the host's immune response by suppressing pro-inflammatory Ub signaling. Human coronaviruses (hCoV), responsible for pandemic outbreaks of severe acute respiratory syndrome (SARS) in 2003 and Middle East respiratory syndrome (MERS) in 2012, are two of such viruses that are interfering with the host's Ub signaling. One way these coronaviruses act is by dampening the immune response through action of viral proteases that possess deubiquitinase and deISGylase functions. Interferon stimulated Gene 15 (ISG15) is a ubiquitin-like (Ubl) modification that has two Ub domains in tandem and is upregulated and attached to substrates during an antiviral response. The papain-like proteases (PLpro's) in both MERS and SARS hCoV are identified to cleave ISG15 and Ub from cellular proteins. In a study comparing MERS and SARS PLpro, it was found that fluorogenic substrates Ub-AMC and ISG15-AMC were both cleaved to a similar extent by MERS PLpro, where SARS PLpro preferred ISG15-AMC (Békés et al. 2015). Covalent capture of these proteases by ABP Ub-propargyl again showed a higher reactivity for the MERS then for the SARS protease. Cleavage assays using native K48-linked tetra-Ub also show a distinct pattern. SARS PLpro cleaves the Ub_4 into Ub_2 and is hardly able to process Ub_2 any further, whereas MERS PLPro cleaves Ub_4 into a mix of Ub_3, Ub_2, and Ub. SARS PLpro seems to have a di-distributive mechanism only making a cut after a Ub_2 moiety, whereas MERS is mono-distributive cleaving after one Ub residue. DiUb ABP's having the reactive element either at the proximal side of the Ub_2 or in-between the Ub moieties help to explain how these two different mechanism function, as MERS PLpro is able to react with the in-between probe and not the proximal probe, and SARS PLpro shows opposite reactivity toward these two probes (Bekes et al. 2016). Hence, activity of SARS PLPro seems to be largely influenced by an S2 site, whereas MERS PLpro does not seem to contain such a site. A crystal structure of the complex between the S2–S1 targeting diUb ABP and SARS PLpro allows identification of the crucial contacts in the S2-site of the DUB and the diUb ABP. All in all these hCoV proteases, although homologous, are shown to have a different mode of action using the three generations of activity-based probes (discussed in Sect. 2.1) and accompanying kinetic parameters could be deduced using fluorogenic mono- and diUb-AMC substrates (discussed in Sect. 2.3).

3.2 Applications in Bacteriology

Similar to viral interference with host immune response, bacteria also have sophisticated mechanisms in place to counteract host immune responses and create optimal conditions for the bacterium to survive and promote replication. Bacterial effector proteins typically are directly secreted into the host cells to interfere with host kinase activities or Ub(l) signaling pathways. Interestingly, those bacteria do not have Ub or Ubl systems themselves and such effector proteins are only in place to promote survival within the host organism. Much of these effector proteases fall in the class of CE clan effector enzymes, which in humans consists of SUMO- and NEDD8-specific proteases. Bacterial CE clan proteases do not only show activity toward Ubl's but some members also show acetyltransferase activity. In a comparative study on such CE clan effectors from intracellular pathogens *Salmonella typhimurium*, *Chlamydia trachomatis*, *Escherichia coli*, *Yersinia pestis*, *Rickettsia bellii*, *Shigella flexneri*, and *Legionella pneumophila*, activity, specificity, and structure were investigated (Pruneda Jonathan et al. 2016). Although all share a common fold most but not all of them were able to react with ABP Ub-Prg. Using Ub-, SUMO1-, NEDD8- and ISG15-FP substrates (discussed in Sect. 2.3), the three effectors unreactive toward Ub-Prg were also shown to be unreactive toward any of the FP-reagents whereas the others showed reactivity toward both Ub and NEDD8 reagents. In contrast to their unreactivity toward Ubl's, the three CE effectors did show acyl transferase activity. Using diUb substrates, the panel of bacterial DUBs was all shown to have a strong preference for K63-linked chains, followed by K48 and K11 at later time points or higher enzyme concentrations. This specificity mostly was shown to be regulated by the S1′-site of these DUBs. Taken together, this study uses both mutational analyses, crystal structure information, and data obtained by Ub(l)-based probes and assay reagents to classify and study the molecular details of a bacterial class of effector proteins in detail.

3.3 *Legionella* Effector Enzymes: The Odd-Ones Out

We thought we understood the basis of the Ub system, but do we actually? Several enzymes have been recently identified to control the Ub system by attaching adenosine diphosphate ribose (ADPr) onto specific positions in Ub, exerting a new layer of control. The intracellular pathogenic bacterium *L. pneumophila*, the cause of Legionnaires disease, releases effector proteins into its host to hijack the host Ub pool (Bhogaraju et al. 2016). These multi-domain SidE family proteins catalyze the ubiquitination of substrates proteins *independent* of the E1-E2-E3 cascade and without the use of energy source ATP. Instead they use NAD^+ (see Fig. 5—structure II) to attach ADPr on the nucleophilic side chain of arginine 42 of Ub using their mono ADPr transferase (mART)-domain (see Fig. 5—structure I (Qiu et al. 2016). Subsequently, the phosphodiesterase (PDE) domain within SidE catalyzes the reaction between a serine-containing substrate (see Fig. 5—structure IV) and the ADP-ribosylated Ub (see Fig. 5—structure III), affording a ribosyl phosphate linkage between Ub and the target protein (see Fig. 5—structure V) (Kotewicz et al. 2017). Besides a mART and PDE domain SidE family members also contain a N-terminal canonical DUB domain that is not necessary for bacterial replication, but is crucial in regulating the extend of ubiquitination at the bacterial vacuolar surface. Cleavage assays using diUb substrates show a preference of K63 over K48 and K11, and crystallization efforts of the formed complex with ABP Ub-VME reveal a different binding modus, not involving the I44 patch on Ub, of the cysteine DUB and Ub (Sheedlo et al. 2015). Besides the multi-domain multi-function SidE's, Legionella also expresses a deubiquitinase SidJ that is able to cleave the formed phosphodiester linkage to Ub, liberating the substrate protein and phosphoribose modified ubiquitin (see Fig. 5—structure VI) (Qiu et al. 2017). This phosphoribose-Ub also interferes with the host Ub machinery as it prevents the canonical E1 to activate and further process the Ub molecule (Puvar et al. 2017). By controlling the construction and deconstruction of modified Ub, without competing with the canonical system, the bacterium tries to hijack the host Ub system.

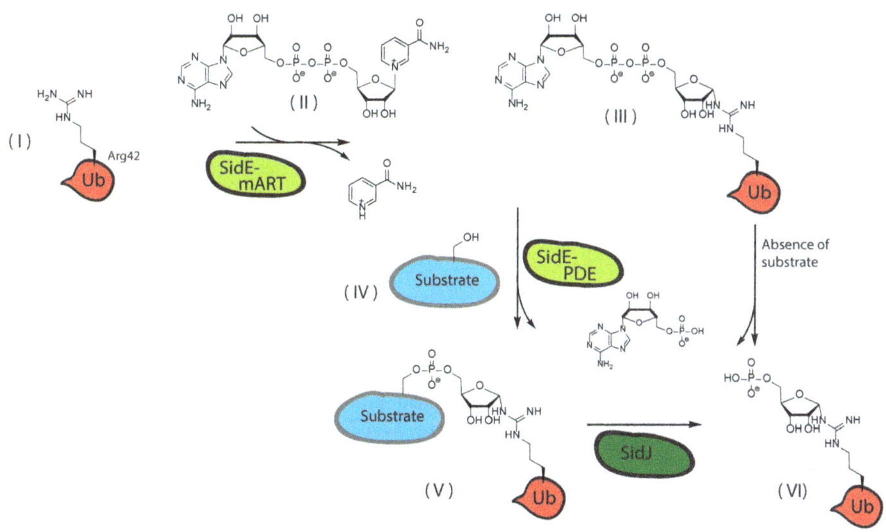

Fig. 5 Actions of SidE effector enzymes ADP-ribosylating Ub(Arg42) and subsequent coupling to substrates forming a phosphoribose linkage between substrate and Ub

Lots of questions on the mechanism of action, substrate recognition, and scope of this E1–E2–E3 independent ubiquitination and attached deubiquitination process call for the development of novel probes specifically targeting such pathways. So far the development of ADP-ribosylated Ub carrying a protease stable triazole linkage between the ribose and Arg42 of ubiquitin has been shown to reflect the native auto-ubiquitination activity of SidE family member SdeA (Liu et al. 2018). No doubtably probes targeting this interesting new way of Ub conjugation to substrates will appear in the near future.

4 Conclusions

The complexity of the ubiquitin system and its involvement in a wide variety of important biological processes makes it a widely and intensively studied field. The roles Ub play in neurodegenerative disease and cancers, for instance, make it a potential target for therapeutic intervention. Unraveling the complexity of this, highly sophisticated system is aided greatly by the development of (activity-based) probes reporting on the dynamics and structural mechanisms used by activating enzymes, ligases, and proteases counteracting the buildup of (poly)-ubiquitinated substrates. Ranging from probes that covalently trap the active cysteines involved in catalysis to probes that have altered fluorescent properties upon enzyme activity and stabilized substrates or poly-Ub chains allowing preference studies and profiling of the Ub interactome have been developed over the past decade or so and already have shed their light on some fundamental question in Ub biology. The more knowledge is gathered on the functioning of those (de)ubiquitinating proteins the more tailored solutions to interrogate their biology, and hence, specific probes involved in such approaches become. Without a doubt will the next generation of Ub-based tools help to increase our knowledge on the system and perhaps ultimately lead to diagnostic tools or therapeutics making it to the clinic.

References

An H, Statsyuk AV (2013) Development of activity-based probes for ubiquitin and ubiquitin-like protein signaling pathways. J Am Chem Soc 135(45):16948–16962. https://doi.org/10.1021/ja4099643

An H, Statsyuk AV (2016) Facile synthesis of covalent probes to capture enzymatic intermediates during E1 enzyme catalysis. Chem Commun 52(12):2477–2480. https://doi.org/10.1039/C5CC08592F

Békés M, Rut W, Kasperkiewicz P, Mulder Monique PC, Ovaa H, Drag M, Lima Christopher D, Huang Tony T (2015) SARS hCoV papain-like protease is a unique Lys^{48} linkage-specific di-distributive deubiquitinating enzyme. Biochem J 468(2):215–226. https://doi.org/10.1042/bj20141170

Bekes M, van Noort GJV, Ekkebus R, Ovaa H, Huang TT, Lima CD (2016) Recognition of Lys48-linked di-ubiquitin and deubiquitinating activities of the SARS coronavirus papain-like protease. Mol Cell 62(4):572–585. https://doi.org/10.1016/j.molcel.2016.04.016

Bhogaraju S, Kalayil S, Liu Y, Bonn F, Colby T, Matic I, Dikic I (2016) Phosphoribosylation of ubiquitin promotes serine ubiquitination and impairs conventional ubiquitination. Cell 167 (6):1636–1649.e1613. https://doi.org/10.1016/j.cell.2016.11.019

Borodovsky A, Kessler BM, Casagrande R, Overkleeft HS, Wilkinson KD, Ploegh HL (2001) A novel active site-directed probe specific for deubiquitylating enzymes reveals proteasome association of USP14. EMBO J 20(18):5187–5196. https://doi.org/10.1093/emboj/20.18.5187

Borodovsky A, Ovaa H, Kolli N, Gan-Erdene T, Wilkinson KD, Ploegh HL, Kessler BM (2002) Chemistry-based functional proteomics reveals novel members of the deubiquitinating enzyme family. Chem Biol 9(10):1149–1159. http://dx.doi.org/10.1016/S1074-5521(02)00248-X

Buetow L, Huang DT (2016) Structural insights into the catalysis and regulation of E3 ubiquitin ligases. Nat Rev Mol Cell Biol 17:626. https://doi.org/10.1038/nrm.2016.91

Dang LC, Melandri FD, Stein RL (1998) Kinetic and mechanistic studies on the hydrolysis of ubiquitin C-terminal 7-amido-4-methylcoumarin by deubiquitinating enzymes. Biochemistry 37(7):1868–1879. https://doi.org/10.1021/bi9723360

de Jong A, Merkx R, Berlin I, Rodenko B, Wijdeven RHM, El Atmioui D, Yalçin Z, Robson CN, Neefjes JJ, Ovaa H (2012) Ubiquitin-based probes prepared by total synthesis to profile the activity of deubiquitinating enzymes. ChemBioChem 13(15):2251–2258. https://doi.org/10.1002/cbic.201200497

de Jong A, Witting K, Kooij R, Flierman D, Ovaa H (2017) Release of enzymatically active deubiquitinating enzymes upon reversible capture by disulfide ubiquitin reagents. Angew Chem Int Ed 56(42):12967–12970. https://doi.org/10.1002/anie.201706738

Ekkebus R, van Kasteren SI, Kulathu Y, Scholten A, Berlin I, Geurink PP, de Jong A, Goerdayal S, Neefjes J, Heck AJR, Komander D, Ovaa H (2013) On terminal alkynes that can react with active-site cysteine nucleophiles in proteases. J Am Chem Soc 135(8):2867–2870. https://doi.org/10.1021/ja309802n

El Oualid F, Merkx R, Ekkebus R, Hameed DS, Smit JJ, de Jong A, Hilkmann H, Sixma TK, Ovaa H (2010) Chemical synthesis of ubiquitin, ubiquitin-based probes, and diubiquitin. Angew Chem Int Ed 49(52):10149–10153. https://doi.org/10.1002/anie.201005995

Faggiano S, Alfano C, Pastore A (2016) The missing links to link ubiquitin: methods for the enzymatic production of polyubiquitin chains. Anal Biochem 492:82–90. http://dx.doi.org/10.1016/j.ab.2015.09.013

Flierman D, van der Heden van Noort GJ, Ekkebus R, Geurink PP, Mevissen TET, Hospenthal MK, Komander D, Ovaa H (2016) Non-hydrolyzable diubiquitin probes reveal linkage-specific reactivity of deubiquitylating enzymes mediated by S2 pockets. Cell Chem Biol 23(4):472–482. https://doi.org/10.1016/j.chembiol.2016.03.009

Geurink PP, El Oualid F, Jonker A, Hameed DS, Ovaa H (2012) A general chemical ligation approach towards isopeptide-linked ubiquitin and ubiquitin-like assay reagents. ChemBioChem 13(2):293–297. https://doi.org/10.1002/cbic.201100706

Geurink PP, van Tol BDM, van Dalen D, Brundel PJG, Mevissen TET, Pruneda JN, Elliott PR, van Tilburg GBA, Komander D, Ovaa H (2016) Development of diubiquitin-based FRET probes to quantify ubiquitin linkage specificity of deubiquitinating enzymes. ChemBioChem 17(9):816–820. https://doi.org/10.1002/cbic.201600017

Haj-Yahya N, Hemantha HP, Meledin R, Bondalapati S, Seenaiah M, Brik A (2014) Dehydroalanine-based diubiquitin activity probes. Org Lett 16(2):540–543. https://doi.org/10.1021/ol403416w

Hassiepen U, Eidhoff U, Meder G, Bulber J-F, Hein A, Bodendorf U, Lorthiois E, Martoglio B (2007) A sensitive fluorescence intensity assay for deubiquitinating proteases using ubiquitin-rhodamine110-glycine as substrate. Anal Biochem 371(2):201–207. https://doi.org/10.1016/j.ab.2007.07.034

Kotewicz KM, Ramabhadran V, Sjoblom N, Vogel JP, Haenssler E, Zhang M, Behringer J, Scheck RA, Isberg RR (2017) A single Legionella effector catalyzes a multistep ubiquitination pathway to rearrange tubular endoplasmic reticulum for replication. Cell Host Microbe 21 (2):169–181. http://dx.doi.org/10.1016/j.chom.2016.12.007

Krist DT, Park S, Boneh GH, Rice SE, Statsyuk AV (2016) UbFluor: a mechanism-based probe for HECT E3 ligases. Chem Sci 7(8):5587–5595. https://doi.org/10.1039/C6SC01167E

Kumar KSA, Haj-Yahya M, Olschewski D, Lashuel HA, Brik A (2009) Highly efficient and chemoselective peptide ubiquitylation. Angew Chem Int Ed 48(43):8090–8094. https://doi.org/10.1002/anie.200902936

Kumar KSA, Bavikar SN, Spasser L, Moyal T, Ohayon S, Brik A (2011) Total chemical synthesis of a 304 amino acid K48-linked tetraubiquitin protein. Angew Chem Int Ed 50(27):6137–6141. https://doi.org/10.1002/anie.201101920

Lam YA, Xu W, DeMartino GN, Cohen RE (1997) Editing of ubiquitin conjugates by an isopeptidase in the 26S proteasome. Nature 385:737. https://doi.org/10.1038/385737a0

Li G, Liang Q, Gong P, Tencer AH, Zhuang Z (2014) Activity-based diubiquitin probes for elucidating the linkage specificity of deubiquitinating enzymes. Chemical Commun (Camb) 50 (2):216–218. https://doi.org/10.1039/c3cc47382a

Liu Q, Kistemaker HAV, Bhogaraju S, Dikic I, Overkleeft HS, van der Marel GA, Ovaa H, van der Heden van Noort GJ, Filippov DV (2018) A general approach towards triazole-linked adenosine diphosphate ribosylated peptides and proteins. Angew Chem Int Ed 57(6):1659–1662. https://doi.org/10.1002/anie.201710527

Lu X, Olsen SK, Capili AD, Cisar JS, Lima CD, Tan DS (2010) Designed semisynthetic protein inhibitors of Ub/Ubl E1 activating enzymes. J Am Chem Soc 132(6):1748–1749. https://doi.org/10.1021/ja9088549

Madiraju C, Welsh K, Cuddy MP, Godoi PH, Pass I, Ngo T, Vasile S, Sergienko EA, Diaz P, Matsuzawa S-I, Reed JC (2012) TR-FRET-based high-throughput screening assay for identification of UBC13 inhibitors. J Biomol Screen 17(2):163–176. https://doi.org/10.1177/1087057111423417

McGouran Joanna F, Gaertner Selina R, Altun M, Kramer Holger B, Kessler Benedikt M (2013) Deubiquitinating enzyme specificity for ubiquitin chain topology profiled by di-ubiquitin activity probes. Chem Biol 20(12):1447–1455. https://doi.org/10.1016/j.chembiol.2013.10.012

Misra M, Kuhn M, Löbel M, An H, Statsyuk AV, Sotriffer C, Schindelin H (2017) Dissecting the specificity of adenosyl sulfamate inhibitors targeting the ubiquitin-activating enzyme. Structure 25(7):1120–1129.e1123. https://doi.org/10.1016/j.str.2017.05.001

Moyal T, Bavikar SN, Karthikeyan SV, Hemantha HP, Brik A (2012) Polymerization behavior of a bifunctional ubiquitin monomer as a function of the nucleophile site and folding conditions. J Am Chem Soc 134(38):16085–16092. https://doi.org/10.1021/ja3078736

Mulder MPC, El Oualid F, ter Beek J, Ovaa H (2014) A native chemical ligation handle that enables the synthesis of advanced activity-based probes: diubiquitin as a case study. ChemBioChem 15(7):946–949. https://doi.org/10.1002/cbic.201402012

Mulder MPC, Witting K, Berlin I, Pruneda JN, Wu K-P, Chang J-G, Merkx R, Bialas J, Groettrup M, Vertegaal ACO, Schulman BA, Komander D, Neefjes J, El Oualid F, Ovaa H (2016) A cascading activity-based probe sequentially targets E1-E2-E3 ubiquitin enzymes. Nat Chem Biol 12(7):523–530. https://doi.org/10.1038/nchembio.2084

Mulder MPC, Merkx R, Witting KF, Hameed DS, El Atmioui D, Lelieveld L, Liebelt F, Neefjes J, Berlin I, Vertegaal ACO, Ovaa H (2018) Total chemical synthesis of SUMO and SUMO-based probes for profiling the activity of SUMO-specific proteases. Angew Chem Int Ed 57 (29):8958–8962. https://doi.org/10.1002/anie.201803483

Ohayon S, Spasser L, Aharoni A, Brik A (2012) Targeting deubiquitinases enabled by chemical synthesis of proteins. J Am Chem Soc 134(6):3281–3289. https://doi.org/10.1021/ja2116712

Orcutt SJ, Wu J, Eddins MJ, Leach CA, Strickler JE (2012) Bioluminescence assay platform for selective and sensitive detection of Ub/Ubl proteases. Biochim Biophys Acta (BBA) – Mol Cell Res 1823(11):2079–2086. https://doi.org/10.1016/j.bbamcr.2012.06.004

Pao K-C, Stanley M, Han C, Lai Y-C, Murphy P, Balk K, Wood NT, Corti O, Corvol J-C, Muqit MMK, Virdee S (2016) Probes of ubiquitin E3 ligases enable systematic dissection of parkin activation. Nat Chem Biol 12:324. https://doi.org/10.1038/nchembio.2045

Park S, Krist DT, Statsyuk AV (2015) Protein ubiquitination and formation of polyubiquitin chains without ATP, E1 and E2 enzymes. Chem Sci 6(3):1770–1779. https://doi.org/10.1039/C4SC02340D

Park S, Foote PK, Krist DT, Rice SE, Statsyuk AV (2017) UbMES and UbFluor: novel probes for ring-between-ring (RBR) E3 ubiquitin ligase PARKIN. J Biol Chem 292(40):16539–16553. https://doi.org/10.1074/jbc.M116.773200

Pasunooti KK, Yang R, Vedachalam S, Gorityala BK, Liu C-F, Liu X-W (2009) Synthesis of 4-mercapto-l-lysine derivatives: potential building blocks for sequential native chemical ligation. Bioorg Med Chem Lett 19(22):6268–6271. http://dx.doi.org/10.1016/j.bmcl.2009.09.107

Pickart CM, Rose IA (1986) Mechanism of ubiquitin carboxyl-terminal hydrolase. Borohydride and hydroxylamine inactivate in the presence of ubiquitin. J Biol Chem 261(22):10210–10217

Pruneda Jonathan N, Durkin Charlotte H, Geurink Paul P, Ovaa H, Santhanam B, Holden David W, Komander D (2016) The molecular basis for ubiquitin and ubiquitin-like specificities in bacterial effector proteases. Mol Cell 63(2):261–276. https://doi.org/10.1016/j.molcel.2016.06.015

Puvar K, Zhou Y, Qiu J, Luo Z-Q, Wirth MJ, Das C (2017) Ubiquitin chains modified by the bacterial ligase SdeA are protected from deubiquitinase hydrolysis. Biochemistry. https://doi.org/10.1021/acs.biochem.7b00664

Qiu J, Sheedlo MJ, Yu K, Tan Y, Nakayasu ES, Das C, Liu X, Luo Z-Q (2016) Ubiquitination independent of E1 and E2 enzymes by bacterial effectors. Nature 533(7601):120–124. https://doi.org/10.1038/nature17657

Qiu J, Yu K, Fei X, Liu Y, Nakayasu ES, Piehowski PD, Shaw JB, Puvar K, Das C, Liu X, Luo Z-Q (2017) A unique deubiquitinase that deconjugates phosphoribosyl-linked protein ubiquitination. Cell Res 27(7):865–881. https://doi.org/10.1038/cr.2017.66

Schneider T, Schneider D, Rösner D, Malhotra S, Mortensen F, Mayer TU, Scheffner M, Marx A (2014) Dissecting ubiquitin signaling with linkage-defined and protease resistant ubiquitin chains. Angew Chem Int Ed 53(47):12925–12929. https://doi.org/10.1002/anie.201407192

Sheedlo MJ, Qiu J, Tan Y, Paul LN, Luo Z-Q, Das C (2015) Structural basis of substrate recognition by a bacterial deubiquitinase important for dynamics of phagosome ubiquitination. Proc Natl Acad Sci 112(49):15090–15095. https://doi.org/10.1073/pnas.1514568112

Spasser L, Brik A (2012) Chemistry and biology of the ubiquitin signal. Angew Chem Int Ed 51(28):6840–6862. https://doi.org/10.1002/anie.201200020

Stanley M, Han C, Knebel A, Murphy P, Shpiro N, Virdee S (2015) Orthogonal thiol functionalization at a single atomic center for profiling transthiolation activity of E1 activating enzymes. ACS Chem Biol 10(6):1542–1554. https://doi.org/10.1021/acschembio.5b00118

Stewart MD, Ritterhoff T, Klevit RE, Brzovic PS (2016) E2 enzymes: more than just middle men. Cell Res 26:423. https://doi.org/10.1038/cr.2016.35

Tang S, Liang L-J, Si Y-Y, Gao S, Wang J-X, Liang J, Mei Z, Zheng J-S, Liu L (2017) Practical chemical synthesis of atypical ubiquitin chains by using an isopeptide-linked Ub isomer. Angew Chem Int Ed 56(43):13333–13337. https://doi.org/10.1002/anie.201708067

Tirat A, Schilb A, Riou V, Leder L, Gerhartz B, Zimmermann J, Worpenberg S, Eidhoff U, Freuler F, Stettler T, Mayr L, Ottl J, Leuenberger B, Filipuzzi I (2005) Synthesis and characterization of fluorescent ubiquitin derivatives as highly sensitive substrates for the deubiquitinating enzymes UCH-L3 and USP-2. Anal Biochem 343(2):244–255. https://doi.org/10.1016/j.ab.2005.04.023

Trang VH, Valkevich EM, Minami S, Chen Y-C, Ge Y, Strieter ER (2012) Nonenzymatic polymerization of ubiquitin: single-step synthesis and isolation of discrete ubiquitin oligomers. Angew Chem Int Ed 51(52):13085–13088. https://doi.org/10.1002/anie.201207171

van der Heden van Noort GJ, Kooij R, Elliott PR, Komander D, Ovaa H (2017) Synthesis of poly-ubiquitin chains using a bifunctional ubiquitin monomer. Org Lett 19(24):6490–6493. https://doi.org/10.1021/acs.orglett.7b03085

van Tilburg GBA, Elhebieshy AF, Ovaa H (2016) Synthetic and semi-synthetic strategies to study ubiquitin signaling. Curr Opin Struct Biol 38:92–101. https://doi.org/10.1016/j.sbi.2016.05.022

Ye Y, Blaser G, Horrocks MH, Ruedas-Rama MJ, Ibrahim S, Zhukov AA, Orte A, Klenerman D, Jackson SE, Komander D (2012) Ubiquitin chain conformation regulates recognition and activity of interacting proteins. Nature 492(7428):266–270. https://doi.org/10.1038/nature11722

Zhang X, Smits AH, van Tilburg GBA, Jansen PWTC, Makowski MM, Ovaa H, Vermeulen M (2017) An interaction landscape of ubiquitin signaling. Mol Cell 65(5):941–955.e948. https://doi.org/10.1016/j.molcel.2017.01.004

Zhao X, Lutz J, Höllmüller E, Scheffner M, Marx A, Stengel F (2017) Identification of proteins interacting with ubiquitin chains. Angew Chem Int Ed 56(49):15764–15768. https://doi.org/10.1002/anie.201705898

ABPP and Host–Virus Interactions

Geneviève F. Desrochers and John Paul Pezacki

Contents

Abstract Successful viral infection, as well as any resultant antiviral response, relies on numerous sequential interactions between host and viral factors. These interactions can take the form of affinity-based interactions between viral and host macromolecules or active, enzyme-based interactions, consisting both of direct enzyme activity performed by viral enzymes and indirect modulation of the activity of the host cell's enzymes via viral interference. This activity has the potential to transform the local microenvironment to the benefit or detriment of both the virus and the host, favouring either the continuation of the viral life cycle or the host's antiviral response. Comprehensive characterisation of enzymatic activity during viral infection is therefore necessary for the understanding of virally induced diseases. Activity-based protein profiling techniques have been established as effective and practicable tools with which to interrogate the regulation of enzymes' catalytic

G. F. Desrochers · J. P. Pezacki (✉)
Department of Chemistry and Biomolecular Sciences, University of Ottawa,
10 Marie Curie Pvt., Ottawa, ON K1N 6N5, Canada
e-mail: john.pezacki@uottawa.ca

Current Topics in Microbiology and Immunology (2019) 420:131–154
DOI 10.1007/82_2018_139

Published Online: 23 September 2018

activity and the roles played by these enzymes in various cell processes. This paper will review the contributions of these techniques in characterising the roles of both host and viral enzymes during viral infection in humans.

1 Introduction

Viruses are infectious biological agents which lack the ability to replicate independently; specifically, they lack translational and metabolic machinery, and they are therefore obligate intracellular parasites. Viral particles consist of either an RNA or DNA genome surrounded by a protein capsid which protects the genetic material. Certain viruses are additionally enveloped by a lipid bilayer derived from host membranes. In order to propagate, viruses must use host cell systems to assemble new RNA, DNA, proteins and lipid envelopes. Though details of viral life cycles vary from virus to virus, there are five general stages: cell entry, translation of viral proteins, replication of the genome, assembly of the viral particle and egress from the cell (Fig. 1). Activity-based profiling has been used broadly to characterise

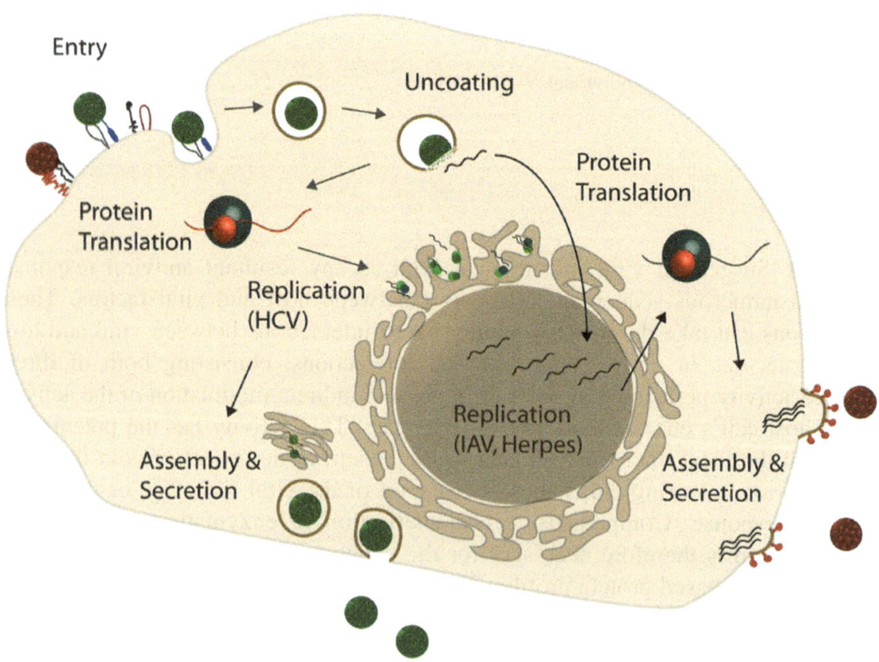

Fig. 1 Viruses utilise multiple host systems throughout their life cycle. Life cycles differ between viruses. Positive-strand RNA viruses (ex: HCV) follow the path illustrated on the left: receptor binding, internalisation, protein translation, replication within membranous structures in the cytosol, assembly and secretion. Viruses such as influenza A and herpesviruses follow the path on the right: replication within the nucleus, export to the cytoplasm, protein translation, assembly and budding from the cell

host–virus interactions. The approaches used and the discoveries of these techniques accorded will be highlighted herein.

The molecular processes involved in virus propagation require the diversion of energy, molecular building blocks and other essential resources. These need to be diverted from the host cell's systems and into pathways and processes needed by the virus. It is important to note that metabolic energy is consumed not only by the synthesis of new viral particles, but also by the remodelling of the cellular environment to meet the demands of the virus. Viruses require specific and often tightly regulated conditions to propagate efficiently. A good example lies in the replication of positive-strand RNA viruses, which occurs within specialised regions called replication organelles (RO) (Neufeldt et al. 2018). These organelles create an optimal environment for replication, allowing the concentration of the required host and viral factors and shielding viral RNA from the innate immune response (Neufeldt et al. 2018). The morphology of these organelles differs from virus to virus: they can be spherical or tubular, originate from the ER, Golgi body, endosome or lysosome, and be composed of single, double or multi-membrane vesicles (Novoa et al. 2005; Neufeldt et al. 2018). These membranes furthermore possess specific lipid profiles unique to each virus. The Rubella virus RO requires elevated saturated fatty acid levels, while West Nile virus, hepatitis C virus and enteroviruses require high cholesterol (Harak and Lohmann 2015; Paul and Bartenschlager 2015). The hepatitis C RO also requires elevated levels of sphingolipids and phosphorylated phosphoinosides (PIPs) (Hsu et al. 2010; Harak and Lohmann 2015; Paul and Bartenschlager 2015). The large amount of metabolic energy required to change the lipid profile of a cell in this manner necessitates significant perturbation of normal lipid homeostasis.

The diversion of metabolic energy and the remodelling of host cell architecture are initiated by viral enzymes; however, direct modification of cellular architecture by viral enzymes is not always feasible due to the small range of functionalities encoded by viruses. Changes in host cell function, through changes in abundance and activity of key enzymes, can be induced by the virus as homeostatic responses to virus-induced changes as well as immunological responses to infection. Viral genomes can be extremely small: the hepatitis B virus, for example, has a genome 3.2 kb long (Liang 2009). As a result, viruses utilise the machinery of the cells they infect to replicate their own genome and assemble new viral particles for release and subsequent reinfection of new host cells.

Functional changes that are induced either directly or indirectly often are the result of changes in enzyme catalytic activity of both viral and host-derived enzymes. Interrogating changes to enzymatic activity during viral infection is therefore essential to form a complete understanding of the mechanisms of virus infection. Activity-based protein profiling (ABPP) techniques (Barglow and Cravatt 2007; Cravatt et al. 2008) are ideally suited to answering questions on the perturbation of enzyme function by host–virus interactions. These techniques use small molecule probes to covalently label active enzymes, while inactive enzymes remain unmodified. By including reporter tags in these probes, it is possible to quantitatively report differences in levels of active enzyme within complex proteomic samples.

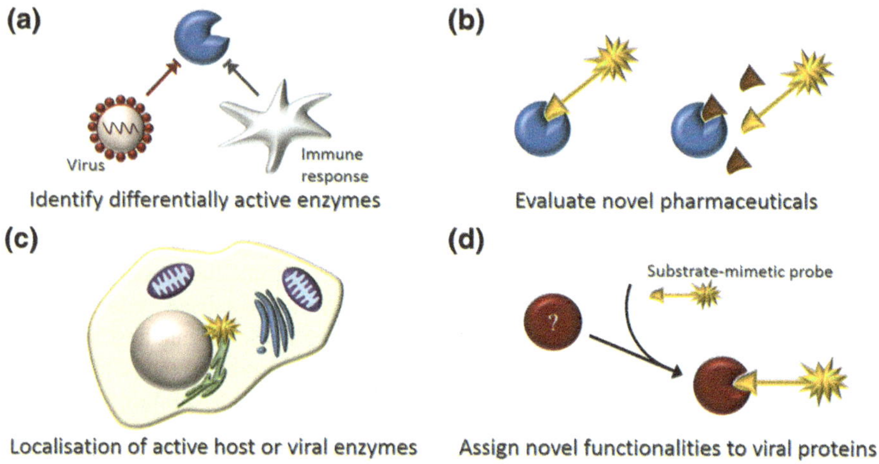

Fig. 2 Contributions of ABPP to the understanding of host–virus interactions. **a** Activity-based profiling identifies and quantifies changes to enzymatic activity caused by the virus or by the host's antiviral defence mechanisms. **b** Competitive activity-based labelling evaluates the effectiveness of novel antiviral compounds. **c** Fluorophore conjugated activity-based probes allow imaging of active enzyme localisation. **d** Activity-based labelling with substrate-mimetic activity-based probes identifies new targets for viral enzymes

This chapter will describe the contributions made by ABPP to the understanding of host–virus interactions. ABPP has supplied a diverse toolbox of activity-based probes which have been applied to study both viral and host enzymes: identifying host enzymes dysregulated during different stages of viral infection, determining how the virus effects these changes and characterizing viral enzymes' functionalities, structure, and catalytic and inhibitory mechanisms (Fig. 2).

2 Viral Entry

For most viruses, the entry process consists of recognition of their target cell, attachment to the cell's membrane, internalisation of the virus into the cell, usually by an endocytic pathway, and the release of viral genetic material into the cytosol of the cell. In the case of enveloped viruses, this requires the fusion of the viral envelope with the endosomal membrane to form a pore through which the genetic material passes. The mechanism of formation of this pore varies between viruses, many of which require significantly different host cell conditions or factors to induce fusion.

A major focus in efforts to design novel pharmaceuticals has been the development of cell entry inhibitors. Targeting cell entry prevents the spread of infection as well as the irreversible damage to tissues caused by infection. Furthermore, as host factors and not viral factors are usually targeted, it can impede the emergence of drug-resistant strains. Antiviral pharmaceuticals targeting cell entry have been developed against hepatitis B virus, human immunodeficiency virus (HIV) and herpesviruses (De Clercq and Li 2016; Sun et al. 2018; de Castro and Camarasa 2018), and entry inhibitors against other viruses such as HCV are currently being developed (Xiao et al. 2015; Qian et al. 2016).

Viruses such as the SARS coronavirus have been shown to require the activity of lysosomal cathepsins, such as cathepsin L, in order to enter the cytosol (Bosch et al. 2008). Cathepsin L cleaves the SARS-CoV spike protein, found on the viral envelope, thereby activating it and allowing membrane fusion to proceed (Bosch et al. 2008). It has furthermore been demonstrated to play an accessory role in Ebola infection, cleaving the GP1 protein in conjunction with cathepsin B and triggering membrane fusion (Chandran et al. 2005). The epoxide-based peptide-mimetic probe DCG-04 was developed to label cysteine proteases, a class which includes cathepsins (Greenbaum et al. 2000).

Shah et al. applied this probe to study novel SARS and Ebola virus entry inhibitors by competitive ABPP (Shah et al. 2010). Cells treated with their novel inhibitors demonstrated significantly reduced labelling by DCG-04, indicating that they were able to target cathepsin L in live cells and suggesting that they inhibited viral entry by targeting cathepsin L (Shah et al. 2010). This study demonstrates how an ABP can be applied in the development of novel pharmaceuticals targeting cell entry.

3 Viral Replication

Members of the *flaviviridae* family significantly remodel their host cell architecture to form an environment favourable for replication. The hepatitis C virus, one of the most studied viruses from this family, induces the formation of double-membrane vesicles inside a membranous web derived from the endoplasmic reticulum (Neufeldt et al. 2018). The formation of the membranous web during HCV infection necessitates significant alteration of the local lipid profile, which results from viral hijacking of lipid metabolic enzymes (Neufeldt et al. 2018). Activity-based protein profiling has been used to identify which enzymes are targeted by the virus and to determine how their regulation is modulated.

3.1 Role of Triglycerides

One of the most pronounced HCV-induced changes to cell architecture is the appearance of larger and more numerous lipid droplets near sites of viral replication (Neufeldt et al. 2018). This is the result of both increases in fatty acid synthesis and decreases in fatty acid oxidation and secretion (Syed et al. 2010).

Fatty acid synthase (FASN), the enzyme responsible for de novo synthesis of fatty acids, has been associated with HCV replication as early as 2002; however, the nature of the interactions between HCV and FASN was not completely characterised (Su et al. 2002). A probe based on the β-lactone inhibitor orlistat, orlistat-alkyne, was shown to strongly and selectively label active FASN (Yang et al. 2010). This probe was used to investigate how HCV altered FASN activity. By comparing changes in protein expression and probe labelling, Nasheri et al. determined that FASN activity was significantly increased during HCV replication by increases both in protein expression and post-translational activation (Nasheri et al. 2013). Imaging of active FASN using the orlistat-alkyne probe demonstrated that HCV does not alter the localisation of FASN, indicating that the post-translational regulation of FASN occurs by another mechanism (Nasheri et al. 2013). Increased activity of FASN during HCV replication was further confirmed using probes containing a β-lactam warhead (Nasheri et al. 2014).

HCV-mediated upregulation of intracellular lipids has been shown to be regulated via numerous mechanisms: in addition to increases in lipid synthesis, decreases in lipolysis and lipid secretion have also been demonstrated to play a significant role (Syed et al. 2010). Singaravelu et al. used a novel activity-based probe with a phenyl sulfonate ester warhead (PS) to identify enzymes displaying differential activity during HCV replication (Singaravelu et al. 2010). The PS probe was shown to label an enzyme involved in the β-oxidation of fatty acids, the electron transfer flavoprotein subunit alpha (ETFA). Comparison of labelling in HCV-replicating and naïve cells shows a decrease in ETFA activity during HCV replication, suggesting that targeting of ETFA may be a mechanism by which HCV reduces lipid oxidation and increases intracellular lipid levels (Singaravelu et al. 2010).

In a more recent study, Nourbakhsh et al. used a fluorescently labelled fluorophosphonate probe (Fig. 3a shows desthiobiotin-labelled variant) to profile serine hydrolase activity in infected vs. naïve hepatoma cells. They identified one enzyme, arylacetamide deacetylase (AADAC), which was significantly down-regulated during HCV infection (Nourbakhsh et al. 2013). It was subsequently demonstrated that AADAC induces lipolysis and decreases VLDL secretion of lipids, thereby decreasing the levels of intracellular triglycerides (Nourbakhsh et al. 2013). Altogether, this study demonstrated that HCV decreases the activity of the serine hydrolase AADAC in order to increase intracellular lipid levels via the inhibition of lipolysis and lipid secretion.

3.2 Role of Phosphatidylinositides

Phosphatidylinositides (PIs) are minor components of the intracellular membranes which play important roles in establishing organelle identity and in propagating cell signalling, and certain species have additionally been shown to be able to induce high membrane curvature (McMahon and Gallop 2005). A phosphorylated species

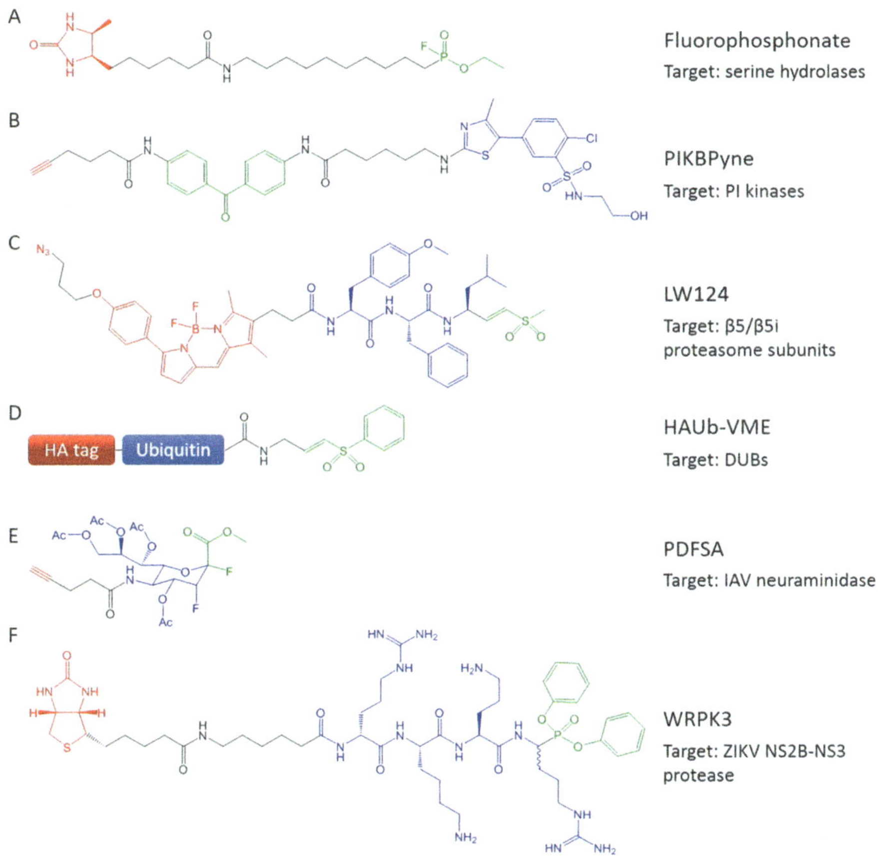

Fig. 3 Examples of types of activity-based probes used to interrogate host–virus interactions. Targeting groups shown in blue, reactive groups shown in green, and reporter tags or handles for attachment of reporter tags shown in red. **a** Inhibitor-based fluorophosphonate conjugated to desthiobiotin possesses a broad specificity towards host cell serine hydrolases. **b** Inhibitor-based PIKBPyne labels specific host cell lipid kinases, phosphatidylinositol kinases type III. **c** Peptide-mimetic LW124 selectively labels proteasomal subunit β5 and immunoproteasomal subunit β5i. The BODIPY group allows LW124 to be used as a fluorescent probe, while the azide handle allows the attachment of affinity tags for purification of labelled targets. **d** Substrate-mimetic HAUb-MVE targets deubiquitinating enzymes from both viruses and host cells. The HA tag allows detection of labelling and purification of labelled enzymes. **e** Substrate-derived PDFSA targets neuraminidases on the envelope of the influenza A virus. **f** Peptide-based WRPK3 contains unnatural amino acids and specifically labels Zika virus NS2B-NS3 protease

of PI, PI4P, has been shown to be upregulated in the membranous web; it has been hypothesised that this it plays a role in establishing the morphology of the membranous web (Hsu et al. 2010; Bishé et al. 2012). PI4P is synthesised from PI by four PI-kinases (PIKs) which are differentially localised within the cell (Bishé et al. 2012). The identity of the kinases which contribute to HCV-mediated upregulation of PI4P levels within the replication complex has been a subject of significant interest (Delang et al. 2012; Bishé et al. 2012). While PIKs such as PIKA are well established as HCV host factors, the role of other PIKs, such as PIKB, during infection remained ambiguous (Delang et al. 2012).

In order to investigate the activity of PI4Ks during HCV infection, Sherratt et al. synthesised a novel activity-based probe, PIKBPyne, derived from the well-known reversible PIK active-site inhibitor PIK93 (Fig. 3b) (Sherratt et al. 2014). The warhead of this probe contained a PIK93 group to enable selective interaction with active PIKs, and a UV-inducible benzophenone crosslinker to form a covalent bond between the probe and its protein target. Using this probe, it was shown that PI4KB activity, but not expression, is upregulated by HCV infection (Sherratt et al. 2014). This finding illustrates the important role ABPP can play in characterising the dysregulation of host factors by viruses.

As the increase in PI4KB activity did not correlate with increased protein expression, activity was hypothesised to be regulated post-translationally. PI4KB activity is known to be regulated by stable protein–protein interactions which recruit PI4KB to membranes and stabilise an activating phosphorylation at Ser294 (Hausser et al. 2005, 2006; Balla 2013). To investigate the role of these interactions in the post-translational regulation of PI4KB activity, Desrochers et al. synthesised new PIKBPyne probes containing flexible linkers of varying length between the targeting PIK93 moiety and the crosslinking benzophenone (Desrochers et al. 2018). These flexible-linker PIKBPyne probes were shown to be able to label proteins interacting with active PI4KB. Labelling with PIKBPyne demonstrated an HCV-induced decrease in the interaction between PI4KB and the Golgi-resident protein ACBD3 (Desrochers et al. 2018). It had previously been reported that HCV induced a change in the localisation of PI4KB from the Golgi towards sites of viral replication (Zhang et al. 2012). This finding suggests that the upregulation of PI4P levels in the membranous web during HCV replication may occur partially through the disruption PI4KB–ACBD3 interaction, allowing the relocalisation of PI4KB.

3.3 *Role of Cholesterol*

The double-membrane vesicles of the HCV replication sites have been shown to contain elevated levels of cholesterol (Neufeldt et al. 2018). Activity-based probes have been applied to characterise the processes by which cholesterol metabolism is hijacked during HCV replication.

Fluorophosphonate-based probes were used to profile serine hydrolase activity during HCV replication, in order to identify enzymes differentially regulated by the virus (Blais et al. 2010a, b). Carboxylesterase 1 (CES1), a liver-abundant serine hydrolase which displays cholesteryl ester hydrolase activity, was identified as a target of activity-based labelling by both fluorophosphonate-based and peptide-based probes containing either a serine or a threonine moiety (Ross et al. 2010; Blais et al. 2010a, b). CES1 activity has been shown to play a role in decreasing the levels of intracellular cholesteryl ester, as well as the secretion of intracellular triglycerides by the VLDL pathway (Zhao et al. 2007; Ross et al. 2010). It has furthermore been demonstrated that CES1 activity is not always correlated to its expression, suggesting that the regulation of CES1 function relies on post-translational modifications and interactions (Ross et al. 2010). Activity-based labelling is therefore ideally suited for interrogating the function of CES1. Both fluorophosphonate and the peptide-based probes demonstrated significantly increased activity during HCV replication, suggesting that CES1 is activated as part of the host cell's antiviral strategy to counteract the virally induced increases in cholesterol and triglycerides (Blais et al. 2010a, b).

β-lactam warhead probes, described above, have also been used to investigate cholesterol metabolic enzymes. These probes have been shown to label hydroxymethylglutaryl-CoA synthase 1 (HMGCS1), an enzyme which catalyses an early step in cholesterol synthesis (Mazein et al. 2013; Nasheri et al. 2014). Labelling with the β-lactam probe showed an increases HMGCS1 activity during HCV replication, suggesting that differential activation of this enzyme is a mechanism by which HCV induces the formation of a favourable replication environment (Nasheri et al. 2014). This finding provided further confirmation of the suggested role of HMGCS1 in HCV infection (Kapadia and Chisari 2005).

Many of the changes in the activity of lipid metabolic enzymes are the result of alterations to their post-translational mechanisms which regulate them. Changes to the activity of signalling pathways which control this post-translational regulation are therefore of great interest. Many signalling pathways rely on kinase activity, and profiling kinase activity therefore has the potential to provide information on the regulation of multiple cell systems and pathways. The probe wortmannin-yne, based on the irreversible kinase inhibitor wortmannin, was designed to be able to broadly profile kinase activity (Desrochers et al. 2015). Comparison of labelling by wortmannin-yne in cells replicating HCV and in naïve cells identified multiple regulatory kinases whose catalytic activity was altered by HCV. Kinases in the MAPK pathway, in particular, displayed significant decreases (Desrochers et al. 2015). This study confirmed previous reports of HCV-mediated decreases in MAPK signalling and demonstrates how ABPs can be used to profile kinase activity.

This section has described ABPs which target lipid metabolic enzymes and how they have been used to profile changes to the regulation of lipids during viral infection. Hijacking of lipid homeostasis is a major contributor to the diversion of metabolic energy during infection. However, other metabolic pathways are also perturbed during infection: glycolysis in particular is targeted by multiple diverse viruses, including HCV, influenza A virus, and several herpesviruses, though the mechanisms by which glycolysis is mis-regulated are not fully understood

(Goodwin et al. 2015). The development of novel probes targeting glucose metabolic enzymes represents an opportunity to increase our understanding of metabolic flux during viral infection.

4 Programmed Cell Death

A fundamental aspect of the innate immune response to infection lies in the activation of apoptotic and cell death pathways upon viral invasion of the cell in order to restrict the spread of infection throughout the host (Barber 2001). Commercial peptide-based probes, FLICA^{TM}, have been established as chemical-based tools to study apoptosis (Furman et al. 2009). These probes are designed to report on the activity of caspaces, enzymes essential to the progression of apoptosis (Furman et al. 2009). Using these probes, Furman et al. determined that norovirus infection induces apoptosis, as demonstrated by the significant increase in caspace activity. Mass spectrometry analysis of enzymes labelled by these probes revealed that they are equally capable of reporting on the activity of the cysteine protease cathepsin B. It was shown that cathepsin B is activated during norovirus infection and acts as an upstream activator of the apoptotic pathway (Furman et al. 2009).

Cathepsin D is a lysosomal aspartic peptidase which has also been shown to induce apoptosis (Benes et al. 2008). Using a peptide-based probe containing a phenylalanine targeting group, Blais et al. demonstrated that HCV significantly down-regulates the activity of cathepsin D, suggesting that HCV decreases cathepsin D activity in order to avoid apoptosis of infected cells (Blais et al. 2010a). Activation of anti-apoptotic genes was also demonstrated during infection of Dengue virus, a close relative of HCV. Novel ATP-based probes containing acyl phosphate warheads were used to profile the activity of kinases during Dengue infection (Vetter et al. 2012). The activity of DNA-dependant protein kinase (DNA-PK), an anti-apoptotic enzyme which recognises and repairs DNA double-stranded breaks, was shown to be significantly upregulated by Dengue infection (Vetter et al. 2012). Altogether, ABPP has shown that both HCV and Dengue virus act to counteract the activation of the apoptotic pathway.

5 Viral Evasion of the Immune Response

Cathepsins also play a role in regulating inflammation and the recruitment of the adaptive immune response during infection. Cathepsin G, for example, has been shown to participate in the degradation of proteins for presentation by the major histocompatibility complex (Burster et al. 2010). It has additionally been shown to negatively regulate STAT5 transcriptional activation (Schuster et al. 2007), as well as inflammation (Burster et al. 2010). Zou et al. reported a novel peptide-based probe, Mars116, specific for cathepsin G (Zou et al. 2012). Labelling of cathepsin G by Mars116 was significantly decreased 12 days after Epstein–Barr virus (EBV) infection (Zou et al. 2012). Labelling of four other cathepsins, cathepsins X, B and S, by DAP22c also showed a significant decrease during EBV infection. This decrease in the activities of immune-linked cathepsins is suggestive of an immune escape strategy by EBV (Zou et al. 2012).

5.1 *Viral Alterations to Proteasome Activity*

The proteasome is a multi-subunit structure responsible for the degradation of damaged or excess proteins within the cell and plays an essential role in protein quality control, cell signalling, immune responses and apoptosis (Kammerl and Meiners 2016). A specialised form of the proteasome, the immunoproteasome, consists of many of the same regulatory subunits in addition to three variant catalytic subunits found only in the immunoproteasome (Ferrington and Gregerson 2012). This specialised form of the proteasome plays a role in directing the activity of the adaptive immune system. It is induced during infection and generates peptides with a hydrophobic C-terminus for cell-surface display in MHC class I molecules. CD8 T cells surveil these peptides and propagate an immune response upon recognition of foreign material (Ferrington and Gregerson 2012). The existence of mixed proteasomes, containing both the constitutively expressed and immunoproteasome subunits, has also been demonstrated, indicating that both categories of catalytic subunits could be implicated in immune response (Kammerl and Meiners 2016).

The catalytically active proteasomal subunits, though not serine hydrolases, containing a nucleophilic threonine as the catalytic residue (Marques et al. 2009; Nasheri et al. 2015), have been shown to be labelled by the activity-based probe fluorophosphonate (Fig. 3a) (Shahiduzzaman et al. 2014). Using a fluorophosphonate-based probe, Shahiduzzaman et al. have shown that the activity of several proteasomal subunits, PSMA2, PSMA3, PSMA6, PSMB3, increased during influenza A infection, possibly due to increases in immunoproteasome activation (Shahiduzzaman et al. 2014). Though fluorophosphate was able to interrogate the activity of a few proteasomal subunits, several other subunits were not detected.

Probes able to specifically and quantitatively assess the activities of all the possible catalytic subunits of the proteasome have been developed by various groups over the past decade. The first to be developed, MV151, is a peptide-mimetic probe conjugated to a BODIPY fluorophore, and labels all catalytically active proteasome subunits (Verdoes et al. 2006). Other more specific probes were later developed: LW124, a probe specific for the β1 subunit and LMP2, its immunoproteasome analogue, and MVB127, a probe specific for β5 and its analogue LMP7 (Fig. 3c) (Li et al. 2013; Keller et al. 2015).

These probes were applied by Keller et al. to investigate perturbation to the proteasome system during viral infection. Proteasome activity was shown to be increased during murine gamma-herpes virus infection using the general-use MV151 probe. LW124 and MVB127 were used to assess the changes to catalytic activity of the proteasomal subunits versus the activity of their immunoproteasomal analogues during infection. The identity of the labelled subunits was determined based on their reactivity and their molecular weight, as determined by in-gel fluorescent imaging. It was demonstrated that while the activity of the proteasomal subunits remained relatively unchanged, the activity of the immunoproteasome subunits increased significantly during the early stages of infection, before decreasing as the infection progressed (Keller et al. 2015). While these probes had been previously demonstrated to be specific proteasome probes (Verdoes et al. 2006; Li et al. 2013), Keller et al. were the first to demonstrate how they can be used to specifically assess regulatory changes to the activity of the immunoproteasome in the context of viral infection (Keller et al. 2015).

These increases in proteasome activity represent the part of the host's response to viral infection wherein proteasomal degradation of proteins is increased in order to generate antigenic peptides for the recruitment of the adaptive immune response. Viral targeting of proteasomal activity could therefore be a potential immune evasion strategy. Recently, Nasheri et al. used a β-lactone-based activity-based probe, orlistat-alkyne, to label active proteasome subunits and thereby interrogate proteasome activity during hepatitis C virus infection (Nasheri et al. 2015). It was shown that the activity of multiple members of the proteasome system displayed decreased activity in HCV infected cells. PSMB5, which displays broad specificity and chymotrypsin-like hydrolase activity (Marques et al. 2009), PSMC6, an ATPase which regulates proteasome activity (Coux et al. 1996), and two immunoproteasomal subunits, PSME1, and PSME2 (de Graaf et al. 2011) were down-regulated between 40 and 80% (Nasheri et al. 2015). Interestingly, changes to protein expression did not match the observed changes in enzymatic activity (Nasheri et al. 2015). Altogether, this suggests that HCV targets proteasomal degradation by altering both protein expression and post-translational regulation as part of its efforts to evade the immune response. Furthermore, this study highlights the advantages of activity-based protein profiling in addition to traditional proteomics to study virally induced perturbation of host cell systems.

5.2 NFκB Signalling During Herpesvirus Infection

Nuclear factor kappa B (NFκB) is a transcription factor, activated during viral infection, which induces the expression of genes involved in the innate immune response (Oeckinghaus and Ghosh 2009). The signalling cascade which is responsible for the activation of NFκB during infection relies in great part on the ubiquitination of key signal transducers (Wertz and Dixit 2010). The auto-ubiquitination of the ubiquitin ligase TRAF6 promotes the activity of the TAK1 kinase (Wertz and Dixit 2010). The IκB kinase (IKK) complex is then activated via TAK1-mediated phosphorylation and TRAF6-mediated ubiquitination (Wertz and Dixit 2010; van Gent et al. 2014). IKK phosphorylation of IκB results in its ubiquitination and subsequent degradation (Oeckinghaus and Ghosh 2009). NFκB is sequestered in the cytoplasm by IκB; upon its degradation, NFκB translocates to the nucleus and induces the expression of antiviral, pro-inflammatory and pro-apoptotic genes (Oeckinghaus and Ghosh 2009).

The activity of enzymes responsible for the regulation of ubiquitination is therefore of interest when investigating host–virus interactions due to the importance of ubiquitination to the immune response to infection. Deubiquitinating enzymes (DUBs) are enzymes responsible for removing ubiquitin from ubiquitinated proteins. As they are cysteine proteases, they are excellent candidates for suicide-based activity probes (Borodovsky et al. 2002).

HAUb probes are based on modified ubiquitin proteins, and mimic the natural substrate of DUBs, allowing them to specifically target DUBs. Ubiquitin is recombinantly expressed with the haemagglutinin tag HA, which functions as a reporter of enzymes labelling. A thiol-reactive group functioning as the warhead is subsequently added to the C-terminus via intein-based chemical ligation. Several variants were produced, containing several different warheads: chloroethylamine (Cl), bromoethylamine (Br2), bromopropylamine (Br3), glycine vinylmethylsulfone (VS), glycine vinyl methylester (VME), glycine vinyl phenylsulfone (VSPh), and glycine vinylcyanide (VCN) (Borodovsky et al. 2002). Immunoblotting and MS/MS profiling of labelled DUBs demonstrated a slight but significant difference in the labelling patterns of the HAUb probes (Borodovsky et al. 2002; Ovaa et al. 2004). The probe containing a vinyl methylester warhead (HAUb-VME, Fig. 3d) displayed the broadest labelling capacity and was used in most subsequent studies (Borodovsky et al. 2002; Ovaa et al. 2004).

After developing these probes, Ovaa et al. applied them to interrogate the activity of host cell deubiquitinating proteins during infection by the herpesvirus Epstein–Barr virus (EBV) (Ovaa et al. 2004). Herpesviruses are large double-stranded DNA viruses and have been shown to modulate the immune response via multiple different mechanisms (Cruz-Muñoz and Fuentes-Pananá 2017). Using the HAUb-VME probe, Ovaa et al. labelled active host cell deubiquitinating enzymes and identified multiple DUBs whose activity was increased during infection: UCH-L3, USP-15, UCH-L1, UCH37, USP7 and USP9X. Several of these enzymes have been shown to regulate the NFκB pathway. USP15 negatively regulates NFκB activation by deubiquitinating IκB (Harhaj and Dixit 2011), while USP7 decreases NFkB signalling by deubiquitinating TRAF6 and IKK (Nanduri et al. 2013). On the contrary, USP37 has been shown to be required for IκB-α degradation and therefore NFkB activity (Mazumdar et al. 2010). Altogether, Ovaa et al. demonstrated that the herpesvirus EBV differentially regulates DUBs implicated in the regulation of the innate immune response, suggesting that this may be a pro-viral immune evasion mechanism.

5.3 Virally Encoded Deubiquitinases

Kattenhorn et al. applied the HAUb-VME probes designed by Borodovsky et al. to study the perturbation of ubiquitination-regulating enzymes during infection of Herpes simplex virus 1 (HHV-1), a herpesvirus related to EBV. MS/MS analysis of labelled proteins identified an enzyme which did not correspond to any known DUB, and whose sequence was found to correspond to a viral protein with unknown function, the large tegument protein deneddylase (UL36) (Kattenhorn et al. 2005).

Perturbation of host cell processes by viruses is usually indirect as the limited size of viral genomes means the virus cannot itself encode the functionalities it requires. Herpesviruses, however, are large, double-stranded DNA viruses and possess larger and more complex genomes than the simpler RNA viruses discussed elsewhere in this chapter. HHV-1, for example, encodes approximately 80 proteins (Macdonald et al. 2012).

UL36 does not share significant sequence homology with any other DUB known at the time (Kattenhorn et al. 2005); tradition bioinformatics could therefore not have predicted its function. This discovery highlights one of the unique advantages of activity-based protein profiling: it reports directly on the functional output of enzymes and allows the identification of novel targets. UL36 has recently been shown to deubiquitinate IκBα, preventing its degradation and promoting the sequestration of NFκB outside of the nucleus and thereby interfering with the activation of the innate immune response (Ye et al. 2017).

Following this discovery, several research groups postulated the existence of homologous deubiquitinating enzymes in closely related viruses whose function had remained undiscovered due to their lack of homology to previously characterised DUBs. Having been established as an effective probe of herpesvirus deubiquitinating enzymes, HAUb-VME was used to interrogate the activity of viral proteins from several different viruses.

Schlieker et al. identified an Epstein–Barr virus protein, BPLF1, and a murine cytomegalovirus protein (MCMV), M48, which possessed significant sequence homology to the HHV-1 UL36 (Schlieker et al. 2005). They subsequently demonstrated that both BPLF1 and M48 were labelled by the HAUb-VME probe, indicating that these proteins were enzymes which targeted and hydrolysed ubiquitinated proteins (Schlieker et al. 2005). It was later shown that BPFL1 acts similarly to HHV-1's UL36, inhibiting the activation of NFκB by catalysing the removal of ubiquitin from the sequestering protein IκBα (van Gent et al. 2014). It has furthermore been demonstrated that BPLF1 exerts additional control over the NFκB pathway, deubiquitinating both TRAF6 and subunits of IKK, and thereby inhibiting the propagation of TLR signalling and the resultant activation of NFκB (Saito et al. 2013; van Gent et al. 2014). Taken together, this demonstrates that EBV uses the virally encoded DUB functions to inhibit the activation of the antiviral innate immune response.

Another herpesvirus, human cytomegalovirus (HCMV), has also been shown to encode a deubiquitinating enzyme (Wang et al. 2006). HAUb-VME labelled a high molecular weight enzyme, pUL48, whose role in infection had previously been unknown. In the absence of pUL48 deubiquitinating activity, virion production was significantly inhibited, though not entirely blocked, indicating that pUL48 deubiquitinating activity contributes to, but is not required for, virion production. HAUb-VME has also been used to identify similar deubiquitinating proteins in murine gamma herpesvirus 68 (MHV-68) and Marek's disease virus (MDV) (Gredmark et al. 2007; Jarosinski et al. 2007).

Assigning novel functionalities is one of the more well-known uses of ABPP, and an essential primary step in characterising poorly understood enzymes which opens up further avenues of investigation. Contextualisation of the reactivity of an enzyme enables the determination of its specific targets. The ubiquitin chains targeted by the DUBs discussed in this section are linked to each other by one of the seven different linkage sub-types, corresponding to the location of the lysine residue to which the next ubiquitin is bound: Lys^{6}, Lys^{11}, Lys^{27}, Lys^{29}, Lys^{33}, Lys^{48} and Lys^{63}. New diUb-VME probes were developed containing two ubiquitin moieties conjugated by one of the seven possible linkage types, thereby allowing the determination of DUB's linkage specificity (Mulder et al. 2014). These probes were used to interrogate the substrate specificity of the SARS coronavirus papain-like protease (PLPro), a deubiquitinating enzyme, thought to play a role in innate immune evasion (Békés et al. 2016). The structural basis of PLPro's specificity was determined by co-crystalising it in conjunction with a diUb-VME probe. Co-crystalisation of enzymes with their substrates is often challenging, as the enzyme will catalyse the substrate and release the products. Using a substrate-mimetic ABP ensures that it is retained within the active site, preserving the interactions which give rise to specificity as well as the active conformation of the enzyme. This showed that PLpro interacts with the two ubiquitin moieties, but has few interactions with the interface between the two (Békés et al. 2016). This suggests that specificity originates from the orientation of the ubiquitin proteins in the S1 and S2 pockets, and not from recognition of the linkage itself. Screening with a panel of Ub-VME probes containing mono-Ub or di-Ub and possessing different linkage types demonstrated that PLpro possessed a similarly low labelling efficiency for mono-Ub and most di-Ub probes (Békés et al. 2016). This lends further support to the idea that interactions with a properly oriented S2 ubiquitin stabilise the enzyme-substrate complex and contribute to the specificity of SARS PLpro.

6 Viral Assembly and Egress

The last stage of the viral life cycle consists of the assembly of the viral particle and its release into the extracellular space. The majority of the contributions made by ABPP towards the understanding of virus infection have focused on viral entry and replication. In some cases, this has been due to limitations in cell culture models of infection: these models could replicate the genome but were unable to efficiently assemble infectious particles (Lohmann and Bartenschlager 2014). As viral assembly and egress do rely on host factors to translocate to the plasma membrane and exit the cell, profiling enzymatic activity during these processes is of significant interest.

Hijacking of cellular trafficking plays a significant role in the packaging and secretion of viruses. The COPI system is essential for correct localisation of IAV particles, while HCV utilises the COPII system to travel from the ER to the Golgi, at which point it can be trafficked through the VLDL secretory pathway (Pohl et al. 2016; Syed et al. 2017). These systems are regulated in part by Rab GTPases, host cell enzymes which play an important role in trafficking throughout the cell (Hutagalung and Novick 2011) and which have been implicated in the egress of multiple viruses (Hogue et al. 2014; Pohl et al. 2016; Takacs et al. 2017). Rab11 has been shown to be essential for the trafficking of IAV RNA from the perinuclear region to the plasma membrane (Pohl et al. 2016). Another Rab protein, Rab1, has been proposed to mediate transport of nascent HCV virions from the ER to the Golgi (Takacs et al. 2017). Rab proteins have also been shown to play a role in DNA virus egress in the case of herpesviruses (Hogue et al. 2014). Altogether, this suggests that Rab GTPase activity could be an interesting target for activity-based profiling during viral infection.

IAV budding has been shown to involve the activity of a class of ATPases, the mitochondrial F1FO-ATPases. These enzymes localise to the plasma membrane and are necessary for efficient budding of new virions, though the precise nature of their contribution has yet to be fully understood (Pohl et al. 2016). ATP-based probes have previously been used to profile host enzyme activity during Dengue infection (Vetter et al. 2012) and applying them to profile the changes in activity of various ATPases during IAV infection could yield new information with regards to how enzymes are recruited and utilised during viral egress.

Lipid regulatory systems, which play an important role in earlier stages of viral life cycles, continue to play an important role in viral egress. HCV provides a good example of this, as its egress relies on the very-low-density-lipoparticle (VLDL) secretory pathway to release the mature virus particle into the bloodstream (Neufeldt et al. 2018). Certain VLDL-associated enzymes, such as the triglyceride-recruiting enzyme ABHD5 or the phospholipid synthesis enzyme LPCAT1, have been identified as pro- or antiviral HCV host factors. However, the interactions between these enzymes and HCV are not fully understood (Vieyres et al. 2016; Beilstein et al. 2017). Profiling enzymes which regulate this pathway could therefore reveal novel information clarifying how it is hijacked by HCV. Even viruses such as influenza, which do not rely on a lipid secretory pathway, still alter the cell's lipid profile in order to exit into the bloodstream (Pohl et al. 2016). Influenza virus egress occurs by budding from the plasma membrane, a process requiring an altered membrane lipid composition (Pohl et al. 2016). Profiling changes to lipid regulatory networks during the assembly and egress of virions could therefore yield new information and improve our understanding of host–virus interactions during late stages of infection.

In general, activity-based profiling on these later stages of viral infection has the potential to yield novel host factors requirements for productive infections. Targeting the secretion of new virions could play an important role within combinatorial antiviral strategies to limit the spread of viral infections.

7 Selective Labelling of Individual Viral Enzymes

Selective labelling of viral enzymes enables the measurement of their activity within a complex proteome. While viral enzymes can be labelled by broadly reactive probes such as fluorophosphonate or HAUb-VME, as discussed in the preceding sections, it can be difficult to detect them amongst the more numerous host cell enzymes which are labelled simultaneously (Blais et al. 2010a). For this reason, specific and selective probes have been developed against multiple viral enzymes.

The infectious particle of the influenza A virus (IAV), the most common cause of the flu, incorporates two viral enzymes on its surface, haemagglutinin (HA) and neuraminidase (NA) (Wohlbold and Krammer 2014). NA is known to play many roles; the most well-established being its sialidase activity which cleaves sialic acid from the nascent virions following virion secretion (Wohlbold and Krammer 2014). NA enzymatic activity has also been suggested to play a role in viral entry, though the mechanism by which this could occur has not been well established (Su et al. 2009; Wohlbold and Krammer 2014). The NA active site is highly conserved between strains of IAV strains, making NA an attractive target for chemical probes as well as therapeutic drugs (Wohlbold and Krammer 2014).

In 2005, Lu et al. reported the synthesis of a mechanism-based neuraminidase probe, consisting of a sialic acid targeting moiety and an *ortho*-difluoromethyl phenyl warhead attached to a biotin reporter tag (Lu et al. 2005). They demonstrated that this probe was able to specifically label neuraminidases and could be used to detect neuraminidase activity. Of particular interest in this study was a demonstration of a more uncommon application of activity-based probes for isolation of viral particles. Immobilised probe bound to NA enzymes present on the particle surface, thereby enabling the isolation of the entire virion (Lu et al. 2005). While this probe was able a covalently label IAV NA, there were several limitations. Its large size and multiple rings necessitated millimolar concentrations to label even purified protein. The probe was also unable to penetrate cell membranes, which precluded the application of this probe in complex systems. Lastly, though the sialic acid did convey some selectivity, the difluoromethyl phenyl warhead nevertheless reacted non-specifically with multiple targets in a complex environment (Tsai et al. 2013).

A more streamlined probe was reported by Tsai et al. Instead of the inclusion of a separate warhead group in addition to the sialic acid moiety, the sialic acid sugar was modified to include fluorine at C2 and C3, displacing a hydroxyl group and a hydrogen, respectively. The electron-withdrawing property of the fluorines transforms the probe into an irreversible mechanism-based inhibitor. Two probes were made: DFSA and an ester-protected probe, PDFSA, both containing the sialic acid warhead and an alkyne handle (Fig. 3e) (Tsai et al. 2013). PDFSA is able to label NA in situ at micromolar concentrations, allowing imaging of sub-cellular localisation of the viral enzyme during infection (Tsai et al. 2013). PDFSA was able to label wild-type NA as well as mutant NA resistant to active-site antivirals, indicating that this probe could be applied to screen for the presence of drug-resistant strains by competitive ABPP (Tsai et al. 2013).

The Zika virus has recently been the subject of international attention following the South American outbreak which caused a dramatic increase in the rates of microencephaly in newborn infants (Lei et al. 2016). The Zika virus, like most positive-strand RNA viruses, encodes a serine protease, NS3, which cleaves the precursor polyprotein into the mature viral proteins. As activity of the NS3 protease is essential to the establishment of a productive infection, the development of tools capable of characterising its activity and regulation is of great interest.

Proteases are capable of displaying high selectivity for the specific amino acid sequences they cleave. While it is possible to design activity-based probes based on these sequences, these probes risk lacking specificity to individual proteases due to the overlapping substrate specificity of closely related enzymes. Recently, this problem was addressed by using unnatural amino acids to expand the chemical space of their peptides' building blocks. The Hybrid Combinatorial Substrate Library (HyCoSuL) uses 102 unnatural amino acids in addition to the twenty canonical ones to build optimal substrates containing a variable 4-amino acid sequence targeting the S1–S4 pockets of the protease active site (Kasperkiewicz et al. 2014). To allow specific activity-based labelling, a fluorogenic coumarin and a diphenyl phosphate warhead are added (Kasperkiewicz et al. 2014). This library has previously been used to design probes targeting human proteases (Kasperkiewicz et al. 2015; Poreba et al. 2016, 2018). Recently, the HyCoSuL methodology was applied to design a probe, WRPK3, which rapidly labels Zika's NS3 protease at low nanomolar concentrations (Fig. 3f) (Rut et al. 2017). This probe has the potential to be applied to interrogate NS3 activity during Zika infection. These studies demonstrate a useful application of unnatural amino acids which has the potential to be used to design activity-based probes against other viral proteases.

8 Future Objectives

The majority of the activity-based profiling reported in this chapter was performed either in vitro on proteomes isolated from animal models or cultured cells, or in situ on live cell monocultures. While this has been able to provide information on targets of viral mediation of host cell systems, the physiological relevance of this information has been subject to some doubt. Cell lysis and protein isolation procedures used in in vitro labelling have the potential to destroy enzyme-cofactor interactions which may result in loss of information about the enzyme's true activity in vivo. Labelling cells in situ preserves the regulation of catalytic activity present in living cells; however, they do not always accurately represent the effects of infection on a live organism. As such, the development of probes which could be administered in vivo would provide the opportunity of characterise changes to enzyme activity in a more complex and physiologically relevant manner.

9 Conclusions

Over the past decade, activity-based protein profiling has proven to be a useful chemical proteomics tool for the interrogation of the role enzymes play in viral infection, providing information not otherwise accessible by traditional genomic and proteomic methods. ABPP has been used to quantify enzymatic activity, detect changes in the localisation of active enzymes, identify protein–protein interactions regulating activity and assign novel functionalities to proteins playing important roles in viral infection. When considered as a part of a larger body of research, they have greatly contributed to our understanding of viral infection and host response.

References

Balla T (2013) Phosphoinositides: tiny lipids with giant impact on cell regulation. Physiol Rev 93:1019–1137

Barber GN (2001) Host defense, viruses and apoptosis. Cell Death Differ 8:113–126. https://doi.org/10.1038/sj.cdd.4400823

Barglow KT, Cravatt BF (2007) Activity-based protein profiling for the functional annotation of enzymes. Nat Methods 4:822–827. https://doi.org/10.1038/nmeth1092

Beilstein F, Lemasson M, Pène V et al (2017) Lysophosphatidylcholine acyltransferase 1 is downregulated by hepatitis C virus: impact on production of lipo-viro-particles. Gut 66:2160–2169. https://doi.org/10.1136/gutjnl-2016-311508

Békés M, van der Heden van Noort GJ, Ekkebus R et al (2016) Recognition of Lys48-Linked Di-ubiquitin and Deubiquitinating Activities of the SARS Coronavirus Papain-like Protease. Mol Cell 62:572–585. https://doi.org/10.1016/j.molcel.2016.04.016

Benes P, Vetvicka V, Fusek M (2008) Cathepsin D—many functions of one aspartic protease. Crit Rev Oncol Hematol 68:12–28. https://doi.org/10.1016/j.critrevonc.2008.02.008

Bishé B, Syed G, Siddiqui A (2012) Phosphoinositides in the hepatitis C virus life cycle. Viruses 4:2340–2358. https://doi.org/10.3390/v4102340

Blais DR, Brûlotte M, Qian Y et al (2010a) Activity-based proteome profiling of hepatoma cells during hepatitis C virus replication using protease substrate probes. J Proteome Res 9:912–923. https://doi.org/10.1021/pr900788a

Blais DR, Lyn RK, Joyce MA et al (2010b) Activity-based protein profiling identifies a host enzyme, carboxylesterase 1, which is differentially active during hepatitis C virus replication. J Biol Chem 285:25602–25612. https://doi.org/10.1074/jbc.M110.135483

Borodovsky A, Ovaa H, Kolli N et al (2002) Chemistry-based functional proteomics reveals novel members of the deubiquitinating enzyme family. Chem Biol 9:1149–1159

Bosch BJ, Bartelink W, Rottier PJM (2008) Cathepsin L functionally cleaves the severe acute respiratory syndrome coronavirus class I fusion protein upstream of rather than adjacent to the fusion peptide. J Virol 82:8887–8890. https://doi.org/10.1128/JVI.00415-08

Burster T, Macmillan H, Hou T et al (2010) Cathepsin G: roles in antigen presentation and beyond. Mol Immunol 47:658–665. https://doi.org/10.1016/j.molimm.2009.10.003

Chandran K, Sullivan NJ, Felbor U et al (2005) Endosomal proteolysis of the Ebola virus glycoprotein is necessary for infection. Science 308:1643–1645. https://doi.org/10.1126/science.1110656

Coux O, Tanaka K, Goldberg AL (1996) Structure and functions of the 20S and 26S proteasomes. Annu Rev Biochem 65:801–847. https://doi.org/10.1146/annurev.bi.65.070196.004101

Cravatt BF, Wright AT, Kozarich JW (2008) Activity-based protein profiling: from enzyme chemistry to proteomic chemistry. Annu Rev Biochem 77:383–414. https://doi.org/10.1146/annurev.biochem.75.101304.124125

Cruz-Muñoz ME, Fuentes-Pananá EM (2017) Beta and gamma human herpesviruses: agonistic and antagonistic interactions with the host immune system. Front Microbiol 8:2521. https://doi.org/10.3389/fmicb.2017.02521

de Castro S, Camarasa M-J (2018) Polypharmacology in HIV inhibition: can a drug with simultaneous action against two relevant targets be an alternative to combination therapy? Eur J Med Chem 150:206–227. https://doi.org/10.1016/J.EJMECH.2018.03.007

De Clercq E, Li G (2016) Approved antiviral drugs over the past 50 years. Clin Microbiol Rev 29:695–747. https://doi.org/10.1128/CMR.00102-15

de Graaf N, van Helden MJG, Textoris-Taube K et al (2011) PA28 and the proteasome immunosubunits play a central and independent role in the production of MHC class I-binding peptides in vivo. Eur J Immunol 41:926–935. https://doi.org/10.1002/eji.201041040

Delang L, Paeshuyse J, Neyts J (2012) The role of phosphatidylinositol 4-kinases and phosphatidylinositol 4-phosphate during viral replication. Biochem Pharmacol 84:1400–1408. https://doi.org/10.1016/j.bcp.2012.07.034

Desrochers GF, Cornacchia C, McKay CS, Pezacki JP (2018) Activity-based phosphatidylinositol kinase probes detect changes to protein–protein interactions during hepatitis C virus replication. ACS Infect Dis. https://doi.org/10.1021/acsinfecdis.8b00047

Desrochers GF, Sherratt AR, Blais DR et al (2015) Profiling kinase activity during hepatitis C virus replication using a wortmannin probe. ACS Infect Dis 1:443–452. https://doi.org/10.1021/acsinfecdis.5b00083

Ferrington DA, Gregerson DS (2012) Immunoproteasomes. In: Progress in molecular biology and translational science, pp 75–112

Furman LM, Maaty WS, Petersen LK et al (2009) Cysteine protease activation and apoptosis in Murine norovirus infection. Virol J 6:139. https://doi.org/10.1186/1743-422X-6-139

Goodwin CM, Xu S, Munger J (2015) Stealing the keys to the kitchen: viral manipulation of the host cell metabolic network. Trends Microbiol 23:789–798. https://doi.org/10.1016/J.TIM.2015.08.007

Gredmark S, Schlieker C, Quesada V et al (2007) A functional ubiquitin-specific protease embedded in the large tegument protein (ORF64) of murine gammaherpesvirus 68 is active during the course of infection. J Virol 81:10300–10309. https://doi.org/10.1128/JVI.01149-07

Greenbaum D, Medzihradszky KF, Burlingame A, Bogyo M (2000) Epoxide electrophiles as activity-dependent cysteine protease profiling and discovery tools. Chem Biol 7:569–581

Harak C, Lohmann V (2015) Ultrastructure of the replication sites of positive-strand RNA viruses. Virology 479–480:418–433. https://doi.org/10.1016/J.VIROL.2015.02.029

Harhaj EW, Dixit VM (2011) Deubiquitinases in the regulation of NF-κB signaling. Cell Res 21:22–39. https://doi.org/10.1038/cr.2010.166

Hausser A, Link G, Hoene M et al (2006) Phospho-specific binding of 14-3-3 proteins to phosphatidylinositol 4-kinase III β protects from dephosphorylation and stabilizes lipid kinase activity. J Cell Sci 119:3613–3621

Hausser A, Storz P, Märtens S et al (2005) Protein kinase D regulates vesicular transport by phosphorylating and activating phosphatidylinositol-4 kinase III β at the Golgi complex. Nat Cell Biol 7:880–887. https://doi.org/10.1038/ncb1289

Hogue IB, Bosse JB, Hu J-R et al (2014) Cellular mechanisms of alpha herpesvirus egress: live cell fluorescence microscopy of pseudorabies virus exocytosis. PLoS Pathog 10:e1004535. https://doi.org/10.1371/journal.ppat.1004535

Hsu N-Y, Ilnytska O, Belov G et al (2010) Viral reorganization of the secretory pathway generates distinct organelles for RNA replication. Cell 141:799–811. https://doi.org/10.1016/j.cell.2010.03.050

Hutagalung AH, Novick PJ (2011) Role of Rab GTPases in membrane traffic and cell physiology. Physiol Rev 91:119–149. https://doi.org/10.1152/physrev.00059.2009

Jarosinski K, Kattenhorn L, Kaufer B et al (2007) A herpesvirus ubiquitin-specific protease is critical for efficient T cell lymphoma formation. Proc Natl Acad Sci 104:20025–20030. https://doi.org/10.1073/pnas.0706295104

Kammerl IE, Meiners S (2016) Proteasome function shapes innate and adaptive immune responses. Am J Physiol Lung Cell Mol Physiol 311:L328–L336. https://doi.org/10.1152/ajplung.00156.2016

Kapadia SB, Chisari FV (2005) Hepatitis C virus RNA replication is regulated by host geranylgeranylation and fatty acids. Proc Natl Acad Sci U S A 102:2561–2566. https://doi.org/10.1073/pnas.0409834102

Kasperkiewicz P, Poreba M, Snipas SJ et al (2014) Design of ultrasensitive probes for human neutrophil elastase through hybrid combinatorial substrate library profiling. Proc Natl Acad Sci 111:2518–2523. https://doi.org/10.1073/pnas.1318548111

Kasperkiewicz P, Poreba M, Snipas SJ et al (2015) Design of a selective substrate and activity based probe for human neutrophil serine protease 4. PLoS ONE 10:e0132818. https://doi.org/10.1371/journal.pone.0132818

Kattenhorn LM, Korbel GA, Kessler BM et al (2005) A deubiquitinating enzyme encoded by HSV-1 belongs to a family of cysteine proteases that is conserved across the family herpesviridae. Mol Cell 19:547–557. https://doi.org/10.1016/j.molcel.2005.07.003

Keller IE, Vosyka O, Takenaka S et al (2015) Regulation of immunoproteasome function in the lung. Sci Rep 5:10230. https://doi.org/10.1038/srep10230

Lei J, Hansen G, Nitsche C et al (2016) Crystal structure of Zika virus NS2B-NS3 protease in complex with a boronate inhibitor. Science 353:503–505. https://doi.org/10.1126/science.aag2419

Li N, Kuo C-L, Paniagua G et al (2013) Relative quantification of proteasome activity by activity-based protein profiling and LC-MS/MS. Nat Protoc 8:1155–1168. https://doi.org/10.1038/nprot.2013.065

Liang TJ (2009) Hepatitis B: the virus and disease. Hepatology 49:S13–S21. https://doi.org/10.1002/hep.22881

Lohmann V, Bartenschlager R (2014) On the history of hepatitis C virus cell culture systems. J Med Chem 57:1627–1642. https://doi.org/10.1021/jm401401n

Lu C-P, Ren C-T, Lai Y-N et al (2005) Design of a mechanism-based probe for neuraminidase to capture influenza viruses. Angew Chemie Int Ed 44:6888–6892. https://doi.org/10.1002/anie.200501738

Macdonald SJ, Mostafa HH, Morrison LA, Davido DJ (2012) Genome sequence of herpes simplex virus 1 strain KOS. J Virol 86:6371–6372. https://doi.org/10.1128/JVI.00646-12

Marques AJ, Palanimurugan R, Matias AC et al (2009) Catalytic mechanism and assembly of the proteasome. Chem Rev 109:1509–1536. https://doi.org/10.1021/cr8004857

Mazein A, Watterson S, Hsieh W-Y et al (2013) A comprehensive machine-readable view of the mammalian cholesterol biosynthesis pathway. Biochem Pharmacol 86:56–66. https://doi.org/10.1016/J.BCP.2013.03.021

Mazumdar T, Gorgun FM, Sha Y et al (2010) Regulation of NF-kappaB activity and inducible nitric oxide synthase by regulatory particle non-ATPase subunit 13 (Rpn13). Proc Natl Acad Sci U S A 107:13854–13859. https://doi.org/10.1073/pnas.0913495107

McMahon HT, Gallop JL (2005) Membrane curvature and mechanisms of dynamic cell membrane remodelling. Nature 438:590–596. https://doi.org/10.1038/nature04396

Mulder MPC, El Oualid F, ter Beek J, Ovaa H (2014) A native chemical ligation handle that enables the synthesis of advanced activity-based probes: diubiquitin as a case study. ChemBioChem 15:946–949. https://doi.org/10.1002/cbic.201402012

Nanduri B, Suvarnapunya AE, Venkatesan M, Edelmann MJ (2013) Deubiquitinating enzymes as promising drug targets for infectious diseases. Curr Pharm Des 19:3234–3247

Nasheri N, Joyce M, Rouleau Y et al (2013) Modulation of fatty acid synthase enzyme activity and expression during hepatitis C virus replication. Chem Biol 20:570–582. https://doi.org/10.1016/j.chembiol.2013.03.014

Nasheri N, McKay CS, Fulton K et al (2014) Hydrophobic triaryl-substituted β-lactams as activity-based probes for profiling eukaryotic enzymes and host-pathogen interactions. ChemBioChem 15:2195–2200. https://doi.org/10.1002/cbic.201402097

Nasheri N, Ning Z, Figeys D et al (2015) Activity-based profiling of the proteasome pathway during hepatitis C virus infection. Proteomics 15:3815–3825. https://doi.org/10.1002/pmic.201500169

Neufeldt CJ, Cortese M, Acosta EG, Bartenschlager R (2018) Rewiring cellular networks by members of the flaviviridae family. Nat Rev Microbiol 16:125–142. https://doi.org/10.1038/nrmicro.2017.170

Nourbakhsh M, Douglas DN, Pu CH et al (2013) Arylacetamide deacetylase: a novel host factor with important roles in the lipolysis of cellular triacylglycerol stores, VLDL assembly and HCV production. J Hepatol 59:336–343. https://doi.org/10.1016/j.jhep.2013.03.022

Novoa RR, Calderita G, Arranz R et al (2005) Virus factories: associations of cell organelles for viral replication and morphogenesis. Biol Cell 97:147–172. https://doi.org/10.1042/BC20040058

Oeckinghaus A, Ghosh S (2009) The NF-κB family of transcription factors and its regulation. Cold Spring Harb Perspect Biol 1:a000034–a000034. https://doi.org/10.1101/cshperspect.a000034

Ovaa H, Kessler BM, Rolen U et al (2004) Activity-based ubiquitin-specific protease (USP) profiling of virus-infected and malignant human cells. Proc Natl Acad Sci 101:2253–2258. https://doi.org/10.1073/pnas.0308411100

Paul D, Bartenschlager R (2015) Flaviviridae replication organelles: oh, what a tangled web we weave. Annu Rev Virol 2:289–310. https://doi.org/10.1146/annurev-virology-100114-055007

Pohl MO, Lanz C, Stertz S (2016) Late stages of the influenza a virus replication cycle—a tight interplay between virus and host. J Gen Virol 97:2058–2072. https://doi.org/10.1099/jgv.0.000562

Poreba M, Rut W, Vizovisek M et al (2018) Selective imaging of cathepsin L in breast cancer by fluorescent activity-based probes. Chem Sci 9:2113–2129. https://doi.org/10.1039/C7SC04303A

Poreba M, Solberg R, Rut W et al (2016) Counter selection substrate library strategy for developing specific protease substrates and probes. Cell Chem Biol 23:1023–1035. https://doi.org/10.1016/j.chembiol.2016.05.020

Qian X-J, Zhu Y-Z, Zhao P, Qi Z-T (2016) Entry inhibitors: new advances in HCV treatment. Emerg Microbes Infect 5:e3. https://doi.org/10.1038/emi.2016.3

Ross MK, Streit TM, Herring KL (2010) Carboxylesterases: dual roles in lipid and pesticide metabolism. J Pestic Sci 35:257–264. https://doi.org/10.1584/jpestics.R10-07

Rut W, Zhang L, Kasperkiewicz P et al (2017) Extended substrate specificity and first potent irreversible inhibitor/activity-based probe design for Zika virus NS2B-NS3 protease. Antiviral Res 139:88–94. https://doi.org/10.1016/j.antiviral.2016.12.018

Saito S, Murata T, Kanda T et al (2013) Epstein-barr virus deubiquitinase downregulates TRAF6-mediated NF-κB signaling during productive replication. J Virol 87:4060–4070. https://doi.org/10.1128/JVI.02020-12

Schlieker C, Korbel GA, Kattenhorn LM, Ploegh HL (2005) A deubiquitinating activity is conserved in the large tegument protein of the herpesviridae. J Virol 79:15582–15585. https://doi.org/10.1128/JVI.79.24.15582-15585.2005
Schuster B, Hendry L, Byers H et al (2007) Purification and identification of the STAT5 protease in myeloid cells. Biochem J 404:81–87. https://doi.org/10.1042/BJ20061877
Shah PP, Wang T, Kaletsky RL et al (2010) A small-molecule oxocarbazate inhibitor of human cathepsin L blocks severe acute respiratory syndrome and ebola pseudotype virus infection into human embryonic kidney 293T cells. Mol Pharmacol 78:319–324. https://doi.org/10.1124/mol.110.064261
Shahiduzzaman M, Ezatti P, Xin G, Coombs KM (2014) Proteasomal serine hydrolases are up-regulated by and required for influenza virus infection. J Proteome Res 13:2223–2238. https://doi.org/10.1021/pr5001779
Sherratt AR, Nasheri N, McKay CS et al (2014) A new chemical probe for phosphatidylinositol kinase activity. ChemBioChem 15:1253–1256. https://doi.org/10.1002/cbic.201402155
Singaravelu R, Blais DR, McKay CS, Pezacki J (2010) Activity-based protein profiling of the hepatitis C virus replication in Huh-7 hepatoma cells using a non-directed active site probe. Proteome Sci 8:5. https://doi.org/10.1186/1477-5956-8-5
Su AI, Pezacki JP, Wodicka L et al (2002) Nonlinear partial differential equations and applications: genomic analysis of the host response to hepatitis C virus infection. Proc Natl Acad Sci 99:15669–15674. https://doi.org/10.1073/pnas.202608199
Su B, Wurtzer S, Rameix-Welti M-A et al (2009) Enhancement of the influenza A hemagglutinin (HA)-mediated cell-cell fusion and virus entry by the viral neuraminidase (NA). PLoS ONE 4: e8495. https://doi.org/10.1371/journal.pone.0008495
Sun D, Zhu L, Yao D et al (2018) Recent progress in potential anti-hepatitis B virus agents: structural and pharmacological perspectives. Eur J Med Chem 147:205–217. https://doi.org/10.1016/j.ejmech.2018.02.001
Syed GH, Amako Y, Siddiqui A (2010) Hepatitis C virus hijacks host lipid metabolism. Trends Endocrinol Metab 21:33–40. https://doi.org/10.1016/j.tem.2009.07.005
Syed GH, Khan M, Yang S, Siddiqui A (2017) Hepatitis C virus lipoviroparticles assemble in the endoplasmic reticulum (ER) and bud off from the ER to the golgi compartment in COPII vesicles. J Virol 91:e00499–17. https://doi.org/10.1128/JVI.00499-17
Takacs CN, Andreo U, Dao Thi VL et al (2017) Differential regulation of lipoprotein and hepatitis C virus secretion by Rab1b. Cell Rep 21:431–441. https://doi.org/10.1016/J.CELREP.2017.09.053
Tsai C-S, Yen H-Y, Lin M-I et al (2013) Cell-permeable probe for identification and imaging of sialidases. Proc Natl Acad Sci U S A 110:2466–2471. https://doi.org/10.1073/pnas.1222183110
van Gent M, Braem SGE, de Jong A et al (2014) Epstein-barr virus large tegument protein BPLF1 contributes to innate immune evasion through interference with toll-like receptor signaling. PLoS Pathog 10:e1003960. https://doi.org/10.1371/journal.ppat.1003960
Verdoes M, Florea BI, Menendez-Benito V et al (2006) A fluorescent broad-spectrum proteasome inhibitor for labeling proteasomes in vitro and in vivo. Chem Biol 13:1217–1226. https://doi.org/10.1016/J.CHEMBIOL.2006.09.013
Vetter ML, Rodgers MA, Patricelli MP, Yang PL (2012) chemoproteomic profiling identifies changes in DNA-PK as markers of early dengue virus infection. ACS Chem Biol 7:2019–2026. https://doi.org/10.1021/cb300420z
Vieyres G, Welsch K, Gerold G et al (2016) ABHD5/CGI-58, the chanarin-dorfman syndrome protein, mobilises lipid stores for hepatitis C virus production. PLoS Pathog 12:e1005568. https://doi.org/10.1371/journal.ppat.1005568
Wang J, Loveland AN, Kattenhorn LM et al (2006) High-molecular-weight protein (pUL48) of human cytomegalovirus is a competent deubiquitinating protease: mutant viruses altered in its active-site cysteine or histidine are viable. J Virol 80:6003–6012. https://doi.org/10.1128/JVI.00401-06

Wertz IE, Dixit VM (2010) Signaling to NF-κB: regulation by ubiquitination. Cold Spring Harb Perspect Biol 2:a003350–a003350. https://doi.org/10.1101/cshperspect.a003350
Wohlbold T, Krammer F (2014) In the shadow of hemagglutinin: a growing interest in influenza viral neuraminidase and its role as a vaccine antigen. Viruses 6:2465–2494. https://doi.org/10.3390/v6062465
Xiao F, Fofana I, Thumann C et al (2015) Synergy of entry inhibitors with direct-acting antivirals uncovers novel combinations for prevention and treatment of hepatitis C. Gut 64:483–494. https://doi.org/10.1136/gutjnl-2013-306155
Yang P-Y, Liu K, Ngai MH et al (2010) Activity-based proteome profiling of potential cellular targets of orlistat—an FDA-approved drug with anti-tumor activities. J Am Chem Soc 132:656–666. https://doi.org/10.1021/ja907716f
Ye R, Su C, Xu H, Zheng C (2017) Herpes simplex virus 1 ubiquitin-specific protease UL36 abrogates NF-κB activation in DNA sensing signal pathway. J Virol 91:e02417–16. https://doi.org/10.1128/JVI.02417-16
Zhang L, Hong Z, Lin W et al (2012) ARF1 and GBF1 generate a PI4P-enriched environment supportive of hepatitis C virus replication. PLoS ONE 7:e32135. https://doi.org/10.1371/journal.pone.0032135
Zhao B, Song J, St. Clair RW, Ghosh S (2007) Stable overexpression of human macrophage cholesteryl ester hydrolase results in enhanced free cholesterol efflux from human THP1 macrophages. Am J Physiol Physiol 292:C405–C412. https://doi.org/10.1152/ajpcell.00306.2006
Zou F, Schmon M, Sienczyk M et al (2012) Application of a novel highly sensitive activity-based probe for detection of cathepsin G. Anal Biochem 421:667–672. https://doi.org/10.1016/j.ab.2011.11.016

Activity-Based Protein Profiling for the Study of Parasite Biology

Henry J. Benns, Edward W. Tate and Matthew A. Child

Contents

Abstract Parasites exist within most ecological niches, often transitioning through biologically and chemically complex host environments over the course of their parasitic life cycles. While the development of technologies for genetic engineering has revolutionised the field of functional genomics, parasites have historically been less amenable to such modification. In light of this, parasitologists have often been at the forefront of adopting new small-molecule technologies, repurposing drugs into biological tools and probes. Over the last decade, activity-based protein profiling (ABPP) has evolved into a powerful and versatile chemical proteomic platform for characterising the function of enzymes. Central to ABPP is the use of activity-based probes (ABPs), which covalently modify the active sites of enzyme classes ranging from serine hydrolases to glycosidases. The application of ABPP to cellular systems has contributed vastly to our knowledge on the fundamental biology of a diverse range of organisms and has facilitated the identification of potential drug targets in many pathogens. In this chapter, we provide a

H. J. Benns · E. W. Tate
Department of Chemistry, Imperial College London, Exhibition Road, South Kensington, London SW7 2AZ, UK

M. A. Child (✉)
Life Sciences, Imperial College London, Exhibition Road, South Kensington, London SW7 2AZ, UK
e-mail: m.child@imperial.ac.uk

Current Topics in Microbiology and Immunology (2019) 420:155–174
DOI 10.1007/82_2018_123

Published Online: 14 August 2018

comprehensive review on the different forms of ABPP that have been successfully applied to parasite systems, and highlight key biological insights that have been enabled through their application.

1 Background

Parasites comprise a large biologically diverse group of eukaryotes that exploit the resources of a host organism to facilitate their own survival and propagation. Broadly, parasites can be taxonomically described as being either microparasitic unicellular protozoa (Kingdom: *Protista*), or complex macroparasitic multicellular helminths and ectoparasites [*Animalia*; (Simner 2017)]. The life cycles of these parasites display remarkable diversity in form and complexity, infecting different host species with a range of host-pathogen interactions, transmission routes and tissue tropism. For instance, while some species exhibit strong preference for specific host cell types, others are more generalist, or exist extracellularly (Fig. 1) (McCall et al. 2016).

Human parasitic infections are a leading cause of morbidity and mortality worldwide. Correspondingly, they represent a major global health and economic burden. In the latest 'Global Burden of Disease' study, parasite-related diseases were estimated to account for over 800,000 deaths in 2016, with most being attributed to protozoan infections (G. B. D. Causes of Death Collaborators 2017). By far the most prevalent of these protozoan infections are those caused by members of the phylum *Apicomplexa,* such as *Plasmodium* spp. and *Toxoplasma gondii. Plasmodium* parasites are responsible for malaria, a devastating mosquito-borne disease that threatens around half of the global population (3.2 billion) (World Health Organization 2017). In 2016, 446,000 malaria-related deaths in 91 countries were reported by the World Health Organisation (WHO), with sub-Saharan Africa carrying 90% of the burden (World Health Organization 2017). Regarded as the most successful parasite on the planet, *T. gondii* has the capacity to infect any vertebrate host, and seroprevalence studies indicate that up to 50% of the human population

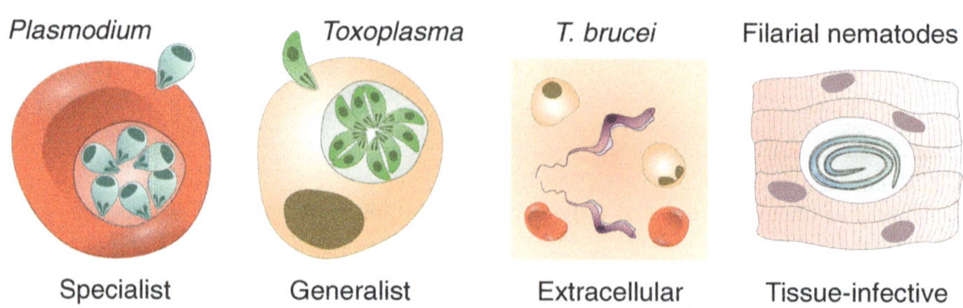

Fig. 1 Major types of infection exhibited by human-infective parasites. Representative parasite species shown are known to infect distinct cell types ('specialists'), a broad range of cell types ('generalists'), reside in extracellular environments (e.g. the bloodstream/lymph), or within tissues

have been infected at some point in their lives (Flegr et al. 2014). *T. gondii* is the aetiological agent of toxoplasmosis, a disease which can lead to life-threatening encephalitis if left untreated. *T. gondii* is also a major veterinary pathogen responsible for the annual loss of over 500,000 lambs, costing the UK sheep industry £12–24 million (Advisory Committee on the Microbiological Safety of Food 2012). Other medically important protozoan parasites include the trypanosomatids, a group of flagellated parasites transmitted by insect vectors and responsible for an array of neglected tropical diseases. Three trypanosomatids of particular concern are *Leishmania* spp., *Trypanosoma brucei* and *Trypanosoma cruzi,* the respective agents of the leishmaniases, human African trypanosomiasis (HAT) and Chagas disease. Infections involving parasitic worms (helminths) are another major contributor to parasite-related deaths. In fact, schistosomiasis caused by trematodes of the *Schistosoma* genus ranks as the second-most-common parasitic disease and claims up to 200,000 lives each year (World Health Organization 2017).

One of the greatest challenges in combating parasitic infection is the development of safe and efficacious anti-parasitic drugs. While chemotherapeutics have been effective in reducing the burden of parasite-related disease, their use has been limited by toxicity, prolonged treatment regimens and/or the emergence of drug-resistant strains (Monzote and Siddiq 2011). For example, resistance to the frontline antimalarial artemisinin has been identified in *Plasmodium falciparum* isolates from across Southeast Asia, compromising the efficacy of well-established artemisinin combination therapies (Haldar et al. 2018). Additionally, drug-resistant *L. donovani* strains in India have limited the effectiveness of pentavalent antimonial compounds for the treatment of visceral leishmaniasis (Croft et al. 2006). On this basis, there is a continuing need to investigate parasite biology and identify novel drug targets for which new efficacious therapeutics can be developed.

While genetic engineering approaches have been powerful for dissecting the basic biology of prokaryotic systems, eukaryotic parasites have historically been less amenable to such modification. For instance, some species exhibit poor transfection efficiencies, or reduced homologous recombination rates due to AT-rich genomes (as is the case with *P. falciparum*; (Suarez et al. 2017). As an alternative to genetic approaches, small-molecule-based chemical biology techniques have provided significant insight into the molecular basis of many essential parasite processes. Activity-based protein profiling (ABPP) has emerged as one particularly versatile strategy for characterising enzyme function on a global scale and has facilitated rapid identification of novel druggable nodes in a range of biological systems, including parasites (Cravatt et al. 2008). Central to this approach is the use of activity-based probes (ABPs). ABPs are small molecules functionalised with fluorescent reporter or affinity-based tags that can be used to profile the activity of specific enzymes classes or residues via covalent interaction with their target(s) (Cravatt et al. 2008). This chapter aims to provide an overview of the diverse applications of ABPP that have advanced our understanding of parasite biology over the last 10 years. For the purpose of clarity, we will focus on a selection of the major biological discoveries that have been made using ABPP, rather than providing a detailed account of all available literature.

2 Activity-Based Probes for Target Discovery in Parasites

2.1 *Profiling Protease Activity*

Perhaps the most widespread use of ABPP in parasites has been for the study of proteases. Proteases are enzymes that catalyse the proteolytic processing of protein substrates via the hydrolysis of peptide bonds (López-Otín and Bond 2008). Based on their catalytic residues or mechanism, proteases are categorised as cysteine, serine, threonine, aspartyl and metalloproteases. Together, they are responsible for a diversity of molecular functions including the regulation of protein activity and localisation, modulation of protein–protein interactions (PPIs) and generation of new active biomolecules (López-Otín and Bond 2008). In parasites, proteases display remarkable functional diversity in order to carry out essential processes such as immune modulation and host cell invasion (McKerrow et al. 2006) with much of our understanding of their contributions to these processes coming through the application of protease-directed ABPs.

2.1.1 Cysteine Proteases

One of the first successful applications of parasite-based ABPP includes the identification and characterisation of the *P. falciparum* papain family of cysteine proteases known as the falcipains (Greenbaum et al. 2002). In two separate studies, it was demonstrated that treatment of intraerythrocytic *P. falciparum* parasites with E-64 inhibits parasite egress from red blood cells (RBCs) and induces the enlargement of parasite food vacuoles (Greenbaum et al. 2002; Salmon et al. 2001). E-64 is an epoxide-containing natural product derived from fungi (*Aspergillus* spp.) and a known covalent inhibitor of clan CA cysteine proteases. Given the clear potential for cysteine proteases to be antimalarial drug targets, a series of E-64-based ABPs were synthesised to profile the proteins associated with the egress and enlarged food vacuole phenotypes (Chandramohanadas et al. 2009; Greenbaum et al. 2000, 2002). Using a biotinylated E-64 derivative, DCG-04 (Fig. 2a), four proteases were identified and found to have distinct activities in the different life cycle stages of the parasite (Greenbaum et al. 2002). For instance, falcipain-1 was shown to be upregulated during the invasive extracellular merozoite stage, highlighting a potential role for this protease in RBC invasion and/or rupture (Greenbaum et al. 2002). To validate a role for falcipain-1 in parasite invasion, Bogyo and colleagues used competitive ABPP of a broader epoxide-based peptide library, identifying several falcipain-1-specific inhibitors (Greenbaum et al. 2002). In this work, merozoite lysates were treated with the epoxides before profiling global cysteine protease activity with a radiolabelled version of DCG-04 (^{125}I-DCG-04; Fig. 2a). The most selective inhibitor, YA29-Eps(S, S), prevented merozoite invasion of erythrocytes without any discernible effects on other parasite processes, indicating that falcipain-1 activity is associated with invasion. Adopting

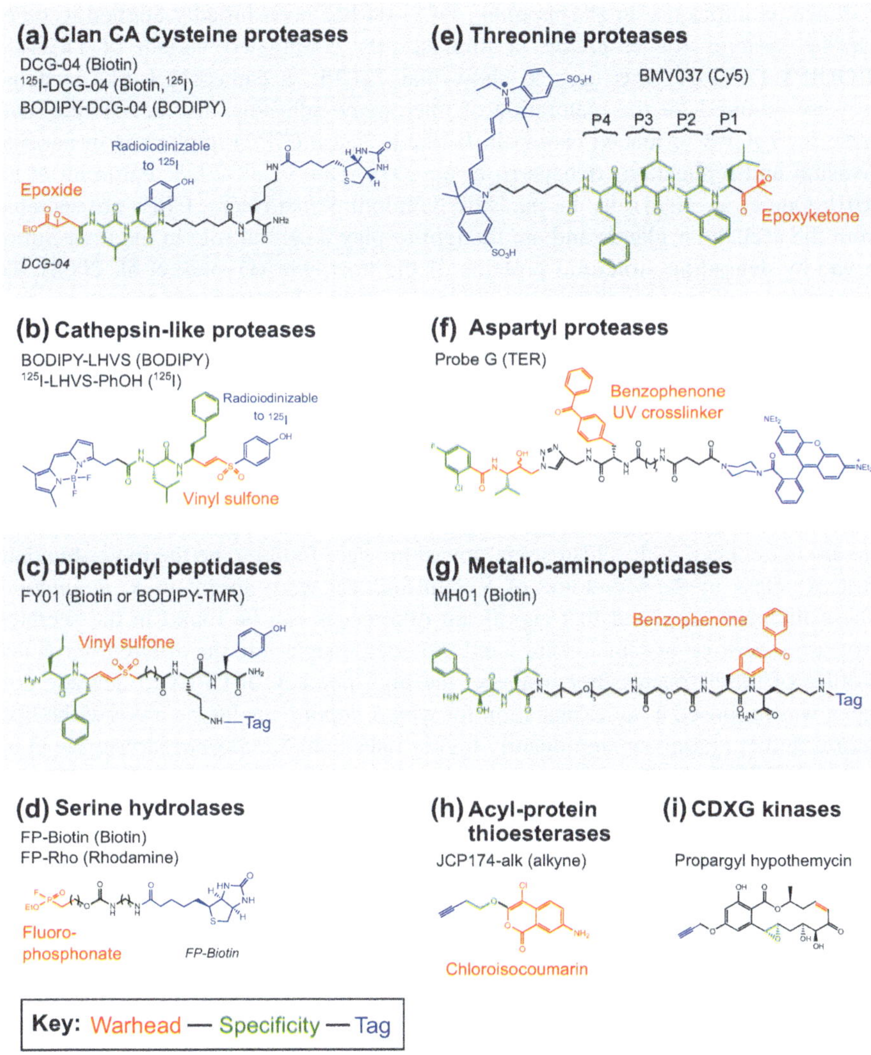

Fig. 2 Representative activity-based probes (ABPs) used to interrogate the function of major enzyme classes in parasites. ABPs have been used for the characterisation of cysteine (**a**–**c**), serine (**d**), threonine (**e**), aspartyl (**f**) and metalloproteases (**g**) of parasites, as well as acyl-protein thioesterases (**h**) and protein kinases that contain a CDXG motif (I). Specific probes are listed under the enzyme class they target, with their associated fluorescent reporter or affinity tag(s) bracketed

a similar ^{125}I-DCG-04 labelling approach, Eksi et al. (2004) later revealed that falcipain-1 is also upregulated in late-stage gametocytes. Here, genetic perturbation of falcipain-1 was shown to ablate oocyst production in the mosquito, suggesting it also plays a role in the sexual stage of the malaria life cycle.

Since its initial use in *Plasmodium*, DCG-04 has been broadly applied to other parasite systems. In *T. gondii*, a fluorescently conjugated variant of DCG-04 (BODIPY-DCG-04) was used to show that *Tg*CPL, a cathepsin L-like cysteine protease required for the maturation of microneme adhesins, also acts as the maturase for *Tg*CPB (a related cathepsin B-like protease (CPB) implicated in parasite invasion and replication) (Chaparro et al. 2018; Dou et al. 2013; Parussini et al. 2010; Que et al. 2004). In the parasitic helminth *Schistosoma*, CPBs are secreted from the acetabular glands and are thought to play a central role in the invasion of larvae by degrading structural proteins of the host skin (Dvořák et al. 2008). To characterise any species differences in terms of the abundance or complement of secreted proteases, Dvořák et al. (2008) used ^{125}I-DCG-04 to compare the activity of CPBs in the pathogenic secretomes of several *Schistosoma* species. Competitive ABPP using a selective inhibitor of CPB activity revealed that CPBs are much more abundant in the cercarial extracts of *Schistosoma japonica* than in *Schistosoma mansoni* (Dvořák et al. 2008). Fluorogenic peptide cleavage assays later demonstrated that the CPB activity in the secretions of *S. japonica* is 40-fold greater than *S. mansoni* (Dvořák et al. 2008). By contrast, serine proteases, which are also associated with schistosomal invasion, were found to be the most abundant protease class in the secretome of *S. mansoni* yet were absent in *S. japoncium.* These findings suggested that significant differences can be found in the secreted protease repertoire of schistosomes and provided insight into the different virulence profiles of these species. For instance, the high potency of CPBs to degrade collagen was proposed as an explanation for why *S. japonicum* larvae traverse the host dermis and epidermis at significantly higher rates than *S. mansoni* larvae (He et al. 2005). DCG-04 labelling was also used to identify 'cruzain' as the major active cysteine protease in the insect-form epimastigotes of *T. cruzi* (Doyle et al. 2011). Subsequent genetic knockout experiments demonstrated that cruzain inhibits the activation of infected macrophages by interrupting signalling pathways relying on NF-κB P65, demonstrating a previously unknown role for this protease in immune evasion (Doyle et al. 2011). This probe has been applied to *Cryptosporidium parvum,* the apicomplexan responsible for cryptosporidiosis, which further emphasises the broad utility of DCG-04 in less tractable parasite systems. *C. parvum* expresses a cysteine protease called cryptopain-1, which is essential for parasites survival in vivo. Through competitive inhibition studies with ^{125}I-DCG-04, the clan CA cysteine inhibitor K11777 was shown to both inhibit recombinant cryptopain-1 in vitro, and cryptopain-1 activity with negligible toxicity in vivo, highlighting cysteine protease inhibitors as potential anti-cryptosporidials (Ndao et al. 2013).

Peptide-based probes bearing vinyl sulfone (VS) reactive groups have also been valuable in delineating the function of parasite papain-family cysteine proteases. In a phenotypic screen using a series of known cysteine protease inhibitors, Teo et al. (2007) identified morpholinourea-leucyl-homophenolalaninyl-phenyl-vinylsulfone (LHVS) as an inhibitor of *T. gondii* host cell invasion, with a half-maximal inhibitory concentration (IC_{50}) of 10 μM. Subsequent biochemical and cell-based studies revealed that LHVS specifically impedes parasite attachment and gliding motility by blocking the secretion of adhesion proteins from specialised apical organelles known as micronemes (Teo et al. 2007). To identify the proteases targeted by LHVS and associated with this phenotype, Carruthers and colleagues performed competitive ABPP using a fluorescently tagged inhibitor derivative (BODIPY-LHVS; Fig. 2b). This study revealed a cathepsin P-like protease (*Tg*CPL) as the principle target of LHVS (Larson et al. 2009). Fluorescence microscopy of BODIPY-LHVS labelling in live extracellular parasites later showed that *Tg*CPL localises to two discrete structures at the apical end of *T. gondii*. While this pointed to a possible role in the proteolytic maturation of invasion-associated microneme proteins, the molecular function of *Tg*CPL remains unclear. Nevertheless, recent pharmacological studies have highlighted *Tg*CPL as a promising therapeutic target in chicken embryo models of acute toxoplasmosis (Chaparro et al. 2018). In similar studies, McKerrow and colleagues used a radioactively iodinated derivative of LHVS (^{125}I-LHVS-PhOH; Fig. 2b) to profile the activity of cysteine proteases in *T. brucei* bloodstream form parasites (Caffrey et al. 2001). Here, ^{125}I-LHVS-PhOH labelling of cell lysates identified brucipain (a cathepsin L-like protease) as an abundant protease in this life cycle stage, which was confirmed through competitive inhibition with two known trypanosomal cysteine protease inhibitors, Z-Phe-Ala-CHN_2 and *N*-Me-pip-Phe-homoPhe-VSPh. Follow-up genetic studies using RNA interference (RNAi) subsequently validated brucipain as an essential secreted virulence factor required for *T. brucei* traversal of the host blood-brain barrier (Abdulla et al. 2008; Nikolskaia et al. 2006).

In *P. falciparum*, an alternative VS-based probe, FY01 (Fig. 2c), was developed and used alongside DCG-04 to characterise a group of CPL proteases known as the dipeptidyl peptidases (DPAPs) (Arastu-Kapur et al. 2008). In this study, Bogyo and colleagues employed competitive ABPP to identify selective inhibitors against DPAP1 and 3, two variants predominantly expressed in the intraerythrocytic merozoite stage. While DPAP1 had previously been associated with the degradation of host haemoglobin in the parasite food vacuole, the precise function of DPAP3 remained unknown. Compared to the broad-spectrum cysteine protease probe DCG-04, FY01 was determined to have high selectivity for DPAP3, as visualised by in-gel fluorescence following labelling with BODIPY TMR-conjugated FY01. Here, selective labelling of DPAP3 was confirmed by mass spectrometry (MS) following affinity purification of proteins with a biotinylated version of FY01. Owing to its inability to efficiently label DPAP3, DCG-04 could be used to assess the specificity of inhibitors against DPAP1. Using competitive ABPP to screen a library of small-molecule irreversible protease inhibitors, vinyl sulfones SAK1 and SAK2 were identified as specific inhibitors of DPAP3 and DPAP1, respectively. In an elegant series of small-molecule inhibition assays, it was demonstrated that DPAP3, but not DPAP1, is required for the rupture of late-stage schizonts during egress, highlighting a novel role for this protease in the asexual phase of the malaria life cycle.

2.1.2 Serine Proteases

Another important class of parasite-derived protease that has been well characterised using ABPP are the serine proteases. Belonging to the serine hydrolase family, such proteases depend on a conserved nucleophilic serine residue positioned within the active site to attack the carbonyl group of an amide bond, forming an acyl-enzyme intermediate that is subsequently hydrolysed to break the bond (Long and Cravatt 2011). The inherent nucleophilicity of the catalytic serine renders it susceptible to covalent modification by a range of broad-spectrum serine hydrolase ABPs including the fluorophosphonates (FPs), aryl phosphonates, isocoumarins, sulfonyl fluorides and carbamates (Cravatt et al. 2008).

In parasites, serine protease activity was first profiled in *Babesia divergens*, a bovine pathogen that can cause lethal babesiosis in immunocompromised humans (Montero et al. 2006). *Babesia* species are apicomplexans with an intraerythrocytic life cycle comparable to that of *Plasmodium*, involving active invasion of RBCs by merozoites. While serine proteases such as *Pf*SUB1 and 2 were thought to play a pivotal role in the entry of *P. falciparum* merozoites (via the proteolytic maturation of parasite and host-derived proteins), whether such enzymes had similar function in *B. divergens* was not known. To identify potential serine hydrolases in *B. divergens*, Montero et al. (2006) labelled the lysates of extracellular merozoites with a biotinylated derivative of the FP probe (FP-biotin; Fig. 2d). Western blot analysis of FP-biotin labelled proteins with an avidin-conjugated secondary antibody identified two distinct protein species. Using antibodies against a related homologue present in *P. falciparum* (*Pf*SUB1), immunoprecipitation experiments demonstrated that these FP-biotin labelled species were the active and precursor forms of a serine subtilisin-like protease, subsequently denoted *Bd*SUB1. Further work found that *Bd*SUB1 localises to invasion-associated secretory organelles known as dense granules and is essential for *B. divergens* invasion, indicating functional conservation with *Pf*SUB1. ABPP has been further used to delineate the roles of several *Plasmodium* serine proteases expressed in the intraerythrocytic parasite stages. In the same chemical biology study that characterised the DPAP1/3 cysteine proteases, Bogyo and co-workers used an isocoumarin-based probe to identify *Pf*SUB1 as a novel regulator of *P. falciparum* merozoite egress (Arastu-Kapur et al. 2008). In this study, a biotinylated chloroisocoumarin from the same protease inhibitor library was identified as an inhibitor of schizont rupture, with a half-maximal effective concentration value (EC_{50}) of 22 μM. This biotinylated inhibitor, JCP104, was used to label protein targets in both intact and permeabilised schizont samples lacking an RBC membrane and with reduced human protein contamination. Western blotting and MS analysis of proteins covalently modified by the biotinylated inhibitor revealed *Pf*SUB1 as the primary target. Furthermore, fluorogenic peptide substrate cleavage assays demonstrated that JCP104 inhibits the proteolytic activity of recombinant *Pf*SUB1 with an IC_{50} consistent with the EC_{50} reported for the schizont rupture assay. These data suggested that JCP104 likely impairs parasite egress by blocking the activity of *Pf*SUB1, supporting a role for this serine protease in the parasite's intraerythrocytic life cycle. To define the mechanism by which *Pf*SUB1 regulates merozoite egress, JCP104 was then used to assess whether inhibition of this protease affects the proteolytic processing of *Pf*SUB1 substrates. For this experiment, Arastu-Kapur et al. focused on the serine repeat antigen 5 (SERA5). SERA5 is an essential pseudoprotease with a role in schizont rupture, and proteolytically processed by *Pf*SUB1 within the parasitophorous vacuole (PV), the membrane-bound structure in which the parasite develops (Collins et al. 2017; Yeoh et al. 2007). As with DPAP3 inhibition, JCP104-mediated inhibition of *Pf*SUB1 resulted in a dose-dependent accumulation of unprocessed SERA5 within intact schizonts concurrent with a reduction in the amount of processed SERA5 in the cell culture media following egress. Intriguingly, the processing of SERA5 was dependent on the presence of functional DPAP3, as small-molecule inhibition of DPAP3 blocked the production of the mature form of *Pf*SUB1 and subsequent merozoite egress. These findings led to the development of a new model for RBC rupture during egress, whereby *Pf*SUB1 is proteolytically matured in exoneme organelles by DPAP3, before being released into the PV where it then processes SERA and matures other merozoite surface proteins including MSP1. This study demonstrated the power of ABPs to characterise the synergy between two mechanistically different proteases in a critical and complex parasite cellular process.

Another *P. falciparum* serine protease that has been characterised with the aid of ABPP is *Pf*ClpP, an orthologue of the cyanobacterial caseinolytic protease P (Rathore et al. 2010). In 2010, Rathore et al. (2010) screened a series of synthetic β-lactone probes to identify compounds that inhibit the activity of *Pf*ClpP. Their best inhibitor, compound U1, had an IC_{50} of 8.4 μM and was converted into an ABP to determine the specificity of this compound in *P. falciparum* proteomes for subsequent use in cell-based studies. Here, a terminal alkyne group was introduced to allow for functionalisation of the probe with an azide-linked rhodamine (Az-Rho) fluorescent reporter by copper-catalysed azide-alkyne cycloaddition reaction, click chemistry (Kolb et al. (2001). Following intact labelling of intraerythrocytic *P. falciparum* trophozoites, in-gel fluorescence analysis of labelled proteins revealed high selectivity of this probe for *Pf*ClpP. Having validated the specificity for the probe, cell-based inhibition studies were conducted with the parent inhibitor (U1) to characterise the function *Pf*ClpP throughout the parasite's intraerythrocytic life cycle. Treatment of *P. falciparum* with U1 resulted in a parasite growth arrest approximately 96 h post-drug treatment, consistent with the well-documented delayed-death phenotype that is associated with the loss of the apicoplast (an essential organelle situated at the anterior of the parasite where isoprenoid, fatty acids and haem are synthesised). Here, U1-treatment induced the formation of abnormally shaped and non-replicative apicoplasts during parasite schizogony. Together, these findings suggested that *Pf*ClpP plays a key role in the functional biogenesis of the apicoplast during intraerythrocytic asexual development.

2.1.3 Threonine Proteases

The proteasome is a large macromolecular protein complex composed of multiple proteolytic subunits that are dependent upon an N-terminal threonine nucleophile for their catalytic activity. As with other eukaryotic systems, the parasite proteasome is critical for a range of essential processes that depend on the proteolytic degradation of polyubiquitinated proteins, such as cell differentiation and replication (Munoz et al. 2015). Small-molecule inhibition studies have highlighted the parasite proteasome as a promising anti-parasitic drug target. For instance, selective inhibition of the catalytic β2 and 5 subunits of the *P. chabaudi* proteasome results in parasite clearance in a mouse model of malaria (Li et al. 2016). Indeed, ABPP of threonine proteases has been integral to the characterisation of the proteasome in various parasites, and its validation as a drug target.

To profile the activity of individual proteasome subunits, Nazif and Bogyo (2001) developed a range of peptidic ABPs that target the catalytic threonine via a vinyl sulfone or epoxyketone warhead. These probes contain variations of a four amino acid specificity sequence, which interact with the specificity pockets of the catalytic β subunits downstream of the site of peptide bond hydrolysis (position P1). Altering the amino acid sequence of the probe's peptide specificity element provided insight into the substrate recognition preferences of each subunit. This approach was used to identify unique substrate recognition properties of the parasite proteasome that can be exploited in drug development. For instance, Wang et al. (2003) used a series of ^{125}I-labelled peptide vinyl sulfones that target the β1, 2 and 5 subunits of the human proteasome to determine the specificity of the equivalent subunits in *T. brucei* (Wang et al. 2003). SDS-PAGE analysis of products radiolabelled with a general probe, ^{125}I-YL$_3$-VS, revealed that β2/5 has similar specificity for hydrophobic residues in the P1 to P4 positions of the substrate. Unlike its human orthologue, the *T. brucei* β1 subunit was not labelled despite its primary sequence being predictive of catalytic activity, suggesting that this subunit has significantly altered peptide specificity in *T. brucei* or may be catalytically inactive. Furthermore, substitution of the P1 leucine to an asparagine in a related probe (^{125}I-NP-L$_2$N-VS) resulted in selective labelling of β2 that was not seen in humans, indicating that this subunit may have different substrate specificity compared with the human proteasome. To define the catalytic profile of the trypanosomal proteasome competitive ABPP was performed, testing the effect of small-molecule inhibitors on subunit labelling with the general ^{125}I-YL$_3$-VS probe. Leupeptin, a specific inhibitor of the trypsin-like activity of the mammalian proteasome, prevents labelling of β2 but not β5, suggesting these subunits are responsible for the trypsin and chymotrypsin-like activity of the *T. brucei* protease, respectively. Bogyo and colleagues screened a positionally scanned library of peptide substrates (P1 to P4) to investigate substrate specificity differences between the *T. brucei* and human proteasome. Using a fluorogenic peptide cleavage assay, they found that the *T. brucei* 20S proteasome has an overall preference for hydrophobic residues in the P1 to P4 positions like its human counterpart. However, unique to the *T. brucei* proteasome was a preference for peptides containing a P1 glutamine, thus identifying a difference in substrate specificity and a possible site to therapeutically target.

Threonine protease-directed ABPs have also been used to validate the specificity of inhibitors against the *P. falciparum* proteasome (Li et al. 2016). Probing substrate specificity with a diverse range of polypeptides, Li et al. (2016) first reported that the *P. falciparum* 20S proteasome, unlike the human proteasome, has strong preference for cleavage of tryptophan residues at P1 and P3. Subsequently, competitive ABPP was used to assess the specificity of inhibitors that contain tryptophan at one or both positions for the different β subunits. The parasite 20S proteasome was treated with the modified inhibitors before labelling with a fluorophore (Cy5)-tagged epoxyketone ABP (BMV037; Fig. 2e) (Li et al. 2014) to detect residual activity. Excitingly, substitutions at both positions resulted in selective inhibition of the parasite β2 catalytic subunit compared to the equivalent human subunit. Structural analysis of the inhibitor-bound proteasome revealed an unusual open conformation around the active site of the *P. falciparum* β2 subunit, providing a basis for future structure-guided drug design. Despite likely having some off-target activity with the P3 position of the human β5 active site, this parasite β2-selective inhibitor effectively perturbed parasite growth with minimal host toxicity in a mouse model of malaria. Overall, this study demonstrated the power of ABPP for the development of selective inhibitors of the *Plasmodium* proteasome, a proven chemically tractable target that could be exploited for future antimalarials.

2.1.4 Aspartyl and Metalloproteases

In contrast to cysteine, serine and threonine proteases, aspartyl and metalloproteases lack a nucleophilic residue and instead depend on the activation of water molecules for their proteolytic activity. Hence, ABPs directed towards these protease classes are typically potent inhibitors of a given target that feature a tag and a photo-crosslinking group for covalent modification (Cravatt et al. 2008). Activity-based profiling of the aspartyl and metalloproteases in parasite systems has been largely restricted to *Plasmodium*. Nevertheless, these proteases are critical for a range of essential processes in diverse macro- and microparasites including food digestion in *Schistosoma* (Goupil et al. 2016), and immune modulation by *Toxoplasma* (Hammoudi et al. 2015).

The plasmepsins (PMs) represent a family of approximately ten functionally-diverse aspartyl proteases expressed in both intra- and exoerythrocytic stages of the *Plasmodium* life cycle (Banerjee et al. 2002; Coombs et al. 2001; Nasamu et al. 2017; Russo et al. 2010). Of these, four (PM-I/II/IV and the histoaspartic protease, HAP) have well-established roles in the digestion of host haemoglobin in the RBC and are essential for parasite growth (Banerjee et al. 2002; Coombs et al. 2001). While PMs have long been considered promising antimalarial targets, their functional redundancy has suggested that each must be simultaneously inhibited in order to effectively clear parasitaemia. To enable assessment of inhibitors that target all four PMs, Liu et al. (2009) developed a series of broad-spectrum hydroxyethyl probes bind to the active site of these proteases adjacent to catalytic aspartic acid residue(s). These probes also contain a terminal azide chemical handle, enabling click-conjugation of alkyne-linked benzophenone (BP) photocrosslinkers and affinity/reporter tags such as the tetraethylrhodamine (TER) fluorophore (Fig. 2f). For one particular probe, consistent labelling of the PMs was confirmed in the lysates of several intraerythrocytic parasite stages by two-dimensional gel electrophoresis (2DGE)-MS and western blotting. Here, the activity of the PMs was present in the insoluble and soluble fractions of trophozoite and schizont stages, respectively. These data supported a change in the subcellular localisation of the PMs from the membrane-bound parasite food vacuole to the soluble compartment of that digestive organelle, consistent with previous reports that PM-II is released in a soluble form from the vacuole during development (Klemba et al. 2004). Competitive ABPP was then used to assay a small library of 152 hydroxyethyl-based inhibitors against PM activity. One compound, G16, caused a dramatic reduction in the fluorescence labelling intensity by the probe ($IC_{50} = 0.84\ \mu M$) and was selected for growth inhibition studies. Treatment of late-stage schizonts with G16 resulted in a decrease in newly formed ring-stage parasites and an accumulation of extracellular merozoites, indicating that this compound affects the development, egress and/or reinvasion of *P. falciparum* erythrocytic stage parasites.

P. falciparum also contains three metallo-aminopeptidases (MAPs) thought to contribute to the proteolytic degradation of host haemoglobin during intraerythrocytic development: aminopeptidase N (*Pf*A-M1), aminopeptidase P (*Pf*APP) and leucyl aminopeptidase (*Pf*-LAP) (McGowan 2013). To elucidate the role of MAPs in the biology of *Plasmodium*, Greenbaum and colleagues established a novel ABPP platform using an ABP scaffold derived from the natural product, bestatin (Harbut et al. 2011). Bestatin is known to inhibit multiple families of MAPs and has been shown to disrupt the growth of *P. falciparum* both in vitro and in vivo, potentially by disrupting the haemoglobin digestion pathway (Naughton et al. 2010). To identify the molecular targets of bestatin, a bestatin-based affinity probe (MH01; Fig. 2g) containing a BP crosslinker and biotin moiety was synthesised and used to label the lysates of asynchronous *P. falciparum* cultures. Western blot detection of probe-labelled biotinylated proteins revealed *Pf*A-M1 and *Pf*-LAP as the principle targets of MH01, with probe labelling successfully outcompeted following pre-treatment of the lysates with the bestatin parent molecule. To gain further insight into the specific functions of the MAPs and thus the mechanism underlying bestatin's effect, the authors generated ABPs specific to *Pf*A-M1 and *Pf*-LAP. This was achieved by screening the inhibitory activity of bestatin-based ABP libraries against each protease, with each probe containing a variation of the two amino acids that govern its interaction with the active sites. Inhibition experiments using the most selective *Pf*A-M1 probe, BTA, demonstrated that *Pf*A-M1 inhibition results in parasite death at the trophozoite stage following disruption of proteolytic digestion of haemoglobin and the phenotypically-associated swelling of the food vacuole. By contrast, inhibition of *Pf*-LAP using a *Pf*-LAP-specific probe (PNAP) correlated with an early death chemotype at the ring-to-trophozoite transition with no obvious morphological features.

2.2 *Profiling Acyl-Protein Thioesterases*

Acyl-protein thioesterases (APTs) are diverse hydrolytic enzymes that catalyse the removal of lipid modifications from protein-associated cysteines through esterase activity. For instance, APTs that specifically cleave palmitate groups are known as palmitoyl-protein thioesterases (PPTs, or depalmitoylases) and play an important role in palmitoylation, a reversible post-translational modification (PTM) that modulates protein function by promoting their membrane localisation, stability and trafficking. As members of the serine hydrolase family, the activity of such enzymes can be profiled using serine hydrolase-directed ABPs, e.g. ABPP was used to characterise the function of a thioesterase in *T. gondii*, *Tg*PPT1 (also known as *Tg*ASH1) (Child et al. 2013; Kemp et al. 2013).

In 2011, Hall et al. (2011) screened a covalent small-molecule library, identifying a related set of substituted chloroisocoumarins that intriguingly enhanced the invasion of host cells by asexual *T. gondii* parasites. Cell-based studies later revealed that these compounds promote an invasive phenotype by inducing microneme secretion and gliding motility, with JCP174 producing the most consistent results (Child et al. 2013). Given the unusual nature of this effect, derivatives of JCP174 were synthesised to facilitate the identification of the targets of this 'enhancer' compound. Using the fluorescently labelled broad-spectrum serine hydrolase probe FP-Rho (Fig. 2d), it was first shown that JCP174 competes for labelling of one specific protein species by in-gel analysis. This ABP competition did not occur when using an inactive analogue of JCP174 that lacked an aromatic amine moiety conserved amongst the enhancer-type compounds. An alkyne-modified variant of JCP174 (JCP174-alk; Fig. 2h) was synthesised to allow probe-labelled proteins to be coupled with a biotin-azide affinity tag and identified by tandem orthogonal proteolysis ABPP (TOP-ABPP) (Weerapana et al. 2007). Mass spectrometry analysis of biotinylated proteins revealed the primary target of JCP174 to be *Tg*PPT1 (Child et al. 2013). A combination of small-molecule and genetic approaches was then used to validate a role for *Tg*PPT1 in microneme secretion and gliding motility, revealing for the first time the importance of dynamic palmitoylation in the regulation of *T. gondii* invasion. The contribution of *Tg*PPT1 to host cell invasion was also demonstrated by Kemp et al. (2013), who used a similar competitive ABPP approach to identify this depalmitoylase as a primary target of various β-lactone-based compounds. These findings initiated the development of a general JCP174-based fluorescent ABP (JCP174-BT) that can be used to profile both parasite and human APTs activity in situ (Garland et al. 2018). Future studies using this probe may provide further insight into the functions of depalmitoylases in other parasite systems and the potential for discovery of novel druggable nodes.

2.3 Profiling Protein Kinases

Protein kinases are implicated in essential functions at virtually every stage of the parasite life cycle and are thus widely acknowledged as drug targets (Doerig 2004). Despite their importance, developing ABPs for profiling protein kinase activity has been a long-standing challenge in chemical proteomics (Rosenblum et al. 2013). Protein kinases catalyse the direct transfer of phosphate from ATP to a protein substrate without any covalent enzyme-ligand intermediate and thus do not typically have a defined nucleophilic residue that can be targeted for covalent modification. Further, as the architecture of the ATP-binding site is similar for many protein kinases (and other enzyme classes), developing competitive inhibitors and probes of ATP-binding with selectivity for the target enzyme(s) remains difficult. Nevertheless, most kinases contain at least one conserved lysine within their active sites that interact with the phosphate backbone of the ATP substrate. This interaction has been exploited in the development of several kinase-directed ABPs including acyl-phosphate ATP probes, Wortmannin-based probes and hypothemycin (Cravatt et al. 2008), some of which have been applied to study kinase function in parasites.

Kinase-based ABPP has been employed to identify the targets of hypothemycin in *T. brucei*, a polyketide natural product inhibitor of CDXG-type kinases with potent trypanocidal activity (Nishino et al. 2013; Schirmer et al. 2006). Kinases play key roles during the life cycle of *T. brucei* and are being increasingly explored as therapeutic targets for African trypanosomiasis (Nett et al. 2009; Parsons et al. 2005). However, while the *T. brucei* kinome comprises 182 potential targets, their interrogation with small-molecule inhibitors has been hampered by the observation that many trypanosomal kinases share high sequence similarity with their human orthologues. Recently, Nishino et al. (2013) produced a semi-synthetic derivative of hypothemycin, a potent inhibitor of CDXG-type kinases, and demonstrated its effectiveness in clearing bloodstream form *T. brucei* parasites in vitro and in vivo. To enable the targets of hypothemycin to be identified, an alkynye-modified hypothemycin-based probe was synthesised for click-conjugation of biotin- or rhodamine-coupled azides (Fig. 2i). Using the biotinylated probe, gel-free competitive ABPP was then conducted with isobaric mass tags to allow for MS-based quantitation of hypothemycin-sensitive proteins in parasite lysates. Using hypothemycin, 11 of 21 total CDXG motif-containing kinases were identified with variable sensitivity to the probe. Here, the previously uncharacterised *Tb*CLK1/2 kinases were shown to have the highest affinity for hypothemycin, followed by other kinases including *Tb*GSK3short and *Tb*MAPK2. Inhibition experiments in live, intact trypanosomes confirmed *Tb*CLK1 as the preferred target of hypothemycin, with its inhibition correlating with loss of cell viability. Although hypothemycin exhibited considerable cytotoxicity in mammalian cells, this study identified *Tb*CLK1 as a potential drug target and introduced hypothemycin-based probes as tools to profile the activity of CDXG-type kinases.

3 Activity-Based Protein Profiling in Parasites: An Exciting Future

Chemical proteomics is a growing field of research that lies at the interface between chemistry and biology. As highlighted in this chapter, the recent development and application of chemical proteomic technologies to parasite systems has contributed vastly to our understanding about the basic biology of a diverse range of pathogens over the last decade. In particular, ABPP has emerged as a powerful technique for functional profiling of diverse enzyme classes and has facilitated the identification of proteins associated with many essential parasitic processes. Additionally, chemical proteomic approaches aimed at profiling PTMs at a systems level have been invaluable for associating PTMs and PTM substrates with critical cellular events. Collectively, these techniques have facilitated the discovery of new potential targets that could be exploited in the development of next-generation therapeutics for many parasite-related diseases. Further, the versatility of ABPP in screening the potency and selectivity of small-molecule inhibitors in complex biological samples has provided a strong foundation for target-based rational drug design. Assessment of the cellular uptake, pharmacokinetic and pharmacodynamic properties of such compounds will provide insight into the full potential of these inhibitors as novel anti-parasitic agents.

Future development of the chemical proteomic technologies described in this review will offer exciting opportunities for parasitologists to explore unchartered territories of parasite biology. Indeed, chemical probes are continually being developed for new enzyme classes, broadening the types of protein that can be experimentally accessed (Chuh et al. 2016; Chuh and Pratt 2015; Yang and Liu 2015). It can therefore be anticipated that further application of these techniques to parasites will aid the functional assignment of previously uncharacterised proteins and thus uncover unique aspects of parasite biology and new drug targets.

This chapter has focused on examples in which chemical biology has been exploited to study parasite proteins with a specific enzymatic activity or PTM. Nevertheless, it is worth mentioning that other chemical proteomic technologies exist, which to our knowledge have yet to be applied to parasites. One well established and rapidly evolving area of ABPP is quantitative reactivity profiling. In this approach, broad-spectrum electrophilic probes are used to quantify the intrinsic reactivity of a specific nucleophilic amino acid type in a complex proteome (Abo et al. 2018). Indeed, this can provide insight into the catalytic and regulatory functions of certain amino acid side chains in proteins, as well as their PTM state and inhibitor occupancy (Abo et al. 2018). While several platforms have been developed for this purpose, perhaps the most widely used is isotopic tandem orthogonal proteolysis activity-based protein profiling (isoTOP-ABPP) (Weerapana et al. 2010). Pioneered by the Cravatt group, isoTOP-ABPP enables the relative abundance of reactive cysteines to be quantified on a global scale using an alkyne-tagged iodoacetamide probe that has specific reactivity towards the cysteine thiol. This technique has been successfully applied to profile cysteine reactivity in diverse biological systems, and since its development has been adapted to study the sensitivity of cysteines to various thiol-dependent PTMs including *S*-sulfenylation and *S*-nitrosylation (Deng et al. 2013; Martell et al. 2016; Yang et al. 2015; Zhou et al. 2016). Furthermore, this platform has recently been expanded to allow for profiling of other nucleophilic hotspots such as lysine side chains using amine-reactive probes (Anderson et al. 2017; Hacker et al. 2017). Although the concept of reactivity profiling remains largely unexplored in parasite systems, the potential for these technologies to reveal the importance of reactive amino acids in parasite proteomes and identify new druggable targets is an exciting prospect.

References

Abdulla MH, O'Brien T, Mackey ZB, Sajid M, Grab DJ, McKerrow JH (2008) RNA interference of *Trypanosoma brucei* cathepsin B and L affects disease progression in a mouse model. PLoS Negl Trop Dis 2(9):e298. https://doi.org/10.1371/journal.pntd.0000298

Abo M, Li C, Weerapana E (2018) Isotopically-labeled iodoacetamide-alkyne probes for quantitative cysteine-reactivity profiling. Mol Pharm 15(3):743–749. https://doi.org/10.1021/acs.molpharmaceut.7b00832

Advisory Committee on the Microbiological Safety of Food (2012) *Ad Hoc* group on vulnerable groups. Risk profile in relation to *Toxoplasma* in the food chain

Anderson KE, To M, Olzmann JA, Nomura DK (2017) Chemoproteomics-enabled covalent ligand screening reveals a thioredoxin-caspase 3 interaction disruptor that impairs breast cancer pathogenicity. ACS Chem Biol 12(10):2522–2528. https://doi.org/10.1021/acschembio.7b00711

Arastu-Kapur S, Ponder EL, Fonovic UP, Yeoh S, Yuan F, Fonovic M, Grainger M, Phillips CI, Powers JC, Bogyo M (2008) Identification of proteases that regulate erythrocyte rupture by the malaria parasite *Plasmodium falciparum*. Nat Chem Biol 4(3):203–213. https://doi.org/10.1038/nchembio.70

Banerjee R, Liu J, Beatty W, Pelosof L, Klemba M, Goldberg DE (2002) Four plasmepsins are active in the *Plasmodium falciparum* food vacuole, including a protease with an active-site histidine. Proc Natl Acad Sci U S A 99(2):990–995. https://doi.org/10.1073/pnas.022630099

Caffrey CR, Hansell E, Lucas KD, Brinen LS, Alvarez Hernandez A, Cheng J, Gwaltney SL 2nd, Roush WR, Stierhof YD, Bogyo M, Steverding D, McKerrow JH (2001) Active site mapping, biochemical properties and subcellular localization of rhodesain, the major cysteine protease of *Trypanosoma brucei* rhodesiense. Mol Biochem Parasitol 118(1):61–73

Chandramohanadas R, Davis PH, Beiting DP, Harbut MB, Darling C, Velmourougane G, Lee MY, Greer PA, Roos DS, Greenbaum DC (2009) Apicomplexan parasites co-opt host calpains to facilitate their escape from infected cells. Science 324(5928):794–797. https://doi.org/10.1126/science.1171085

Chaparro JD, Cheng T, Tran UP, Andrade RM, Brenner SBT, Hwang G, Cohn S, Hirata K, McKerrow JH, Reed SL (2018) Two key cathepsins, TgCPB and TgCPL, are targeted by the vinyl sulfone inhibitor K11777 in in vitro and in vivo models of toxoplasmosis. PLoS ONE 13(3):e0193982. https://doi.org/10.1371/journal.pone.0193982

Child MA, Hall CI, Beck JR, Ofori LO, Albrow VE, Garland M, Bowyer PW, Bradley PJ, Powers JC, Boothroyd JC, Weerapana E, Bogyo M (2013) Small-molecule inhibition of a depalmitoylase enhances *Toxoplasma* host-cell invasion. Nat Chem Biol 9(10):651–656. https://doi.org/10.1038/nchembio.1315

Chuh KN, Pratt MR (2015) Chemical methods for the proteome-wide identification of posttranslationally modified proteins. Curr Opin Chem Biol 24:27–37. https://doi.org/10.1016/j.cbpa.2014.10.020

Chuh KN, Batt AR, Pratt MR (2016) Chemical methods for encoding and decoding of posttranslational modifications. Cell Chem Biol 23(1):86–107. https://doi.org/10.1016/j.chembiol.2015.11.006

Collins CR, Hackett F, Atid J, Tan MSY, Blackman MJ (2017) The Plasmodium falciparum pseudoprotease SERA5 regulates the kinetics and efficiency of malaria parasite egress from host erythrocytes. PLoS Pathog 13(7):e1006453. https://doi.org/10.1371/journal.ppat.1006453

Coombs GH, Goldberg DE, Klemba M, Berry C, Kay J, Mottram JC (2001) Aspartic proteases of *Plasmodium falciparum* and other parasitic protozoa as drug targets. Trends Parasitol 17(11):532–537

Cravatt BF, Wright AT, Kozarich JW (2008) Activity-based protein profiling: from enzyme chemistry to proteomic chemistry. Annu Rev Biochem 77:383–414. https://doi.org/10.1146/annurev.biochem.75.101304.124125

Croft SL, Sundar S, Fairlamb AH (2006) Drug resistance in leishmaniasis. Clin Microbiol Rev 19(1):111–126. https://doi.org/10.1128/CMR.19.1.111-126.2006

Deng X, Weerapana E, Ulanovskaya O, Sun F, Liang H, Ji Q, Ye Y, Fu Y, Zhou L, Li J, Zhang H, Wang C, Alvarez S, Hicks LM, Lan L, Wu M, Cravatt BF, He C (2013) Proteome-wide quantification and characterization of oxidation-sensitive cysteines in pathogenic bacteria. Cell Host Microbe 13(3):358–370. https://doi.org/10.1016/j.chom.2013.02.004

Doerig C (2004) Protein kinases as targets for anti-parasitic chemotherapy. Biochim Biophys Acta 1697(1–2):155–168. https://doi.org/10.1016/j.bbapap.2003.11.021

Dou Z, Coppens I, Carruthers VB (2013) Non-canonical maturation of two papain-family proteases in *Toxoplasma gondii*. J Biol Chem 288(5):3523–3534. https://doi.org/10.1074/jbc.M112.443697

Doyle PS, Zhou YM, Hsieh I, Greenbaum DC, McKerrow JH, Engel JC (2011) The *Trypanosoma cruzi* protease cruzain mediates immune evasion. PLoS Pathog 7(9):e1002139. https://doi.org/10.1371/journal.ppat.1002139

Dvořák J, Mashiyama ST, Braschi S, Sajid M, Knudsen GM, Hansell E, Lim KC, Hsieh I, Bahgat M, Mackenzie B, Medzihradszky KF, Babbitt PC, Caffrey CR, McKerrow JH (2008) Differential use of protease families for invasion by *Schistosome cercariae*. Biochimie 90(2):345–358. https://doi.org/10.1016/j.biochi.2007.08.013

Eksi S, Czesny B, Greenbaum DC, Bogyo M, Williamson KC (2004) Targeted disruption of *Plasmodium falciparum* cysteine protease, falcipain 1, reduces oocyst production, not erythrocytic stage growth. Mol Microbiol 53(1):243–250. https://doi.org/10.1111/j.1365-2958.2004.04108.x

Flegr J, Prandota J, Sovickova M, Israili ZH (2014) Toxoplasmosis–a global threat. Correlation of latent toxoplasmosis with specific disease burden in a set of 88 countries. PLoS ONE 9(3):e90203. https://doi.org/10.1371/journal.pone.0090203

G. B. D. Causes of Death Collaborators (2017) Global, regional, and national age-sex specific mortality for 264 causes of death, 1980–2016: a systematic analysis for the Global Burden of Disease Study 2016. Lancet 390(10100):1151–1210. https://doi.org/10.1016/s0140-6736(17)32152-9

Garland M, Schulze CJ, Foe IT, van der Linden WA, Child MA, Bogyo M (2018) Development of an activity-based probe for acyl-protein thioesterases. PLoS ONE 13(1):e0190255. https://doi.org/10.1371/journal.pone.0190255

Goupil LS, Ivry SL, Hsieh I, Suzuki BM, Craik CS, O'Donoghue AJ, McKerrow JH (2016) Cysteine and aspartyl proteases contribute to protein digestion in the gut of freshwater planaria. PLoS Negl Trop Dis 10(8):e0004893. https://doi.org/10.1371/journal.pntd.0004893

Greenbaum D, Medzihradszky KF, Burlingame A, Bogyo M (2000) Epoxide electrophiles as activity-dependent cysteine protease profiling and discovery tools. Chem Biol 7(8):569–581

Greenbaum DC, Baruch A, Grainger M, Bozdech Z, Medzihradszky KF, Engel J, DeRisi J, Holder AA, Bogyo M (2002) A role for the protease falcipain 1 in host cell invasion by the human malaria parasite. Science 298(5600):2002–2006. https://doi.org/10.1126/science.1077426

Hacker SM, Backus KM, Lazear MR, Forli S, Correia BE, Cravatt BF (2017) Global profiling of lysine reactivity and ligandability in the human proteome. Nat Chem 9(12):1181–1190. https://doi.org/10.1038/nchem.2826

Haldar K, Bhattacharjee S, Safeukui I (2018) Drug resistance in *Plasmodium*. Nat Rev Microbiol 16(3):156–170. https://doi.org/10.1038/nrmicro.2017.161

Hall CI, Reese ML, Weerapana E, Child MA, Bowyer PW, Albrow VE, Haraldsen JD, Phillips MR, Sandoval ED, Ward GE, Cravatt BF, Boothroyd JC, Bogyo M (2011) Chemical genetic screen identifies *Toxoplasma* DJ-1 as a regulator of parasite secretion, attachment, and invasion. Proc Natl Acad Sci U S A 108(26):10568–10573. https://doi.org/10.1073/pnas.1105622108

Hammoudi PM, Jacot D, Mueller C, Di Cristina M, Dogga SK, Marq JB, Romano J, Tosetti N, Dubrot J, Emre Y, Lunghi M, Coppens I, Yamamoto M, Sojka D, Pino P, Soldati-Favre D (2015) Fundamental roles of the golgi-associated *Toxoplasma* aspartyl protease, ASP5, at the host-parasite interface. PLoS Pathog 11(10):e1005211. https://doi.org/10.1371/journal.ppat.1005211

Harbut MB, Velmourougane G, Dalal S, Reiss G, Whisstock JC, Onder O, Brisson D, McGowan S, Klemba M, Greenbaum DC (2011) Bestatin-based chemical biology strategy reveals distinct roles for malaria M1- and M17-family aminopeptidases. Proc Natl Acad Sci U S A 108(34):E526–E534. https://doi.org/10.1073/pnas.1105601108

He YX, Salafsky B, Ramaswamy K (2005) Comparison of skin invasion among three major species of *Schistosoma*. Trends Parasitol 21(5):201–203. https://doi.org/10.1016/j.pt.2005.03.003

Kemp LE, Rusch M, Adibekian A, Bullen HE, Graindorge A, Freymond C, Rottmann M, Braun-Breton C, Baumeister S, Porfetye AT, Vetter IR, Hedberg C, Soldati-Favre D (2013) Characterization of a serine hydrolase targeted by acyl-protein thioesterase inhibitors in *Toxoplasma gondii*. J Biol Chem 288(38):27002–27018. https://doi.org/10.1074/jbc.M113.460709

Klemba M, Beatty W, Gluzman I, Goldberg DE (2004) Trafficking of plasmepsin II to the food vacuole of the malaria parasite *Plasmodium falciparum*. J Cell Biol 164(1):47–56. https://doi.org/10.1083/jcb200307147

Kolb HC, Finn MG, Sharpless KB (2001) Click chemistry: diverse chemical function from a few good reactions. Angew Chem Int Ed Engl 40(11):2004–2021. https://doi.org/10.1002/15213773(20010601)40:11

Larson ET, Parussini F, Huynh MH, Giebel JD, Kelley AM, Zhang L, Bogyo M, Merritt EA, Carruthers VB (2009) *Toxoplasma gondii* cathepsin L is the primary target of the invasion-inhibitory compound morpholinurea-leucyl-homophenyl-vinyl sulfone phenyl. J Biol Chem 284(39):26839–26850. https://doi.org/10.1074/jbc.M109.003780

Li H, van der Linden WA, Verdoes M, Florea BI, McAllister FE, Govindaswamy K, Elias JE, Bhanot P, Overkleeft HS, Bogyo M (2014) Assessing subunit dependency of the *Plasmodium* proteasome using small molecule inhibitors and active site probes. ACS Chem Biol 9(8):1869–1876. https://doi.org/10.1021/cb5001263

Li H, O'Donoghue AJ, van der Linden WA, Xie SC, Yoo E, Foe IT, Tilley L, Craik CS, da Fonseca PC, Bogyo M (2016) Structure- and function-based design of *Plasmodium*-selective proteasome inhibitors. Nature 530(7589):233–236. https://doi.org/10.1038/nature16936

Liu K, Shi H, Xiao H, Chong AG, Bi X, Chang YT, Tan KS, Yada RY, Yao SQ (2009) Functional profiling, identification, and inhibition of plasmepsins in intraerythrocytic malaria parasites. Angew Chem Int Ed Engl 48(44):8293–8297. https://doi.org/10.1002/anie.200903747

Long JZ, Cravatt BF (2011) The metabolic serine hydrolases and their functions in mammalian physiology and disease. Chem Rev 111(10):6022–6063. https://doi.org/10.1021/cr200075y

López-Otín C, Bond JS (2008) Proteases: multifunctional enzymes in life and disease. J Biol Chem 283(45):30433–30437. https://doi.org/10.1074/jbc.R800035200

Martell J, Seo Y, Bak DW, Kingsley SF, Tissenbaum HA, Weerapana E (2016) Global cysteine-reactivity profiling during impaired insulin/IGF-1 signaling in *C. elegans* identifies uncharacterized mediators of longevity. Cell Chem Biol 23(8):955–966. https://doi.org/10.1016/j.chembiol.2016.06.015

McCall LI, Siqueira-Neto JL, McKerrow JH (2016) Location, location, location: five facts about tissue tropism and pathogenesis. PLoS Pathog 12(5):e1005519. https://doi.org/10.1371/journal.ppat.1005519

McGowan S (2013) Working in concert: the metalloaminopeptidases from *Plasmodium falciparum*. Curr Opin Struct Biol 23(6):828–835. https://doi.org/10.1016/j.sbi.2013.07.015

McKerrow JH, Caffrey C, Kelly B, Loke P, Sajid M (2006) Proteases in parasitic diseases. Annu Rev Pathol 1:497–536. https://doi.org/10.1146/annurev.pathol.1.110304.100151

Montero E, Gonzalez LM, Rodriguez M, Oksov Y, Blackman MJ, Lobo CA (2006) A conserved subtilisin protease identified in *Babesia divergens* merozoites. J Biol Chem 281(47):35717–35726. https://doi.org/10.1074/jbc.M604344200

Monzote L, Siddiq A (2011) Drug development to protozoan diseases. Open Med Chem J 5:1–3. https://doi.org/10.2174/1874104501105010001

Munoz C, San Francisco J, Gutierrez B, Gonzalez J (2015) Role of the ubiquitin-proteasome systems in the biology and virulence of protozoan parasites. Biomed Res Int 2015:141526. https://doi.org/10.1155/2015/141526

Nasamu AS, Glushakova S, Russo I, Vaupel B, Oksman A, Kim AS, Fremont DH, Tolia N, Beck JR, Meyers MJ, Niles JC, Zimmerberg J, Goldberg DE (2017) Plasmepsins IX and X are essential and druggable mediators of malaria parasite egress and invasion. Science 358 (6362):518–522. https://doi.org/10.1126/science.aan1478

Naughton JA, Nasizadeh S, Bell A (2010) Downstream effects of haemoglobinase inhibition in *Plasmodium falciparum*-infected erythrocytes. Mol Biochem Parasitol 173(2):81–87. https://doi.org/10.1016/j.molbiopara.2010.05.007

Nazif T, Bogyo M (2001) Global analysis of proteasomal substrate specificity using positional-scanning libraries of covalent inhibitors. Proc Natl Acad Sci U S A 98(6):2967–2972. https://doi.org/10.1073/pnas.061028898

Ndao M, Nath-Chowdhury M, Sajid M, Marcus V, Mashiyama ST, Sakanari J, Chow E, Mackey Z, Land KM, Jacobson MP, Kalyanaraman C, McKerrow JH, Arrowood MJ, Caffrey CR (2013) A cysteine protease inhibitor rescues mice from a lethal *Cryptosporidium parvum* infection. Antimicrob Agents Chemother 57(12):6063–6073. https://doi.org/10.1128/AAC.00734-13

Nett IR, Martin DM, Miranda-Saavedra D, Lamont D, Barber JD, Mehlert A, Ferguson MA (2009) The phosphoproteome of bloodstream form *Trypanosoma brucei*, causative agent of African sleeping sickness. Mol Cell Proteomics 8(7):1527–1538. https://doi.org/10.1074/mcp.M800556-MCP200

Nikolskaia OV, de ALAP, Kim YV, Lonsdale-Eccles JD, Fukuma T, Scharfstein J, Grab DJ (2006) Blood-brain barrier traversal by African trypanosomes requires calcium signaling induced by parasite cysteine protease. J Clin Invest 116(10):2739–2747. https://doi.org/10.1172/jci27798

Nishino M, Choy JW, Gushwa NN, Oses-Prieto JA, Koupparis K, Burlingame AL, Renslo AR, McKerrow JH, Taunton J (2013) Hypothemycin, a fungal natural product, identifies therapeutic targets in *Trypanosoma brucei* [corrected]. Elife 2:e00712. https://doi.org/10.7554/eLife.00712

Parsons M, Worthey EA, Ward PN, Mottram JC (2005) Comparative analysis of the kinomes of three pathogenic trypanosomatids: *Leishmania* major, *Trypanosoma brucei* and *Trypanosoma cruzi*. BMC Genom 6:127. https://doi.org/10.1186/1471-2164-6-127

Parussini F, Coppens I, Shah PP, Diamond SL, Carruthers VB (2010) Cathepsin L occupies a vacuolar compartment and is a protein maturase within the endo/exocytic system of *Toxoplasma gondii*. Mol Microbiol 76(6):1340–1357. https://doi.org/10.1111/j.1365-2958.2010.07181.x

Que X, Wunderlich A, Joiner KA, Reed SL (2004) Toxopain-1 is critical for infection in a novel chicken embryo model of congenital toxoplasmosis. Infect Immun 72(5):2915–2921

Rathore S, Sinha D, Asad M, Bottcher T, Afrin F, Chauhan VS, Gupta D, Sieber SA, Mohmmed A (2010) A cyanobacterial serine protease of *Plasmodium falciparum* is targeted to the apicoplast and plays an important role in its growth and development. Mol Microbiol 77(4):873–890. https://doi.org/10.1111/j.1365-2958.2010.07251.x

Rosenblum JS, Nomanbhoy TK, Kozarich JW (2013) Functional interrogation of kinases and other nucleotide-binding proteins. FEBS Lett 587(13):1870–1877. https://doi.org/10.1016/j.febslet.2013.05.008

Russo I, Babbitt S, Muralidharan V, Butler T, Oksman A, Goldberg DE (2010) Plasmepsin V licenses *Plasmodium* proteins for export into the host erythrocyte. Nature 463(7281):632–636. https://doi.org/10.1038/nature08726

Salmon BL, Oksman A, Goldberg DE (2001) Malaria parasite exit from the host erythrocyte: a two-step process requiring extraerythrocytic proteolysis. Proc Natl Acad Sci U S A 98(1):271–276. https://doi.org/10.1073/pnas.011413198

Schirmer A, Kennedy J, Murli S, Reid R, Santi DV (2006) Targeted covalent inactivation of protein kinases by resorcylic acid lactone polyketides. Proc Natl Acad Sci U S A 103 (11):4234–4239. https://doi.org/10.1073/pnas.0600445103

Simner PJ (2017) Medical parasitology taxonomy update: january 2012 to december 2015. J Clin Microbiol 55(1):43–47. https://doi.org/10.1128/JCM.01020-16

Suarez CE, Bishop RP, Alzan HF, Poole WA, Cooke BM (2017) Advances in the application of genetic manipulation methods to apicomplexan parasites. Int J Parasitol 47(12):701–710. https://doi.org/10.1016/j.ijpara.2017.08.002

Teo CF, Zhou XW, Bogyo M, Carruthers VB (2007) Cysteine protease inhibitors block *Toxoplasma gondii* microneme secretion and cell invasion. Antimicrob Agents Chemother 51(2):679–688. https://doi.org/10.1128/AAC.01059-06

Wang CC, Bozdech Z, Liu CL, Shipway A, Backes BJ, Harris JL, Bogyo M (2003) Biochemical analysis of the 20 S proteasome of *Trypanosoma brucei*. J Biol Chem 278(18):15800–15808. https://doi.org/10.1074/jbc.M300195200

Weerapana E, Speers AE, Cravatt BF (2007) Tandem orthogonal proteolysis-activity-based protein profiling (TOP-ABPP)–a general method for mapping sites of probe modification in proteomes. Nat Protoc 2(6):1414–1425. https://doi.org/10.1038/nprot.2007.194

Weerapana E, Wang C, Simon GM, Richter F, Khare S, Dillon MB, Bachovchin DA, Mowen K, Baker D, Cravatt BF (2010) Quantitative reactivity profiling predicts functional cysteines in proteomes. Nature 468(7325):790–795. https://doi.org/10.1038/nature09472

World Health Organization (2017) World Malaria Report 2017. World Health Organization, Geneva

Yang P, Liu K (2015) Activity-based protein profiling: recent advances in probe development and applications. ChemBioChem 16(5):712–724. https://doi.org/10.1002/cbic.201402582

Yang J, Gupta V, Tallman KA, Porter NA, Carroll KS, Liebler DC (2015) Global, in situ, site-specific analysis of protein S-sulfenylation. Nat Protoc 10(7):1022–1037. https://doi.org/10.1038/nprot.2015.062

Yeoh S, O'Donnell RA, Koussis K, Dluzewski AR, Ansell KH, Osborne SA, Hackett F, Withers-Martinez C, Mitchell GH, Bannister LH, Bryans JS, Kettleborough CA, Blackman MJ (2007) Subcellular discharge of a serine protease mediates release of invasive malaria parasites from host erythrocytes. Cell 131(6):1072–1083. https://doi.org/10.1016/j.cell.2007.10.049

Zhou Y, Wynia-Smith SL, Couvertier SM, Kalous KS, Marletta MA, Smith BC, Weerapana E (2016) Chemoproteomic strategy to quantitatively monitor transnitrosation uncovers functionally relevant S-nitrosation sites on Cathepsin D and HADH2. Cell Chem Biol 23(6):727–737. https://doi.org/10.1016/j.chembiol.2016.05.008

Deciphering T Cell Immunometabolism with Activity-Based Protein Profiling

Adam L. Borne, Tao Huang, Rebecca L. McCloud, Boobalan Pachaiyappan, Timothy N. J. Bullock and Ku-Lung Hsu

Contents

Adam L. Borne, Tao Huang, Rebecca L. McCloud and Boobalan Pachaiyappan contributed equally

---

T. Huang · R. L. McCloud · B. Pachaiyappan · K.-L. Hsu (✉)
Department of Chemistry, University of Virginia, McCormick Road,
P.O. Box 400319, Charlottesville, VA 22904, USA
e-mail: kenhsu@virginia.edu

A. L. Borne · K.-L. Hsu
Department of Pharmacology, University of Virginia School of Medicine,
Charlottesville, VA 22908, USA

T. N. J. Bullock
Department of Pathology, University of Virginia School of Medicine,
Charlottesville, VA 22908, USA

Current Topics in Microbiology and Immunology (2019) 420:175–210
DOI 10.1007/82_2018_124

Published Online: 21 August 2018

Abstract As a major sentinel of adaptive immunity, T cells seek and destroy diseased cells using antigen recognition to achieve molecular specificity. Strategies to block checkpoint inhibition of T cell activity and thus reawaken the patient's antitumor immune responses are rapidly becoming standard of care for treatment of diverse cancers. Adoptive transfer of patient T cells genetically engineered with tumor-targeting capabilities is redefining the field of personalized medicines. The diverse opportunities for exploiting T cell biology in the clinic have prompted new efforts to expand the scope of targets amenable to immuno-oncology. Given the complex spatiotemporal regulation of T cell function and fate, new technologies capable of global molecular profiling in vivo are needed to guide selection of appropriate T cell targets and subsets. In this chapter, we describe the use of activity-based protein profiling (ABPP) to illuminate different aspects of T cell metabolism and signaling as fertile starting points for investigation. We highlight the merits of ABPP methods to enable target, inhibitor, and biochemical pathway discovery of T cells in the burgeoning field of immuno-oncology.

1 Introduction

1.1 Cellular Interactions that Regulate T Cell Responses

Thymus-educated lymphocytes (T cells) contribute substantially to our overall health: on the one hand, by protecting the body from pathogenic events including infections and cancer; while on the other hand helping potentiate inflammatory events and autoimmune diseases. T cells recognize their targets via their T cell receptors (TCRs), which are exquisitely sensitive and generally very specific for their ligands [a combination of a short 8-20aa peptides derived from protein proteolytic products presented by the major histocompatibility complex (MHC) Class I or Class II molecules at the cell surface; referred to as their "antigen"; Fig. 1]. Mature T cells encounter their ligands initially in secondary lymphoid tissues such as lymph nodes (LNs) and the spleen. "Priming," or activation, of T cells is initiated by dendritic cells (DCs) that have acquired the proteins that T cells recognize from the extracellular environment. These proteins are acquired by phagocytosis of dying or excreted materials in the periphery (Yatim et al. 2017), which precedes DCs migration via the lymphatics to LN (Randolph et al. 2008). DCs proteolytically degrade proteins to peptides (Alloatti et al. 2016), which then bind to the MHC molecules within DC and are transported out to the surface of the DC where they can be surveilled by T cells (Fig. 1). For tumors, the antigens are commonly

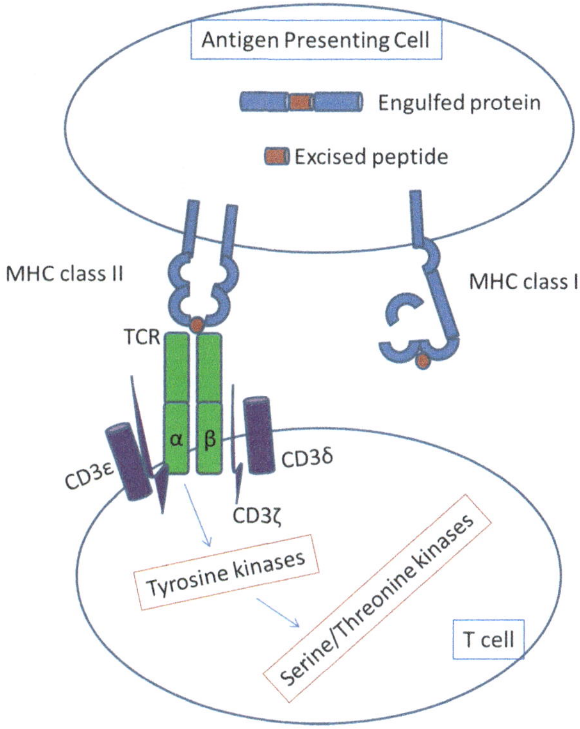

Fig. 1 Activation of T cells via MHC–peptide complexes on professional APC triggering the T cell receptor and cascades of tyrosine and serine–threonine kinases

over-expressed self-proteins; proteins derived from the viruses that drive tumorigenesis; or "new" neo-antigens that arise due to mutations within the genome. Within the LN, DCs secrete chemokines that attract naïve T cells (Luster 2002), providing the opportunity for the T cells to survey the MHC–peptide complexes on the surface of the DC with their TCRs. Once activated, effector T cells undergo rounds of proliferation and exit the secondary lymphoid organ into the circulation. Trafficking T cells receive signals from chemokines and homing receptor ligands expressed on the surface of endothelial cells at sites of inflammation which cause them to adhere to the endothelial cells, and then extravasate from the vasculature into tissue (Jalkanen et al. 1986). Once in the peripheral tissue, chemokine gradients again cause T cells to move to infected or cancerous cells, where their TCR can be engaged by MHC–peptide complexes on the surface of the target cell. Engagement of the TCR can lead to a variety of functions directly performed by the T cell, including proliferation, production of cell–cell communication molecules called cytokines that can promote or limit inflammation, and the release of lytic granules that can kill target cells. These functions drive the ability of T cells to destroy or modify their target cells directly or cooperate, educate, and regulate other cells of the immune system, such as antibody-producing B cells and myeloid cells that engulf damaged and diseased tissues (Fig. 2). Most of the activated T cells die after a short period in the periphery once the infection and the derived antigen are

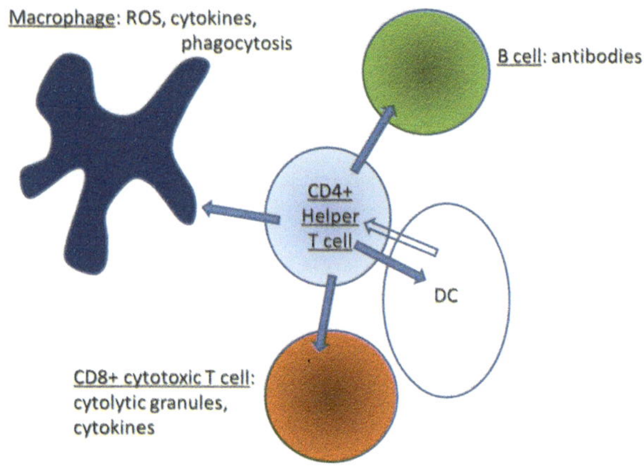

Fig. 2 Interactivity between immune cell subsets. Helper T cells are activated by DC and in turn provide survival and differentiation signals to cytotoxic CD8+ T cells, B cells, and macrophages

removed, but some live on as long-lived memory T cells residing in the tissue, draining lymph nodes or circulating in the blood in a quiescent state, awaiting re-exposure to the antigen that originally promoted their expansion (Reading et al. 2018). These interactions underlie the ability of the immune system to counter pathogens and cancer and yet are also the underpinnings of many cardiovascular and neurological diseases, autoimmunity, and transplant engraftment. Thus, understanding how T cell activation is regulated is anticipated to provide ample opportunities for therapeutic intervention.

In addition to TCR engagement, T cells receive signals from a variety of other receptors on their cell surface. Prominent among these are costimulatory molecules that are members of the immunoglobulin superfamily (IgSF), particularly CD28, and the tumor necrosis factor receptor superfamily (TNFRSF), which supports the cycling and survival of activated T cells without which nascent T cell responses abort. Further, cytokine receptors can have striking influence on T cell fate decisions with respect to both their differentiation into specialized subsets with discrete functions and their long-term survival. On the one hand, pro-inflammatory cytokines such as the type 1 interferon family members or IL-12 can promote the differentiation of helper T cell (CD4+) subsets (Hsieh et al. 1993) and effector function and terminal differentiation in cytotoxic (CD8+) T cells (Cui et al. 2009). Conversely, IL-10 and IL-21 can induce transcription factors that support the ability of T cells to differentiate into long-lived memory T cells (Cui et al. 2011).

1.2 Cell Signaling in T Cell Responses

The initial signals from the TCR are mediated by tyrosine kinases that ultimately activate a collection of serine–threonine kinases that are intrinsically involved in activating/sustaining activities needed for the proliferation and survival of the expanding T cell population (Brownlie and Zamoyska 2013). Like most signaling networks, second messengers play a prominent role in the translation of ligand binding to functional outcomes. In the case of T cells, TCR-mediated activation of the phospholipase C-γ results in the production of inositol-(1,4,5)-trisphosphate (IP3) and diacylglycerols (DAGs). IP3 leads to the activation of protein kinase B/ Akt via activating PDK1 (Rosse et al. 2010). Akt phosphorylates many members of the FoxO transcription factor family, leading to their exclusion from the nucleus (Eijkelenboom and Burgering 2013). FoxO transcription factors regulate the expression of cytokine receptors that contribute to T cell survival and, via regulating KLF2, influence the expression of homing receptors and chemokine receptors that direct the trafficking of activating T cells (Kerdiles et al. 2009). Akt additionally regulates the expression of cytolytic effector molecules in cytotoxic T cells and cytokine receptors such as IL12Rb (Macintyre et al. 2011), which supports the expression of T-bet and Eomesodermin, transcription factors that engrain effector differentiation. DAG can bind and activate many members of the protein kinase C (PKC) serine–threonine kinase family (Quann et al. 2011), guanine nucleotide exchange factors [RasGRPs (Jun et al. 2013)] and subsequently Erk1 and Erk2 via the mitogen-activated protein kinase (MAPK) pathway (Fischer et al. 2005), and protein kinase D (PKD) members (Spitaler et al. 2006), with ensuing impact on the formation of the immunological synapse that connects T cells to antigen-presenting cells, cell motility, and cytokine production.

1.3 Cell Metabolism in T Cell Responses

The tremendous expansion in T cells after TCR engagement requires substantial alterations to their metabolism to support proliferation and differentiation. This involves the increase in oxidative phosphorylation (OXPHOS) and the induction of glycolysis. Glycolysis is activated by either AKT [in helper T cells (Rathmell et al. 2003)] or Erk-mediated regulation of c-Myc (cytotoxic T cells) (Wang et al. 2011). mTORC1, the target of the immunosuppressant rapamycin, is critical kinase involved in metabolic regulation via its regulation of p70-S6 kinase and eIF4E-binding protein (contributing to protein transcription and translation) and the SREBP fatty acid synthesis regulators (Zeng and Chi 2017). Furthermore, mTORC1 links to the HIF-1 transcription factor which has a major role in promoting the sustained expression of glucose transporters and the major enzymes of the glycolytic pathway and concomitantly to many genes that are reflected in the "effector" activities of T cells [such as the expression of lytic granules and cytokines (Finlay et al. 2012)]. Conversely, nutrient-sensing serine–threonine kinases, exemplified by the AMP-activated kinase α1 (AMPKα1) which works with liver kinase B1 (LKB1) to sense the availability of nutrients within the T cell, can attenuate mTORC1 and arrest glycolysis once nutrients and ATP become limiting; this allows OXPHOS to become more prominent and elicits a greater use of fatty acids rather than glucose as the T cells transition into a more quiescent state (Tamas et al. 2006, 2010). Thus, glycolysis is commonly linked to the effector functions of T cells, and its induction and sustained activation are regulated by kinases at critical points.

1.4 Fate Decisions in T Cells

As mentioned, once actively proliferating, T cells can be polarized into different effector subsets (Th1, Th2, etc.) that secrete different patterns of cytokines. Further, the capacity for T cells to become long-lived memory cells resides within this primary T cell response, as a subpopulation either that has not received signals that drive to terminal effector differentiation or has received signals that preserve homeostatic functions. These fate decisions are commonly influenced by the amount of antigen and cytokines (interferons; interleukins; colony-stimulating factors) present during the early expansion phases of T cells, and the signals transmitted by their respective receptors (type I and type II) induce chromatin remodeling that durably influences the gene sets transcribed in these subpopulations. Many of the influential cytokine receptors are linked to Janus kinase (JAK) non-receptor tyrosine kinases that auto-phosphorylate and control the STAT family of transcription factors, which in turn promote the expression of genes as homo- or hetero-dimers binding to their target promoters (Yamaoka et al. 2004). For example, interleukin-2, a critical T cell growth factor, binds to the IL-2 receptor complex, activating JAK1 and JAK3, with the ensuing phosphorylation of STAT5A and STAT5B promoting the expression of the IL-4 receptor, which in turn promotes TH2 subset differentiation. However, these signaling pathways are not generally linear, as IL-2 is also a key determinant in the differentiation of immunosuppressive regulatory T cells (Treg). Similarly, IL-12, which activates Jak1 and Tyk2 and STAT4, drives TH1 differentiation and substantially promotes the differentiation of T cells into terminally differentiated effectors by enhancing T-bet expression. Conversely, IL-23 and IL-10, via STAT3, promote memory T cell differentiation at the expense of effectors. How these different outcomes are controlled, from the same receptor kinase–transcription factor complexes, is an area of intense focus, being partially revealed by phosphoproteomic studies (Ross et al. 2016; Hukelmann et al. 2016), and likely involves unequal activity of STATs (and thus JAKs) and feedback loops between STATs in transcriptional regulation (Hirahara et al. 2015). The contribution of this family of kinases to these fate decisions provides substantial opportunity for therapeutic intervention, potentially ameliorating autoimmune disease by inflating Treg or polarizing vaccine-induced T cell responses to subtypes that are most effective with respect to the purpose of the vaccine. Given that JAKs can also influence chromatin structure, the development of assays that interrogate, and potentially inhibit, the function of these kinases during T cell responses will likely empower much greater understanding of how a relatively constrained family of kinases may drive such complex outcomes.

2 Features of Dysfunctional T Cells

The immune system is a powerful defense against pathogens and tumors, but, as evidenced by autoimmune diseases, toxic shock syndromes and other chronic inflammatory diseases can be deleterious to the host if left unchecked. Consequentially, a multitude of negative feedback mechanisms has evolved to limit immune cell function. This is particularly apparent for T cells in cases of chronic exposure to antigen, such as unresolved pathogen infections (e.g., HIV; HCV; malaria), and these mechanisms for restraining T cell responses are commonly hijacked by tumors.

2.1 Anergy/Tolerance

The dendritic cells that acquire antigen from infected tissues also engulf self-antigens from dying cells. As inflammation from infection is an important element in activating DC to drive T cell responses to a pathogen, it is important that T cells do not also respond to the self-antigens. The majority of self-reactive T cells are culled in the thymus by the processes of negative selection. However, this process is not absolute, and many proteins are expressed late in development and thus avoid "central" tolerance. In the periphery, self-tolerance can be maintained by unactivated DC presenting self-antigen, derived from cells continually undergoing turnover, to T cells in the absence of the costimulatory molecules that are induced by inflammation. Under these circumstances, T cells undergo abortive proliferation and enter a state of anergy, whereby their response to re-exposure to the same antigen is blunted. Anergy is regulated by both ubiquitin ligases and transcription factors that limit the expression of genes that would normally promote T cell expansion and differentiation. Recently, it has become apparent that T cell responses to tumor antigens can be similarly blunted. Although tumors commonly express neo-antigens that would not be the subject of central tolerance, the limited amount of inflammation within the tumor microenvironment can limit DC activation, leading to presentation of antigen in the absence of costimulation.

2.2 Dysfunction/Exhaustion

In many cases of chronic antigen exposure, T cells differentiate but become unresponsive to further stimulation, a situation referred to as exhaustion. This state of exhaustion reflects intrinsic mechanisms of limiting responsiveness to further TCR engagement. T cell receptor activation of tyrosine kinases is highly regulated, and effective signal transduction requires sustained stimulation due to the constitutive activation of protein tyrosine phosphatases (PTPs) that counteract kinase activity. It has been shown in tumor-infiltrating lymphocytes (TILs) that TCR stimulation can be attenuated downstream of antigen engagement (Koneru et al. 2005) by the recruitment of src homology region 2 domain-containing phosphatase (SHP)-1 to the immunological synapse where the T cell engages the target cell (Monu and Frey 2007). SHP-1 limits further signaling, Ca^{2+} flux, and integrin-mediated binding. This is an active inhibition as removal of TILs from the tumor environment, and brief in vitro culture can restore function. Additionally, after engagement of the TCR, inhibitory molecules are expressed that contain immunoreceptor tyrosine-based inhibitory motifs (ITIMs) and are subsequently phosphorylated on their ITIMs, recruiting and activating PTP such as SHP-1 and SHP-2 (Pauls and Marshall 2017; Srivastava et al. 2013). Activated PTP in turn dephosphorylates proximal substrates: These range from kinases to antigen receptors to adaptor molecules. Supporting this notion, polymorphisms in PTPN22 have been shown to promote autoimmune diseases and anticancer responses due to an inability to constrain TCR signaling (Salmond et al. 2015; Brownlie et al. 2017), and short hairpin RNA (shRNA) screens in vivo identified Ppp2r2d, a regulatory subunit of the PP2A phosphatase, as limiting T cell function in the tumor microenvironment (Zhou et al. 2014). Among the inhibitory molecules that recruit PTPs are those being clinically evaluated with considerable success in immuno-oncology: PD1; CTLA-4; LAG3; TIGIT; TIM3 (Vazquez-Cintron et al. 2010). In addition to SHP recruitment to TCR, signaling can be attenuated by recruiting suppressors of cytokine signaling (SOCS) family member cytokine-inducible SH2-containing protein (CISH), which normally blocks STAT5 activation to support the degradation of PLC-γ1 (Palmer et al. 2015). Interestingly, RNAi or CRISPR/Cas9 screens have also implicated kinases in tumor cells for their ability to indirectly impede antitumor immunity by desensitizing responsiveness to cytokine signaling (Patel et al. 2017; Manguso et al. 2017).

2.3 Extrinsic Regulation of T Cell Metabolism

A third approach to limiting T cell function in the tumor microenvironment has recently become clear. Tumors can be considered "competitors" for immune cells for the resources that both need to maintain function. For example, studies have implicated the consumption of glucose by tumors (commonly referred to as Warburg metabolism as glycolysis is used even though oxygen may not be limiting) as a mechanism by which T cells can lose their function, given the importance of glycolysis to T cell effector activity (Chang et al. 2013, 2015a). Other essential metabolites have been demonstrated to be critical T cell function, and their availability is attenuated in the TME. The amino acids arginine and tryptophan are actively depleted by myeloid cells expressing arginase-1 [Arg 1 (Vasquez-Dunddel et al. 2013)] and a variety of cells expressing indoleamine 2,3-dioxygenase 1 [IDO-1 (Balachandran et al. 2011; Liu et al. 2010)], respectively. Interestingly, IDO-1 was originally identified as a mechanism of maintaining maternal-fetal tolerance (Munn et al. 1998). The availability of serine has also shown to be critical for T cell effector function (Ma et al. 2017). Thus, these "extrinsic" regulators of T cell metabolism are also current targets for cancer immunotherapy.

2.4 New Opportunities for ABPP Discovery in T Cell Biology

Multiple tyrosine and serine–threonine kinase pathways regulate T cell responses from inception to termination. Conversely, tyrosine phosphatases and ubiquitin ligases play critical roles in attenuating signals that emanate from the TCR and cytokine receptors. Thus, considerable opportunities exist to regulate fate decisions and metabolic activity of T cells, as evidenced by the clinical implementation of rapamycin, the inhibitor of mTOR, and JAK inhibitors. The challenge at hand is that many of these enzymes are active in many different pathways in many normal cells, so selecting the most appropriate target for the physiological condition is critical to prevent on-target toxicities. Further identifying inhibitors that are truly specific for each of the wide variety of isoforms of different enzymes will be a major thrust of future work. In Sect. 3, we describe examples of using ABPP methods to probe enzymes and pathways that have substantial influence on T cell function.

3 Functional Insights into T Cell Biology Using ABPP Methods

3.1 ABPP Technology

Activity-based protein profiling (ABPP) is a global proteomic method for measuring functional state of enzymes directly in native biological systems (Cravatt et al. 2008; Niphakis and Cravatt 2014). An activity-based probe contains at least two key features: (1) a reactive group that binds and covalently modifies active sites of a large number of enzymes, which share conserved mechanistic and/or structural features and (2) a reporter tag (e.g., fluorophore or biotin) to enable detection, enrichment, and identification of probe-labeled enzyme targets by gel electrophoresis (Jessani et al. 2002) (in-gel analysis) and liquid chromatography–mass spectrometry (LC-MS) (Jessani et al. 2005; Adibekian et al. 2011; Inloes et al. 2014) (LC-MS, Fig. 3a). ABPP probes have been generated for more than a dozen enzymes classes including serine hydrolases (SHs) (Liu et al. 1999), kinases (Patricelli et al. 2011), glycosidases (Vocadlo and Bertozzi 2004; Hekmat et al. 2005), oxidoreductases (Wright and Cravatt 2007; Speers and Cravatt 2005), and cysteine proteases (Greenbaum et al. 2000). ABPP probes label active enzymes but not their inactive forms, enabling detection of changes in enzyme activity independent of alterations in gene or protein expression (Jessani et al. 2002, 2004).

ABPP methods can be configured for inhibitor discovery using a competitive format (Fig. 3b). The potency and selectivity of small molecules can be profiled against many enzymes in parallel by performing competitive ABPP in complex proteomes, where inhibitors compete probe labeling of enzyme targets. Competitive ABPP offers several advantages over conventional substrate assays. First, enzymes can be tested in native proteomes without the need for recombinant protein expression and purification (Adibekian et al. 2011; Long et al. 2009; Hsu et al. 2013a; Inloes et al. 2014; Chang et al. 2015b; Nagano et al. 2013). Second, inhibitors can be developed for uncharacterized enzymes that lack any known substrates (Adibekian et al. 2011; Hsu et al. 2013b; Inloes et al. 2014). Since competitive ABPP is compatible with any enzyme, regardless of its degree of functional annotation, optimized inhibitors can be developed for enzymes that are biologically characterized but devoid of inhibitors, as well as orphan enzymes. In most cases, the compounds developed from these medicinal chemistry programs represent the first pharmacological probes for studying the function of their respective enzyme targets in vivo. Finally, inhibitors are evaluated against many enzymes in parallel, enabling identification of selective agents from non-specific compounds (Bachovchin et al. 2010; Adibekian et al. 2011; Hsu et al. 2013a, b) (Fig. 3b).

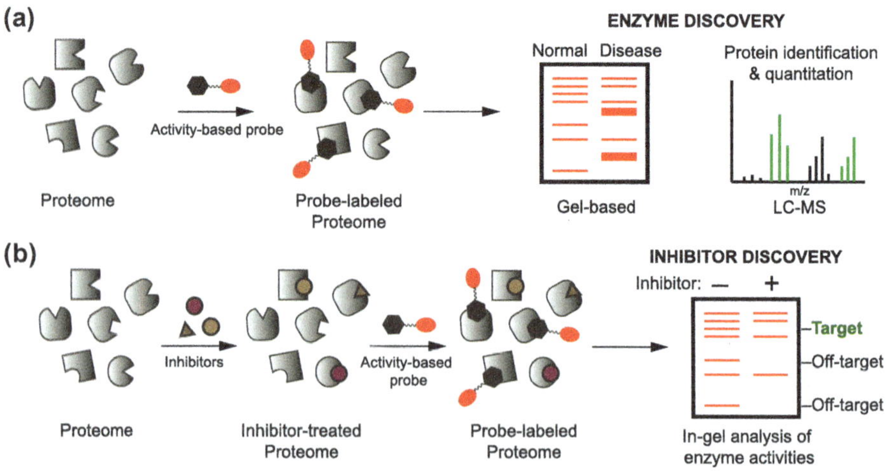

Fig. 3 **a** Activity-based protein profiling (ABPP) utilizes active-site directed chemical probes to measure the functional state of large numbers of enzymes in native biological systems (e.g., cells or tissues). Activity-based probes consist of a reactive group (hexagon) for targeting a specific enzyme class and a reporter tag (e.g., fluorophore or biotin denoted by oval) for detection by in-gel fluorescence scanning or by avidin-enrichment coupled with liquid chromatography–mass spectrometry (LC-MS), respectively. **b** For inhibitor discovery, the potency and selectivity of small molecules can be profiled against many enzymes in parallel by performing competitive ABPP in complex proteomes, where inhibitors compete for probe labeling of enzyme targets

3.2 *ABPP Discovery of Dimethyl Fumarate Mode of Action in T Cells*

ABPP has been used to determine mode of action of immunosuppressive drugs to reveal protein targets important for T cell activation. A recent study used isotopic tandem orthogonal proteolysis-activity-based protein profiling [isoTOP-ABPP (Weerapana et al. 2010; Wang et al. 2014)] to identify protein targets and site of binding for dimethyl fumarate (DMF) in T cells (Blewett et al. 2016). DMF is an FDA-approved immunosuppressive drug used to treat autoimmune disorders including psoriasis (Mrowietz et al. 2007) and multiple sclerosis (MS) (Linker and Gold 2013). DMF exhibits both neuroprotective and immunomodulatory activities through a poorly understood mechanism of action (Scannevin et al. 2012; Albrecht et al. 2012; Gillard et al. 2015). DMF contains a Michael acceptor functional group that mediates covalent reaction with cysteine residues in proteins (Fig. 4). Based on its cysteine reactivity, several pathways have been proposed to explain DMF activity including Nrf-KEAP1 complex dissociation through cysteine modifications (Taguchi et al. 2011; Scannevin et al. 2012) or perturbation of redox hemostasis by changing cellular GSH concentrations (Mrowietz and Asadullah 2005; Lehmann et al. 2007; Lin et al. 2011; Albrecht et al. 2012). DMF is also proposed to function as a prodrug that is metabolized to monomethyl fumarate (MMF, Fig. 4), which acts as a ligand for the G protein-coupled receptor, hydroxycarboxylic acid receptor 2 (HCAR2) (Chen et al. 2014).

To determine DMF mode of action, the Cravatt Lab applied isoTOP-ABPP (Weerapana et al. 2010; Wang et al. 2014) for global cysteine reactivity profiling to identify and quantify DMF-sensitive cysteine residues in primary human and mouse T cell proteomes (Blewett et al. 2016). In this study, the general cysteine-reactive probe iodoacetamide (IA)-alkyne was used to evaluate the reactivity of DMF across >2400 cysteine residues in a competitive ABPP experiment. Sensitivity of probe-modified sites to DMF treatment was compared with the non-electrophilic analogs MMF and DMS (Fig. 4) to determine DMF-specific covalent reactions with T cell protein targets. The results revealed that DMF, but not MMF or DMS, inhibited T cell activation at an early stage, and the observed cell biology was independent of Nrf2 activation or GSH depletion. The unbiased nature of the isoTOP-ABPP approach revealed that DMF mode of action is via covalent modification of several proteins with known biochemical or genetic links to T cell

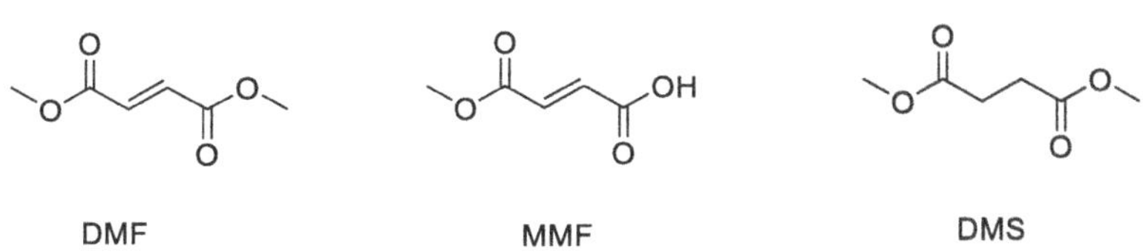

Fig. 4 Chemical structures of dimethyl fumarate (DMF), monomethyl fumarate (MMF), and the non-electrophilic analog dimethyl succinate (DMS)

function including protein kinase C theta (PKCθ). DMF blocked association of PKCθ with the costimulatory receptor CD28 by perturbing a cysteine site in the C2 domain of this kinase (Blewett et al. 2016).

3.3 ABPP Discovery of New Metabolic Regulators of T Cell Proliferation

In contrast to the DMF example above, ABPP can also facilitate studies of protein families without established biology in T cells. Here, we highlight an example of using ABPP methods to discover a new serine hydrolase (SH) target involved in T cell proliferation (Adibekian et al. 2011). SHs represent a large and diverse enzyme class that includes ~200 enzymes or 1% of the genes in the human proteome (Simon and Cravatt 2010; Long and Cravatt 2011). The defining feature of these enzymes is the activated serine nucleophile, as part of a catalytic triad, in their active sites to hydrolyze ester, thioester, or amide bonds in a wide range of substrates including small molecules, peptides, and proteins (Simon and Cravatt 2010; Long and Cravatt 2011). The important biological role of SHs has been exemplified by several clinical drugs targeting individual SHs to treat human diseases including obesity (Henness and Perry 2006), diabetes (Thornberry and Weber 2007), Alzheimer's disease (Birks et al. 2009), and microbial infections (Kluge and Petter 2010). Despite the disease relevance and the successful bench-to-clinic translations, the majority of the metabolic SHs remain poorly characterized (Simon and Cravatt 2010; Long and Cravatt 2011). In recent years, chemical proteomic methods such as ABPP have emerged as a powerful tool to study these unexplored enzymes.

A case in point is a recent study that identified 1,2,3-triazole ureas (1,2,3-TUs, Fig. 5) as an exceptional chemotype for SH inhibition that offers broad activity against the SH class with tunable selectivity for individual enzymes (Adibekian et al. 2011; Hsu et al. 2012, 2013a, b; Yun et al. 2014; Ogasawara et al. 2016; Shin et al. 2018b). Reporter-tagged fluorophosphonates (FPs) were used as general SH activity-based probes for these studies because these probes detect a large fraction of mammalian SHs with negligible cross-reactivity against other enzyme classes that contain nucleophilic residues (Bachovchin et al. 2010; Jessani et al. 2005; Okerberg et al. 2005). Competitive ABPP using FP probes was used to discover a broad-spectrum 1,2,3-TU inhibitor AA26-9, which showed inhibitory activity against approximately 1/3 of the 40+ SHs detected in immortalized T cell lines. AA26-9 targets included serine peptidases, lipases, amidases, esterases, and thioesterases, which suggests that 1,2,3-TUs represent a versatile starting point for developing inhibitors to probe T cell biology.

Next, a library of 1,2,3-TUs based on AA26-9 was expanded using efficient and accessible click chemistry and screened for novel SH inhibitors. Among those

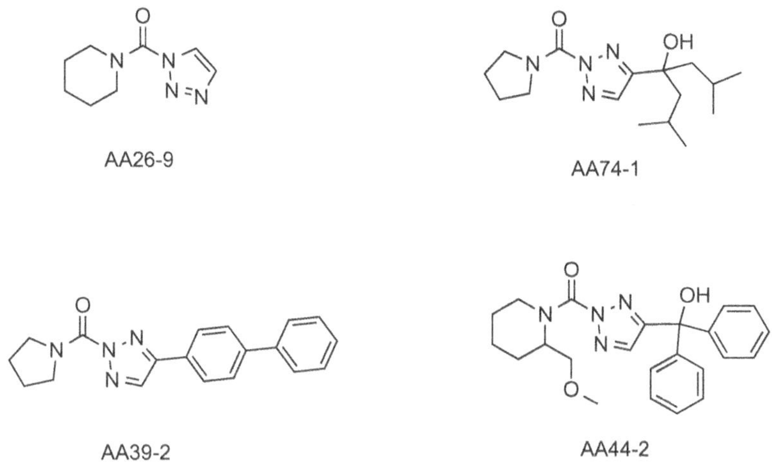

Fig. 5 Chemical structures of 1,2,3-TUs that have been optimized by ABPP-guided medicinal chemistry into selective SH chemical probes for in vivo analysis

analogs, three inhibitors AA74-1, AA39-2, and AA44-2 stood out with remarkable potency and specificity for their respective SH targets in mouse T cell proteomes (Fig. 5). All three compounds exhibited potent inhibitory activity and high selectivity in live T cells; AA74-1 and AA44-2 blocked more than 95% of acyl peptide hydrolase (APEH) and α,β-hydrolase-11 (ABHD11) activities, respectively, without affecting any of the other 40+ SHs in T cells. AA39-2 displayed similar inhibitory activity against PAFAH2 in T cells and cross-reacted with a single SH off-target, α,β-hydrolase-6 (ABHD6). Furthermore, treatment with AA74-1 resulted in near-complete blockade of APEH activity in brain and heart at doses as low as 0.4 mg kg^{-1} and exhibited high selectivity against the 44 SHs detected in the mouse brain. Treatment with AA74-1 resulted in accumulation of a panel of N-acetylated proteins important for T cell proliferation, which helped annotate the substrate specificity of APEH and positions this enzyme as a promising T cell metabolic target.

3.4 *ABPP Discovery of Proteasome Regulation of T Cell Fate*

ABPP can also be used to elucidate underlying mechanisms that control T cell lineage fate. It has been recognized that T cells undergo polarized reorganization of proteins upon activation and this polarization may persist through cell division (Chang et al. 2007; Yeh et al. 2008). Therefore, asymmetric cell division has been suggested in T cells, which enables a single-parent T cell to produce two daughter cells that have distinct fates toward the effector and memory lineages (Chang et al. 2007). It remains poorly understood, however, what critical proteins undergo polarized segregation and unequal partition into progenies for a selected T cell type and what are the underlying mechanisms.

A recent study provided evidence of a lineage-determining transcription factor, T-bet, that was unequally partitioned between the two daughter cells in activated naïve T cells undergoing division (Chang et al. 2011; Lin et al. 2015). This disparity of T-bet is thought to result from asymmetric destruction during mitosis in the setting of asymmetric distribution of the proteasome, the protein degradation machinery. The progeny that received less proteasome acquired more T-bet, and inhibition of polarized segregation of the proteasome or proteasome-dependent degradation prevented the asymmetric partitioning of T-bet, suggesting that a cell may unevenly deploy cellular activities during mitosis so as to pass a different abundance of cell fate regulators to daughter cells.

To examine whether asymmetric localization of the proteasome is associated with differential rates of T-bet degradation within a mitotic cell, the activity-based probe, MVB003, was used to probe the degradative activity of the proteasome (Fig. 6). The MVB003 probe was designed based on the natural product epoxomicin, which potently inhibits the 20S proteasome via covalent reaction of the amino-terminal threonine with the epoxyketone electrophile (Groll et al. 2000; Meng et al. 1999; Sin et al. 1999). The addition of a BODIPY fluorescent tag converts epoxomicin into a probe for activity-based profiling of the proteasome (Li et al. 2013) (Fig. 6). The authors discovered using MVB003 and immunofluorescence microscopy that activated T cells exhibited unequal proteasome activity within mitotic T cells. Costaining experiments with activated CD8+ T lymphocytes indicated that a greater share of T-bet was partitioned into the daughter cell that received less proteasome. The same asymmetric partitioning of T-bet and proteasome was also observed in activated CD4+ T lymphocytes.

The findings in this study suggest that both CD8+ and CD4+ daughter T cells display differential T-bet abundance, which will influence their subsequent fates. By a similar mechanism of distinct localization of the degradation machinery, other biomolecules such as transcription factors and cell proliferation regulators could be inherited by daughter cells with disparity because of asymmetric distribution of the proteasome. The full landscape of the disparities caused by unequal segregation of the proteasome remains to be extensively explored. In this regard, ABPP probes capable of measuring PI3K and mTOR activity [e.g., kinase activity-based probes (Patricelli et al. 2007, 2011; Shin et al. 2018a)] may help further elucidate out how unequal transduction of PI3K/AKT/mTOR signaling during cell division bifurcates transcriptional networks and fates of effector T cells (Lin et al. 2015).

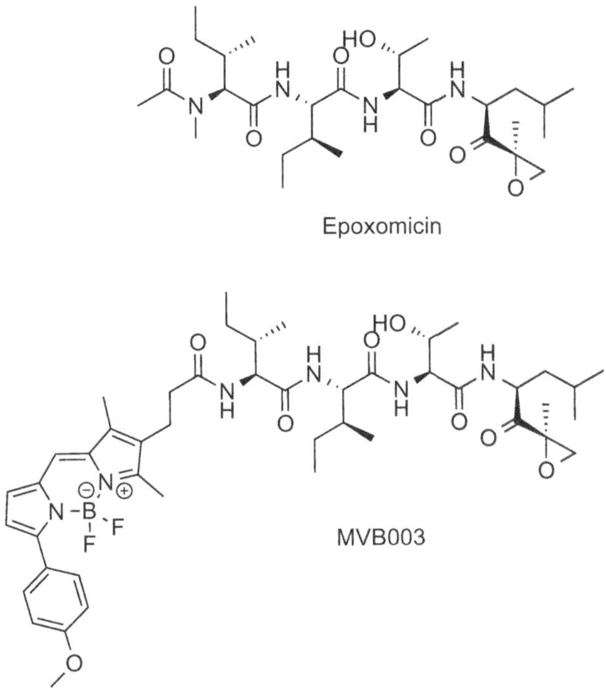

Fig. 6 Chemical structure of a proteasome activity-based probe, MVP003 based on epoxomicin, a natural product that functions as a cell-permeable, potent, selective, and irreversible proteasome inhibitor. The epoxyketone functional group enables covalent reaction with the amino-terminal threonine of the 20S proteasome

4 Using ABPP to Probe Immunometabolism in the Tumor Microenvironment

The field of cancer immunotherapy has seen great excitement recently, primarily due to the success of immune checkpoint inhibitors, cell therapy with tumor-infiltrating lymphocytes, and chimeric antigen receptor T cell (CAR-T) therapy. As therapies for oncology continue to witness a paradigm shift, small molecules or biological agents that target clinically under-explored immunomodulatory proteins are being actively pursued for potential therapeutic intervention. In the subsequent sections, we focus on metabolic pathways operating in the tumor microenvironment that restrain antitumor immunity via extrinsic and intrinsic mechanisms. Tumors are a metabolically demanding environment that can perturb metabolism and function of infiltrating T cells through competition for substrates required to fuel optimal antitumor effector functions (Buck et al. 2017; Chang and Pearce 2016; Ho et al. 2015; Nakaya et al. 2014). Metabolites produced from these pathways not only provide energy and building blocks for growth, but also serve as cellular signals to regulate transcriptional programs that tune T cell function and fate. Thus, defining the metabolic pathways and enzymes differentially utilized by cancer cells and tumor-specific T cells could uncover new "checkpoints" for immuno-oncology. ABPP can facilitate these efforts because metabolism is regulated by enzyme chemistry that is highly amenable to activity-based profiling methods.

4.1 Diacylglycerol Signaling as an Intrinsic Regulator of TCR Activation

Studies have demonstrated that the balance between T cell activation and anergy (explained in Sect. 2.1) can be controlled by activity of DGKs. There is compelling evidence in the literature showing that DGKs are key regulators of T cell effector function (Mérida et al. 2015; Riese et al. 2011) due at least in part to regulation of Ras signaling. DGKα expression increases in the absence of costimulation and is high in anergic T cells, and ectopic DGKα overexpression causes an anergic state. Conversely, anergy can be relieved either by ectopic overexpression of a dominant negative DGKα mutant or by pharmacological inhibition of DGKα (Zha et al. 2006; Olenchock et al. 2006). Moreover, T cells from DGKα knockout mice are resistant to induction of anergy and are hyper-responsive to immune stimulation. Blocking the activity of DGKα increases Ras activation in T cells, as measured by GTP loading; overexpression of RasGRP1 or dominant negative Ras overcomes anergy. An alternative isoform, DGKζ, has also been implicated in restraining effector T cell function directly (Riese et al. 2011), or by promoting Treg development (Joshi et al. 2013), offering an additional target.

4.1.1 Diacylglycerol Kinase Family and Biochemistry

Diacylglycerol kinases (DGKs) are a family of lipid kinases that regulate cell biology through ATP-dependent phosphorylation of diacylglycerol (DAG) to form phosphatidic acid (PA, Fig. 7). Humans express 10 DGK isoforms that are further classified into five subtypes based on the number and type of regulatory domains (Mérida et al. 2008; Sakane et al. 2016; Shulga et al. 2011). All DGKs contain at

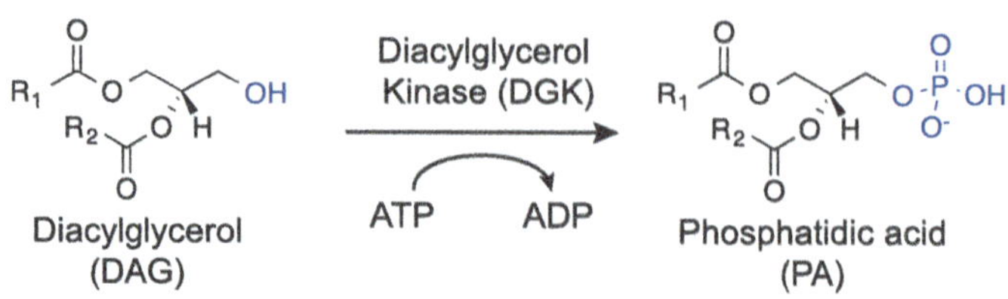

Fig. 7 Diacylglycerol kinases (DGKs) phosphorylate DAGs to regulate cellular signaling

least two zinc finger domains (C1A and C1B, Fig. 8), which have been shown in DGKγ and β to respond to phorbol ester (DAG mimetics) stimulation (Colón-González and Kazanietz 2006). The role of C1 domains in other DGK isoforms remains poorly understood with respect to catalytic and signaling functions (Shindo et al. 2003).

Type 1 DGKs (α, β, γ) contain a lipid kinase domain split into catalytic (DAGKc) and accessory (DAGKa) regions, a pair of EF-hands, a recoverin homology domain (RVH), and the C1A/B domains (Fig. 8). The EF-hands bind Ca^{2+} and have been shown to activate kinase activity, but various mechanisms of Ca^{2+}-independent activation of DGKs have been demonstrated including lipid binding at the C1 domains (Fanani et al. 2004). Type 1 DGK C1s are classified as typical (DGKβ and DGKγ) or atypical (DGKα) based on protein sequence homology with prototypical C1s found in the protein kinase C (PKC) family of Ser/Thr kinases (Merino et al. 2007). DGKα is highly expressed in T cells, DGKβ in the caudate nucleus, and DGKγ in cerebellum according to bio-GPS global mRNA expression analysis (Wu et al. 2016).

The type 2 DGKs (δ, η, κ) have the same C1A/B pair, but the DAGKc and DAGKa regions are separated by a short peptide linker of unknown function. They are grouped based on the presence of a pleckstrin homology (PH) domain which bind phosphoinositides (PIP3) and localize proteins to membrane (Hurley and Misra 2000) (Fig. 8). DGKδ also contains a sterile alpha motif (SAM) domain allowing for interactions with other SAM domain-containing proteins (Stapleton et al. 1999). DGKκ has a region of 33 repeats of the amino acid sequence Glu-Pro-Ala-Pro referred to as an EPAP domain of poorly defined function (Shulga et al. 2011). DGKη lacks both the SAM and EPAP domains. DGKδ is highly expressed in T cells as well as B cells and natural killer (NK) cells. DGKκ and DGKη are expressed at low levels in cardiac myocytes.

DGKε is the only type 3 DGK and the smallest of the human DGKs. DGKε contains a combined DAGKc/DAGKa domain, C1A/B, and a hydrophobic domain (HD, Fig. 8). The HD domain at the N terminus is proposed to localize DGKε to the inner membrane of the cell (Jennings et al. 2015). While other DGKs are expressed as soluble proteins that can associate with membranes, DGKε is expressed exclusively in the membrane fraction (Franks et al. 2017). DGKε is highly expressed in the cervical cortical ganglion and the only DGK to show substrate specificity for DAGs that contain arachidonic acid chains (Lung et al. 2009).

Type 4 DGKs (ζ, ι) contain the combined catalytic DAGKc/DAGKa domain, C1A/B, a myristoylated alanine-rich C kinase substrate (MARCKS) domain, a series of four ankyrin repeats, and a PDZ protein–protein interaction domain (Fig. 8). The MARCKS domain can be phosphorylated by PKC, and this domain in DGKζ also contains a nuclear localization sequence that may facilitate nuclear localization (Aderem 1992). The ankyrin repeats and the PDZ domain mediate protein–protein interactions and are often associated with proteins that function as scaffolds in signaling cascades (Mosavi et al. 2009; Lee and Zheng 2010). DGKζ is

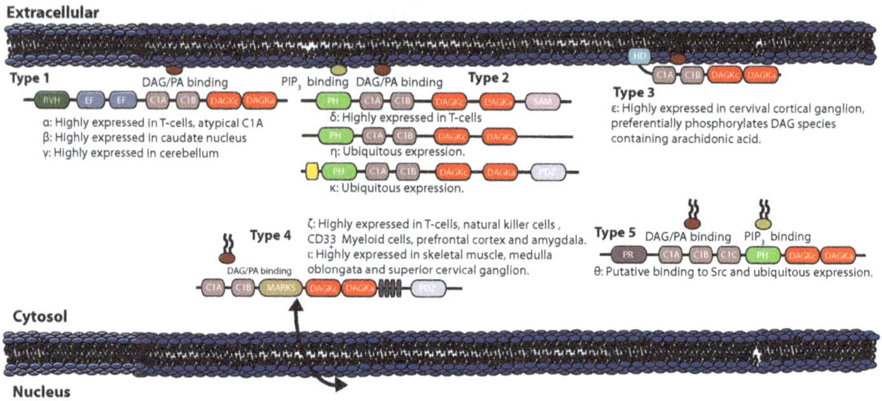

Fig. 8 Domain architecture of the mammalian diacylglycerol kinase (DGK) superfamily. All DGKs contain at least 2 C1 domains (C1A and C1B) and a catalytic domain (divided into DAGKc and DAGKa subdomains). Additional DGK domains permit control of when and where DGKs are active in cell signaling. Only type 3 DGKs are expressed as transmembrane proteins by virtue of a hydrophobic domain (HD). The majority of DGKs are expressed in soluble and membrane fractions and translocated between subcellular locales in an activation-dependent manner. Distinct cell type and tissue expression of DGKs support specific biology regulated by this enzyme family and therapeutic potential for various human diseases

highly expressed in T cells, NK cells, CD33+ myeloid cells, prefrontal cortex, and the amygdala. DGKι is highly expressed in skeletal muscle, medulla oblongata, and the superior cervical ganglion.

DGKθ is the only member of the type 5 subtype with a combined catalytic DAGKc/DAGKa domain, three C1 domains (C1A/C1B/C1C), PH domain, and a proline-rich domain (Fig. 8). Proline-rich regions can frequently bind proteins containing Src homology 3 domains, but it is unclear whether this domain in DGKθ is mediating protein–protein interactions (Tu-Sekine et al. 2016). DGKθ is expressed ubiquitously throughout normal human tissues.

4.1.2 ABPP Discovery of Novel Druggable Sites of DGKs

Development of DGK inhibitors has been hampered by the lack of crystal structures, chemical scaffolds for medicinal chemistry, and functional assays suited for measuring DGK activity in vitro and in vivo. Recent efforts using ABPP methods are revealing new features of the DGK active site that can be leveraged for development of highly potent and selective inhibitors. Functional insights into ligand binding in DGK active sites can also clarify how DGKs mediate substrate recognition and specificity, which will be critical to elucidate isoform-specific biology in vivo.

To expand knowledge of molecular recognition of DGK active sites, the Hsu laboratory established a chemoproteomic strategy using kinase ATP acyl phosphate

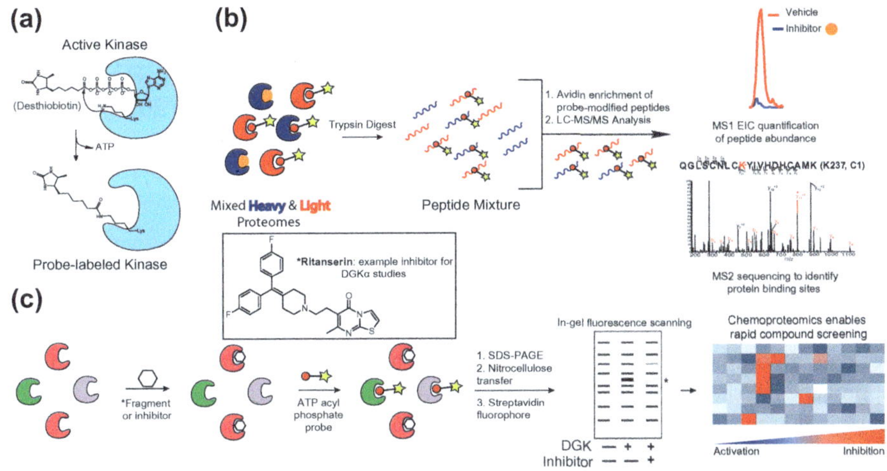

Fig. 9 ABPP analysis of the DGK superfamily. **a** Mechanism of ATP acyl phosphate probe labeling. Nucleophilic lysines in kinase active sites attack the acyl carbon of the ATP probe. Once ATP has been displaced, the kinase is labeled with a desthiobiotin tag for protein identification by in-gel fluorescence scanning or LC-MS/MS. **b** The binding site of DGK inhibitors can be determined by performing a trypsin digest of proteomes from cells cultured in isotopically heavy (^{13}C, ^{15}N) and light (^{12}C, ^{14}N) lysine and arginine amino acids, and enrichment for probe-modified, active-site peptides by affinity chromatography. Mixed heavy and light samples can be directly compared by their MS1 peaks for quantification. Identification of peptides is achieved by MS2 fragmentation and sequencing using bioinformatics. LC-MS/MS studies using the ATP acyl phosphate probe have uncovered the C1 and DAGKa domains as novel druggable sites. **c** Fragment inhibitors can be screened by competitive ABPP. Screening for DGK inhibitors using ABPP methods provides information on potency and selectivity to identify ligand-efficient small molecule binders. ABPP screening of fragment compounds in complex proteomes can accelerate the medicinal chemistry process by guiding selection of hit compounds based on both potency and selectivity. Inset box shows chemical structure of a non-selective DGKα inhibitor ritanserin that has been used for biochemical and biological studies

activity-based probes to map substrate and inhibitor binding sites of all five DGK subtypes (Franks et al. 2017; McCloud et al. 2018) (Fig. 9). Identification of probe-modified, active-site peptides by quantitative LC-MS indicated that the DAGKc and DAGKa domains are important components of the DGK active-site in complex proteomes (Franks et al. 2017). Competition with free ATP resulted in near-complete blockade of probe binding at both DAGKc and DAGKa domains, designating DAGKc/DAGKa as a primary ATP binding site (Franks et al. 2017). A surprising result from these ABPP studies was discovery of a probe-modified, active-site peptide derived from the atypical C1A domain of DGKα. Competitive ABPP LC-MS using a non-selective DGK inhibitor (ritanserin, inset box, Fig. 9) confirmed that C1A and DAGKa were principal ligand-binding sites for ritanserin and fragments derived from this compound (Franks et al. 2017; McCloud et al. 2018). These findings raised the intriguing question of whether C1A and DAGKc/DAGKa were independent ligand-binding sites or if they interact to form a single contiguous binding site. Competitive ABPP LC-MS using a non-selective DGK fragment inhibitor (RLM001) showed equipotent competition at all three binding sites, providing evidence in support of the C1A, DAGKc, and DAGKa forming a single contiguous binding site (McCloud et al. 2018).

4.1.3 ABPP Strategies for Discovery of DGK Inhibitors

While high-throughput screening (HTS) has been pursued for developing DGKα inhibitors (Liu et al. 2016), one of the challenges facing DGK inhibitor development is optimization of selectivity against the 500+ human kinases that also utilize ATP as a common substrate. An alternative to HTS is the use of fragment-based approaches (Erlanson et al. 2016) for development of DGK inhibitors (Fig. 9). The power of fragment-based lead discovery includes identification of fragment compounds (<300 Da) that possess ligand-efficient interactions with protein binding sites that can produce highly selective compounds (Kuntz et al. 1999). A fragment-based approach for developing DGK inhibitors was first pursued by deconstruction of ritanserin into a hydrophobic fragment (RF001) and a thiazolopyrimidinone (RLM001) (Franks et al. 2017; McCloud et al. 2018). RF001 was discovered to be more selective than ritanserin, which helped identify a minimal pharmacophore for selective DGKα inactivation (McCloud et al. 2018). In contrast, another ritanserin-derived fragment, RLM001, which contained nucleotide-like structural features, functioned as a general binding element to kinase ATP pockets and helps explain why the parent molecule ritanserin shows non-specific binding across the kinome (McCloud et al. 2018). Given the ability of competitive ABPP to rapidly identify fragment leads (Backus et al. 2016; Hacker et al. 2017; Parker et al. 2017), future efforts will benefit from identification of new DGK fragment binders that can be optimized along with RF001 by linking, merging, and/or elaboration strategies to develop potent isoform-selective DGK inhibitors (Erlanson et al. 2016).

4.2 *Extrinsic Regulators of T Cell Metabolism*

4.2.1 Metabolic Checkpoints in Glycolysis

Activated T cells also undergo a metabolic switch akin to tumor cells and upregulate aerobic glycolysis to allow differentiation into functional effector T cells. In addition to supplying energy and building blocks, the glycolytic metabolite phosphoenolpyruvate (PEP) is required for sustaining TCR-mediated Ca^{2+} nuclear factor of activated T cells (NFAT) signaling and effector functions by suppressing sarco/ER Ca^{2+}-ATPase (SERCA) activity (Ho et al. 2015). A key glycolytic enzyme that regulates PEP production is the enolase family (EC 4.2.1.11), which comprises a family of metal ion-activated enzymes that catalyze the penultimate step in

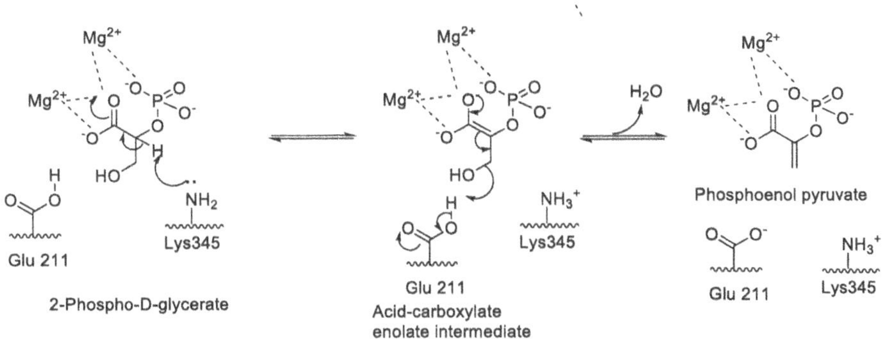

Scheme 1 Enzymatic reaction of ENO1. Residues are numbered based on X-ray crystal structure information obtained from yeast variant. The homologous catalytic residues in human ENO1 are Glu 209 and Lys 342

glycolysis, which is the conversion of 2-phospho-D-glycerate (2PG) to PEP (Liu et al. 2000) (Scheme 1). Humans express three isoforms of enolase, namely non-neuronal enolase (ENO1 or α-ENO) that is expressed in almost all tissues, neuronal enolase (ENO2 or γ-ENO) that is expressed in neuron and neuroendocrine tissues, and muscle-specific enolase (ENO3 or β-ENO) that is predominantly expressed in muscle tissues (Lohman and Meyerhof 1934). Enolases are expressed as homo- or hetero-dimers of α-, β-, or γ-subunits with molecular weights between 82 and 100 kDa (Capello et al. 2011).

Interestingly, ENO1 is also expressed as a shorter isoform (myc-binding protein-1, MBP-1), which localizes to the nucleus and functions as a transcription factor that regulates gene expression of c-Myc (Subramanian and Miller 2000) and FOXP3 (De Rosa et al. 2015). MBP-1 binds to the c-Myc promoter and down-regulates expression of this proto-oncogene (Subramanian and Miller 2000). In contrast, binding of MBP-1 (and potentially other ENO1 isoforms) is required for FOXP3 and specifically the FOXP3-E2 splice variant, expression that drives generation and immunosuppressive activity of human regulatory T cells (Treg cells) (De Rosa et al. 2015). Thus, enolase has adopted both catalytic (glycolysis) and non-catalytic (DNA-binding) functions that will be difficult to distinguish using standard biochemical techniques. Here, the development of activity-based probes that report on ENO1 activity may permit monitoring of active enolase pools in T cell subsets and tumor cells, and guide appropriate strategies to perturb the multifunctional features of this protein. For example, enolase activity-based probes may identify the appropriate ENO1 isoform to selectively inactivate immunosuppressive Tregs in tumors as a potential immunotherapy strategy.

Several enolase inhibitors have been reported (Fig. 10) and serve as starting points for development of enolase activity-based probes. Classical ENO1 inhibitors were designed largely to mimic either the 2PG substrate or its transition state in the

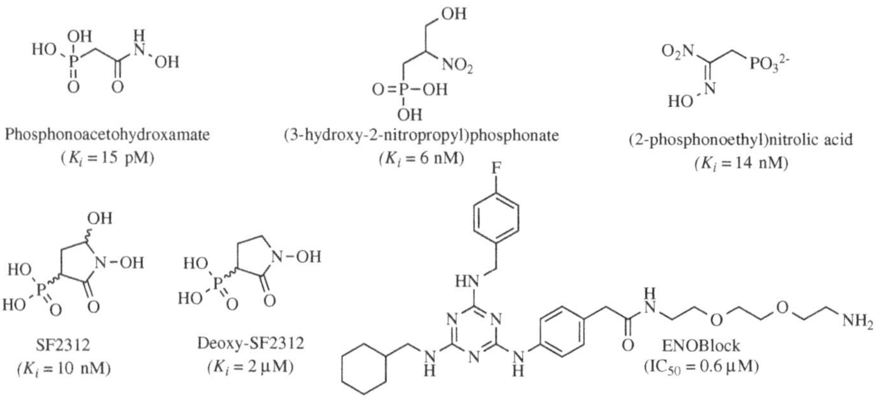

Fig. 10 Chemical structures of enolase inhibitors

dehydration mechanism (Scheme 1). For example, phosphonoacetohydroxamate (PhAH; K_i = 15 pM) is the most potent ENO1 inhibitor known to date (Anderson et al. 1984), and it is thought to mimic the acid–carboxylate form of the intermediate carbanion in the dehydration reaction. Close analogs of PhAH, (3-hydroxy-2-nitropropyl)phosphonate (K_i = 6 nM) and (2-phosphonoethyl)nitrolic acid (K_i = 14 nM), were studied biochemically but not in the context of therapeutic intervention (Anderson et al. 1984). Racemic SF2312, a natural product-based "ring-stabilized" analog of PhAH identified using computational modeling of PEP and 2-PGA in the ENO2 binding site, displays inhibition against both ENO1 (recombinant hENO1 IC_{50} = 38 nM) and ENO2 (recombinant hENO2 IC_{50} = 43 nM) (Leonard et al. 2016). A deoxy-SF2312 (a deoxy analog of SF2312) displayed poor potency (IC_{50} = 2 μM) compared to SF2312, possibly due to the loss of a key hydrogen-bonding interaction with a glutamic acid [Glu 166 in the case of human ENO1 (Leonard et al. 2016)].

As an alternative to designing inhibitors based on substrates, Jung et al. disclosed a novel small molecule named ENOblock (Fig. 10) that was identified through a phenotypic screen to identify molecules that selectively killed cancer cells in a hypoxic environment (Jung et al. 2013). ENOblock inhibited ENO1 (IC_{50} = 0.6 μM) and also reduced cancer cell invasion, migration, and metastasis in vivo (Jung et al. 2013) when tested in HCT116 colon cancer cells. Mechanistically, ENOblock treatment decreased the expression of AKT and Bcl-xL (target proteins for cellular responses to hypoxia) and thus reduced the ability of cancer cells to adapt to hypoxic condition. Further investigations using ABPP methods are needed to carefully define the selectivity profile of ENOblock [which has recently been questioned (Satani et al. 2016)] in order to fully understand its target profile and suitability for probing T cell and/or tumor cell metabolism in vivo.

4.2.2 Glutamine Dependence for Effector Function

T cells also rely on amino acids, in addition to glucose, for proper function and survival (Chang and Pearce 2016). Glutamine serves as a reservoir for nitrogen atoms and a key nutrient for intracellular processes including metabolism, generation of ATP, and biosynthesis of biomolecules including proteins (Cluntun et al. 2017). Akin to glucose, competition for glutamine in the tumor microenvironment can block proliferation and cytokine production of T cells. For example, TCR activation of naïve T cells is coupled with enhanced glutamine uptake that is dependent on the amino acid transporter ASCT2. In addition, ASCT2 couples TCR activation with mTORC1 regulation of metabolic signaling, which supports glutamine as an immunomodulatory metabolite (Nakaya et al. 2014).

Glutamine is synthesized from glutamate by glutamine synthetase, whereas the mitochondrial glutaminase (EC 3.5.1.2) catabolizes glutamine to glutamate. A simplified pathway for glutamine metabolism is shown in Scheme 2. In the first step, L-glutamine is hydrolyzed to L-glutamate by glutaminase. Glutamate is subsequently deaminated by glutamate dehydrogenase to α-ketoglutarate (α-KG), a key intermediate that revives the tricarboxylic acid (TCA) cycle (citric acid cycle or Krebs cycle) when the concentrations of citrate fall below the threshold (Lukey et al. 2013). In tumor cells, the glutamine pathway is upregulated by enhanced expression of glutaminase and the c-Myc-regulated transporter SLC1A5 that supplies exogenous glutamine. The net effect is acceleration of the glutamine pathway, which results in excess generation of α-KG (via the formation of glutamate) to drive the TCA cycle and promote cancer cell proliferation by increasing the production of ATP, proteins, lipids, and nucleic acids. Thus, glutaminase through its role as a rate-limiting enzyme in the degradation of glutamine can regulate TCA cycle anaplerosis (Lukey et al. 2013).

Two isoforms of glutaminase exist in mammals: kidney-type glutaminase (KGA; encoded by *GLS*) and liver-type glutaminase (LGA; encoded by *GLS2*). KGA is

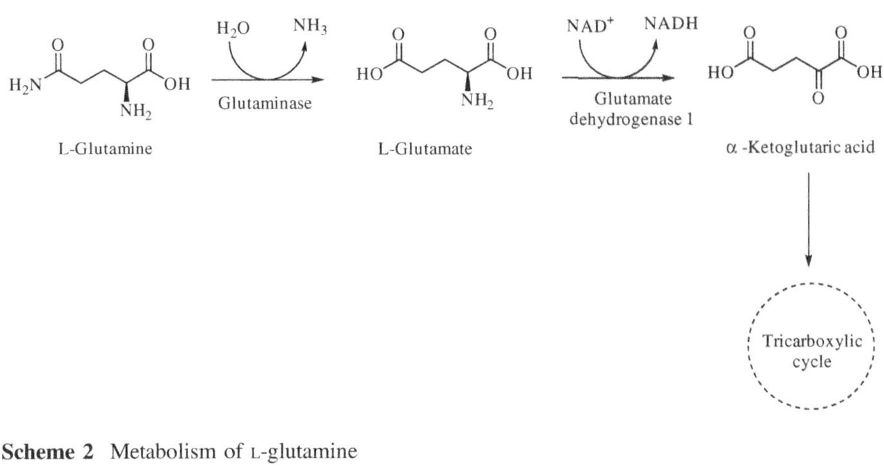

Scheme 2 Metabolism of L-glutamine

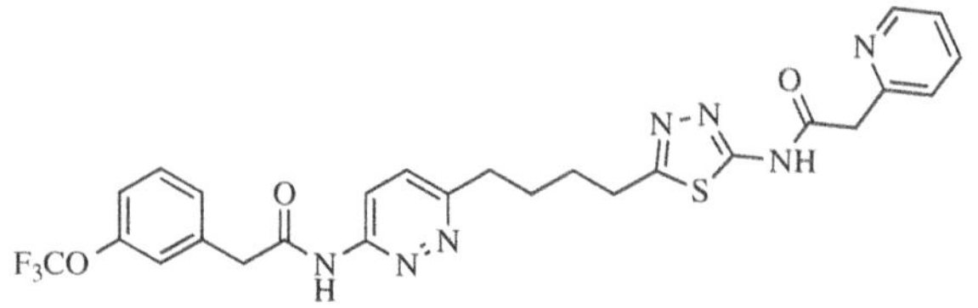

CB-839
(IC_{50} = 50 nM)

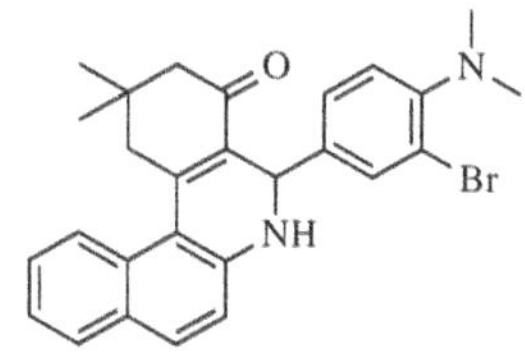

Compound 968
(IC_{50} = 2.5 μM)

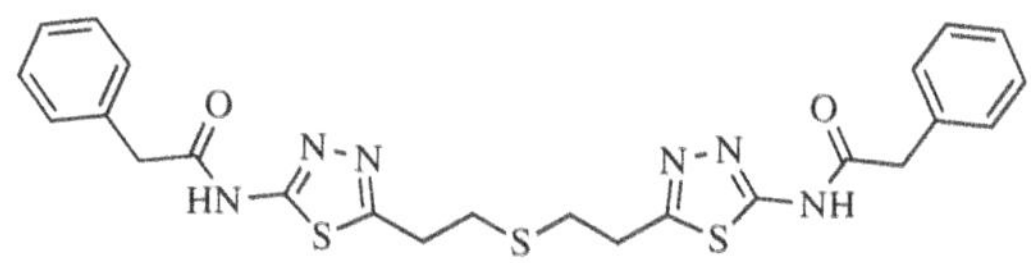

Compound BPTES
(IC_{50} = 60 nM)

Fig. 11 Chemical structures of glutaminase allosteric inhibitors

believed to be an oncogene that is regulated by c-Myc expression, whereas LGA functions as a tumor suppressor and is a target gene for p53 (Lukey et al. 2013). KGA is widely distributed in kidney, brain, intestine, fetal liver, lymphocytes, and transformed cells, whereas LGA is expressed only in periportal hepatocytes of the postnatal liver. Interestingly, a shorter splice variant of KGA, also called GAC, is upregulated in many cancers. A correlation between the expression of GAC and the degree of malignancy, as well as the tumor grade, in breast cancer samples has been established (Cassago et al. 2012; Wang et al. 2010).

First-generation glutaminase inhibitors (Fig. 11) focused on the identification of glutamine-mimetic metabolites; however, the clinical utility of these compounds was limited due to toxicity (Ahluwalia et al. 1990). Compound CB-839 selectively inhibits both the splice variants of *GLS*, namely KGA and GAC (recombinant GAC IC_{50} = 50 nM), and is over 1000-fold less active toward *GLS2*. It exerts antiproliferative effects against triple-negative breast cancer (TNBC) cell lines HCC1806 and MDA-MB-231CB-839, as well as against AML and lung cancer cell lines (Gregory et al. 2018; Gross et al. 2014; Momcilovic et al. 2017). Because of its promising in vivo effects, CB-839 is under investigation in clinical trials for possible treatment of solid and hematological tumors. The compound BPTES (Fig. 11) is identified as an allosteric glutaminase inhibitor with an IC_{50} value of ~60 nM in vitro and ~20 μM in cells (Seltzer et al. 2010). X-ray crystallography data revealed that two molecules of BPTES bind to the glutaminase tetramer and convert it to an inactive conformation (DeLaBarre et al. 2011). Compound 968, a cell-permeable allosteric inhibitor with an IC_{50} of 2.5 μM, blocks the conversion of inactive glutaminase to active glutaminase by stabilizing the conformation of the former state (Wang et al. 2010; Katt et al. 2012). Molecular docking experiments offered additional insights that this inhibitor binds at the hydrophobic pocket between the N- and the C-termini (Katt et al. 2012).

The reported glutaminase inhibitors provide a good starting point for developing novel glutaminase activity-based probes that can serve as reporters of T cell health in the tumor microenvironment. Given that glutamine is essential for metabolism of both immune and tumor cells, a combination of ABPP and in vivo tumor immunology will be needed to determine how cell metabolism can be tuned in support of T cell effector functions in the complex tumor environment. For example, glutaminase activity-based probes may enable selection of appropriate T cell subsets with metabolic phenotypes suitable for vaccination as well as serving as a biomarker for current immunotherapies.

4.2.3 Immunosuppression by Tryptophan Depletion

Tumors hijack enzymes that contribute to an immunosuppressive tumor microenvironment to evade immune responses (Cheong and Sun 2018). Several lines of study suggest that proliferating tumor cells exploit kynurenine metabolism to support immune evasion (Cheong and Sun 2018; Hornyak et al. 2018). The heme-containing enzyme indoleamine 2,3-dioxygenase 1 (IDO1) catalyzes conversion of L-tryptophan (Trp) to N-formylkynurenine (NFK) in the presence of molecular oxygen as the first and rate-limiting step in formation of L-Kynurenine (Efimov et al. 2011; Cheong and Sun 2018) (Scheme 3). The mechanisms of how kynurenine pathway metabolites contribute to immunosuppression remain an active area of investigation. Two pathways are proposed. In the first mechanism, which is based on Trp depletion, a link between Trp depletion and T cells has been established: a greater Trp local concentration results in greater activity and proliferation of T cells; in contrast, lower Trp levels result in T lymphocyte arrest in the G1 phase of the cell cycle. A second mechanism focused on the role of accumulated Trp catabolites such as 3-hydroxykynurenine, 3-hydroxyanthranilic acid, quinolinic

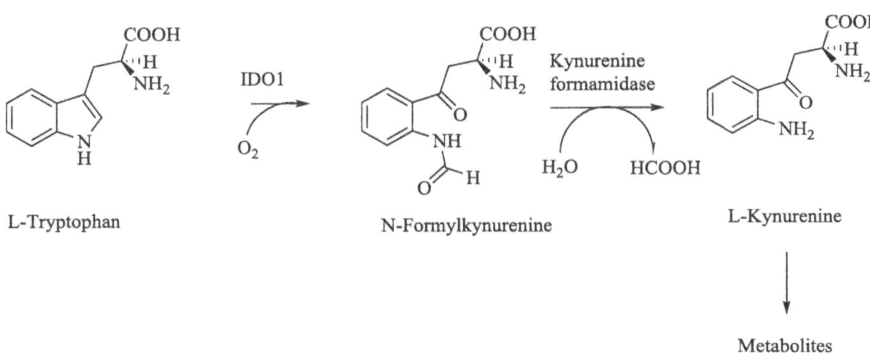

Scheme 3 IDO1 enzymatic pathway for metabolism of L-tryptophan to produce L-Kynurenine and associated metabolites

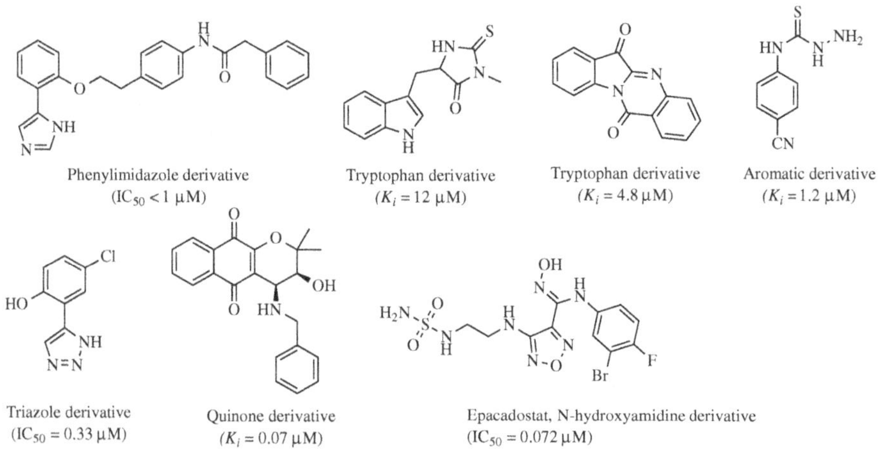

Fig. 12 Representative IDO1 inhibitors published in the literature. For a detailed review, see reference Rohrig et al. (2015)

acid, and picolinic acid which induce differentiation of Treg cells and apoptosis of effector T cells (Hornyak et al. 2018).

Several classes of IDO1 inhibitors that elicit antitumor effects are published in the literature (Rohrig et al. 2015) (Fig. 12). Specifically, epacadostat, a hydroxyamidine IDO1 inhibitor that is orally bioavailable, potently suppresses tryptophan metabolism in vitro in CT26 colon carcinoma and PAN02 pancreatic carcinoma cells (Koblish et al. 2010). In Phase I/II study, this investigational drug showed favorable overall response rate, disease control rate, and progression-free survival (Gajewski et al. 2016). The combination of epacadostat and pembrolizumab (monoclonal anti-PD-1 antibody) was found to be well tolerated, evaluated for efficacy in patients with unresectable or metastatic melanoma (Gajewski et al. 2016), but ultimately failed in the clinic (NCT02752074). Although the cause of failure is currently under investigation, development of a IDO1-tailored activity-based probe could serve as a valuable biomarker for confirming target engagement with epacadostat in vivo to guide future clinical efforts.

5 Conclusions

In summary, we illustrate how ABPP methods can provide new tools to study T cell metabolism, signaling, and cell fate. The use of ABPP to identify lipid-sensing modules of key protein kinases as a principal mechanism of DMF activity in T cells should help guide development of new immunosuppressive drugs (Blewett et al. 2016). The use of universal serine hydrolase activity-based probes provides opportunities to discover new protein mediators of T cell proliferation (Adibekian et al. 2011). Proteasome activity-based probes are capable of tracking asymmetric distribution of active proteasomes in daughter cells to uncover regulatory mechanisms of T cell fate (Chang et al. 2011). Thus, ABPP enables interrogation of specific biology while exploiting conserved enzyme chemistry for global discovery efforts. Projecting forward, ABPP is poised to tackle the challenge of identifying aberrant metabolic remodeling events that limit antitumor immunity (e.g., metabolic checkpoints in glycolysis and lipid phosphorylation) and the development of corresponding agents to reprogram T cell metabolism for beneficial and ultimately therapeutic purposes.

References

Aderem A (1992) The Marcks brothers: a family of protein kinase C substrates. Cell 71:713–716. https://doi.org/10.1016/0092-8674(92)90546-O

Adibekian A, Martin BR, Wang C, Hsu KL, Bachovchin DA, Niessen S, Hoover H, Cravatt BF (2011) Click-generated triazole ureas as ultrapotent in vivo-active serine hydrolase inhibitors. Nat Chem Biol 7(7):469–478. https://doi.org/10.1038/nchembio.579

Ahluwalia GS, Grem JL, Hao Z, Cooney DA (1990) Metabolism and action of amino-acid analog anticancer agents. Pharmacol Therapeut 46(2):243–271. https://doi.org/10.1016/0163-7258(90)90094-I

Albrecht P, Bouchachia I, Goebels N, Henke N, Hofstetter HH, Issberner A, Kovacs Z, Lewerenz J, Lisak D, Maher P, Mausberg AK, Quasthoff K, Zimmermann C, Hartung HP, Methner A (2012) Effects of dimethyl fumarate on neuroprotection and immunomodulation. J Neuroinflamm 9:163. https://doi.org/10.1186/1742-2094-9-163

Alexander JW (1990) Mechanism of immunologic suppression in burn injury. J Trauma 30(12 Suppl):S70–S75

Alloatti A, Kotsias F, Magalhaes JG, Amigorena S (2016) Dendritic cell maturation and cross-presentation: timing matters! Immunol Rev 272(1):97–108. https://doi.org/10.1111/imr.12432

Anderson VE, Weiss PM, Cleland WW (1984) Reaction intermediate analogs for enolase. Biochemistry 23(12):2779–2786. https://doi.org/10.1021/bi00307a038

Bachovchin DA, Ji T, Li W, Simon GM, Blankman JL, Adibekian A, Hoover H, Niessen S, Cravatt BF (2010) Superfamily-wide portrait of serine hydrolase inhibition achieved by library-versus-library screening. Proc Natl Acad Sci U S A 107(49):20941–20946. https://doi.org/10.1073/pnas.1011663107

Backus KM, Correia BE, Lum KM, Forli S, Horning BD, Gonzalez-Paez GE, Chatterjee S, Lanning BR, Teijaro JR, Olson AJ, Wolan DW, Cravatt BF (2016) Proteome-wide covalent ligand discovery in native biological systems. Nature 534(7608):570–574. https://doi.org/10.1038/nature18002

Balachandran VP, Cavnar MJ, Zeng S, Bamboat ZM, Ocuin LM, Obaid H, Sorenson EC, Popow R, Ariyan C, Rossi F, Besmer P, Guo T, Antonescu CR, Taguchi T, Yuan J, Wolchok JD, Allison JP, DeMatteo RP (2011) Imatinib potentiates antitumor T cell responses in gastrointestinal stromal tumor through the inhibition of Ido. Nat Med 17(9):1094–1100. https://doi.org/10.1038/nm.2438

Birks J, Grimley Evans J, Iakovidou V, Tsolaki M, Holt FE (2009) Rivastigmine for Alzheimer's disease. Cochrane Database Syst Rev (2):Cd001191. https://doi.org/10.1002/14651858.cd001191.pub2

Blewett MM, Xie J, Zaro BW, Backus KM, Altman A, Teijaro JR, Cravatt BF (2016) Chemical proteomic map of dimethyl fumarate–sensitive cysteines in primary human T cells. Sci Signal 9(445):rs10. https://doi.org/10.1126/scisignal.aaf7694

Brownlie RJ, Zamoyska R (2013) T cell receptor signalling networks: branched, diversified and bounded. Nat Rev Immunol 13(4):257–269. https://doi.org/10.1038/nri3403

Brownlie RJ, Garcia C, Ravasz M, Zehn D, Salmond RJ, Zamoyska R (2017) Resistance to TGFbeta suppression and improved anti-tumor responses in CD8(+) T cells lacking PTPN22. Nat Commun 8(1):1343. https://doi.org/10.1038/s41467-017-01427-1

Buck MD, Sowell RT, Kaech SM, Pearce EL (2017) Metabolic instruction of immunity. Cell 169 (4):570–586. https://doi.org/10.1016/j.cell.2017.04.004

Capello M, Ferri-Borgogno S, Cappello P, Novelli F (2011) Alpha-Enolase: a promising therapeutic and diagnostic tumor target. FEBS J 278(7):1064–1074. https://doi.org/10.1111/j.1742-4658.2011.08025.x

Cassago A, Ferreira APS, Ferreira IM, Fornezari C, Gomes ERM, Greene KS, Pereira HM, Garratt RC, Dias SMG, Ambrosio ALB (2012) Mitochondrial localization and structure-based phosphate activation mechanism of Glutaminase C with implications for cancer metabolism. Proc Natl Acad Sci U S A 109(4):1092–1097. https://doi.org/10.1073/pnas.1112495109

Chang C-H, Pearce EL (2016) Emerging concepts of T cell metabolism as a target of immunotherapy. Nat Immunol 17(4):364–368. https://doi.org/10.1038/ni.3415

Chang JT, Palanivel VR, Kinjyo I, Schambach F, Intlekofer AM, Banerjee A, Longworth SA, Vinup KE, Mrass P, Oliaro J, Killeen N, Orange JS, Russell SM, Weninger W, Reiner SL (2007) Asymmetric T lymphocyte division in the initiation of adaptive immune responses. Science (New York, NY) 315(5819):1687–1691. https://doi.org/10.1126/science.1139393

Chang JT, Ciocca ML, Kinjyo I, Palanivel VR, McClurkin CE, DeJong CS, Mooney EC, Kim JS, Steinel NC, Oliaro J, Yin CC, Florea BI, Overkleeft HS, Berg LJ, Russell SM, Koretzky GA, Jordan MS, Reiner SL (2011) Asymmetric proteasome segregation as a mechanism for unequal partitioning of the transcription factor T-bet during T lymphocyte division. Immunity 34(4): 492–504. https://doi.org/10.1016/j.immuni.2011.03.017

Chang CH, Curtis JD, Maggi LB Jr, Faubert B, Villarino AV, O'Sullivan D, Huang SC, van der Windt GJ, Blagih J, Qiu J, Weber JD, Pearce EJ, Jones RG, Pearce EL (2013) Posttranscriptional control of T cell effector function by aerobic glycolysis. Cell 153(6): 1239–1251. https://doi.org/10.1016/j.cell.2013.05.016

Chang CH, Qiu J, O'Sullivan D, Buck MD, Noguchi T, Curtis JD, Chen Q, Gindin M, Gubin MM, van der Windt GJ, Tonc E, Schreiber RD, Pearce EJ, Pearce EL (2015a) Metabolic competition in the tumor microenvironment is a driver of cancer progression. Cell 162(6):1229–1241. https://doi.org/10.1016/j.cell.2015.08.016

Chang JW, Zuhl AM, Speers AE, Niessen S, Brown SJ, Mulvihill MM, Fan YC, Spicer TP, Southern M, Scampavia L, Fernandez-Vega V, Dix MM, Cameron MD, Hodder PS, Rosen H, Nomura DK, Kwon O, Hsu KL, Cravatt BF (2015b) Selective inhibitor of platelet-activating factor acetylhydrolases 1b2 and 1b3 that impairs cancer cell survival. ACS Chem Biol 10 (4):925–932. https://doi.org/10.1021/cb500893q

Chen H, Assmann JC, Krenz A, Rahman M, Grimm M, Karsten CM, Kohl J, Offermanns S, Wettschureck N, Schwaninger M (2014) Hydroxycarboxylic acid receptor 2 mediates dimethyl fumarate's protective effect in EAE. J Clin Investig 124(5):2188–2192. https://doi.org/10.1172/jci72151

Cheong JE, Sun L (2018) Targeting the IDO1/TDO2-KYN-AhR pathway for cancer immunotherapy—challenges and opportunities. Trends Pharmacol Sci 39(3):307–325. https://doi.org/10.1016/j.tips.2017.11.007

Cluntun AA, Lukey MJ, Cerione RA, Locasale JW (2017) Glutamine metabolism in cancer: understanding the heterogeneity. Trends Cancer 3(3):169–180. https://doi.org/10.1016/j.trecan.2017.01.005

Colón-González F, Kazanietz MG (2006) C1 domains exposed: from diacylglycerol binding to protein–protein interactions. Biochim Biophys Acta 1761:827–837. https://doi.org/10.1016/j.bbalip.2006.05.001

Cravatt BF, Wright AT, Kozarich JW (2008) Activity-based protein profiling: from enzyme chemistry to proteomic chemistry. Annu Rev Biochem 77:383–414. https://doi.org/10.1146/annurev.biochem.75.101304.124125

Cui W, Joshi NS, Jiang A, Kaech SM (2009) Effects of Signal 3 during CD8 T cell priming: bystander production of IL-12 enhances effector T cell expansion but promotes terminal differentiation. Vaccine 27(15):2177–2187. https://doi.org/10.1016/j.vaccine.2009.01.088

Cui W, Liu Y, Weinstein JS, Craft J, Kaech SM (2011) An interleukin-21-interleukin-10-STAT3 pathway is critical for functional maturation of memory CD8+ T cells. Immunity 35(5): 792–805. https://doi.org/10.1016/j.immuni.2011.09.017

De Rosa V, Galgani M, Porcellini A, Colamatteo A, Santopaolo M, Zuchegna C, Romano A, De Simone S, Procaccini C, La Rocca C, Carrieri PB, Maniscalco GT, Salvetti M, Buscarinu MC, Franzese A, Mozzillo E, La Cava A, Matarese G (2015) Glycolysis controls the induction of human regulatory T cells by modulating the expression of FOXP3 exon 2 splicing variants. Nat Immunol 16(11):1174–1184. https://doi.org/10.1038/ni.3269

DeLaBarre B, Gross S, Fang C, Gao Y, Jha A, Jiang F, Song JJ, Wei WT, Hurov JB (2011) Full-length human glutaminase in complex with an allosteric inhibitor. Biochemistry 50(50): 10764–10770. https://doi.org/10.1021/bi201613d

Efimov I, Basran J, Thackray SJ, Handa S, Mowat CG, Raven EL (2011) Structure and reaction mechanism in the heme dioxygenases. Biochemistry 50(14):2717–2724. https://doi.org/10.1021/bi101732n

Eijkelenboom A, Burgering BM (2013) FOXOs: signalling integrators for homeostasis maintenance. Nat Rev Mol Cell Biol 14(2):83–97. https://doi.org/10.1038/nrm3507

Erlanson DA, Fesik SW, Hubbard RE, Jahnke W, Jhoti H (2016) Twenty years on: the impact of fragments on drug discovery. Nat Rev Drug Discov 15(9):605–619. https://doi.org/10.1038/nrd.2016.109

Fanani ML, Topham MK, Walsh JP, Epand RM (2004) Lipid modulation of the activity of diacylglycerol kinase α- and ζ-isoforms: activation by phosphatidylethanolamine and cholesterol. Biochemistry 43:14767–14777. https://doi.org/10.1021/bi049145z

Finlay DK, Rosenzweig E, Sinclair LV, Feijoo-Carnero C, Hukelmann JL, Rolf J, Panteleyev AA, Okkenhaug K, Cantrell DA (2012) PDK1 regulation of mTOR and hypoxia-inducible factor 1 integrate metabolism and migration of CD8+ T cells. J Exp Med 209(13):2441–2453. https://doi.org/10.1084/jem.20112607

Fischer AM, Katayama CD, Pages G, Pouyssegur J, Hedrick SM (2005) The role of erk1 and erk2 in multiple stages of T cell development. Immunity 23(4):431–443. https://doi.org/10.1016/j.immuni.2005.08.013

Franks CE, Campbell ST, Purow BW, Harris TE, Hsu KL (2017) The ligand binding landscape of diacylglycerol kinases. Cell Chem Biol 24(7):870–880 e875. https://doi.org/10.1016/j.chembiol.2017.06.007

Gajewski TF, Hamid O, Smith DC, Bauer TM, Wasser JS, Luke JJ, Balmanoukian AS, Kaufman DR, Zhao YF, Maleski J, Leopold L, Gangadhar TC (2016) Preliminary safety and efficacy data from a phase 1/2 study of epacadostat (INCB024360) in combination with pembrolizumab in patients with advanced/metastatic melanoma. J Transl Med 14

Gillard GO, Collette B, Anderson J, Chao J, Scannevin RH, Huss DJ, Fontenot JD (2015) DMF, but not other fumarates, inhibits NF-kappaB activity in vitro in an Nrf2-independent manner. J Neuroimmunol 283:74–85. https://doi.org/10.1016/j.jneuroim.2015.04.006

Greenbaum D, Medzihradszky KF, Burlingame A, Bogyo M (2000) Epoxide electrophiles as activity-dependent cysteine protease profiling and discovery tools. Chem Biol 7(8):569–581

Gregory MA, Nemkov T, Reisz JA, Zaberezhnyy V, Hansen KC, D'Alessandro A, DeGregori J (2018) Glutaminase inhibition improves FLT3 inhibitor therapy for acute myeloid leukemia. Exp Hematol 58:52–58. https://doi.org/10.1016/j.exphem.2017.09.007

Groll M, Kim KB, Kairies N, Huber R, Crews CM (2000) Crystal structure of epoxomicin: 20S proteasome reveals a molecular basis for selectivity of α', β'-epoxyketone proteasome inhibitors. J Am Chem Soc 122(6):1237–1238. https://doi.org/10.1021/ja993588m

Gross MI, Demo SD, Dennison JB, Chen L, Chernov-Rogan T, Goyal B, Janes JR, Laidig GJ, Lewis ER, Li J, Mackinnon AL, Parlati F, Rodriguez ML, Shwonek PJ, Sjogren EB, Stanton TF, Wang T, Yang J, Zhao F, Bennett MK (2014) Antitumor activity of the glutaminase inhibitor CB-839 in triple-negative breast cancer. Mol Cancer Ther 13(4): 890–901. https://doi.org/10.1158/1535-7163.MCT-13-0870

Hacker SM, Backus KM, Lazear MR, Forli S, Correia BE, Cravatt BF (2017) Global profiling of lysine reactivity and ligandability in the human proteome. Nat Chem 9(12):1181–1190. https://doi.org/10.1038/nchem.2826

Hekmat O, Kim YW, Williams SJ, He S, Withers SG (2005) Active-site peptide "fingerprinting" of glycosidases in complex mixtures by mass spectrometry. Discovery of a novel retaining beta-1,4-glycanase in Cellulomonas fimi. J Biol Chem 280(42):35126–35135. https://doi.org/10.1074/jbc.M508434200

Henness S, Perry CM (2006) Orlistat: a review of its use in the management of obesity. Drugs 66(12):1625–1656

Hirahara K, Onodera A, Villarino AV, Bonelli M, Sciume G, Laurence A, Sun HW, Brooks SR, Vahedi G, Shih HY, Gutierrez-Cruz G, Iwata S, Suzuki R, Mikami Y, Okamoto Y, Nakayama T, Holland SM, Hunter CA, Kanno Y, O'Shea JJ (2015) Asymmetric action of STAT transcription factors drives transcriptional outputs and cytokine specificity. Immunity 42(5):877–889. https://doi.org/10.1016/j.immuni.2015.04.014

Ho PC, Bihuniak JD, Macintyre AN, Staron M, Liu X, Amezquita R, Tsui YC, Cui G, Micevic G, Perales JC, Kleinstein SH, Abel ED, Insogna KL, Feske S, Locasale JW, Bosenberg MW, Rathmell JC, Kaech SM (2015) Phosphoenolpyruvate is a metabolic checkpoint of anti-tumor T cell responses. Cell 162(6):1217–1228. https://doi.org/10.1016/j.cell.2015.08.012

Hornyak L, Dobos N, Koncz G, Karanyi Z, Pall D, Szabo Z, Halmos G, Szekvolgyi L (2018) The role of indoleamine-2,3-dioxygenase in cancer development, diagnostics, and therapy. Front Immunol 9:151. https://doi.org/10.3389/fimmu.2018.00151

Hsieh CS, Macatonia SE, Tripp CS, Wolf SF, O'Garra A, Murphy KM (1993) Development of TH1 CD4+ T cells through IL-12 produced by Listeria-induced macrophages. Science (New York, NY) 260(5107):547–549

Hsu KL, Tsuboi K, Adibekian A, Pugh H, Masuda K, Cravatt BF (2012) DAGLbeta inhibition perturbs a lipid network involved in macrophage inflammatory responses. Nat Chem Biol 8(12):999–1007. https://doi.org/10.1038/nchembio.1105

Hsu KL, Tsuboi K, Chang JW, Whitby LR, Speers AE, Pugh H, Cravatt BF (2013a) Discovery and optimization of piperidyl-1,2,3-triazole ureas as potent, selective, and in vivo-active inhibitors of alpha/beta-hydrolase domain containing 6 (ABHD6). J Med Chem 56(21):8270–8279. https://doi.org/10.1021/jm400899c

Hsu KL, Tsuboi K, Whitby LR, Speers AE, Pugh H, Inloes J, Cravatt BF (2013b) Development and optimization of piperidyl-1,2,3-triazole ureas as selective chemical probes of endocannabinoid biosynthesis. J Med Chem 56(21):8257–8269. https://doi.org/10.1021/jm400898x

Hukelmann JL, Anderson KE, Sinclair LV, Grzes KM, Murillo AB, Hawkins PT, Stephens LR, Lamond AI, Cantrell DA (2016) The cytotoxic T cell proteome and its shaping by the kinase mTOR. Nat Immunol 17(1):104–112. https://doi.org/10.1038/ni.3314

Hurley JH, Misra S (2000) Signaling and subcellular targeting by membrane-binding domains. Annu Rev Biophys Biomol Struct 29:49–79. https://doi.org/10.1146/annurev.biophys.29.1.49

Inloes JM, Hsu KL, Dix MM, Viader A, Masuda K, Takei T, Wood MR, Cravatt BF (2014) The hereditary spastic paraplegia-related enzyme DDHD2 is a principal brain triglyceride lipase. Proc Natl Acad Sci U S A 111(41):14924–14929. https://doi.org/10.1073/pnas.1413706111

Jalkanen S, Reichert RA, Gallatin WM, Bargatze RF, Weissman IL, Butcher EC (1986) Homing receptors and the control of lymphocyte migration. Immunol Rev 91:39–60

Jennings W, Doshi S, D'Souza K, Epand RM (2015) Molecular properties of diacylglycerol kinase-epsilon in relation to function. Chem Phys Lipid 192:100–108. https://doi.org/10.1016/j.chemphyslip.2015.06.003

Jessani N, Liu Y, Humphrey M, Cravatt BF (2002) Enzyme activity profiles of the secreted and membrane proteome that depict cancer cell invasiveness. Proc Natl Acad Sci U S A 99(16): 10335–10340. https://doi.org/10.1073/pnas.162187599

Jessani N, Humphrey M, McDonald WH, Niessen S, Masuda K, Gangadharan B, Yates JR 3rd, Mueller BM, Cravatt BF (2004) Carcinoma and stromal enzyme activity profiles associated with breast tumor growth in vivo. Proc Natl Acad Sci U S A 101(38):13756–13761. https://doi.org/10.1073/pnas.0404727101

Jessani N, Niessen S, Wei BQ, Nicolau M, Humphrey M, Ji Y, Han W, Noh DY, Yates JR 3rd, Jeffrey SS, Cravatt BF (2005) A streamlined platform for high-content functional proteomics of primary human specimens. Nat Methods 2(9):691–697. https://doi.org/10.1038/nmeth778

Joshi RP, Schmidt AM, Das J, Pytel D, Riese MJ, Lester M, Diehl JA, Behrens EM, Kambayashi T, Koretzky GA (2013) The zeta isoform of diacylglycerol kinase plays a predominant role in regulatory T cell development and TCR-mediated ras signaling. Sci Signal 6(303):ra102. https://doi.org/10.1126/scisignal.2004373

Jun JE, Rubio I, Roose JP (2013) Regulation of ras exchange factors and cellular localization of ras activation by lipid messengers in T cells. Front Immunol 4:239. https://doi.org/10.3389/fimmu.2013.00239

Jung DW, Kim WH, Park SH, Lee J, Kim J, Su D, Ha HH, Chang YT, Williams DR (2013) A unique small molecule inhibitor of enolase clarifies its role in fundamental biological processes. ACS Chem Biol 8(6):1271–1282. https://doi.org/10.1021/cb300687k

Katt WP, Ramachandran S, Erickson JW, Cerione RA (2012) Dibenzophenanthridines as inhibitors of glutaminase C and cancer cell proliferation. Mol Cancer Ther 11(6):1269–1278. https://doi.org/10.1158/1535-7163.MCT-11-0942

Kerdiles YM, Beisner DR, Tinoco R, Dejean AS, Castrillon DH, DePinho RA, Hedrick SM (2009) Foxo1 links homing and survival of naive T cells by regulating L-selectin, CCR7 and interleukin 7 receptor. Nat Immunol 10(2):176–184. https://doi.org/10.1038/ni.1689

Kluge AF, Petter RC (2010) Acylating drugs: redesigning natural covalent inhibitors. Curr Opin Chem Biol 14(3):421–427. https://doi.org/10.1016/j.cbpa.2010.03.035

Koblish HK, Hansbury MJ, Bowman KJ, Yang GJ, Neilan CL, Haley PJ, Burn TC, Waeltz P, Sparks RB, Yue EW, Combs AP, Scherle PA, Vaddi K, Fridman JS (2010) Hydroxyamidine inhibitors of indoleamine-2,3-dioxygenase potently suppress systemic tryptophan catabolism and the growth of IDO-expressing tumors. Mol Cancer Ther 9(2):489–498. https://doi.org/10.1158/1535-7163.MCT-09-0628

Koneru M, Schaer D, Monu N, Ayala A, Frey AB (2005) Defective proximal TCR signaling inhibits CD8+ tumor-infiltrating lymphocyte lytic function. J Immunol 174(4):1830–1840

Kuntz ID, Chen K, Sharp KA, Kollman PA (1999) The maximal affinity of ligands. Proc Natl Acad Sci U S A 96(18):9997–10002

Lee H-J, Zheng JJ (2010) PDZ domains and their binding partners: structure, specificity, and modification. Cell Commun Signal 8:8. https://doi.org/10.1186/1478-811X-8-8

Lehmann JC, Listopad JJ, Rentzsch CU, Igney FH, von Bonin A, Hennekes HH, Asadullah K, Docke WD (2007) Dimethylfumarate induces immunosuppression via glutathione depletion and subsequent induction of heme oxygenase 1. J Invest Dermatol 127(4):835–845. https://doi.org/10.1038/sj.jid.5700686

Leonard PG, Satani N, Maxwell D, Lin YH, Hammoudi N, Peng Z, Pisaneschi F, Link TM, Lee GRt, Sun D, Prasad BAB, Di Francesco ME, Czako B, Asara JM, Wang YA, Bornmann W, DePinho RA, Muller FL (2016) SF2312 is a natural phosphonate inhibitor of enolase. Nat Chem Biol 12(12):1053–1058. https://doi.org/10.1038/nchembio.2195

Li N, Kuo C-L, Paniagua G, van den Elst H, Verdoes M, Willems LI, van der Linden WA, Ruben M, van Genderen E, Gubbens J, van Wezel GP, Overkleeft HS, Florea BI (2013) Relative quantification of proteasome activity by activity-based protein profiling and LC-MS/MS. Nat Protoc 8:1155. https://doi.org/10.1038/nprot.2013.065

Lin SX, Lisi L, Dello Russo C, Polak PE, Sharp A, Weinberg G, Kalinin S, Feinstein DL (2011) The anti-inflammatory effects of dimethyl fumarate in astrocytes involve glutathione and haem oxygenase-1. ASN Neuro 3(2). https://doi.org/10.1042/an20100033

Lin WH, Adams WC, Nish SA, Chen YH, Yen B, Rothman NJ, Kratchmarov R, Okada T, Klein U, Reiner SL (2015) Asymmetric PI3K signaling driving developmental and regenerative cell fate bifurcation. Cell Rep 13(10):2203–2218. https://doi.org/10.1016/j.celrep.2015.10.072

Linker RA, Gold R (2013) Dimethyl fumarate for treatment of multiple sclerosis: mechanism of action, effectiveness, and side effects. Curr Neurol Neurosci Rep 13(11):394. https://doi.org/10.1007/s11910-013-0394-8

Liu Y, Patricelli MP, Cravatt BF (1999) Activity-based protein profiling: the serine hydrolases. Proc Natl Acad Sci U S A 96(26):14694–14699

Liu H, Zhang Y, Yang W (2000) How is the active site of enolase organized to catalyze two different reaction steps? J Am Chem Soc 122(28):6560–6570. https://doi.org/10.1021/ja9936619

Liu X, Shin N, Koblish HK, Yang G, Wang Q, Wang K, Leffet L, Hansbury MJ, Thomas B, Rupar M, Waeltz P, Bowman KJ, Polam P, Sparks RB, Yue EW, Li Y, Wynn R, Fridman JS, Burn TC, Combs AP, Newton RC, Scherle PA (2010) Selective inhibition of IDO1 effectively regulates mediators of antitumor immunity. Blood 115(17):3520–3530. https://doi.org/10.1182/blood-2009-09-246124

Liu K, Kunii N, Sakuma M, Yamaki A, Mizuno S, Sato M, Sakai H, Kado S, Kumagai K, Kojima H, Okabe T, Nagano T, Shirai Y, Sakane F (2016) A novel diacylglycerol kinase alpha-selective inhibitor, CU-3, induces cancer cell apoptosis and enhances immune response. J Lipid Res 57(3):368–379. https://doi.org/10.1194/jlr.M062794

Lohman K, Meyerhof O (1934) Über die enzymatische Umwandlung von Phosphoglyzerinsäure in Brenztraubensäure und Phosphorsäure (Enzymatic transformation of phosphoglyceric acid into pyruvic and phosphoric acid). Biochem Z 273:60–72

Long JZ, Cravatt BF (2011) The metabolic serine hydrolases and their functions in mammalian physiology and disease. Chem Rev 111(10):6022–6063. https://doi.org/10.1021/cr200075y

Long JZ, Li W, Booker L, Burston JJ, Kinsey SG, Schlosburg JE, Pavon FJ, Serrano AM, Selley DE, Parsons LH, Lichtman AH, Cravatt BF (2009) Selective blockade of 2-arachidonoylglycerol hydrolysis produces cannabinoid behavioral effects. Nat Chem Biol 5(1):37–44. https://doi.org/10.1038/nchembio.129

Lukey MJ, Wilson KF, Cerione RA (2013) Therapeutic strategies impacting cancer cell glutamine metabolism. Future Med Chem 5(14):1685–1700. https://doi.org/10.4155/fmc.13.130

Lung M, Shulga YV, Ivanova PT, Myers DS, Milne SB, Brown HA, Topham MK, Epand RM (2009) Diacylglycerol kinase ϵ is selective for both acyl chains of phosphatidic acid or diacylglycerol. J Biol Chem 284:31062–31073. https://doi.org/10.1074/jbc.M109.050617

Luster AD (2002) The role of chemokines in linking innate and adaptive immunity. Curr Opin Immunol 14(1):129–135

Ma EH, Bantug G, Griss T, Condotta S, Johnson RM, Samborska B, Mainolfi N, Suri V, Guak H, Balmer ML, Verway MJ, Raissi TC, Tsui H, Boukhaled G, Henriques da Costa S, Frezza C, Krawczyk CM, Friedman A, Manfredi M, Richer MJ, Hess C, Jones RG (2017) Serine is an essential metabolite for effector T cell expansion. Cell Metab 25(2):345–357. https://doi.org/10.1016/j.cmet.2016.12.011

Macintyre AN, Finlay D, Preston G, Sinclair LV, Waugh CM, Tamas P, Feijoo C, Okkenhaug K, Cantrell DA (2011) Protein kinase B controls transcriptional programs that direct cytotoxic T cell fate but is dispensable for T cell metabolism. Immunity 34(2):224–236. https://doi.org/10.1016/j.immuni.2011.01.012

Manguso RT, Pope HW, Zimmer MD, Brown FD, Yates KB, Miller BC, Collins NB, Bi K, LaFleur MW, Juneja VR, Weiss SA, Lo J, Fisher DE, Miao D, Van Allen E, Root DE, Sharpe AH, Doench JG, Haining WN (2017) In vivo CRISPR screening identifies Ptpn2 as a cancer immunotherapy target. Nature 547(7664):413–418. https://doi.org/10.1038/nature23270

McCloud RL, Franks CE, Campbell ST, Purow BW, Harris TE, Hsu KL (2018) Deconstructing lipid kinase inhibitors by chemical proteomics. Biochemistry 57(2):231–236. https://doi.org/10.1021/acs.biochem.7b00962

Meng L, Mohan R, Kwok BH, Elofsson M, Sin N, Crews CM (1999) Epoxomicin, a potent and selective proteasome inhibitor, exhibits in vivo antiinflammatory activity. Proc Natl Acad Sci U S A 96(18):10403–10408

Mérida I, Ávila-Flores A, Merino E (2008) Diacylglycerol kinases: at the hub of cell signalling. Biochem J 409:1–18. https://doi.org/10.1042/BJ20071040

Mérida I, Andrada E, Gharbi SI, Avila-Flores A (2015) Redundant and specialized roles for diacylglycerol kinases alpha and zeta in the control of T cell functions. Sci Signal 8(374):re6. https://doi.org/10.1126/scisignal.aaa0974

Merino E, Sanjuan MA, Moraga I, Cipres A, Merida I (2007) Role of the diacylglycerol kinase alpha-conserved domains in membrane targeting in intact T cells. J Biol Chem 282(48):35396–35404. https://doi.org/10.1074/jbc.M702085200

Momcilovic M, Bailey ST, Lee JT, Fishbein MC, Magyar C, Braas D, Graeber T, Jackson NJ, Czernin J, Emberley E, Gross M, Janes J, Mackinnon A, Pan A, Rodriguez M, Works M, Zhang W, Parlati F, Demo S, Garon E, Krysan K, Walser TC, Dubinett SM, Sadeghi S, Christofk HR, Shackelford DB (2017) Targeted inhibition of EGFR and glutaminase induces metabolic crisis in EGFR mutant lung cancer. Cell reports 18(3):601–610. https://doi.org/10.1016/j.celrep.2016.12.061

Monu N, Frey AB (2007) Suppression of proximal T cell receptor signaling and lytic function in CD8+ tumor-infiltrating T cells. Cancer Res 67(23):11447–11454. https://doi.org/10.1158/0008-5472.CAN-07-1441

Mosavi LK, Cammett TJ, Desrosiers DC, Peng Z-y (2009) The ankyrin repeat as molecular architecture for protein recognition. Protein Sci 13:1435–1448. https://doi.org/10.1110/ps.03554604

Mrowietz U, Asadullah K (2005) Dimethylfumarate for psoriasis: more than a dietary curiosity. Trends Mol Med 11(1):43–48. https://doi.org/10.1016/j.molmed.2004.11.003

Mrowietz U, Altmeyer P, Bieber T, Röcken M, Schopf RE, Sterry W (2007) Treatment of psoriasis with fumaric acid esters (Fumaderm®). JDDG: J Dtsch Dermatol Ges 5(8):716–717. https://doi.org/10.1111/j.1610-0387.2007.06346.x

Munn DH, Zhou M, Attwood JT, Bondarev I, Conway SJ, Marshall B, Brown C, Mellor AL (1998) Prevention of allogeneic fetal rejection by tryptophan catabolism. Science (New York, NY) 281(5380):1191–1193. https://doi.org/10.1126/science.281.5380.1191

Nagano JM, Hsu KL, Whitby LR, Niphakis MJ, Speers AE, Brown SJ, Spicer T, Fernandez-Vega V, Ferguson J, Hodder P, Srinivasan P, Gonzalez TD, Rosen H, Bahnson BJ, Cravatt BF (2013) Selective inhibitors and tailored activity probes for lipoprotein-associated phospholipase A(2). Bioorg Med Chem Lett 23(3):839–843. https://doi.org/10.1016/j.bmcl.2012.11.061

Nakaya M, Xiao Y, Zhou X, Chang JH, Chang M, Cheng X, Blonska M, Lin X, Sun SC (2014) Inflammatory T cell responses rely on amino acid transporter ASCT2 facilitation of glutamine uptake and mTORC1 kinase activation. Immunity 40(5):692–705. https://doi.org/10.1016/j.immuni.2014.04.007

Niphakis MJ, Cravatt BF (2014) Enzyme inhibitor discovery by activity-based protein profiling. Annu Rev Biochem 83(1):341–377. https://doi.org/10.1146/annurev-biochem-060713-035708

Ogasawara D, Deng H, Viader A, Baggelaar MP, Breman A, den Dulk H, van den Nieuwendijk AM, Soethoudt M, van der Wel T, Zhou J, Overkleeft HS, Sanchez-Alavez M, Mori S, Nguyen W, Conti B, Liu X, Chen Y, Liu QS, Cravatt BF, van der Stelt M (2016) Rapid and profound rewiring of brain lipid signaling networks by acute diacylglycerol lipase inhibition. Proc Natl Acad Sci U S A 113(1):26–33. https://doi.org/10.1073/pnas.1522364112

Okerberg ES, Wu J, Zhang B, Samii B, Blackford K, Winn DT, Shreder KR, Burbaum JJ, Patricelli MP (2005) High-resolution functional proteomics by active-site peptide profiling. Proc Natl Acad Sci U S A 102(14):4996–5001. https://doi.org/10.1073/pnas.0501205102

Olenchock BA, Guo R, Carpenter JH, Jordan M, Topham MK, Koretzky GA, Zhong XP (2006) Disruption of diacylglycerol metabolism impairs the induction of T cell anergy. Nat Immunol 7 (11):1174–1181. https://doi.org/10.1038/ni1400

Palmer DC, Guittard GC, Franco Z, Crompton JG, Eil RL, Patel SJ, Ji Y, Van Panhuys N, Klebanoff CA, Sukumar M, Clever D, Chichura A, Roychoudhuri R, Varma R, Wang E,

Gattinoni L, Marincola FM, Balagopalan L, Samelson LE, Restifo NP (2015) Cish actively silences TCR signaling in CD8+ T cells to maintain tumor tolerance. J Exp Med 212(12): 2095–2113. https://doi.org/10.1084/jem.20150304

Parker CG, Galmozzi A, Wang Y, Correia BE, Sasaki K, Joslyn CM, Kim AS, Cavallaro CL, Lawrence RM, Johnson SR, Narvaiza I, Saez E, Cravatt BF (2017) Ligand and target discovery by fragment-based screening in human cells. Cell 168(3):527–541 e529. https://doi.org/10.1016/j.cell.2016.12.029

Patel SJ, Sanjana NE, Kishton RJ, Eidizadeh A, Vodnala SK, Cam M, Gartner JJ, Jia L, Steinberg SM, Yamamoto TN, Merchant AS, Mehta GU, Chichura A, Shalem O, Tran E, Eil R, Sukumar M, Guijarro EP, Day CP, Robbins P, Feldman S, Merlino G, Zhang F, Restifo NP (2017) Identification of essential genes for cancer immunotherapy. Nature 548 (7669):537–542. https://doi.org/10.1038/nature23477

Patricelli MP, Szardenings AK, Liyanage M, Nomanbhoy TK, Wu M, Weissig H, Aban A, Chun D, Tanner S, Kozarich JW (2007) Functional interrogation of the kinome using nucleotide acyl phosphates. Biochemistry 46(2):350–358. https://doi.org/10.1021/bi062142x

Patricelli MP, Nomanbhoy Tyzoon K, Wu J, Brown H, Zhou D, Zhang J, Jagannathan S, Aban A, Okerberg E, Herring C, Nordin B, Weissig H, Yang Q, Lee J-D, Gray Nathanael S, Kozarich John W (2011) In situ kinase profiling reveals functionally relevant properties of native kinases. Chem Biol 18(6):699–710

Pauls SD, Marshall AJ (2017) Regulation of immune cell signaling by SHIP1: a phosphatase, scaffold protein, and potential therapeutic target. Eur J Immunol 47(6):932–945. https://doi.org/10.1002/eji.201646795

Quann EJ, Liu X, Altan-Bonnet G, Huse M (2011) A cascade of protein kinase C isozymes promotes cytoskeletal polarization in T cells. Nat Immunol 12(7):647–654. https://doi.org/10.1038/ni.2033

Randolph GJ, Ochando J, Partida-Sanchez S (2008) Migration of dendritic cell subsets and their precursors. Annu Rev Immunol 26:293–316. https://doi.org/10.1146/annurev.immunol.26.021607.090254

Rathmell JC, Elstrom RL, Cinalli RM, Thompson CB (2003) Activated Akt promotes increased resting T cell size, CD28-independent T cell growth, and development of autoimmunity and lymphoma. Eur J Immunol 33(8):2223–2232. https://doi.org/10.1002/eji.200324048

Reading JL, Galvez-Cancino F, Swanton C, Lladser A, Peggs KS, Quezada SA (2018) The function and dysfunction of memory CD8(+) T cells in tumor immunity. Immunol Rev 283 (1):194–212. https://doi.org/10.1111/imr.12657

Riese MJ, Grewal J, Das J, Zou T, Patil V, Chakraborty AK, Koretzky GA (2011) Decreased diacylglycerol metabolism enhances ERK activation and augments CD8+ T cell functional responses. J Biol Chem 286(7):5254–5265. https://doi.org/10.1074/jbc.M110.171884

Rohrig UF, Majjigapu SR, Vogel P, Zoete V, Michielin O (2015) Challenges in the discovery of indoleamine 2,3-dioxygenase 1 (IDO1) inhibitors. J Med Chem 58(24):9421–9437. https://doi.org/10.1021/acs.jmedchem.5b00326

Ross SH, Rollings C, Anderson KE, Hawkins PT, Stephens LR, Cantrell DA (2016) Phosphoproteomic analyses of interleukin 2 signaling reveal integrated JAK kinase-dependent and -independent networks in CD8(+) T cells. Immunity 45(3):685–700. https://doi.org/10.1016/j.immuni.2016.07.022

Rosse C, Linch M, Kermorgant S, Cameron AJ, Boeckeler K, Parker PJ (2010) PKC and the control of localized signal dynamics. Nat Rev Mol Cell Biol 11(2):103–112. https://doi.org/10.1038/nrm2847

Sakane F, Mizuno S, Komenoi S (2016) Diacylglycerol kinases as emerging potential drug targets for a variety of diseases: an update. Front Cell Dev Biol 4. https://doi.org/10.3389/fcell.2016.00082

Salmond RJ, Brownlie RJ, Zamoyska R (2015) Multifunctional roles of the autoimmune disease-associated tyrosine phosphatase PTPN22 in regulating T cell homeostasis. Cell Cycle 14(5):705–711. https://doi.org/10.1080/15384101.2015.1007018

Satani N, Lin YH, Hammoudi N, Raghavan S, Georgiou DK, Muller FL (2016) ENOblock does not inhibit the activity of the glycolytic enzyme enolase. PLoS ONE 11(12):e0168739. https://doi.org/10.1371/journal.pone.0168739

Scannevin RH, Chollate S, Jung MY, Shackett M, Patel H, Bista P, Zeng W, Ryan S, Yamamoto M, Lukashev M, Rhodes KJ (2012) Fumarates promote cytoprotection of central nervous system cells against oxidative stress via the nuclear factor (erythroid-derived 2)-like 2 pathway. J Pharmacol Exp Ther 341(1):274–284. https://doi.org/10.1124/jpet.111.190132

Seltzer MJ, Bennett BD, Joshi AD, Gao P, Thomas AG, Ferraris DV, Tsukamoto T, Rojas CJ, Slusher BS, Rabinowitz JD, Dang CV, Riggins GJ (2010) Inhibition of glutaminase preferentially slows growth of glioma cells with mutant IDH1. Cancer Res 70(22):8981–8987. https://doi.org/10.1158/0008-5472.CAN-10-1666

Shin M, Franks CE, Hsu KL (2018a) Isoform-selective activity-based profiling of ERK signaling. Chem Sci 9(9):2419–2431. https://doi.org/10.1039/c8sc00043c

Shin M, Snyder HW, Donvito G, Schurman LD, Fox TE, Lichtman AH, Kester M, Hsu KL (2018b) Liposomal delivery of diacylglycerol lipase-beta inhibitors to macrophages dramatically enhances selectivity and efficacy in vivo. Mol Pharm 15(3):721–728. https://doi.org/10.1021/acs.molpharmaceut.7b00657

Shindo M, Irie K, Masuda A, Ohigashi H, Shirai Y, Miyasaka K, Saito N (2003) Synthesis and phorbol ester binding of the cysteine-rich domains of diacylglycerol kinase (DGK) isozymes. DGKgamma and DGKbeta are new targets of tumor-promoting phorbol esters. J Biol Chem 278(20):18448–18454

Shulga YV, Topham MK, Epand RM (2011) Regulation and functions of diacylglycerol kinases. Chem Rev 111:6186–6208. https://doi.org/10.1021/cr1004106

Simon GM, Cravatt BF (2010) Activity-based proteomics of enzyme superfamilies: serine hydrolases as a case study. J Biol Chem 285(15):11051–11055. https://doi.org/10.1074/jbc.R109.097600

Sin N, Kim KB, Elofsson M, Meng L, Auth H, Kwok BH, Crews CM (1999) Total synthesis of the potent proteasome inhibitor epoxomicin: a useful tool for understanding proteasome biology. Bioorg Med Chem Lett 9(15):2283–2288

Speers AE, Cravatt BF (2005) A tandem orthogonal proteolysis strategy for high-content chemical proteomics. J Am Chem Soc 127(28):10018–10019. https://doi.org/10.1021/ja0532842

Spitaler M, Emslie E, Wood CD, Cantrell D (2006) Diacylglycerol and protein kinase D localization during T lymphocyte activation. Immunity 24(5):535–546. https://doi.org/10.1016/j.immuni.2006.02.013

Srivastava N, Sudan R, Kerr WG (2013) Role of inositol poly-phosphatases and their targets in T cell biology. Front Immunol 4:288. https://doi.org/10.3389/fimmu.2013.00288

Stapleton D, Balan I, Pawson T, Sicheri F (1999) The crystal structure of an Eph receptor SAM domain reveals a mechanism for modular dimerization. Nat Struct Biol 6(1):44–49. https://doi.org/10.1038/4917

Subramanian A, Miller DM (2000) Structural analysis of alpha-enolase. Mapping the functional domains involved in down-regulation of the c-myc protooncogene. J Biol Chem 275(8):5958–5965

Taguchi K, Motohashi H, Yamamoto M (2011) Molecular mechanisms of the Keap1-Nrf2 pathway in stress response and cancer evolution. Genes Cells: Devoted Mol Cell Mech 16(2):123–140. https://doi.org/10.1111/j.1365-2443.2010.01473.x

Tamas P, Hawley SA, Clarke RG, Mustard KJ, Green K, Hardie DG, Cantrell DA (2006) Regulation of the energy sensor AMP-activated protein kinase by antigen receptor and Ca^{2+} in T lymphocytes. J Exp Med 203(7):1665–1670. https://doi.org/10.1084/jem.20052469

Tamas P, Macintyre A, Finlay D, Clarke R, Feijoo-Carnero C, Ashworth A, Cantrell D (2010) LKB1 is essential for the proliferation of T-cell progenitors and mature peripheral T cells. Eur J Immunol 40(1):242–253. https://doi.org/10.1002/eji.200939677

Thornberry NA, Weber AE (2007) Discovery of JANUVIA (Sitagliptin), a selective dipeptidyl peptidase IV inhibitor for the treatment of type 2 diabetes. Curr Top Med Chem 7(6):557–568

Tu-Sekine B, Goldschmidt HL, Raben DM (2016) DGK-θ: structure, enzymology, and physiological roles. Front Cell Dev Biol 4. https://doi.org/10.3389/fcell.2016.00101

Vasquez-Dunddel D, Pan F, Zeng Q, Gorbounov M, Albesiano E, Fu J, Blosser RL, Tam AJ, Bruno T, Zhang H, Pardoll D, Kim Y (2013) STAT3 regulates arginase-I in myeloid-derived suppressor cells from cancer patients. J Clin Investig 123(4):1580–1589. https://doi.org/10.1172/JCI60083

Vazquez-Cintron EJ, Monu NR, Frey AB (2010) Tumor-induced disruption of proximal TCR-mediated signal transduction in tumor-infiltrating CD8+ lymphocytes inactivates antitumor effector phase. J Immunol 185(12):7133–7140. https://doi.org/10.4049/jimmunol.1001157

Vocadlo DJ, Bertozzi CR (2004) A strategy for functional proteomic analysis of glycosidase activity from cell lysates. Angew Chem Int Ed Engl 43(40):5338–5342. https://doi.org/10.1002/anie.200454235

Wang JB, Erickson JW, Fuji R, Ramachandran S, Gao P, Dinavahi R, Wilson KF, Ambrosio AL, Dias SM, Dang CV, Cerione RA (2010) Targeting mitochondrial glutaminase activity inhibits oncogenic transformation. Cancer Cell 18(3):207–219. https://doi.org/10.1016/j.ccr.2010.08.009

Wang R, Dillon CP, Shi LZ, Milasta S, Carter R, Finkelstein D, McCormick LL, Fitzgerald P, Chi H, Munger J, Green DR (2011) The transcription factor Myc controls metabolic reprogramming upon T lymphocyte activation. Immunity 35(6):871–882. https://doi.org/10.1016/j.immuni.2011.09.021

Wang C, Weerapana E, Blewett MM, Cravatt BF (2014) A chemoproteomic platform to quantitatively map targets of lipid-derived electrophiles. Nat Methods 11(1):79–85. https://doi.org/10.1038/nmeth.2759

Weerapana E, Wang C, Simon GM, Richter F, Khare S, Dillon MBD, Bachovchin DA, Mowen K, Baker D, Cravatt BF (2010) Quantitative reactivity profiling predicts functional cysteines in proteomes. Nature 468(7325):790–795. https://doi.org/10.1038/nature09472

Wright AT, Cravatt BF (2007) Chemical proteomic probes for profiling cytochrome p450 activities and drug interactions in vivo. Chem Biol 14(9):1043–1051. https://doi.org/10.1016/j.chembiol.2007.08.008

Wu C, Jin X, Tsueng G, Afrasiabi C, Su AI (2016) BioGPS: building your own mash-up of gene annotations and expression profiles. Nucleic Acids Res 44:D313–D316. https://doi.org/10.1093/nar/gkv1104

Yamaoka K, Saharinen P, Pesu M, Holt VE 3rd, Silvennoinen O, O'Shea JJ (2004) The janus kinases (jaks). Genome Biol 5(12):253. https://doi.org/10.1186/gb-2004-5-12-253

Yatim N, Cullen S, Albert ML (2017) Dying cells actively regulate adaptive immune responses. Nat Rev Immunol 17(4):262–275. https://doi.org/10.1038/nri.2017.9

Yeh JH, Sidhu SS, Chan AC (2008) Regulation of a late phase of T cell polarity and effector functions by Crtam. Cell 132(5):846–859. https://doi.org/10.1016/j.cell.2008.01.013

Yun B, Lee H, Ghosh M, Cravatt BF, Hsu KL, Bonventre JV, Ewing H, Gelb MH, Leslie CC (2014) Serine hydrolase inhibitors block necrotic cell death by preventing calcium overload of the mitochondria and permeability transition pore formation. J Biol Chem 289(3):1491–1504. https://doi.org/10.1074/jbc.M113.497651

Zeng H, Chi H (2017) mTOR signaling in the differentiation and function of regulatory and effector T cells. Curr Opin Immunol 46:103–111. https://doi.org/10.1016/j.coi.2017.04.005

Zha Y, Marks R, Ho AW, Peterson AC, Janardhan S, Brown I, Praveen K, Stang S, Stone JC, Gajewski TF (2006) T cell anergy is reversed by active Ras and is regulated by diacylglycerol kinase-alpha. Nat Immunol 7(11):1166–1173. https://doi.org/10.1038/ni1394

Zhou P, Shaffer DR, Alvarez Arias DA, Nakazaki Y, Pos W, Torres AJ, Cremasco V, Dougan SK, Cowley GS, Elpek K, Brogdon J, Lamb J, Turley SJ, Ploegh HL, Root DE, Love JC, Dranoff G, Hacohen N, Cantor H, Wucherpfennig KW (2014) In vivo discovery of immunotherapy targets in the tumour microenvironment. Nature 506(7486):52–57. https://doi.org/10.1038/nature12988

Small-Molecule Inhibitors of PARPs: From Tools for Investigating ADP-Ribosylation to Therapeutics

Ilsa T. Kirby and Michael S. Cohen

Contents

I. T. Kirby · M. S. Cohen
Program in Chemical Biology, Oregon Health & Science University, Portland, OR 97210, USA

I. T. Kirby · M. S. Cohen (✉)
Department of Physiology and Pharmacology, Oregon Health & Science University, Portland, OR 97210, USA
e-mail: cohenmic@ohsu.edu

Current Topics in Microbiology and Immunology (2019) 420:211–232
DOI 10.1007/82_2018_137

Published Online: 22 September 2018

Abstract Over the last 60 years, poly-ADP-ribose polymerases (PARPs, 17 family members in humans) have emerged as important regulators of physiology and disease. Small-molecule inhibitors have been essential tools for unraveling PARP function, and recently the first PARP inhibitors have been approved for the treatment of various human cancers. However, inhibitors have only been developed for a few PARPs and in vitro profiling has revealed that many of these exhibit polypharmacology across the PARP family. In this review, we discuss the history, development, and current state of the field, highlighting the limitations and opportunities for PARP inhibitor development.

1 Introduction

1.1 The PARP Family

Poly-ADP-ribose polymerases (PARPs1-16; also known as ADP-ribosyltransferases or ARTDs) catalyze the transfer of ADP-ribose (ADPr) from nicotinamide adenine dinucleotide (NAD^+) to their target. Most PARPs (PARP3, 4, 6–8, 10–12, 14–16) transfer a single unit of ADPr onto targets, a process known as mono-ADP-ribosylation (MARylation) (Fig. 1). Only four PARPs (PARP1, 2, 5a, 5b) have been shown to transfer multiple units of ADPr onto targets, a process known as poly-ADP-ribosylation (PARylation) (Fig. 1). In most cases, ADPr is transferred onto amino acids in proteins; however, recent studies demonstrate that PARP-mediated ADPr transfer can occur on DNA (Munnur and Ahel 2017; Belousova et al. 2018). Similar to phosphorylation and other more well-characterized posttranslational modifications (PTMs), both MARylation and PARylation are reversible suggesting that these PTMs are dynamic.

Though best known for their role in DNA damage repair, recent studies have revealed that PARPs play much wider roles in cells: from transcriptional regulation, miRNA processing, mRNA stability, and nuclear core complex biology to the unfolded protein response (Bock et al. 2015). Regarding the physiological role of PARPs, there is a wealth of evidence demonstrating the involvement of PARPs in the immune system and T cell regulation, which has been well reviewed (Rosado et al. 2013; Krishnakumar and Kraus 2010; Giansanti et al. 2010). For example, PARP1, 2, and 14 have been shown to mediate pro-inflammatory responses (Mehrotra et al. 2013; Zingarelli et al. 1998; Andreone et al. 2003; Bai and Virág 2012). A recent study demonstrates a role of PARP6 in neurodevelopment (Huang et al. 2016). Lastly, several PARPs are implicated in human diseases; for

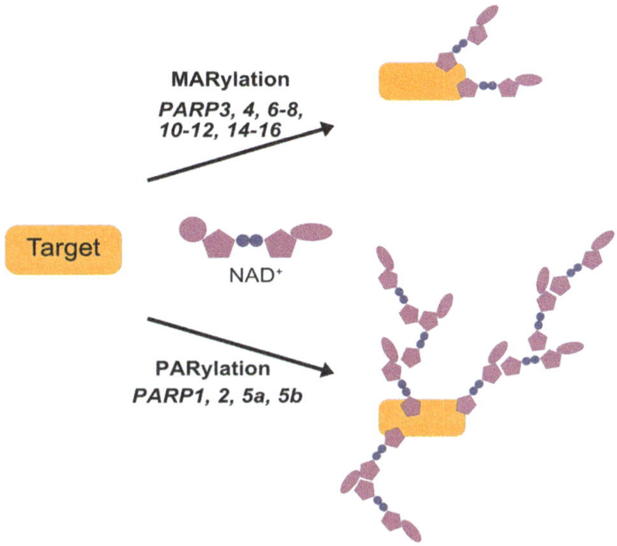

Fig. 1 PARPs catalyze the transfer of ADPr from NAD^+ to target proteins. The majority of PARPs transfer either a single unit of ADPr, a process known as MARylation, whereas only four PARP family members transfer multiple units of ADPr, a process known as PARylation

example, PARP14 mediates allergic responses in asthma (Mehrotra et al. 2013) and is a survival factor in multiple myeloma (Barbarulo et al. 2013) and hepatocellular carcinoma (Iansante et al. 2015).

1.2 Summary of Review

Despite the rapid expansion of the PARP field in recent years, most PARP family members that catalyze MARylation are poorly understood. This is due, in part, to the dearth of small-molecule inhibitors that selectively inhibit PARPs that catalyze MARylation. In contrast, there are numerous selective inhibitors for PARPs that catalyze PARylation and these have been useful not only as tools to reveal the functions of PARylation in the cell, but also validate PARPs as therapeutic targets. There are currently three FDA-approved PARP inhibitors (olaparib, rucaparib, and niraparib) that exhibit good selectivity for PARP1/2.

In this review, we discuss the history of PARP inhibitor development, structural features of PARP inhibitors, PARP inhibitor screening assays, and the current challenges and opportunities for PARP inhibitor development. For a general review of ADP-ribosylation and PARP biology, we direct the reader to the following references (Bock et al. 2015; Bai 2015; Vyas et al. 2014; Kraus 2015; Barkauskaite et al. 2015; Cohen and Chang 2018; Hottiger et al. 2010). To begin, we will describe structural features within the PARP catalytic domain.

2 Common and Distinct Features in the Catalytic Domain of PARPs

2.1 The NAD⁺ Binding Domain

While the PARP family is diverse in regard to their domain architecture, they all share a highly conserved catalytic domain known as the ADP-ribosyltransferase (ART) fold, which binds NAD^+ in a conformation optimal for ADPr transfer. Nearly all PARP inhibitors target the ART fold and are competitive with NAD^+. The key interaction between NAD^+ and PARPs has been gleaned from crystal structures of related ART bacterial toxins (diphtheria, pertussis, cholera, and certain clostridial toxins) (Hottiger et al. 2010) bound to NAD^+(Tsurumura et al. 2013) (Fig. 2a), and a recent structure of PARP1 bound to benzamide adenine dinucleotide (BAD) (Langelier et al. 2018), a non-hydrolyzable NAD^+ analog (Fig. 2b). Two key interactions between NAD^+ and PARPs are the interaction of the exocyclic amide of the nicotinamide moiety with the main chain of a conserved glycine

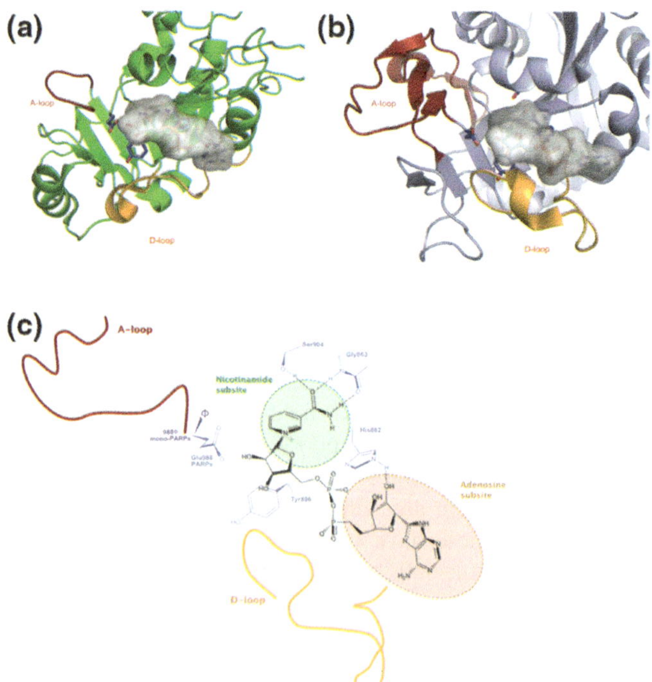

Fig. 2 Structural features of the NAD^+ binding site. **a** The bacterial toxin ExoA (PDB 2ZIT) bound to endogenous NAD^+ shows how the H-Y-E catalytic triad, A-loop, and D-loop hold NAD^+ in the binding site of a protein that catalyzes MARylation; **b** PARP1 in complex with a non-hydrolyzable NAD^+ analog (PDB 6BHV) closely mirrors the structural interactions of endogenous NAD^+ with ExoA; **c** Based on these crystal structures a simplified model of key interactions between NAD^+ and the PARP catalytic domain reveals two important regions: the nicotinamide subsite (green) and the adenosine subsite (orange)

(Gly863, human PARP1 numbering) and the side chain of a conserved serine (Ser 904, human PARP1 numbering) (Fig. 2c). These interactions are commonly exploited by PARP inhibitors, as will be illustrated below.

2.2 *HYE Versus HYΦ PARPs*

Active PARPs can be divided into two subfamilies based on the sequence of an active site triad motif in the ART fold (Fig. 2c)—the histidine–tyrosine–glutamate (HYE) PARPs (PARP1, 2, 3, 4, 5a and 5b) and the histidine–tyrosine–hydrophobic amino acid (HYΦ) PARPs (PARP 6–8, 10–12, 14–16) (Hottiger et al. 2010). In both the HYE and HYΦ, the conserved histidine forms a hydrogen bond with the 2-OH of the adenosine ribose and the conserved tyrosine π-stacks with the nicotinamide moiety (Fig. 2c). The glutamate in the HYE PARPs is necessary but not sufficient for the PARylation activity of several HYE PARPs (Marsischky et al. 1995; Rolli et al. 1997). HYΦ PARPs were recently shown to exclusively catalyze MARylation (Vyas et al. 2014).

2.3 *Non-Conserved Loops*

In addition to these conserved amino acids, there are two loop regions in the ART fold that are less well conserved yet thought to be critical for catalysis and NAD^+ binding: the acceptor loop (A-loop) and the donor-loop (D-loop) (Fig. 2a–c). The A-loop varies widely in terms of length and amino acid composition across the PARP family and is thought to interact with PARP substrates (Fig. 2c). The D-loop is a variable region that interacts with the ADPr moiety of NAD^+ (Fig. 2a–c). Based on mutagenesis studies of the related bacterial ARTs, the D-loop in PARPs is thought to be necessary for NAD^+ binding and ADPr transfer (Pinto and Schüler 2015). The high variability of the A-loop and D-loop of PARPs could be exploited for the design of isoform-selective inhibitors.

3 Early Developments in the Field: Initial Focus on Small-Molecule Inhibitors of PARP1

3.1 *The Founding PARP Inhibitors*

Interest in PARP inhibitors emerged in the early 1980s following the discovery of PARP1, the most ubiquitous and abundant PARP family member (Fig. 3). The first-described PARP1 inhibitors focused on targeting the nicotinamide subsite and were simple biomimetics of nicotinamide such as 3-aminobenzamide (3-AB) (Purnell and Whish 1980) (Fig. 4), which inhibits PARP1 with a half-maximal inhibitory concentration (IC_{50}) ~ 10 μM. These inhibitors played a crucial role in

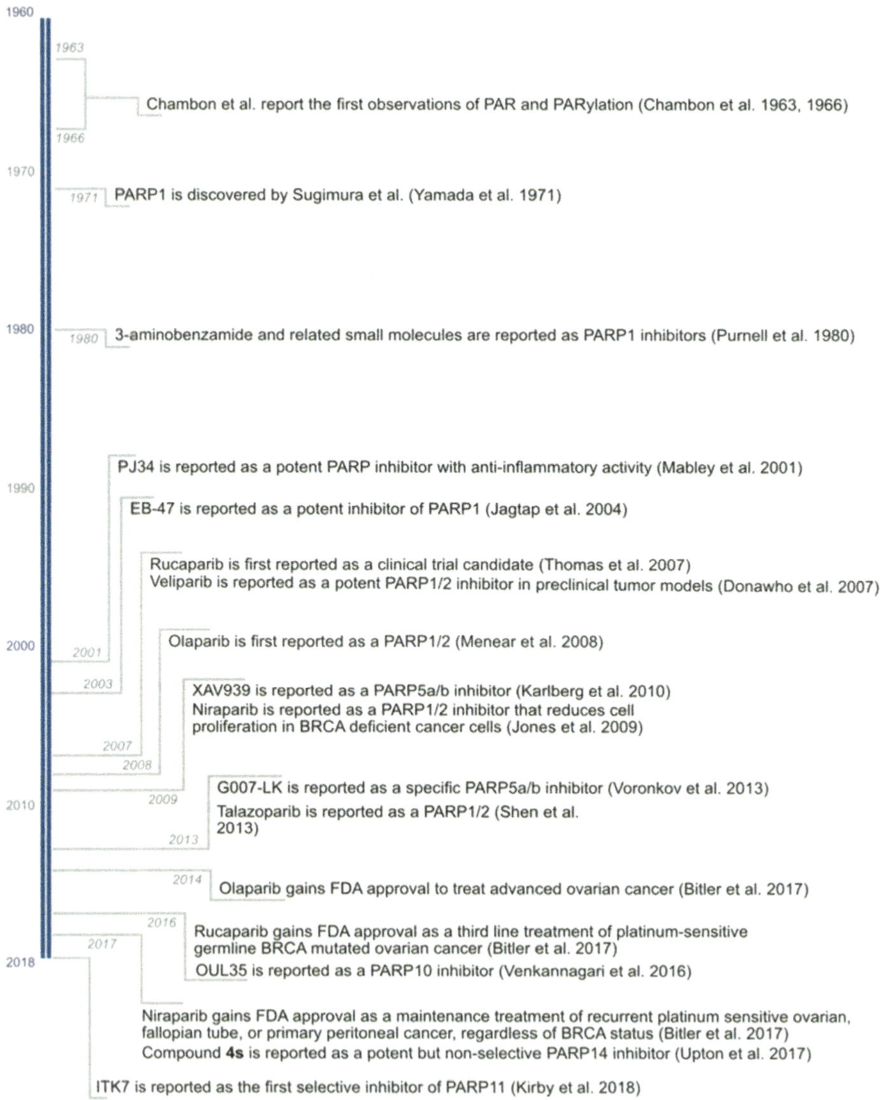

Fig. 3 Timeline of PARP inhibitor development

elucidating the role of PARP1 in DNA damage repair. Consistent with their role in DNA damage repair, these first-generation PARP inhibitors were shown to potentiate the cytotoxicity induced by DNA damaging agents in cancer cells (Purnell and Whish 1980; Nduka et al. 1980; Durrant and Boyle 1982). During the next two decades, significant advances were made in PARP inhibitor development, and by the mid-2000s inhibitors with nanomolar IC_{50} values against PARP1, and the closely related PARP2, were identified.

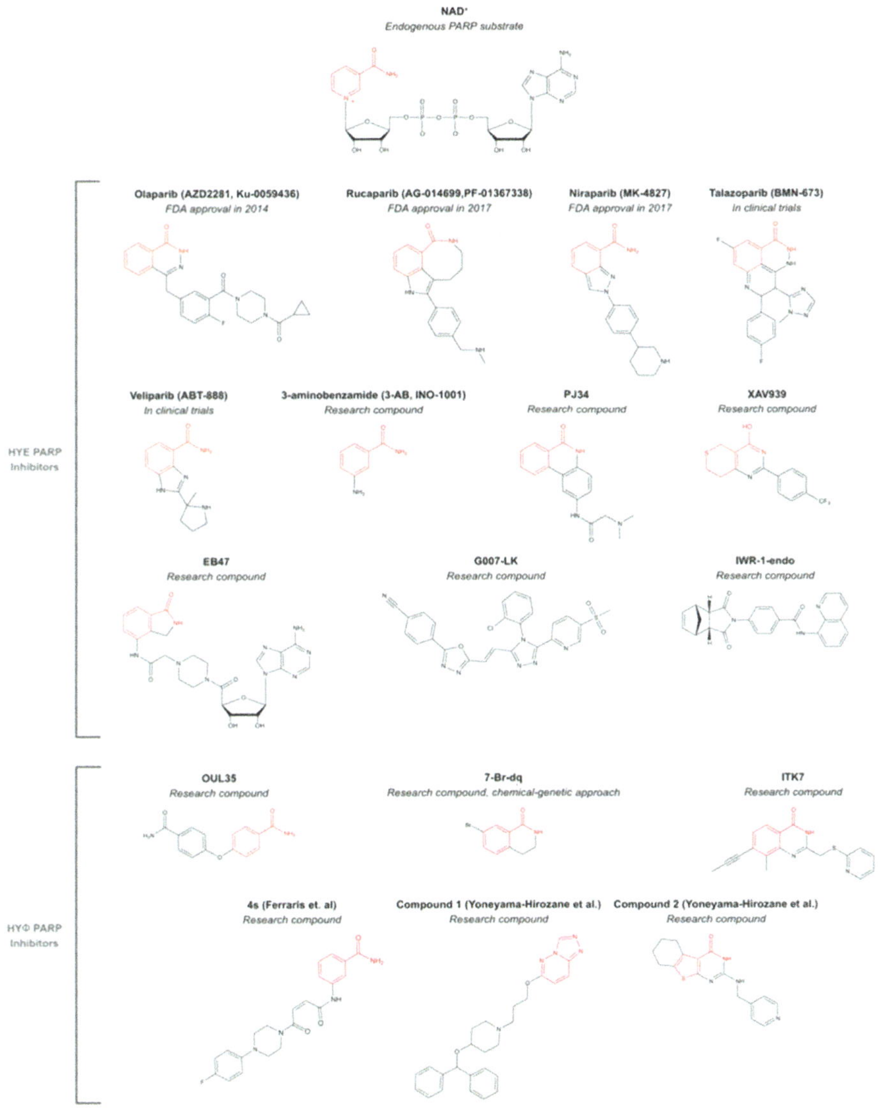

Fig. 4 Structures of common PARP inhibitors and their classifications

3.2 *PARP1/2 Identified as Therapeutic Targets in Cancer*

In 2005, two seminal papers demonstrated that potent PARP1/2 inhibitors induce synthetic lethality in BRCA-deficient cancers by blocking PARP1 mediated DNA repair pathways (Farmer et al. 2005; Bryant et al. 2005). These studies motivated the development of PARP1/2 inhibitors as adjuvants with DNA damaging reagents or as single agents in cancers that have defects in the DNA repair machinery. First disclosed in 2008, the potent PARP1/2 inhibitor olaparib (Fig. 4) effectively killed BRCA-deficient cancer cells at low nanomolar concentrations (Menear et al. 2008). In 2009, Fong et al. published the first Phase 1 trial of olaparib reporting its antitumor effects in BRCA1/2 mutated cancer (Fong et al. 2009), and in 2014 olaparib became the first FDA-approved PARP inhibitor. Since then off-label and approved uses of olaparib have expanded beyond its original approval as a monotherapy for BRCA-deficient ovarian cancers to include prostate cancer and germ line mutated metastatic breast cancer. Following the FDA approval of olaparib, two more potent PARP1/2 inhibitors have entered the clinic: rucaparib in 2016 and niraparib in 2017 (Fig. 4).

These FDA-approved PARP inhibitors have shifted the paradigm for cancer treatment. Because they exploit synthetic lethality they are relatively non-toxic to normal cells, thus avoiding many of the side effects of frontline chemotherapeutics. For a thorough review of PARP1 history, biology, and inhibitors see the following references (Kraus 2015; Ferraris 2010; Feng et al. 2015). For a recent review of the clinical applications of PARP1/2 inhibitors including olaparib, rucaparib, and niraparib, see the following references (Ohmoto and Yachida 2017; Mariappan et al. 2017).

4 Moving Beyond PARP1/2

4.1 *Forward Chemical Genetic Screen Identifies a Small-Molecule Inhibitor of PARP5a/b*

In contrast to the targeted approach used to identify potent and selective inhibitors of PARP1/2, the first inhibitor of the other PAR-generating PARPs, PARP5a/b (also known as tankyrase 1/2), was identified in a forward chemical genetic screen focused on identifying inhibitors of Wnt/β-catenin signaling (Huang et al. 2009). A small molecule, XAV939 (Fig. 4), was found to induce degradation of β-catenin by stabilizing axin, thereby blocking β-catenin-mediated transcription (Huang et al. 2009). Using an immobilized, active analog of XAV939, it was shown that the target of XAV939 is PARP5a/b (Huang et al. 2009). XAV939 potently inhibits the activity of PARP5a/b-mediated PARylation, and later structural studies demonstrate that XAV939 binds to the nicotinamide subsite in PARP5a (Fig. 5f) (Karlberg et al. 2010). Knockdown of PARP5a/b phenocopied the effects of XAV939 on axin stabilization, providing evidence that PARP5a/b are the targets of XAV939 (Huang et al. 2009). In this same study, the authors showed that a previously described axin-stabilizing compound, IWR-1-endo, also potently inhibited PARP5a/b-mediated PARylation, suggesting that this compound stabilizes axin by inhibiting PARP5a/b catalytic activity (Huang et al. 2009). Intriguingly, structural studies show IWR-1-endo binds exclusively to the adenosine subsite, which at the time, was the first example of this type if binding mode to a PARP (Narwal et al. 2012).

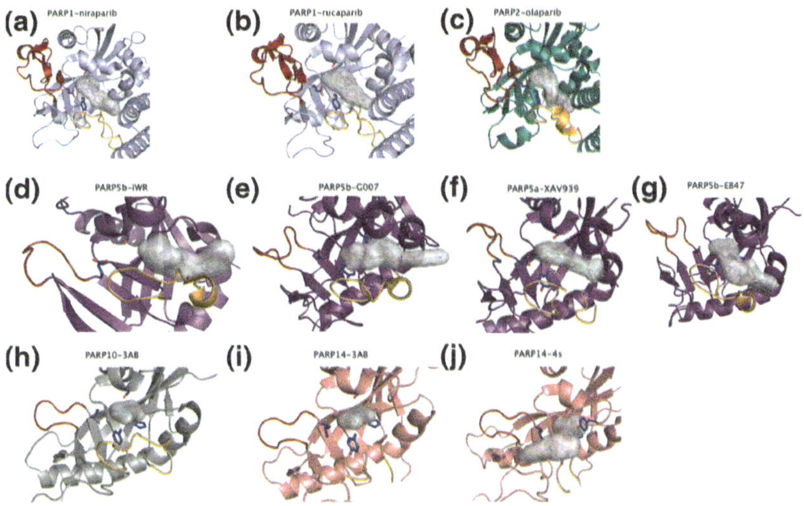

Fig. 5 Different small molecules access different regions of the NAD^+ binding pocket. **a** PARP1 in complex with niraparib (PDB 4R6E); **b** PARP1 in complex with rucaparib (PDB 4RV6); **b** PARP2 in complex with olaparib (PDB 4TVJ); **d** PARP5b in complex with EB-47 (PDB 4TK5); **e** PARP5b in complex with IWR-1-endo (PDB 3UA9); **f** PARP5b in complex with G007-LK; **g** PARP10 in complex with 3-aminobenzamide (PDB 3HKV); **h** PARP14 in complex with 3-aminobenzamide (PDB 3GOY); **i** PARP14 in complex with compound **4s** (PDB 5NQE)

Since this study, several academic and industry efforts have led to the development of more potent and selective PARP5a/b inhibitors, including G007-LK (Fig. 4) (Menear et al. 2008), which targets the adenosine subsite of the NAD^+ binding pocket (Fig. 5). This compound as well as structurally unrelated PARP5a/b inhibitors have been invaluable tools for uncovering the roles of PARP5a/b-mediated PARylation in cells and show promise as anticancer drugs (Vyas et al. 2014; Zhan et al. 2014; Kamal et al. 2014).

4.2 A Dearth of Selective and Potent Inhibitors for the HYΦ PARP Subfamily

Historically, PARP inhibitor design has focused on PARP1/2 and PARP5a/b with comparatively little attention given to the rest of the family, and in particular to the HYΦ PARPs. Recent years have seen a growing interest in the HYΦ PARP subfamily and a subsequent increase in HYΦ PARP inhibitor development. Of particular interest to many in drug discovery research has been PARP14 because of its role in several pathologies including asthma (Mehrotra et al. 2013), multiple myeloma (Barbarulo et al. 2013), and hepatocellular carcinoma (Iansante et al. 2015). Promising inhibitors of PARP14 have been published by Upton et al. (e.g. **4s**, Fig. 4) (Upton et al. 2017) and Yoneyama-Hirozane et al. at Takeda Pharmaceuticals

(Compound 1 and 2, Fig. 3) (Yoneyama-Hirozane et al. 2017). Upton et al. expanded on a previously identified small-molecule inhibitor of PARP14 (Andersson et al. 2012; Ekblad et al. 2015) based on 3-AB to develop potent but non-selective PARP14 inhibitors containing cis-maleic amide substituents emanating from the 3-amino group of the 3-AB scaffold (Upton et al. 2017). Though originally designed to interact with the D-loop and adenosine subsite, crystal structure analysis of **4s** shows that the cis-maleic amide substituent accesses an induced pocket adjacent to the NAD^+ binding site (further discussion below). Yoneyama-Hirozane et al. screened a small-molecule library to identify two compounds (Compound 1 and 2, Fig. 4) based on divergent scaffolds that showed potent PARP14 inhibition and no activity against PARP1 up to 25 μM (Yoneyama-Hirozane et al. 2017). Though untested against the majority of the PARP family, these scaffolds are promising starting points for further development of PARP14 inhibitors.

There have been some efforts to generate selective inhibitors for other HYΦ PARPs. For example, a mono-selective inhibitor of PARP10 was developed using a chemical genetic strategy (Fig. 4; Morgan et al. 2015). A screen of a small-molecule library from the National Cancer Institute led to the discovery of a PARP10 inhibitor (OUL35, Fig. 4) (Venkannagari et al. 2016). OUL35, an ether linked dimer of benzamide, appears to exhibit some selectivity for PARP10 when tested against several other PARP family members. Recently, a PARP11 inhibitor (ITK7, Fig. 4) with low nano-molar potency and greater than 200-fold selectivity over the entire PARP family was reported. This inhibitor was used to elucidate the connection between the catalytic activity of PARP11 its localization to the nuclear pore (Kirby et al. 2018).

A number of non-selective PARP inhibitors and promising scaffolds have been described, but broadly speaking these compounds lack the selectivity or potency necessary for use as probes for exploring the individual role of HYΦ PARPs in cells (Wahlberg et al. 2012; Thorsell et al. 2017).

5 Structural Studies of PARP Inhibitors: Insights into Inhibitor Potency and Selectivity

Structural studies of PARPs bound to various inhibitors have provided insight into inhibitor potency and selectivity and have been useful for structure-based design efforts. Here we discuss a few of these structures. The majority of structural studies have focused on HYE PARPs, such as PARP1 and PARP5a/b (Fig. 5).

5.1 Structural Studies of HYE PARP Inhibitors

The three FDA-approved inhibitors olaparib, rucaparib, and niraparib all share similar binding modes to PARP1/2 despite their divergent scaffolds (Figs. 4, 5a–c). A common and seemingly essential feature among these inhibitors is the interaction of an amide moiety (nicotinamide mimic) with the backbone glycine and side chain of an active site serine, which mimics the hydrogen bonding interaction observed with the nicotinamide of NAD^+ (Figs. 2c, 5a–c). Additionally, these inhibitors contain various substituents that engage with, to varying degrees, the D-loop and adenosine subsite (Fig. 5a–c).

Another potent PARP inhibitor, EB47, which is designed to mimic the full NAD^+ structure, occupies both the nicotinamide and adenosine subsites of PARP5b (Fig. 5d). Unsurprisingly, this molecule is not specific for PARP5b and has been shown to bind to several PARPs (Wahlberg et al. 2012). By contrast, two PARP5a/b inhibitors, IWR-1-endo and G007-LK, which do not contain nicotinamide mimics, exclusively occupy the adenosine subsite (Fig. 5e, f). PARP inhibitors that target the adenosine subsite but not the nicotinamide subsite are still fairly uncommon, and their efficacy against HYΦ PARPs has not been explored.

5.2 Structural Studies of HYΦ PARP Inhibitors

In recent years, greater attention has been given to the HYΦ PARPs, though inhibitors and crystal structures remain relatively rare by comparison to the better characterized HYE PARP family members. Some of the first reported crystal structures of HYΦ PARPs were PARP10 and PARP14 bound to 3-aminobenzamide (3-AB). As expected, 3-AB binds in the nicotinamide subsite in PARP10 and PARP14 and makes the same interactions with the conserved glycine backbone as the benzamide moiety of BAD bound to PARP1 (Fig. 5h, i). The PARP14 inhibitors mentioned above, which extend substituents from the 3-amino position of the 3-AB scaffold, occupy a unique, induced pocket in PARP14. The crystal structure of one of these compounds (**4s**) shows that 3-AB binds in the nicotinamide subsite as expected, but that the cis-maleic amide substituent emanating from 3-amino position reaches into a nascent pocket (near the A-loop) that likely results from a compound-induced movement of the D-loop (Fig. 5j). This finding reveals a new pocket in PARP14, and perhaps other HYΦ PARPs that could be targeted by other inhibitors.

5.3 Pharmacophore for PARP Inhibitors

From these structures, we can construct a model to summarize known inhibitor interactions with the NAD^+ binding pocket, which can guide future PARP inhibitor development (Fig. 6). A crucial feature of many PARP inhibitors is mimicking the interaction between the nicotinamide moiety of NAD^+ and the backbone glycine and side chain of an active site serine. Indeed, a wide range of scaffolds exploiting this interaction have been developed into successful PARP inhibitors. Targeting the nicotinamide site in concert with the adenosine subsite and D-loop appears optimal for potent PARP inhibition; however, many of these inhibitors are not selective (more on this below). Exclusively targeting the adenosine subsite, as has been

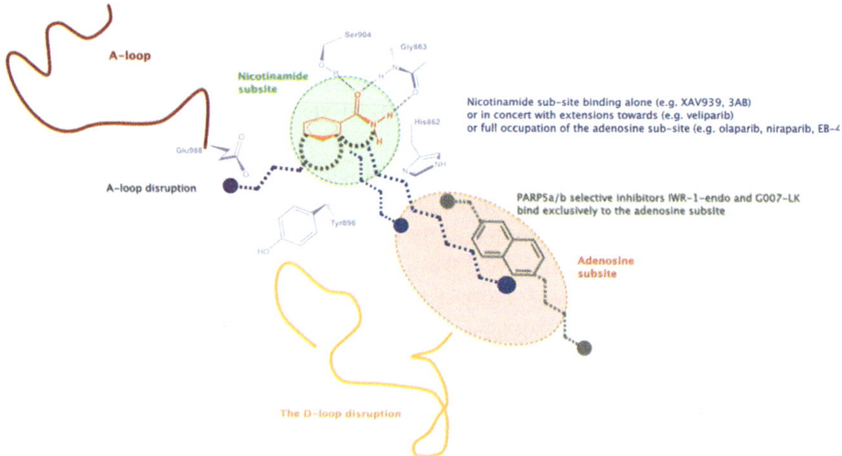

Fig. 6 Major structural interactions of known PARP inhibitors with the NAD^+ binding pocket. Crystal structures of known PARP inhibitors demonstrate how various small molecules can exploit different features in the nicotinamide and adenosine subsites in the NAD^+ binding pocket

shown for PARP5a/b, may result in more selective PARP inhibitors. Lastly, D-loop disrupting compounds (e.g., **4s**) that can induce unique pockets outside the nicotinamide and adenosine subsites may turn out to be a generalizable approach for generating potent and selective PARP inhibitors.

6 Chemical and Biological Reagents for Measuring PARP Activity

6.1 NAD^+ Analogs for In Vitro Analysis

As described above, PARPs use NAD^+ as a substrate to mediate PARylation or MARylation. Historically, PARP activity was measured using radioactive NAD^+ (e.g., adenylate phosphate-[^{32}P] NAD^+) (Surowy and Berger 1985) or using biotin-NAD^+ (Zhang and Snyder 1992) (Fig. 7).Compared to [^{32}P] NAD^+, biotin-NAD^+ can be used for identifying P/MARylated targets (Narendja and Sauermann 1994). Other NAD^+ analogs include *N*-6-etheno-NAD^+ (Barrio et al. 1972), a fluorescent NAD^+ analog, and ADP-ribose-*p*-nitrophenoxy (Oei et al. 1999), an analog in which the nicotinamide is replace with a p-nitrophenol for use in colorimetric assays (Fig. 7). Additionally, various "clickable" NAD^+ analogs have recently been developed, which contain an alkyne at various positions on the adenosine ring of NAD^+ (Fig. 7) (Jiang et al. 2010; Wallrodt et al. 2016; Wang et al. 2014; Carter-O'Connell et al. 2014). These clickable NAD^+ analogs can be coupled to a fluorescent-azide (visualize) or biotin-azide (visualize and identify) via the Huisgen 1,3-Dipolar Cycloaddition ("click reaction").

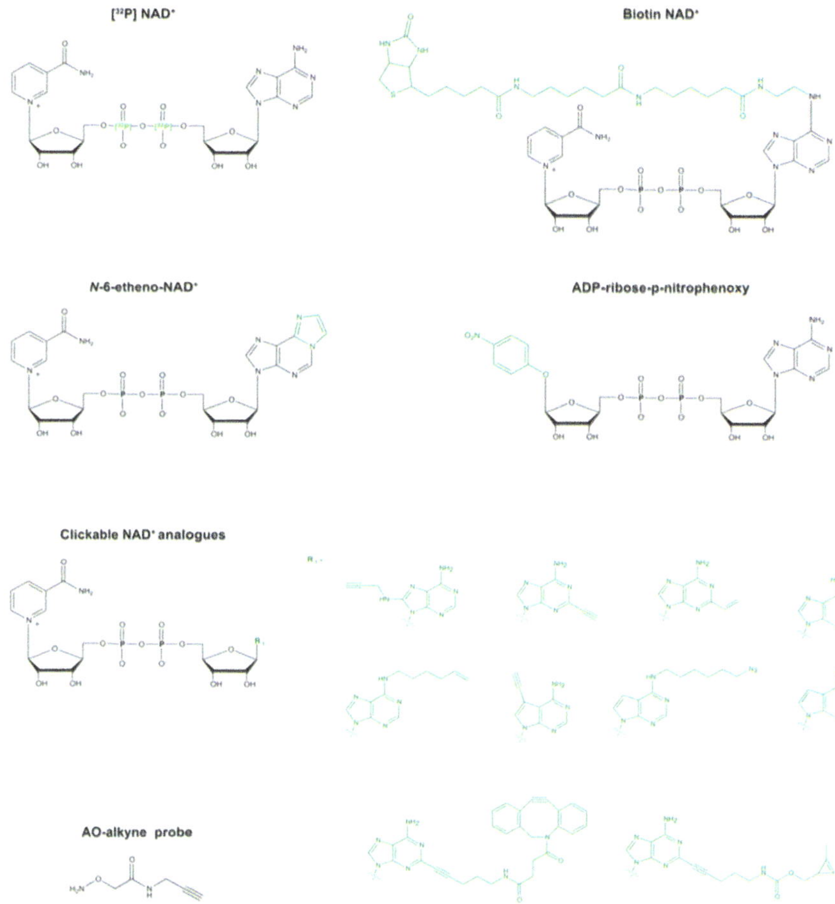

Fig. 7 Various NAD^+ analogs and probes have been developed to monitor PARP activity. Salient modifications to NAD^+ are highlighted in green

6.2 Detection Methods with Endogenous NAD^+

In addition to using NAD^+ analogs, several other strategies for detecting PARylation or MARylation have been described. For PARylation detection, the most commonly used reagents are antibodies that specifically recognize PARylated substrates (e.g. 10H) (Kawamitsu et al. 1984; Meyer and Hilz 1986; Küpper et al. 1996). Recently, protein-based reagents for detecting both PARylated and MARylated proteins have been described. These reagents consist of domains that recognize either ADPr (e.g., macro domain) or poly-ADPr (e.g., WWE domain) fused to Fc (Gibson et al. 2017). These reagents have been used in Western blot experiments as well as plate-based assays, as well as pull-down experiments. Lastly, an aminooxy-alkyne probe (AO-alkyne, Fig. 7) was described, which can readily detect proteins that are P/MARylated on acidic amino acids. AO-alkyne can also be used in cells for detecting cellular PARylation and MARylation (Morgan and Cohen 2015).

7 Assessing PARP Inhibitor Selectivity Across the PARP Family

7.1 Profiling Using Protein Stabilization Reveals Lack of Selectivity of Many PARP Inhibitors

Arguably the most important aspect of inhibitor development is assessing its target selectivity. Among other things, this is essential for understanding any cell-based or in vivo studies conducted with an inhibitor. Unfortunately, there are few studies that assess PARP inhibitor selectivity across multiple PARP family members, let alone the entire PARP family. One of the first examples of profiling inhibitor selectivity across multiple PARP family members was described in 2012. In this study, known and potential PARP inhibitors were screened against 13 PARPs using differential scanning fluorimetry (DSF), which assesses whether a compound can stabilize proteins (Wahlberg et al. 2012). One of the main findings of this study was that compounds previously described as selective PARP1/2 inhibitors, such as veliparib, rucaparib, and olaparib, stabilize several other HYE PARPs suggesting that these compounds may not be as selective as previously thought. In general, most of the known PARP inhibitors did not stabilize HYΦ PARPs, suggesting that they would not inhibit these PARPs. While this study provided the first insight into PARP inhibitor selectivity, DSF only assesses whether these PARP inhibitors can stabilize PARPs, which does not necessarily correlate with inhibition of catalytic activity.

7.2 Polypharmacology Among PARP Inhibitors

Recently, a high throughput 96-well-plate-based ADP-ribosylation assay using biotin-NAD^+ was used for screening known PARP inhibitors against 11 PARP family members (Thorsell et al. 2017), most of which were the same PARPs used in the DSF study. In general, there was reasonable agreement between the ADP-ribosylation activity study and the DSF study, although the DSF study slightly overestimated PARP inhibitor selectivity. The selectivity profile of various PARP inhibitors is summarized in a heat map shown in Fig. 8 (Voronkov et al. 2013; Upton et al. 2017; Thorsell et al. 2017; Huang et al. 2009; Ishida et al. 2006; Papeo et al. 2014; Kirby et al. 2018). There are several important findings worth noting: 1. veliparib appears to be the most potent and selective PARP1/2 inhibitor, exhibiting greater than 100-fold selectivity for PARP1/2 versus other PARPs; 2.

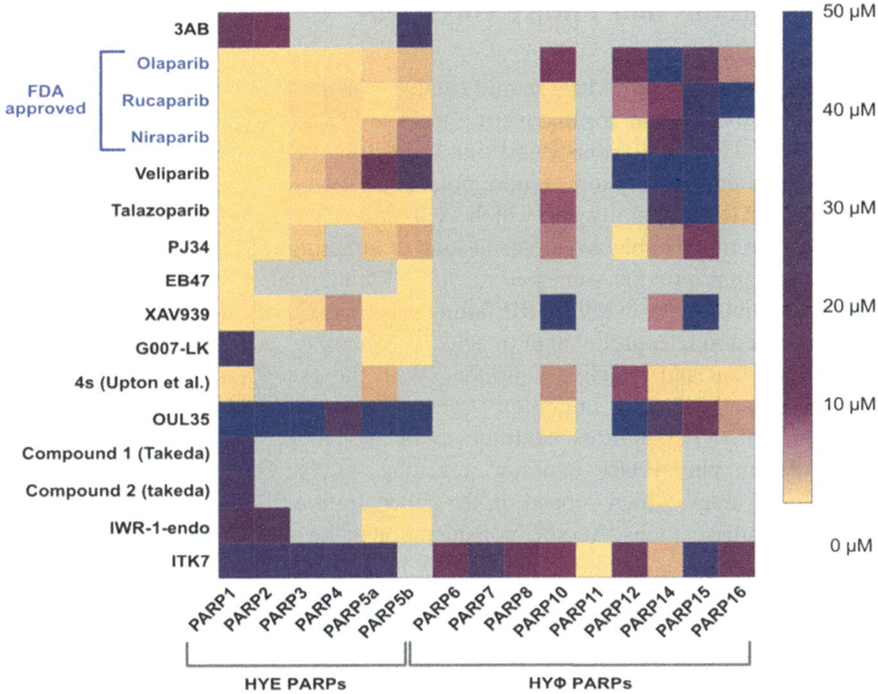

Fig. 8 Heat map showing the known IC_{50} values of PARP inhibitors shown in Fig. 3. The IC_{50} values used to generate this heat map derived from serval sources, as referenced in the text. In gray: values unknown

XAV939, which was previously described as selective inhibitor of PARP5a/b, potently inhibits PARP1 and PARP2, whereas IWR-1 is highly selective for PARP5a/b; 3. rucaparib and olaparib, while most selective for PARP1 and PARP2, inhibit several other PARP family members with sub-micromolar IC_{50} values.

While a comprehensive assay for screening inhibitors across the entire PARP family is still needed to fully assess family-wide PARP inhibitor selectivity, these findings have several important implications for interpreting results from cell-based experiments using these compounds. For example, a PubMed search reveals that many papers describe studies using XAV939 in cell-based assays at concentrations that also inhibit PARP1 and PARP2, making it difficult to conclude that the effects of the compound were in fact due to PARP5a/b inhibition. For selective inhibition of PARP5a/b in cells the adenosine pocket binders IWR-1 or G007-LK are better options in our opinion.

Lastly, these findings have important implications for evaluating PARP inhibitors in a clinical setting for cancer treatment. The different selectivity profiles of the three FDA-approved PARP inhibitors (Fig. 8) could potentially contribute to efficacy and/or toxicity. Knowing the selectivity profiles will help assess the effectiveness of polypharmacology for certain cancers.

8 Conclusions and Future Directions

The growing interest in PARPs should catalyze the development of selective PARP inhibitors to use as tools for uncovering the role of PARPs in cells and as potential therapeutics. The multifarious PARP-inhibitor structures available should guide the design of the next generation of more potent and selective inhibitors, especially for the HYΦ PARP subfamily for which is there a dearth of inhibitors. Exploiting differences in the variable A-loop or D-loop or targeting regions outside the NAD^+-binding site may provide strategies for the ultimate goal of generating potent and selective inhibitors for every PARP family member. Covalent inhibitors that target non-conserved nucleophilic amino acids are another potential strategy for the design of potent and selective inhibitors. With the exception of some early work (Watson et al. 1998), this inhibition strategy has not been pursued.

As new PARP inhibitors continue to be developed, it will be important to standardize in vitro PARP inhibitor screening assays. This is essential for comparing IC_{50} values, which depend on the concentration of NAD^+ or NAD^+ analog, obtained in different labs. Another consideration is the use of co-activators. PARP1, for example, requires single-stranded DNA for activation, whereas as PARP2 and PARP3 are optimally activated by 5'-phosphorylated double strand nicked DNA (Langelier et al. 2014). Most assays assessing PARP activity have focused on auto-P/MARylation; however, PARPs can also trans-P/MARylate targets in a cellular context (Carter-O'Connell et al. 2016; Gibson et al. 2016). It will be important to incorporate this into in vitro assays as auto-modification versus trans-modification may yield different inhibition profiles. In some cases, co-activators may be required for optimal trans-P/MARylation. For example, recently, it was shown that histone PARylation factor 1 (HPF1), which binds to PARP1, promotes PARP1 trans-PARylation of histones on serines (Gibbs-Seymour et al. 2016). Whether protein co-activators exist for other PARP family members is unclear, but as we learn more about PARP activation mechanisms these will need to be incorporated into in vitro PARP inhibitor screening assays.

Approaches to broadly assess PARP selectivity in a cellular context are desperately needed. Chemical proteomics approaches using resin bound PARP inhibitors is one potential strategy. This approach has the ability to identify potential non-PARP targets. Indeed, a recent study using resin bound olaparib, veliparib, rucaparib, and niraparib revealed that rucaparib and niraparib also target hexose-6-Phosphate Dehydrogenase (H6PD) and Deoxycytidine kinase (DCK), and inhibition of these targets may be clinically relevant (Knezevic et al. 2016). This chemical proteomics approach could also be used to profile PARP inhibitors in cell lysates, similar to the way Kinobeads have been used to profile kinase inhibitors (Golkowski et al. 2014); however, this approach requires a pan-PARP inhibitor with good potency, which unfortunately does not currently exist. Another approach could be to use activity-based protein profiling (ABPP), which exploits a conserved nucleophile in an enzyme active site for the development of a broad-spectrum probe for screening inhibitors across an enzyme family in cell lysates or cells (Cravatt et al. 2008). Such an approach for PARPs could be quite useful for profiling PARP inhibitors in a cellular context.

References

Andersson CD, Karlberg T, Ekblad T, Lindgren AEG, Thorsell A-G, Spjut S et al (2012) Discovery of ligands for ADP-ribosyltransferases via docking-based virtual screening. J Med Chem 55(17):7706–7718

Andreone TL, O'Connor M, Denenberg A, Hake PW, Zingarelli B (2003) Poly(ADP-Ribose) polymerase-1 regulates activation of activator protein-1 in murine fibroblasts. J Immunol 170 (4):2113–2120. 15 Feb 2003. (American Association of Immunologists)

Bai P (2015) Biology of poly(ADP-Ribose) polymerases: the factotums of cell maintenance. Mol Cell 58(6):947–958

Bai P, Virág L (2012) Role of poly(ADP-ribose) polymerases in the regulation of inflammatory processes. FEBS Lett 586(21):3771–3777. 26 Sep 2012. (Wiley-Blackwell)

Barbarulo A, Iansante V, Chaidos A, Naresh K, Rahemtulla A, Franzoso G et al (2013) Poly (ADP-ribose) polymerase family member 14 (PARP14) is a novel effector of the JNK2-dependent pro-survival signal in multiple myeloma. Oncogene 32(36):4231–4242

Barkauskaite E, Jankevicius G, Ahel I (2015) Structures and mechanisms of enzymes employed in the synthesis and degradation of PARP-dependent protein ADP-ribosylation. Mol Cell 58 (6):935–946. 18 Jun 2015. (Elsevier)

Barrio JR, Secrist JA, Leonard NJ (1972) A fluorescent analog of nicotinamide adenine dinucleotide. Proc Natl Acad Sci 69(8):2039–2042

Belousova EA, Ishchenko AA, Lavrik OI (2018) DNA is a new target of PARP3. Sci Rep. Nature Publishing Group 8(1):101. 8 Mar 2018

Bitler BG, Gynecologic ZW (2017) PARP inhibitors: clinical utility and possibilities of overcoming resistance. Gynecologiconcology

Bock FJ, Todorova TT, Chang P (2015) RNA Regulation by Poly(ADP-Ribose) Polymerases. Mol Cell 58(6):959–969

Bryant HE, Schultz N, Thomas HD, Parker KM, Flower D, Lopez E et al (2005) Specific killing of BRCA2-deficient tumours with inhibitors of poly(ADP-ribose) polymerase. Nature 434 (7035):913–917 Nature Publishing Group

Carter-O'Connell I, Jin H, Morgan RK, David LL, Cohen MS (2014) Engineering the substrate specificity of ADP-ribosyltransferases for identifying direct protein targets. J Am Chem Soc 136(14):5201–5204. 9 Apr 2014

Carter-O'Connell I, Jin H, Morgan RK, Zaja R, David LL, Ahel I et al Identifying Family-Member-Specific Targets of Mono-ARTDs by Using a Chemical Genetics Approach. Cell Rep 14(3):621–631

Chambon P, Weill JD, Doly J, Strosser MT, Mandel P (1966) On the formation of a novel adenylic compound by enzymatic extracts of liver nuclei. Biochem Biophys Res Commun 25(6):638–643

Chambon P, Weill JD, Mandel P (1963) Nicotinamide mononucleotide activation of a new DNA-dependent polyadenylic acid synthesizing nuclear enzyme. Biochem Biophys Res Commun 11(1):39–43

Cohen MS, Chang P (2018) Insights into the biogenesis, function, and regulation of ADP-ribosylation. Nat Chem Biol 14(3):236–243 Nature Publishing Group

Cravatt BF, Wright AT, Kozarich JW (2008) Activity-based protein profiling: from enzyme chemistry to proteomic chemistry. Annu Rev Biochem 77(1):383–414

Donawho CK, Luo Y, Penning TD, Bauch JL, Bouska JJ, Bontcheva-Diaz VD et al (2007) ABT-888, an orally active poly(ADP-Ribose) polymerase inhibitor that potentiates DNA-damaging agents in preclinical tumor models. Clini Cancer Res 13(9):2728–2737 American Association for Cancer Research

Durrant LG, Boyle JM (1982) Potentiation of cell killing by inhibitors of poly (ADP-ribose) polymerase in four rodent cell lines exposed to N-methyl-N-nitrosourea or UV light. Chem Biol Interact 38(3):325–338

Ekblad T, Lindgren AEG, Andersson CD, Caraballo R, Thorsell A-G, Karlberg T et al (2015) Towards small molecule inhibitors of mono-ADP-ribosyltransferases. Eur J Med Chem 95:546–551

Farmer H, McCabe N, Lord CJ, Tutt ANJ, Johnson DA, Richardson TB et al (2005) Targeting the DNA repair defect in BRCA mutant cells as a therapeutic strategy. Nature 434(7035):917–921 Nature Publishing Group

Feng FY, de Bono JS, Rubin MA, Knudsen KE (2015) Chromatin to clinic: the molecular rationale for PARP1 inhibitor function. Mol Cell 58(6):925–934

Ferraris DV (2010) Evolution of poly(ADP-ribose) polymerase-1 (PARP-1) inhibitors. From concept to clinic. J Med Chem 53(12):4561–4584

Fong PC, Boss DS, Yap TA, Tutt A, Wu P, Mergui-Roelvink M et al (2009) Inhibition of poly (ADP-Ribose) polymerase in tumors from BRCAMutation carriers. N Engl J Med 361(2):123–134

Giansanti V, Donà F, Tillhon M, Scovassi AI (2010) PARP inhibitors: new tools to protect from inflammation. Biochem Pharmacol 80(12):1869–1877

Gibbs-Seymour I, Fontana P, Rack JGM, Ahel I (2016) HPF1/C4orf27 Is a PARP-1-interacting protein that regulates PARP-1 ADP-ribosylation activity. Mol Cell 62(3):432–442

Gibson BA, Conrad LB, Huang D, Kraus WL (2017) Generation and characterization of recombinant antibody-like ADP-ribose binding proteins. Biochemistry 56(48):6305–6316

Gibson BA, Zhang Y, Jiang H, Hussey KM, Shrimp JH, Lin H et al (2016) Chemical genetic discovery of PARP targets reveals a role for PARP-1 in transcription elongation. Science 353 (6294):45–50

Golkowski M, Brigham JL, Perera BGK, Romano GS, Maly DJ, Ong S-E (2014) Rapid profiling of protein kinase inhibitors by quantitative proteomics. Med Chem Commun 5(3):363–369 The Royal Society of Chemistry

Hottiger MO, Hassa PO, Lüscher B, Schüler H, Koch-Nolte F (2010a) Toward a unified nomenclature for mammalian ADP-ribosyltransferases. Trends Biochem Sci 35(4):208–219

Hottiger MO, Hassa PO, Lüscher B, Schüler H, Koch-Nolte F (2010) Toward a unified nomenclature for mammalian ADP-ribosyltransferases. Trends Biochem Sci 35(4):208–219. (Elsevier)

Huang JY, Wang K, Vermehren Schmaedick A, Adelman JP, Cohen MS (2016) PARP6 is a regulator of hippocampal dendritic morphogenesis. Sci Rep 6(1):208. 4 Jan 2016. (Nature Publishing Group)

Huang S-MA, Mishina YM, Liu S, Cheung A, Stegmeier F, Michaud GA et al (2009) Tankyrase inhibition stabilizes axin and antagonizes Wnt signalling. Nature 461(7264):614–620. 16 Sept 2009. (Nature Publishing Group)

Huang S-MA, Mishina YM, Liu S, Cheung A, Stegmeier F, Michaud GA et al (2009) Tankyrase inhibition stabilizes axin and antagonizes Wnt signalling. Nature 461(7264):614–620. 1 Oct 2009. (Nature Publishing Group)

Iansante V, Choy PM, Fung SW, Liu Y, Chai J-G, Dyson J et al (2015) PARP14 promotes the Warburg effect in hepatocellular carcinoma by inhibiting JNK1-dependent PKM2 phosphorylation and activation. Nat Commun 10(6):7882

Ishida J, Yamamoto H, Kido Y, Kamijo K, Murano K, Miyake H et al (2006) Discovery of potent and selective PARP-1 and PARP-2 inhibitors: SBDD analysis via a combination of X-ray structural study and homology modeling. Bioorg Med Chem 14(5):1378–1390

Jagtap PG, Southan GJ, Baloglu E, Ram S, Mabley JG, Marton A et al (2004) The discovery and synthesis of novel adenosine substituted 2,3-Dihydro-1H-isoindol-1-ones: potent inhibitors of poly(ADP-ribose) polymerase-1 (PARP-1). ChemInform 35(18):81 4 May 2004. (WILEY-VCH Verlag)

Jiang H, Kim JH, Frizzell KM, Kraus WL, Lin H (2010) Clickable NAD Analogues for Labeling Substrate Proteins of Poly(ADP-ribose) Polymerases. J Am Chem Soc 132(27):9363–9372

Jones P, Altamura S, Boueres J, Ferrigno F, Fonsi M, Giomini C et al (2009) Discovery of 2-{4-[(3 S)-Piperidin-3-yl]phenyl}-2 H-indazole-7-carboxamide (MK-4827): A Novel Oral Poly (ADP-ribose)polymerase (PARP) Inhibitor Efficacious in BRCA-1 and -2 Mutant Tumors. J Med Chem 52(22):7170–7185

Kamal A, Riyaz S, Srivastava AK, Rahim A (2014) Tankyrase inhibitors as therapeutic targets for cancer. Curr Top Med Chem 14(17):1967–1976

Karlberg T, Markova N, Johansson I, Hammarström M, Schütz P, Weigelt J et al (2010) Structural basis for the interaction between tankyrase-2 and a potent Wnt-signaling inhibitor. J Med Chem 53(14):5352–5355

Kawamitsu H, Hoshino H, Okada H, Miwa M, Momoi H, Sugimura T (1984) Monoclonal antibodies to poly(adenosine diphosphate ribose) recognize different structures. Biochemistry 23(16):3771–3777

Kirby IT, Kojic A, Arnold MR, Thorsell A-G, Karlberg T, Vermehren Schmaedick A et al (2018) A potent and selective PARP11 inhibitor suggests coupling between cellular localization and catalytic activity. Cell Chem Biol

Knezevic CE, Wright G, Remsing Rix LL, Kim W, Kuenzi BM, Luo Y et al (2016) Proteome-wide profiling of clinical PARP inhibitors reveals compound-specific secondary targets. Cell Chem Biol 23(12):1490–1503

Kraus WL (2015) PARPs and ADP-Ribosylation: 50 Years … and Counting. Mol Cell 58(6):902–910

Krishnakumar R, Kraus WL (2010) The PARP side of the nucleus: molecular actions, physiological outcomes, and clinical targets. Mol Cell 39(1):8–24

Küpper J-H, van Gool L, Müller M, Bürkle A (1996) Detection of poly(ADP-ribose) polymerase and its reaction product poly(ADP-ribose) by immunocytochemistry. Histochem J 28(5):391–395. (Kluwer Academic Publishers)

Langelier MF, Riccio AA, Pascal JM (2014) PARP-2 and PARP-3 are selectively activated by 5' phosphorylated DNA breaks through an allosteric regulatory mechanism shared with PARP-1. Nucleic Acids Res 42(12):7762–75

Langelier M-F, Zandarashvili L, Aguiar PM, Black BE, Pascal JM (2018) NAD+ analog reveals PARP-1 substrate-blocking mechanism and allosteric communication from catalytic center to DNA-binding domains. Nat Commun 9(1):844 Nature Publishing Group

Mabley JG, Jagtap P, Perretti M, Getting SJ, Salzman AL, Virág L et al (2001) Anti-inflammatory effects of a novel, potent inhibitor of poly (ADP-ribose) polymerase. Inflamm res 50(11):561–569. (Birkhäuser Verlag)

Mariappan L, Jiang XY, Jackson J, Drew Y (2017) Emerging treatment options for ovarian cancer: focus on rucaparib. IJWH 9:913–924 Dove Press

Marsischky GT, Wilson BA, Collier RJ (1995) Role of glutamic acid 988 of human poly-ADP-ribose polymerase in polymer formation. Evidence for active site similarities to the ADP-ribosylating toxins. J Biol Chem 270(7):3247–3254

Mehrotra P, Hollenbeck A, Riley JP, Li F, Patel RJ, Akhtar N et al (2013) Poly (ADP-ribose) polymerase 14 and its enzyme activity regulates TH2 differentiation and allergic airway disease. J Allergy Clin Immunol 131(2):521–531.e12. (Elsevier)

Menear KA, Adcock C, Boulter R, Cockcroft X-L, Copsey L, Cranston A et al (2008) 4-[3-(4-cyclopropanecarbonylpiperazine-1-carbonyl)-4-fluorobenzyl]-2H-phthalazin-1-one: a novel bioavailable inhibitor of poly(ADP-ribose) polymerase-1. J Med Chem 51(20):6581–6591

Meyer T, Hilz H (1986) Production of anti-(ADP-ribose) antibodies with the aid of a dinucleotide-pyrophosphatase-resistant hapten and their application for the detection of mono(ADP-ribosyl)ated polypeptides. Eur J Biochem 155(1):157–165. (Blackwell Publishing Ltd)

Morgan RK, Carter-OConnell I, Cohen MS (2015) Selective inhibition of PARP10 using a chemical genetics strategy. Bioorg Med Chem Lett

Morgan RK, Cohen MS (2015) A clickable aminooxy probe for monitoring cellular ADP-ribosylation. ACS Chem Biol 10(8):1778–1784. 27 May 2015. (American Chemical Society)

Munnur D, Ahel I (2017) Reversible mono-ADP-ribosylation of DNA breaks. FEBS J. Wiley/ Blackwell (10.1111); 2017 Nov 8;284(23):4002–16

Narendja FM, Sauermann G (1994) The use of biotinylated poly(ADP-ribose) for studies on poly (ADP-ribose)-protein interaction. Anal Biochem 220:415–419. (Vienna)

Narwal M, Venkannagari H, Lehtiö L (2012) Structural basis of selective inhibition of human tankyrases. J Med Chem 55(3):1360–1367

Nduka N, Skidmore CJ, Shall S (1980) The enhancement of cytotoxicity of *N*-Methyl-*N*-nitrosourea and of y-radiation by inhibitors of poly(ADP-ribose) Polymerase. Eur J Biochem 105(3):525–530. (Blackwell Publishing Ltd)

Oei SL, Griesenbeck J, Buchlow G, Jorcke D, Mayer-Kuckuk P, Wons T et al (1999) NAD^+ analogs substituted in the purine base as substrates for poly(ADP-ribosyl) transferase. FEBS Lett 397(1):17–21

Ohmoto A, Yachida S (2017) Current status of poly(ADP-ribose) polymerase inhibitors and future directions. OTT 10:5195–5208 Dove Press

Papeo G, Avanzi N, Bettoni S, Leone A, Paolucci M, Perego R et al (2014) Insights into PARP inhibitors' selectivity using fluorescence polarization and surface plasmon resonance binding assays. J Biomol Screen 19(8):1212–1219

Pinto AF, Schüler H (2015) Comparative structural analysis of the putative mono-ADP-ribosyltransferases of the ARTD/PARP family. Curr Top Microbiol Immunol 384 (Chapter 417):153–166. (Springer International Publishing, Cham)

Purnell MR, Whish WJ. Novel inhibitors of poly(ADP-ribose) synthetase. Biochem J 185(3):775–777. 1 Mar 1980. (Portland Press Limited)

Rolli V, O'Farrell M, Ménissier-de Murcia J, de Murcia G (1997) Random mutagenesis of the poly (ADP-ribose) polymerase catalytic domain reveals amino acids involved in polymer branching †. Biochemistry 36(40):12147–12154

Rosado MM, Bennici E, Novelli F, Pioli C (2013) Beyond DNA repair, the immunological role of PARP-1 and its siblings. Immunology (10.1111); 139(4):428–437. 2 Jul 2013. (Wiley/ Blackwell)

Shen Y, Rehman FL, Feng Y, Boshuizen J, Bajrami I, Elliott R et al (2013) BMN 673, a novel and highly potent PARP1/2 inhibitor for the treatment of human cancers with DNA repair deficiency. Clin Cancer Res 19(18):5003–5015 American Association for Cancer Research

Surowy CS, Berger NA (1985) A, 3-aminobenzamide-resistant labeled protein in [32P]NAD^+-labeled cells. Biochimica et Biophysica Acta (BBA)—Molecular. Cell Res 847(3):309–315

Thomas HD, Calabrese CR, Batey MA, Canan S, Hostomsky Z, Kyle S et al (2007) A simple, sensitive, and generalizable plate assay for screening PARP inhibitors. In: Methods in molecular biology. American Association for Cancer Research, pp 945–56

Thorsell A-G, Ekblad T, Karlberg T, Löw M, Pinto AF, Trésaugues L et al (2017a) Structural basis for potency and promiscuity in poly(ADP-ribose) polymerase (PARP) and tankyrase inhibitors. J Med Chem 60(4):1262–1271

Thorsell A-G, Ekblad T, Karlberg T, Löw M, Pinto AF, Trésaugues L et al (2017b) Structural basis for potency and promiscuity in poly(ADP-ribose) polymerase (PARP) and tankyrase inhibitors. J Med Chem 60(4):1262–1271

Tsurumura T, Tsumori Y, Qiu H, Oda M, Sakurai J, Nagahama M et al (2013) Arginine ADP-ribosylation mechanism based on structural snapshots of iota-toxin and actin complex. Proc Natl Acad Sci 110(11):4267–4272. 12 Mar 2013

Upton K, Meyers M, Thorsell A-G, Karlberg T, Holechek J, Lease R et al (2017) Design and synthesis of potent inhibitors of the mono(ADP-ribosyl)transferase, PARP14. Bioorg Med Chem Lett 27(13):2907–2911

Venkannagari H, Verheugd P, Koivunen J, Haikarainen T, Obaji E, Ashok Y et al (2016) Small-molecule chemical probe rescues cells from mono-ADP-ribosyltransferase ARTD10/PARP10-induced apoptosis and sensitizes cancer cells to DNA damage. Cell Chem Biol 23 (10):1251–1260. (Elsevier)

Voronkov A, Holsworth DD, Waaler J, Wilson SR, Ekblad B, Perdreau-Dahl H et al (2013) Structural basis and SAR for G007-LK, a lead stage 1,2,4-triazole based specific tankyrase 1/2 inhibitor. J Med Chem 56(7):3012–3023

Vyas S, Matic I, Uchima L, Rood J, Zaja R, Hay RT et al (2014) Family-wide analysis of poly (ADP-ribose) polymerase activity. Nat Commun 5:4426

Wahlberg E, Karlberg T, Kouznetsova E, Markova N, Macchiarulo A, Thorsell A-G et al (2012) Family-wide chemical profiling and structural analysis of PARP and tankyrase inhibitors. Nat Biotechnol 30(3):283–288

Wallrodt S, Buntz A, Wang Y, Zumbusch A, Marx A (2016) Bioorthogonally functionalized NAD^+ analogues for in-cell visualization of poly(ADP-ribose) formation. Angew Chem Int Ed 55(27):7660–7664

Wang Y, Rösner D, Grzywa M, Marx A (2014) Chain-terminating and clickable NAD^+ analogues for labeling the target proteins of ADP-ribosyltransferases. Angew Chem Int Ed Engl 53 (31):8159–8162. 28 July 2014. (WILEY-VCH Verlag)

Watson CY, Whish WJD, Threadgill MD (1998) Synthesis of 3-substituted benzamides and 5-substituted isoquinolin-1(2H)-ones and preliminary evaluation as inhibitors of poly (ADP-ribose)polymerase (PARP). Bioorg Med Chem 6(6):721–734

Yamada M, Miwa M, Sugimura T (1971) Studies on poly (adenosine diphosphate-ribose): X. Properties of a partially purified poly (adenosine diphosphate-ribose) polymerase. Arch Biochem Biophy 146(2):579–586

Yoneyama-Hirozane M, Matsumoto S-I, Toyoda Y, Saikatendu KS, Zama Y, Yonemori K et al (2017) Identification of PARP14 inhibitors using novel methods for detecting auto-ribosylation. Biochem Biophys Res Commun 486(3):626–631

Zhan P, Song Y, Itoh Y, Suzuki T, Liu X (2014) Recent advances in the structure-based rational design of TNKSIs. Mol BioSyst 10(11):2783–2799 The Royal Society of Chemistry

Zhang J, Snyder SH (1992) Nitricoxidestimulatesauto-ADP-ribosylationof glyceraldehyde-3-phosphatedehydrogenase. Proc Natl Acad Sci U S A 89:9382–9385

Zingarelli B, Salzman AL, Szabó C (1998) Genetic disruption of poly (ADP-Ribose) synthetase inhibits the expression of *P*-Selectin and intercellular adhesion molecule-1 in myocardial ischemia/reperfusion injury. Circ Res 13;83(1):85–94. 13 Jul 1998. (American Heart Association, Inc)

Development of Activity-Based Proteomic Probes for Protein Citrullination

Venkatesh V. Nemmara and Paul R. Thompson

Contents

Abstract Protein arginine deiminases (PADs) catalyze the post-translational deimination of peptidyl arginine to form peptidyl citrulline. This modification is increased in multiple inflammatory diseases and in certain cancers. PADs regulate a variety of signaling pathways including apoptosis, terminal differentiation, and transcriptional regulation. Activity-based protein profiling (ABPP) probes have been developed to understand the role of the PADs in vivo and to investigate the effect of protein citrullination in various pathological conditions. Furthermore, these ABPPs have been utilized as a platform for high-throughput inhibitor discovery.

V. V. Nemmara · P. R. Thompson (✉)
Department of Biochemistry and Molecular Pharmacology, University of Massachusetts Medical School, LRB 826, 364 Plantation Street, Worcester, MA 01605, USA
e-mail: paul.thompson@umassmed.edu

V. V. Nemmara · P. R. Thompson
Program in Chemical Biology, University of Massachusetts Medical School, LRB 826, 364 Plantation Street, Worcester, MA 01605, USA

Current Topics in Microbiology and Immunology (2019) 420:233–252
DOI 10.1007/82_2018_132

Published Online: 11 September 2018

This review will showcase the development of ABPPs targeting the PADs. In addition, it provides a brief overview of PAD structure and function along with recent advances in PAD inhibitor development.

1 Introduction

Protein citrullination was first described by Rogers and Simmonds in 1958 (Fig. 1a) (Rogers and Simmonds 1958). This post-translational modification (PTM) converts an arginine residue into a citrulline via hydrolysis of the guanidinium group in a so-called deimination reaction.

This PTM is generated by a family of calcium-dependent enzymes called protein arginine deiminases (PADs) (Bicker and Thompson 2013; Fuhrmann et al. 2015; Fuhrmann and Thompson 2016). There are five human PADs, i.e., PAD1, PAD2, PAD3, PAD4, and PAD6, of which only PADs 1–4 are catalytically active (Raijmakers et al. 2007; Taki et al. 2011; Witalison et al. 2015). These isozymes,

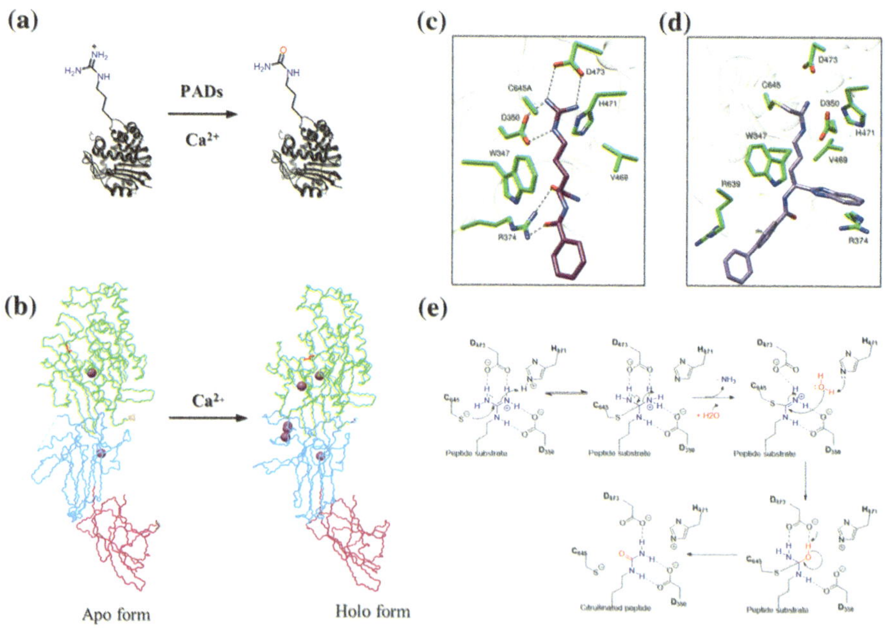

Fig. 1 **a** PAD-catalyzed hydrolysis of peptidyl arginine to peptidyl citrulline. **b** Backbone conformation of PAD2 showing both the apo and holo forms that is generated upon calcium binding. The structural change due to calcium binding is clearly evident in the catalytic domain (green), which harbors the catalytic cysteine C647 (shown in red in the catalytic domain). **c** Crystal structure of PAD4 C654A protomer bound to the substrate BAA (PDB code 1WDA). **d** Co-crystal structure of BB-F-amidine (5a) bounds to PAD4 (PDB code 5N0 M). **e** Proposed catalytic mechanism for PAD4

which share 50–55% sequence identity within the same species (Arita et al. 2004), are found in a myriad of cells and tissue types. PAD1 is highly expressed in epidermis and uterus and is thought to citrullinate keratins and filaggrins during skin keratinization (Ishida-Yamamoto et al. 2002; Senshu et al. 1996). PAD2 is distributed in various tissues and is abundant in muscles and brain (Moscarello et al. 2002). PAD3 is found in hair follicles and epidermis (Rogers et al. 1997), whereas PAD4 is expressed in neutrophils, granulocytes and macrophages (Nakashima et al. 1999). PAD6 is only found in oocytes and embryos (Esposito et al. 2007). In addition to their localization in the cytoplasm, PAD1, PAD2, and PAD4 also localize to the nucleus where they citrullinate histones and other chromatin-associated proteins (Cherrington et al. 2010; Fuhrmann et al. 2015; Jang et al. 2011; Kan et al. 2012; Lewallen et al. 2015).

Citrullination can have profound effects on the primary, secondary, and tertiary structures of proteins due to the loss of a positive charge. Additionally, this PTM can result in the loss of protein–protein or protein–DNA interactions with consequent effects on cell signaling (Clancy et al. 2017; Fuhrmann et al. 2015; Lewis and Nacht 2016). Notably, the PADs regulate various biological processes, including myelination, cell differentiation, gene regulation, and the innate immune response (Bicker and Thompson, 2013; Christophorou et al. 2014; Li et al. 2010; Nauseef and Borregaard, 2014; Senshu et al. 1996; Slade et al. 2014). Additionally, aberrant PAD activity can lead to protein hypercitrullination, which is a hallmark of various inflammatory and neurodegenerative disorders (Jang et al. 2008; Jones et al. 2009; Khandpur et al. 2013; Knight et al. 2013, 2014; Leffler et al. 2012; Musse et al. 2008).

Specifically, citrullination occurs during NETosis, a proinflammatory form of cell death that is aberrantly upregulated in numerous autoimmune diseases including RA, atherosclerosis, and lupus (Khandpur et al. 2013; Knight et al. 2013, 2014). Protein citrullination is also elevated in luminal breast cancer, multiple sclerosis and certain inflammatory diseases (Jones et al. 2009; McElwee et al. 2012; Moscarello et al. 2002; Zhang et al. 2012). Importantly, inhibition of PAD2 in breast cancer cell lines decreases disease progression by increasing apoptosis (McElwee et al. 2012). Recently, PAD1 was shown to be overexpressed in human triple-negative breast cancer lines (e.g., MDA-MB-231 cells) as well as in xenograft mouse models and its inhibition resulted in reduced cell proliferation and metastasis (Qin et al. 2017). PAD2 and PAD4 are also activated in the central nervous system (CNS) during neurodegenerative processes and are observed to be co-localized in regions of degraded neurons in Alzheimer's patients (Acharya et al. 2012; Ishigami et al. 2005). The distinct roles of the PADs in various pathophysiologic states are quite alarming, and there is a pressing need to develop isozyme-specific inhibitors to be used as therapeutics and probes to decipher the physiological roles of these enzymes in various disease states. This review will focus on the discovery and development of PAD-targeted ABPPs. It will also provide a brief overview of the structure and function of the PADs and recent progress in PAD inhibitor development.

2 PAD Structure and Function

2.1 PAD Structures

One of the major advances that aided the development of PAD inhibitors and activity-based probes was the determination of high-resolution X-ray crystal structures for PAD1, PAD2, and PAD4 (Arita et al. 2004; Saijo et al. 2016; Slade et al. 2015). PADs contain an α/β propeller domain located in the C-terminal lobe that harbors the active site. The N-terminal domain is comprised of two immunoglobulin-like subdomains that are proposed to be important for protein–protein interactions and for substrate selection (Fig. 1b) (Arita et al. 2004; Fuhrmann et al. 2015). Crystal structures of both the inactive (apo, calcium-free) and active (holo, calcium-bound) states of PAD2 and PAD4 have been reported confirming that calcium binding is required to form a catalytically competent active site (Fig. 1b). A significant difference between the apo and holo states is the position of the catalytic cysteine (C647 in PAD2 and C645 in PAD4), which moves ≥ 10 Å into the active site upon calcium binding (Fig. 1b–d). From a careful analysis of the crystal structures, we know that PAD1, PAD2, and PAD4 bind four, six, and five calcium ions, respectively—PAD1 lacks Ca5 and PAD2 has a unique sixth site, i.e., Ca6. Based on calcium titration experiments with PAD2, we know that Ca1 and Ca6 are tightly and constitutively bound to PAD2. Ca3, Ca4, and Ca5 bind next and trigger a conformational change that generates the Ca2-binding site. Ca2 binding then triggers a conformational change that moves C647, as well as W347, into catalytically competent conformations (Slade et al. 2015).

2.2 PAD Substrate Recognition and Catalytic Mechanism

PADs are highly selective in citrullinating peptidyl arginine over free arginine (Knuckley et al. 2010a). However, not all arginine residues are equally citrullinated by the PADs. Evidence suggests that PADs preferentially catalyze the citrullination of arginine residues present in a β-turn or in specific loops as compared to other secondary structure elements in a protein (Knuckley et al. 2010a; Tarcsa et al. 1996). PADs are also known to catalyze the citrullination of arginine containing small peptides. A structure of the PAD4 (C645A) bound to benzoyl-L-arginine amide (**BAA**) depicts the key residues that interact with the substrate (Arita et al. 2004). As shown in Fig. 1c, D473 and D350 make salt bridges with the guanidinium group of the substrate, C645A is positioned for nucleophilic attack, and H471 is involved in general acid/base catalysis. The methylene groups of the substrate side chain make hydrophobic interactions with W347 and V469. R374 makes hydrogen bonding interactions with the backbone carbonyl group of the substrate. This is a notable interaction as it confers specificity toward peptidyl arginine as opposed to free arginine, which lacks an amide bond. This observation was further confirmed by crystal structures of PAD4 bound to various histone substrate peptides, which illustrated that most of the interactions occur between the backbone carbonyl groups of the substrate and the side chain of the active-site residues (Arita et al. 2006). The lack of relevant interactions between the side chain of protein substrates and PAD4 is consistent with the broad substrate scope of PAD4. The substrate scope of the other PADs is similarly broad suggesting that substrate recognition is driven primarily by sequence context rather than key side chain interactions.

Based on the available crystal structures and biochemical experiments (Arita et al. 2004; Knuckley et al. 2007, 2010a, b), we proposed that PAD4 catalyzes citrullination via attack on the guanidinium carbon by a nucleophilic cysteine thiolate, which results in the formation of a tetrahedral intermediate (Fig. 1e). Proton transfer from H471 stabilizes the positively charged intermediate via interactions with D350 and D473 (Knuckley et al. 2010b). Collapse of the intermediate generates an *S*-alkylthiouronium intermediate that is ultimately hydrolyzed by water via a second tetrahedral intermediate. The thiol group in the catalytic cysteine needs to be deprotonated for optimal enzyme activity. Based on pH-dependent kinetic inactivation studies, the p*Ka* of this cysteine was calculated to be ~8.3 when iodoacetamide was used as the thiol-reactive reagent (Dreyton et al. 2014b; Knuckley et al. 2007, 2010a, b). This data coupled with the observation of an inverse solvent isotope effect on *k*cat/*K*m supports the notion that PAD4 (as well as PADs 1 and 3) uses a reverse protonation mechanism. Further supporting this mechanism is the fact that 2-chloroacetamidine, a positively charged thiol-reactive compound, does not depress the p*K*a of C645 in PAD4 (Knuckley et al. 2010b). By contrast, 2-chloroacetamidine does depress the p*K*a of C647 in PAD2 (p*K*a ~ 7.2) (Dreyton et al. 2014b). This data along with a normal solvent isotope effect indicates that PAD2 uses a substrate assisted mechanism of thiol deprotonation (Dreyton et al. 2014b).

3 PAD Inhibitors

Over the past several years, numerous PAD inhibitors have been developed. These compounds include both reversible and irreversible inhibitors with decent potencies and selectivities. The mechanisms of action have been extensively studied using biochemical and X-ray crystallographic techniques. This section will provide an overview of both reversible (non-covalent) and irreversible (covalent) PAD inhibitors.

3.1 Reversible PAD Inhibitors

Taxol (pacitaxel) was discovered as a weak non-competitive inhibitor (*K*i = 4–10 mM) of PAD4 by Pritzker and Moscarello two decades ago. This finding sparked the beginning of further PAD inhibitor development (Pritzker and

Moscarello 1998). Knuckley et al. later discovered streptomycin (Ki ~ 0.56 mM), minocycline (Ki ~ 0.63 mM), and chlorotetracycline (Ki ~ 0.11 mM) as relatively potent inhibitors of PAD4 (Knuckley et al. 2008). Recently, Lewis et al. reported the development of highly potent reversible inhibitors that are selective for the apo form of PAD4 (Lewis et al. 2015). A DNA-encoded small molecule library screen for PAD inhibitors in the presence and absence of calcium yielded GSK121 as a modest inhibitor of apo PAD4. A detailed SAR for the primary hit resulted in optimized leads GSK199 and GSK484, which inhibited PAD4 with IC50 values of 250 nM and 80 nM, respectively, in the presence of 0.2 mM Ca^{2+} (Fig. 2a). The potencies of these compounds were, however, reduced by >fivefold at higher calcium concentrations. Detailed kinetic analyses and a co-crystal structure indicate that these inhibitors are competitive with respect to calcium binding and

(a)

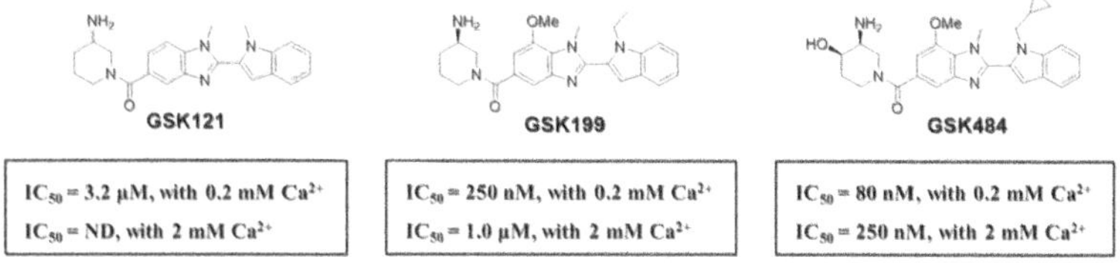

(b)

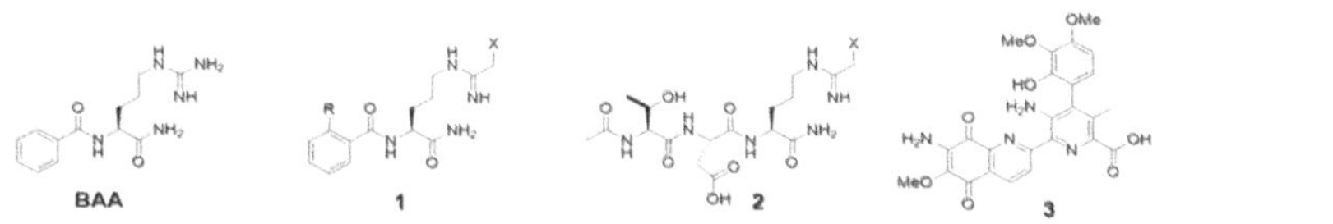

Compounds	X	R	PAD1 k_{inact}/K_I ($M^{-1}min^{-1}$)	PAD2 k_{inact}/K_I ($M^{-1}min^{-1}$)	PAD3 k_{inact}/K_I ($M^{-1}min^{-1}$)	PAD4 k_{inact}/K_I ($M^{-1}min^{-1}$)
1a	F	H	2800	380	170	3000
1b	Cl	H	37000	1200	2000	13000
1c	F	COOH	181000	7500	6700	32500
1d	Cl	COOH	106000	14100	10400	38000
2a	F	-	1700	500	400	26000
2b	Cl	-	21000	300	920	24000
3	-	-	3700	12000	3500	440000

Compounds	X	R1	PAD1 k_{inact}/K_I ($M^{-1}min^{-1}$)	PAD2 k_{inact}/K_I ($M^{-1}min^{-1}$)	PAD3 k_{inact}/K_I ($M^{-1}min^{-1}$)	PAD4 k_{inact}/K_I ($M^{-1}min^{-1}$)
4a	F	-	1220	160	3	130
4b	Cl	-	13500	270	65	1400
5a	F	-	900	1200	3400	3750
5b	Cl	-	16100	4100	6800	13300
6a	-	H	129100	210300	4430	14300
6b	-	Me	59760	212700	2010	2300
7	-	-	360	1110	15600	1460

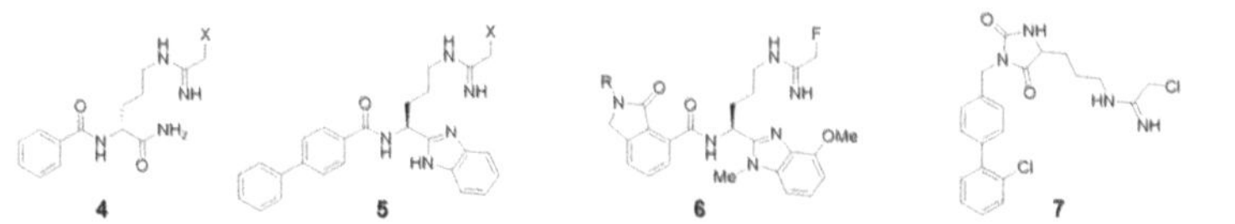

Fig. 2 **a** Structure of reversible PAD4 inhibitors. The IC50 values for each compound with PAD4 are shown underneath. These compounds are >15-fold selective for PAD4. **b** Structure of irreversible PAD inhibitors. The generic warhead on the irreversible inhibitors is colored in pink. The second-order rate constants for PAD inactivation by covalent inhibitors are shown in the table

preferentially bind the apo form of PAD4, thereby preventing the calcium-induced movement of Cys645 into the active site. Not only did these inhibitors demonstrate more than 15-fold selectivity for PAD4, but they also revealed a unique approach for targeting the apo state of a specific PAD isoform. Importantly, GSK199 shows efficacy in the murine collagen induced arthritis model of RA (Willis et al. 2017).

3.2 *Irreversible PAD Inhibitors*

The search for irreversible covalent inhibitors of PAD isoforms began with the discovery of 2-chloroacetamidine, a time-dependent inactivator of PAD4 with an IC50 > 0.5 mM. In parallel, Luo et al. replaced the guanidinium group in the PAD substrate benzoyl-L-arginine amide (**BAA**) with either a fluoro- or chloroacetamidine-based warhead, thereby generating F-amidine (**1a**) or Cl-amidine (**1b**), respectively (Fig. 2b) (Luo et al. 2006a, c). Kinetic studies revealed that both F-amidine and Cl-amidine function as mechanism-based inhibitors that irreversibly inactivate PAD4 by modifying the catalytic cysteine in a time-dependent manner. Interestingly, Cl-amidine was more potent than F-amidine when evaluated against all PAD isoforms (except PAD6, which is not active), consistent with the increased reactivity of the chloro-warhead over the fluoro-warhead. More importantly, Cl-amidine showed efficacy in treating a variety of animal models, including mouse models of RA, lupus, atherosclerosis, ulcerative colitis, neuron injury, and breast cancer (Willis et al. 2011; Knight et al. 2013, 2014, 2015; Lange et al. 2011) (Chumanevich et al. 2011; McElwee et al. 2012).

Optimization of F-amidine and Cl-amidine resulted in the identification of the more potent *ortho*-carboxylate derivatives *o*-F-amidine (**1c**) and *o*-Cl-amidine (**1d**) (Causey et al. 2011). The increased potency of **1c** and **1d** against PAD4 was found to be due to enhanced interactions between the *o*-carboxylate groups and the indole NH of W347 as well as a water-mediated hydrogen bond with the side chain of Q346 (Causey et al. 2011). From an extensive tripeptide library screen, the haloacetamidine-based peptides, TDFA (**2a**) and TDCA (**2b**) (Fig. 2b), were discovered to be potent and selective PAD4 inactivators (Jones et al. 2012). The improved inhibition of these peptides was attributed to their interaction with Q346, R374, and R639 in PAD4. Interaction with R639 was looked upon as key for selectivity against PAD4 as this residue is unique to this isozyme. From a fluorescence polarization activity-based high-throughput screen (HTS), assay streptonigrin (**3**) was also found to be a selective inhibitor of PAD4 (Dreyton et al. 2014a; Knuckley et al. 2010c).

Despite these early advances, the peptidic nature of these first-generation haloacetamidine-based inhibitors possessed various limitations, including reduced metabolic stability due to proteolytic cleavage and poor membrane permeability (Knight et al. 2015). This led to the development of second-generation inhibitors that were predicated on improving the lipophilicity of the core structure by substituting the C-terminal carboxamide in **1a** and **1b** with a benzimidazole moiety and the N-terminal phenyl ring with a biphenyl substituent resulting in BB-F-amidine (**5a**) and BB-Cl-amidine (**5b**) (Figs. 1d and 2b) (Knight et al. 2015; Muth et al. 2017). This led to a dramatic increase in the lipophilicity of BB-Cl-amidine (CLogP = 4.17) as compared to Cl-amidine (CLogP = −0.23), which was predicted to aid cell entry (Knight et al. 2015). Consistent with this prediction, BB-Cl-amidine was far superior when compared to Cl-amidine in a variety of cell-based assays and animal models but showed similar in vitro efficacy (Ghari et al. 2016; Horibata et al. 2015; Kawalkowska et al. 2016; Knight et al. 2015). Recently, Muth et al. reported a detailed SAR for **5a** and **5b** with a methyl benzimidazole scaffold that included a lactam ring in place of N-terminal phenyl group (Muth et al. 2017). The most potent inhibitors **6a** and **6b** exhibited excellent PAD2 selectivity (up to 106-fold) and proved to be potent in several cell-based assays (Fig. 2b) (Muth et al. 2017).

The search for other PAD isoform-selective inhibitors led to the discovery of D-F-amidine (**4a**) and D-Cl-amidine (**4b**) (Fig. 2b), the D-ornithine derivatives of **1a** and **1b**, which exhibited remarkable selectivity against PAD1 (400-fold by **4a** and 200-fold by **4b**) (Bicker et al. 2012a). D-Cl-amidine also demonstrated improved metabolic stability as compared to Cl-amidine (Bicker et al. 2012a). Furthermore, Jamali et al. identified a novel scaffold (**7**) through a substrate-based fragment screen, inhibitor, which was potent and selective for PAD3 (Fig. 2b) (Jamali et al. 2015, 2016). In addition to the in vitro selectivity, the hydantoin-based inhibitor (**7**) also demonstrated efficacy in a cellular model of thapsigargin induced cell death in PAD3-expressing HEK293T cells (Jamali et al. 2016).

4 Activity-Based Proteomic Probes Targeting the PADs

The ability of haloacetamidine-based inhibitors to covalently modify PADs in an irreversible manner was considered to be a prerequisite for the development of activity-based probes. In addition to a chemical warhead that reacts with functional sites in proteins, activity-based probes also bear chemical handles, such as fluorophores, biotin, or alkynes for subsequent fluorescent visualization, streptavidin enrichment and mass spectrometry-based quantification or bio-orthogonal conjugation, respectively, for subsequent analysis of protein activities (Fig. 3a). An overview of the development of PAD-targeted ABPPs is described below.

4.1 First-Generation PAD-Targeted ABPPs

First-generation PAD-targeted ABPPS were based on the structure of Cl-amidine and F-amidine. Replacement of the benzoyl group with a *p*-benzylic azide facilitated their conjugation to rhodamine–alkyne via the formation of a triazole linker to generate rhodamine-conjugated F-amidine (RFA) and rhodamine-conjugated

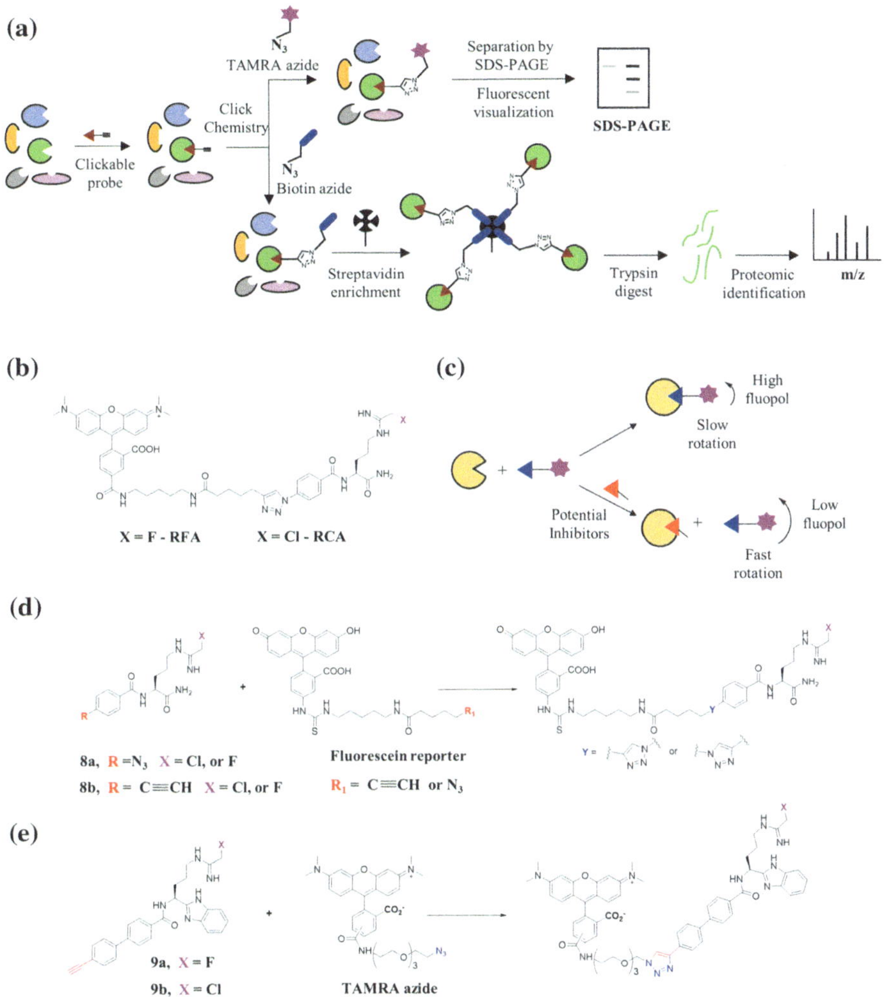

Fig. 3 **a** Schematic representation depicting the use of clickable ABPPs. **b** Chemical structures of RFA and RCA. **c** Schematic overview of the fluorescence polarization assay using RFA as an ABPP for high-throughput inhibitor discovery. **d** Structures of first-generation 'clickable' PAD-targeted ABPPs. Labeled probes are subjected to click-chemistry using fluorescein reporters through a post-inactivation strategy. **e** Structures of second-generation 'clickable' PAD-targeted ABPPs. Probe-labeled proteins are subjected to click-chemistry using TAMRA azide reporter through a post-inactivation strategy

Cl-amidine (RCA) (Fig. 3b) (Luo et al. 2006b). RFA and RCA preferentially labeled the calcium-bound holo form of PAD4 and do not modify the C645S active-site mutant (Luo et al. 2006b). These probes are as potent as the parent compounds suggesting that the fluorophore does not alter enzyme binding (Luo et al. 2006b). The versatility of RFA and RCA was further demonstrated through the development of a competitive gel-based ABPP-assay, which was used to identify novel PAD inhibitors from diverse chemical libraries (Knuckley et al. 2008). Using this strategy, Knuckley et al. discovered streptomycin as a competitive inhibitor and chlorotetracycline as a mixed inhibitor of PAD4 (Knuckley et al. 2008).

The success of this RFA-based screening strategy led to the development of an RFA-based high-throughput screen (HTS), termed a fluorescence polarization activity-based protein profiling (fluopol-ABPP)-based HTS assay (Fig. 3c) (Knuckley et al. 2010c). This assay monitors changes in fluorescence polarization (fluopol) as a result of RFA binding to PAD4. Since the RFA-PAD4 complex rotates slowly, there is a larger change in the fluorescent polarization signal. In contrast, unbound RFA rotates faster and leads to a smaller change in the fluorescence polarization signal. In the case of an inhibitor bound to PAD4, RFA-PAD4 binding will be hampered, resulting in a lower fluopol signal (Knuckley et al. 2010c). Using this strategy, Knuckley et al. screened more than 2000 compounds and identified 10 inhibitors, from which streptonigrin (**3**) was discovered as a potent and selective inhibitor of PAD4. Despite the utility of this RFA-based HTS assay, the methodology suffered limitations including a strong bias toward irreversible inhibitors that preferentially targeted the active holo form of PAD4.

To overcome this limitation, Lewallen et al. developed a modified approach to identify inhibitors that targeted the apo, calcium-free, form of PAD2 (Lewallen et al. 2014). Specifically, they performed the screen at lower Ca^{2+} concentrations (350 μM $CaCl_2$) and screened the 1280 compound LOPAC library (Sigma-Aldrich Library of Pharmacologically Active Compounds). Using this approach, Lewallen et al. identified ruthenium red as a calcium competitive inhibitor that binds to the apo form of PAD2 (Lewallen et al. 2014). In the future, this RFA-based screening strategy could be used to identify isoform-specific PAD inhibitors from various compound libraries. Furthermore, this approach will be useful for the discovery of allosteric inhibitors that target alternate sites in PAD isoforms, similarly to the PAD4-selective inhibitors GSK199 and GSK484.

4.2 Second-Generation PAD-Targeted ABPPs

A major drawback of the rhodamine-based probes RFA and RCA is that they are not ideal cellular probes due to the presence of a bulky fluorophore, which may alter their cell permeability. To circumvent this issue, Slack et al. employed a bio-orthogonal strategy that involved the conjugation of fluorescein and biotin-based reporters to F-amidine and Cl-amidine using the copper(I)-catalyzed azide–alkyne cycloaddition (CuAAC) reaction (Slack et al. 2011). Development of fluorescein-based ABPPs, in addition to the rhodamine-based ABPPs, in principle, generates tools that allow for the optimal visualization of labeled proteins across the entire pH spectrum. To achieve this, Slack et al. synthesized azide and alkyne containing F-amidine and Cl-amidine (**8a** and **8b**) and used them to label recombinant PAD4 and PAD4 in *Escherichia coli* cell extracts (Fig. 3d). The labeled samples were then subjected to click-chemistry using the corresponding azide or alkyne fluorescein reporter tags to obtain post-inactivated ABPPs bound to PAD4 that were visualized by fluorescence. F-amidine and Cl-amidine containing an alkyne handle demonstrated efficient post-inactivation click-chemistry as compared to their azide counterparts and proved to be selective for PAD4 in *E. coli* cell lysates (Fig. 3a, d). Despite their utility in the efficient detection of PAD4 by fluorescent visualization, these fluorescein-based ABPPs were not useful for isolating PAD4 or to identify PAD4-interacting proteins in biological samples (Slack et al. 2011).

To demonstrate efficient isolation of PAD4 from bacterial (*E. coli*) and mammalian (e.g., MCF-7 and HL60) cells, Slack et al. treated the cell lysates with F-amidine-Y-ne and Cl-amidine-Y-ne and employed a TEV-tagged biotin–azide reporter to perform post-inactivation click-chemistry (Slack et al. 2011). The subsequent pulldown of the probe-modified proteins using streptavidin–agarose and Western blot established the utility of these biotin–azide reporter tags in isolating the active form of PAD4 and identifying several known PAD4-binding proteins, including HDAC1, p53, and histone H3. The approach was also used to isolate PAD4 from HL60 granulocytes, suggesting a potential future role for these ABPPs in identifying the post-translational modifications that occur to the PADs under various physiological conditions.

Building on the success of the second-generation inhibitors BB-F-amidine and BB-Cl-amidine as cell-permeable and metabolically stable pan-PAD inhibitors, we next developed clickable probes based on this scaffold (Nemmara et al. 2018). Specifically, Nemmara et al. developed alkyne containing derivatives of BB-F-amidine and BB-Cl-amidine, which are termed BB-F-Yne (**9a**) and BB-Cl-Yne (**9b**). These probes are as potent as the parent compounds (**5a** and **5b**) and can label PAD2 in HEK cells (Fig. 3a, e). The versatility of these probes was verified by their ability to pull down PAD2 from HEK cell lysates after post-inactivation click-chemistry with biotin–azide. Furthermore, using a chemoproteomic approach, we demonstrated that BB-F-Yne, which possesses a fluoroacetamidine warhead, is remarkably selective for PAD2 in HEK cells—the only protein isolated was PAD2. By contrast, BB-Cl-Yne has a few 'off-targets' in addition to PAD2. Most of the off-targets are highly abundant proteins with reactive cysteines, i.e., 'the usual suspects' that are found in other proteomic screens that use cysteine-reactive electrophiles (Weerapana et al. 2007, 2010). Projecting forward, it will be worthwhile to assess the potential of future ABPPs harboring a fluoroacetamidine warhead, especially, for the selective detection of PAD isoforms in complex biological systems as well as in identifying novel proteins and signaling pathways associated with PAD biology.

4.3 Phenylglyoxal-Based Probes to Detect Protein Citrullination

Given that aberrant citrullination is a hallmark of various autoimmune diseases and certain tumor types, it is important to develop protein detection methods to identify citrullinated proteins in complex biological systems. Due to the low abundance of this PTM in biological samples, the enrichment of citrullinated proteins by chemical derivatization improves detection. It should be noted that the detection of citrullinated proteins by mass spectrometry is also challenging because the 1 Da mass change is easily confused with a deamidation event or an isotope effect. To improve the MS-based detection of citrullinated proteins, Holm et al. showed that selective derivatization of the urea group in citrulline is possible with 2,3-butanedione alone or in combination with antipyrine (De Ceuleneer et al. 2011; Holm et al. 2006). These modifications resulted in a mass increase of 50 or 238 Da, which is readily detected by MS, thereby differentiating citrullinated peptides from other arginine containing ones. However, this methodology does not facilitate the enrichment of citrullinated proteins or peptides.

Building upon this concept, Bicker et al. developed a citrulline-specific probe by linking rhodamine and phenylglyoxal via a triazole linker. This compound rhodamine–phenylglyoxal (Rh-PG) specifically labels citrullinated proteins under acidic conditions (Bicker et al. 2012b) (Fig. 4a, b). Rh-PG was able to monitor the kinetics of protein citrullination of PAD4 substrates (e.g., histone H3) due to its low limit of detection (LOD $\sim$ 600 fmol for citrullinated histone H3). In addition, Bicker et al. used Rh-PG to detect citrullinated proteins in a mouse model of ulcerative colitis treated with and without a PAD inhibitor. Notably, several proteins showed reduced citrullination in the presence of inhibitor, suggesting they could be useful biomarkers of ulcerative colitis (Fig. 4a). Using Rh-PG, Bawadekar et al. also quantified citrullinated proteins in lysates in a mouse model of TNF-α induced lung inflammation and showed that PADs other than PAD4 are responsible for inflammation-mediated protein citrullination in lungs (Bawadekar et al. 2016). Moreover, Carmona-Rivera et al. used Rh-PG to show that citrullinated proteins are abundantly generated during NET formation induced by either IgM or rheumatoid factor (Carmona-Rivera et al. 2017). In total, Rh-PG is a powerful probe for the robust quantification of citrullinated proteins and can be used to characterize unique disease biomarkers in various autoimmune diseases.

To extend the versatility of these phenylglyoxal-based probes, Lewallen et al. developed biotin-conjugated phenylglyoxal (BPG). This probe serves as an antibody surrogate for Western blotting and as a chemical handle to isolate citrullinated proteins from biological mixtures (Fig. 4a, b) (Lewallen et al. 2015). Using this probe, numerous citrullinated proteins were identified for the first time from PAD2-expressing HEK293T cells treated with ionomycin. These novel PAD substrates included multiple proteins involved in RNA splicing, suggesting a potential role for the PADs in this process (Lewallen et al. 2015). More recently, Tilvawala et al. used BPG to map the RA-associated citrullinome in serum, synovial fluid,

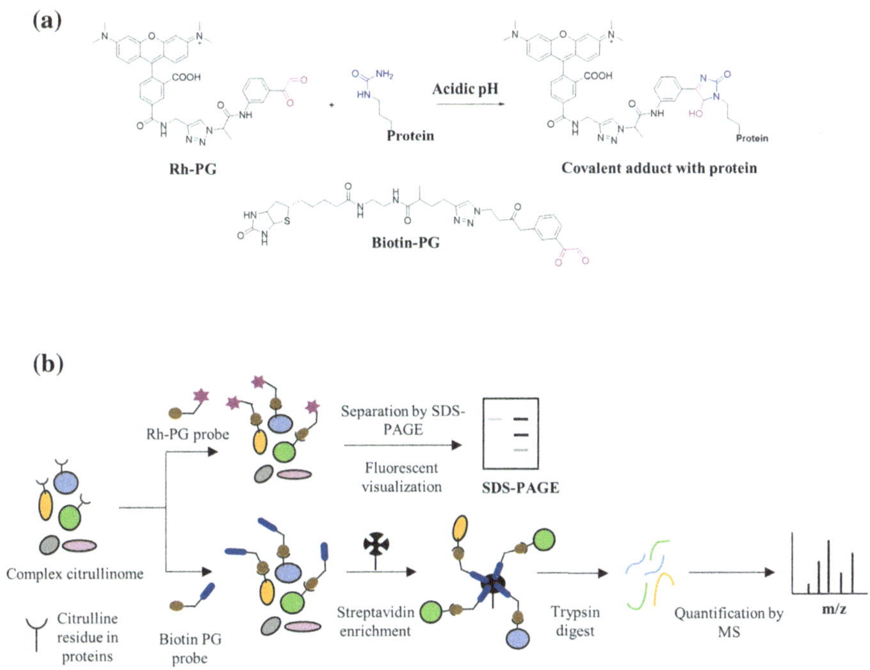

Fig. 4 **a** Chemical structures of Rh-PG and BPG and the chemoselective reaction of Rh-PG with citrulline under acidic conditions. **b** Schematic representation of the strategy for the chemical derivatization of peptidyl citrulline using Rh-PG and biotin PG

and synovial tissue samples (Tilvawala et al. 2018). In addition to identifying more than 150 novel citrullinated proteins, they developed a BPG-based ELISA assay to validate the proteomic data. Among the various proteins identified, there were numerous SERPINs (SERine Protease INhibitors) and metabolic enzymes. Fascinatingly, Tilvawala et al. demonstrated that citrullination inactivates a subset of SERPINs and modulates the activity of a range of metabolic enzymes. Specifically, citrullination of antiplasmin, C1-inhibitor and tissue plasminogen inhibitor (t-PAI) abolished their inhibitory activity against their cognate proteases, i.e., plasmin, kallikrein, and tissue plasminogen (t-PA).

An important limitation of this probe is that in its current form, it is not possible to identify the exact site of citrullination in target proteins. Moreover, it does not distinguish between citrullination and lysine carbamylation. In the future, development of phenylglyoxal-based probes that can identify endogenous sites of citrullination should be a priority.

5 Conclusions

Over the past several years, chemical labeling has produced a wealth of novel technologies to identify the biological roles of a variety of proteins. Activity-based protein profiling (ABPP) is one such technique that has become a centerpiece of chemical biology. The discovery of RFA, a haloacetamidine-based ABPP, has led to the development of high-throughput assay platforms for identifying isoform-selective PAD inhibitors. Moreover, the first-generation PAD-targeted ABPPs, F-amidine-Yne (**8a**) and Cl-amidine-Yne (**8b**), were able to label PAD4 in cells and isolate PAD4-binding proteins. Remarkably, the second-generation PAD-targeted ABPPs, BB-F-Yne (**9a**) and BB-Cl-Yne (**9b**), showed enhanced cellular uptake and efficient labeling of PADs in cells (Fig. 3e). Perhaps the most exciting discovery is the superior selectivity of fluroacetamidine-based probes toward PADs in cell systems, a finding that is counterintuitive to the historical perception that covalent inhibitors are non-selective in nature. We predict that these fluoroacetamidine-based probes will be used to detect the PADs in complex biological milieus and in identifying pathways that regulate PAD activity. Recent advances in chemical labeling techniques have also led to the development of citrulline-specific probes to detect protein citrullination in a variety of disease states. The citrulline-specific probes Rh-PG and BPG have proved to be powerful tools to detect and quantify protein citrullination in complex systems. We predict that these probes have the potential to transform our understanding of PAD biology by uncovering novel citrullinated biomarkers associated with a range of pathophysiological states. Finally, we expect the next-generation citrulline-specific probes will enable the identification of specific sites of citrullination.

References

Acharya NK, Nagele EP, Han M, Coretti NJ, DeMarshall C, Kosciuk MC, Boulos PA, Nagele RG (2012) Neuronal PAD4 expression and protein citrullination: possible role in production of autoantibodies associated with neurodegenerative disease. J Autoimmun 38(4):369–380. https://doi.org/10.1016/j.jaut.2012.03.004

Arita K, Hashimoto H, Shimizu T, Nakashima K, Yamada M, Sato M (2004) Structural basis for Ca(2+)-induced activation of human PAD4. Nat Struct Mol Biol 11(8):777–783. https://doi.org/10.1038/nsmb799

Arita K, Shimizu T, Hashimoto H, Hidaka Y, Yamada M, Sato M (2006) Structural basis for histone N-terminal recognition by human peptidylarginine deiminase 4. Proc Natl Acad Sci U S A 103(14):5291–5296. https://doi.org/10.1073/pnas.0509639103

Bawadekar M, Gendron-Fitzpatrick A, Rebernick R, Shim D, Warner TF, Nicholas AP, Lundblad LK, Thompson PR, Shelef MA (2016) Tumor necrosis factor alpha, citrullination, and peptidylarginine deiminase 4 in lung and joint inflammation. Arthritis Res Ther 18(1):173. https://doi.org/10.1186/s13075-016-1068-0

Bicker KL, Anguish L, Chumanevich AA, Cameron MD, Cui X, Witalison E, Subramanian V, Zhang X, Chumanevich AP, Hofseth LJ et al (2012a) D-amino acid based protein arginine deiminase inhibitors: Synthesis, pharmacokinetics, and in cellulo efficacy. ACS Med Chem Lett 3(12):1081–1085. https://doi.org/10.1021/ml300288d

Bicker KL, Subramanian V, Chumanevich AA, Hofseth LJ, Thompson PR (2012b) Seeing citrulline: development of a phenylglyoxal-based probe to visualize protein citrullination. J Am Chem Soc 134(41):17015–17018. https://doi.org/10.1021/ja308871v

Bicker KL, Thompson PR (2013) The protein arginine deiminases: Structure, function, inhibition, and disease. Biopolymers 99(2):155–163. https://doi.org/10.1002/bip.22127

Carmona-Rivera C, Carlucci PM, Moore E, Lingampalli N, Uchtenhagen H, James E, Liu Y, Bicker KL, Wahamaa H, Hoffmann V et al (2017) Synovial fibroblast-neutrophil interactions promote pathogenic adaptive immunity in rheumatoid arthritis. Sci Immunol 2(10). https://doi.org/10.1126/sciimmunol.aag3358

Causey CP, Jones JE, Slack JL, Kamei D, Jones LE, Subramanian V, Knuckley B, Ebrahimi P, Chumanevich AA, Luo Y et al (2011) The development of N-alpha-(2-carboxyl)benzoyl-N(5)-(2-fluoro-1-iminoethyl)-L-ornithine amide (*o*-F-amidine) and *N*-alpha-(2-carboxyl)benzoyl-*N*(5)-(2-chloro-1-iminoethyl)-L-ornithine amide (*o*-Cl-amidine) as second generation protein arginine deiminase (PAD) inhibitors. J Med Chem 54(19):6919–6935. https://doi.org/10.1021/jm2008985

Cherrington BD, Morency E, Struble AM, Coonrod SA, Wakshlag JJ (2010) Potential role for peptidylarginine deiminase 2 (PAD2) in citrullination of canine mammary epithelial cell histones. PLoS ONE 5(7):e11768. https://doi.org/10.1371/journal.pone.0011768

Christophorou MA, Castelo-Branco G, Halley-Stott RP, Oliveira CS, Loos R, Radzisheuskaya A, Mowen KA, Bertone P, Silva JC, Zernicka-Goetz M et al (2014) Citrullination regulates pluripotency and histone H1 binding to chromatin. Nature 507(7490):104–108. https://doi.org/10.1038/nature12942

Chumanevich AA, Causey CP, Knuckley BA, Jones JE, Poudyal D, Chumanevich AP, Davis T, Matesic LE, Thompson PR, Hofseth LJ (2011) Suppression of Colitis in Mice by Cl-Amidine: a novel Peptidylarginine Deiminase (Pad) Inhibitor. Am J Physiol Gastrointest Liver Physiol 300(6):G929–G938. https://doi.org/10.1152/ajpgi.00435.2010

Clancy KW, Russell AM, Subramanian V, Nguyen H, Qian Y, Campbell RM, Thompson PR (2017) Citrullination/methylation crosstalk on Histone H3 regulates ER-target gene transcription. ACS Chem Biol 12(6):1691–1702. https://doi.org/10.1021/acschembio.7b00241

De Ceuleneer M, De Wit V, Van Steendam K, Van Nieuwerburgh F, Tilleman K, Deforce D (2011) Modification of citrulline residues with 2,3-butanedione facilitates their detection by liquid chromatography/mass spectrometry. Rapid Commun Mass Spectrom 25(11):1536–1542. https://doi.org/10.1002/rcm.5015

Dreyton CJ, Anderson ED, Subramanian V, Boger DL, Thompson PR (2014a) Insights into the mechanism of streptonigrin-induced protein arginine deiminase inactivation. Bioorg Med Chem 22(4):1362–1369. https://doi.org/10.1016/j.bmc.2013.12.064

Dreyton CJ, Knuckley B, Jones JE, Lewallen DM, Thompson PR (2014b) Mechanistic studies of protein arginine deiminase 2: evidence for a substrate-assisted mechanism. Biochemistry 53(27):4426–4433. https://doi.org/10.1021/bi500554b

Esposito G, Vitale AM, Leijten FP, Strik AM, Koonen-Reemst AM, Yurttas P, Robben TJ, Coonrod S, Gossen JA (2007) Peptidylarginine deiminase (PAD) 6 is essential for oocyte cytoskeletal sheet formation and female fertility. Mol Cell Endocrinol 273(1–2):25–31. https://doi.org/10.1016/j.mce.2007.05.005

Fuhrmann J, Clancy KW, Thompson PR (2015) Chemical biology of protein arginine modifications in epigenetic regulation. Chem Rev 115(11):5413–5461. https://doi.org/10.1021/acs.chemrev.5b00003

Fuhrmann J, Thompson PR (2016) Protein arginine methylation and citrullination in epigenetic regulation. ACS Chem Biol 11(3):654–668. https://doi.org/10.1021/acschembio.5b00942

Ghari F, Quirke AM, Munro S, Kawalkowska J, Picaud S, McGouran J, Subramanian V, Muth A, Williams R, Kessler B et al (2016) Citrullination-acetylation interplay guides E2F-1 activity during the inflammatory response. Sci Adv 2(2):e1501257. https://doi.org/10.1126/sciadv.1501257

Holm A, Rise F, Sessler N, Sollid LM, Undheim K, Fleckenstein B (2006) Specific modification of peptide-bound citrulline residues. Anal Biochem 352(1):68–76. https://doi.org/10.1016/j.ab.2006.02.007

Horibata S, Vo TV, Subramanian V, Thompson PR, Coonrod SA (2015) Utilization of the soft agar colony formation assay to identify inhibitors of tumorigenicity in breast cancer cells. J Vis Exp (99):e52727. https://doi.org/10.3791/52727

Ishida-Yamamoto A, Senshu T, Eady RA, Takahashi H, Shimizu H, Akiyama M, Iizuka H (2002) Sequential reorganization of cornified cell keratin filaments involving filaggrin-mediated compaction and keratin 1 deimination. J Invest Dermatol 118(2):282–287. https://doi.org/10.1046/j.0022-202x.2001.01671.x

Ishigami A, Ohsawa T, Hiratsuka M, Taguchi H, Kobayashi S, Saito Y, Murayama S, Asaga H, Toda T, Kimura N et al (2005) Abnormal accumulation of citrullinated proteins catalyzed by peptidylarginine deiminase in hippocampal extracts from patients with Alzheimer's disease. J Neurosci Res 80(1):120–128. https://doi.org/10.1002/jnr.20431

Jamali H, Khan HA, Stringer JR, Chowdhury S, Ellman JA (2015) Identification of multiple structurally distinct, nonpeptidic small molecule inhibitors of protein arginine deiminase 3 using a substrate-based fragment method. J Am Chem Soc 137(10):3616–3621. https://doi.org/10.1021/jacs.5b00095

Jamali H, Khan HA, Tjin CC, Ellman JA (2016) Cellular activity of new small molecule protein arginine deiminase 3 (PAD3) inhibitors. ACS Med Chem Lett 7(9):847–851. https://doi.org/10.1021/acsmedchemlett.6b00215

Jang B, Kim E, Choi JK, Jin JK, Kim JI, Ishigami A, Maruyama N, Carp RI, Kim YS, Choi EK (2008) Accumulation of citrullinated proteins by up-regulated peptidylarginine deiminase 2 in brains of scrapie-infected mice: a possible role in pathogenesis. Am J Pathol 173(4):1129–1142. https://doi.org/10.2353/ajpath.2008.080388

Jang B, Shin HY, Choi JK, du Nguyen PT, Jeong BH, Ishigami A, Maruyama N, Carp RI, Kim YS, Choi EK (2011) Subcellular localization of peptidylarginine deiminase 2 and citrullinated proteins in brains of scrapie-infected mice: nuclear localization of PAD2 and membrane fraction- enriched citrullinated proteins. J Neuropathol Exp Neurol 70(2):116–124. https://doi.org/10.1097/NEN.0b013e318207559e

Jones JE, Causey CP, Knuckley B, Slack-Noyes JL, Thompson PR (2009) Protein arginine deiminase 4 (PAD4): current understanding and future therapeutic potential. Curr Opin Drug Discov Devel 12(5):616–627

Jones JE, Slack JL, Fang P, Zhang X, Subramanian V, Causey CP, Coonrod SA, Guo M, Thompson PR (2012) Synthesis and screening of a haloacetamidine containing library to identify PAD4 selective inhibitors. ACS Chem Biol 7(1):160–165. https://doi.org/10.1021/cb200258q

Kan R, Jin M, Subramanian V, Causey CP, Thompson PR, Coonrod SA (2012) Potential role for PADI-mediated histone citrullination in preimplantation development. BMC Dev Biol 1219. https://doi.org/10.1186/1471-213x-12-19

Kawalkowska J, Quirke AM, Ghari F, Davis S, Subramanian V, Thompson PR, Williams RO, Fischer R, La Thangue NB, Venables PJ (2016) Abrogation of collagen–induced arthritis by a peptidyl arginine deiminase inhibitor is associated with modulation of T cell-mediated immune responses. Sci Rep 626430. https://doi.org/10.1038/srep26430

Khandpur R, Carmona-Rivera C, Vivekanandan-Giri A, Gizinski A, Yalavarthi S, Knight JS, Friday S, Li S, Patel RM, Subramanian V et al (2013) NETs are a source of citrullinated autoantigens and stimulate inflammatory responses in rheumatoid arthritis. Sci Transl Med 5(178):178ra140. https://doi.org/10.1126/scitranslmed.3005580

Knight JS, Luo W, O'Dell AA, Yalavarthi S, Zhao W, Subramanian V, Guo C, Grenn RC, Thompson PR, Eitzman DT et al (2014) Peptidylarginine deiminase inhibition reduces vascular damage and modulates innate immune responses in murine models of atherosclerosis. Circ Res 114(6):947–956. https://doi.org/10.1161/CIRCRESAHA.114.303312

Knight JS, Subramanian V, O'Dell AA, Yalavarthi S, Zhao W, Smith CK, Hodgin JB, Thompson PR, Kaplan MJ (2015) Peptidylarginine deiminase inhibition disrupts NET formation and protects against kidney, skin and vascular disease in lupus-prone MRL/lpr mice. Ann Rheum Dis 74(12):2199–2206. https://doi.org/10.1136/annrheumdis-2014-205365

Knight JS, Zhao W, Luo W, Subramanian V, O'Dell AA, Yalavarthi S, Hodgin JB, Eitzman DT, Thompson PR, Kaplan MJ (2013) Peptidylarginine deiminase inhibition is immunomodulatory and vasculoprotective in murine lupus. J Clin Invest 123(7):2981–2993. https://doi.org/10.1172/JCI67390

Knuckley B, Bhatia M, Thompson PR (2007) Protein arginine deiminase 4: evidence for a reverse protonation mechanism. Biochemistry 46(22):6578–6587. https://doi.org/10.1021/bi700095s

Knuckley B, Causey CP, Jones JE, Bhatia M, Dreyton CJ, Osborne TC, Takahara H, Thompson PR (2010a) Substrate specificity and kinetic studies of PADs 1, 3, and 4 identify potent and selective inhibitors of protein arginine deiminase 3. Biochemistry 49(23):4852–4863. https://doi.org/10.1021/bi100363t

Knuckley B, Causey CP, Pellechia PJ, Cook PF, Thompson PR (2010b) Haloacetamidine- based inactivators of protein arginine deiminase 4 (PAD4): evidence that general acid catalysis promotes efficient inactivation. ChemBioChem 11(2):161–165. https://doi.org/10.1002/cbic.200900698

Knuckley B, Jones JE, Bachovchin DA, Slack J, Causey CP, Brown SJ, Rosen H, Cravatt BF, Thompson PR (2010c) A fluopol-ABPP HTS assay to identify PAD inhibitors. Chem Commun (Camb) 46(38):7175–7177. https://doi.org/10.1039/c0cc02634d

Knuckley B, Luo Y, Thompson PR (2008) Profiling Protein Arginine Deiminase 4 (PAD4): a novel screen to identify PAD4 inhibitors. Bioorg Med Chem 16(2):739–745. https://doi.org/10.1016/j.bmc.2007.10.021

Lange S, Gogel S, Leung KY, Vernay B, Nicholas AP, Causey CP, Thompson PR, Greene ND, Ferretti P (2011) Protein deiminases: new players in the developmentally regulated loss of neural regenerative ability. Dev Biol 355(2):205–214. https://doi.org/10.1016/j.ydbio.2011.04.015

Leffler J, Martin M, Gullstrand B, Tyden H, Lood C, Truedsson L, Bengtsson AA, Blom AM (2012) Neutrophil extracellular traps that are not degraded in systemic lupus erythematosus activate complement exacerbating the disease. J Immunol 188(7):3522–3531. https://doi.org/10.4049/jimmunol.1102404

Lewallen DM, Bicker KL, Madoux F, Chase P, Anguish L, Coonrod S, Hodder P, Thompson PR (2014) A FluoPol-ABPP PAD2 high-throughput screen identifies the first calcium site inhibitor targeting the PADs. ACS Chem Biol 9(4):913–921. https://doi.org/10.1021/cb400841k

Lewallen DM, Bicker KL, Subramanian V, Clancy KW, Slade DJ, Martell J, Dreyton CJ, Sokolove J, Weerapana E, Thompson PR (2015) Chemical proteomic platform to identify citrullinated proteins. ACS Chem Biol 10(11):2520–2528. https://doi.org/10.1021/acschembio.5b00438

Lewis HD, Liddle J, Coote JE, Atkinson SJ, Barker MD, Bax BD, Bicker KL, Bingham RP, Campbell M, Chen YH et al (2015) Inhibition of PAD4 activity is sufficient to disrupt mouse and human NET formation. Nat Chem Biol 11(3):189–191. https://doi.org/10.1038/nchembio.1735

Lewis HD, Nacht M (2016) iPAD or PADi-'tablets' with therapeutic disease potential? Curr Opin Chem Biol 33169-178. https://doi.org/10.1016/j.cbpa.2016.06.020

Li P, Wang D, Yao H, Doret P, Hao G, Shen Q, Qiu H, Zhang X, Wang Y, Chen G et al (2010) Coordination of PAD4 and HDAC2 in the regulation of p53-target gene expression. Oncogene 29(21):3153–3162. https://doi.org/10.1038/onc.2010.51

Luo Y, Arita K, Bhatia M, Knuckley B, Lee YH, Stallcup MR, Sato M, Thompson PR (2006a) Inhibitors and inactivators of protein arginine deiminase 4: functional and structural characterization. Biochemistry 45(39):11727–11736. https://doi.org/10.1021/bi061180d

Luo Y, Knuckley B, Bhatia M, Thompson PR (2006b) Activity based protein profiling reagents for Protein Arginine Deiminase 4 (PAD4): synthesis and in vitro evaluation of a fluorescently-labeled probe. J Am Chem Soc 128(45):14468–14469. https://doi.org/10.1021/ja0656907

Luo Y, Knuckley B, Lee YH, Stallcup MR, Thompson PR (2006c) A fluoro-acetamidine based inactivator of protein arginine deiminase 4 (PAD4): design, synthesis, and in vitro and in vivo evaluation. J Am Chem Soc 128(45):1092–1093. https://doi.org/10.1021/ja0576233

McElwee JL, Mohanan S, Griffith OL, Breuer HC, Anguish LJ, Cherrington BD, Palmer AM, Howe LR, Subramanian V, Causey CP et al (2012) Identification of PADI2 as a potential breast cancer biomarker and therapeutic target. BMC Cancer 12(1):500. https://doi.org/10.1186/1471-2407-12-500

Moscarello MA, Pritzker L, Mastronardi FG, Wood DD (2002) Peptidylarginine deiminase: a candidate factor in demyelinating disease. J Neurochem 81(2): 335–343. https://doi.org/10.1191/1352458502ms776oa

Musse AA, Li Z, Ackerley CA, Bienzle D, Lei H, Poma R, Harauz G, Moscarello MA, Mastronardi FG (2008) Peptidylarginine deiminase 2 (PAD2) overexpression in transgenic mice leads to myelin loss in the central nervous system. Dis Model Mech 1(4–5):229–240. https://doi.org/10.1242/dmm.000729

Muth A, Subramanian V, Beaumont E, Nagar M, Kerry P, McEwan P, Srinath H, Clancy K, Parelkar S, Thompson PR (2017) Development of a selective inhibitor of protein arginine deiminase 2. J Med Chem 60(7):3198–3211. https://doi.org/10.1021/acs.jmedchem.7b00274

Nakashima K, Hagiwara T, Ishigami A, Nagata S, Asaga H, Kuramoto M, Senshu T, Yamada M (1999) Molecular characterization of peptidylarginine deiminase in HL-60 cells induced by retinoic acid and 1-alpha,25-dihydroxyvitamin D(3). J Biol Chem 274(39):27786–27792. https://doi.org/10.1074/jbc.274.39.27786

Nauseef WM, Borregaard N (2014) Neutrophils at work. Nat Immunol 15(7):602–611. https://doi.org/10.1038/ni.2921

Nemmara VV, Subramanian V, Muth A, Mondal S, Salinger AJ, Maurais AJ, Tilvawala R, Weerapana E, Thompson PR (2018) The Development of benzimidazole-based clickable probes for the efficient labeling of cellular Protein Arginine Deiminases (PADs). ACS Chem Bioldoi. https://doi.org/10.1021/acschembio.7b00957

Pritzker LB, Moscarello MA (1998) A novel microtubule independent effect of paclitaxel: the inhibition of peptidylarginine deiminase from bovine brain. Biochim Biophys Acta 1388 (1):154–160. https://doi.org/10.1016/S0167-4838(98)00175-7

Qin H, Liu X, Li F, Miao L, Li T, Xu B, An X, Muth A, Thompson PR, Coonrod SA et al (2017) PAD1 promotes epithelial-mesenchymal transition and metastasis in triple-negative breast cancer cells by regulating MEK1-ERK1/2-MMP2 signaling. Cancer Lett 40930-41. https://doi.org/10.1016/j.canlet.2017.08.019

Raijmakers R, Zendman AJ, Egberts WV, Vossenaar ER, Raats J, Soede-Huijbregts C, Rutjes FP, van Veelen PA, Drijfhout JW, Pruijn GJ (2007) Methylation of arginine residues interferes with citrullination by peptidylarginine deiminases in vitro. J Mol Biol 367(4):1118–1129. https://doi.org/10.1016/j.jmb.2007.01.054

Rogers G, Winter B, McLaughlan C, Powell B, Nesci T (1997) Peptidylarginine deiminase of the hair follicle: characterization, localization, and function in keratinizing tissues. J Invest Dermatol 108(5):700–707. https://doi.org/10.1111/1523-1747.ep12292083

Rogers GE, Simmonds DH (1958) Content of citrulline and other amino-acids in a protein of hair follicles. Nature 182(4629):186–187. https://doi.org/10.1038/182186a0

Saijo S, Nagai A, Kinjo S, Mashimo R, Akimoto M, Kizawa K, Yabe-Wada T, Shimizu N, Takahara H, Unno M (2016) Monomeric Form of peptidylarginine deiminase Type I revealed by X-ray crystallography and small-angle X-ray scattering. J Mol Biol 428(15):3058–3073. https://doi.org/10.1016/j.jmb.2016.06.018

Senshu T, Kan S, Ogawa H, Manabe M, Asaga H (1996) Preferential deimination of keratin K1 and filaggrin during the terminal differentiation of human epidermis. Biochem Biophys Res Commun 225(3):712–719. https://doi.org/10.1006/bbrc.1996.1240

Slack JL, Causey CP, Luo Y, Thompson PR (2011) Development and use of clickable activity based protein profiling agents for protein arginine deiminase 4. ACS Chem Biol 6(5):466–476. https://doi.org/10.1021/cb1003515

Slade DJ, Fang P, Dreyton CJ, Zhang Y, Fuhrmann J, Rempel D, Bax BD, Coonrod SA, Lewis HD, Guo M et al (2015) Protein arginine deiminase 2 binds calcium in an ordered fashion: implications for inhibitor design. ACS Chem Biol 10(4):1043–1053. https://doi.org/10.1021/cb500933j

Slade DJ, Subramanian V, Thompson PR (2014) Pluripotency: citrullination unravels stem cells. Nat Chem Biol 10(5):327–328. https://doi.org/10.1038/nchembio.1504

Taki H, Gomi T, Knuckley B, Thompson PR, Vugrek O, Hirata K, Miyahara T, Shinoda K, Hounoki H, Sugiyama E et al (2011) Purification of enzymatically inactive peptidylarginine deiminase type 6 from mouse ovary that reveals hexameric structure different from other dimeric isoforms. Adv Biosci Biotechnol 02(04):304–310. https://doi.org/10.4236/abb.2011.24044

Tarcsa E, Marekov LN, Mei G, Melino G, Lee SC, Steinert PM (1996) Protein unfolding by peptidylarginine deiminase. Substrate specificity and structural relationships of the natural substrates trichohyalin and filaggrin. J Biol Chem 271(48):30709–30716. https://doi.org/10.1074/jbc.271.48.30709

Tilvawala R, Nguyen SH, Maurais AJ, Nemmara VV, Nagar M, Salinger AJ, Nagpal S, Weerapana E, Thompson PR (2018) The rheumatoid arthritis-associated citrullinome. Cell Chem Biol 25(6):691–704. https://doi.org/10.1016/j.chembiol.2018.03.002

Weerapana E, Speers AE, Cravatt BF (2007) Tandem orthogonal proteolysis-activity-based protein profiling (TOP-ABPP)—A general method for mapping sites of probe modification in proteomes. Nat Protoc 2(6):1414–1425. https://doi.org/10.1038/nprot.2007.194

Weerapana E, Wang C, Simon GM, Richter F, Khare S, Dillon MB, Bachovchin DA, Mowen K, Baker D, Cravatt BF (2010) Quantitative reactivity profiling predicts functional cysteines in proteomes. Nature 468(7325):790–795. https://doi.org/10.1038/nature09472

Willis VC, Banda NK, Cordova KN, Chandra PE, Robinson WH, Cooper DC, Lugo D, Mehta G, Taylor S, Tak PP et al (2017) PAD4 inhibition is sufficient for the amelioration of collagen-induced arthritis. Clin Exp Immunoldoi. https://doi.org/10.1111/cei.12932

Willis VC, Gizinski AM, Banda NK, Causey CP, Knuckley B, Cordova KN, Luo Y, Levitt B, Glogowska M, Chandra P et al (2011) *N*-alpha;-Benzoyl-N5-(2-Chloro-1-Iminoethyl)-L-Ornithine Amide, a protein arginine deiminase inhibitor, reduces the severity of murine collagen-induced arthritis. J Immunol 186(7):4396–4404. https://doi.org/10.4049/jimmunol.1001620

Witalison EE, Thompson PR, Hofseth LJ (2015) Protein arginine deiminases and associated citrullination: physiological functions and diseases associated with dysregulation. Curr Drug Targets 16(7):700–710. https://doi.org/10.2174/1389450116666150202160954

Zhang X, Bolt M, Guertin MJ, Chen W, Zhang S, Cherrington BD, Slade DJ, Dreyton CJ, Subramanian V, Bicker KL et al (2012) Peptidylarginine deiminase 2-catalyzed histone H3 arginine 26 citrullination facilitates estrogen receptor alpha target gene activation. Proc Natl Acad Sci U S A 10913331–13336. https://doi.org/10.1073/pnas.1203280109

Recent Advances in Activity-Based Protein Profiling of Proteases

Suravi Chakrabarty, Jan Pascal Kahler, Merel A. T. van de Plassche, Roeland Vanhoutte and Steven H. L. Verhelst

Contents

Suravi Chakrabarty, Jan Pascal Kahler, Merel A. T. van de Plassche, Roeland Vanhoutte—equal contribution

S. Chakrabarty · J. P. Kahler · M. A. T. van de Plassche · R. Vanhoutte · S. H. L. Verhelst (✉)
Laboratory of Chemical Biology, Department of Cellular & Molecular Medicine, KU Leuven —University of Leuven, Herestraat 49, Box 802, 3000 Louvain, Belgium
e-mail: steven.verhelst@kuleuven.be

S. H. L. Verhelst
Leibniz Institute for Analytical Sciences ISAS, Otto-Hahn-Str. 6b, 44227 Dortmund, Germany

Current Topics in Microbiology and Immunology (2019) 420:253–282
DOI 10.1007/82_2018_138

Published Online: 23 September 2018

Abstract The activity of proteases is tightly regulated, and dysregulation is linked to a variety of human diseases. For this reason, ABPP is a well-suited method to study protease biology and the design of protease probes has pushed the boundaries of ABPP. The development of highly selective protease probes is still a challenging task. After an introduction, the first section of this chapter discusses several strategies to enable detection of a single active protease species. These range from the usage of non-natural amino acids, combination of probes with antibodies, and engineering of the target proteases. A next section describes the different types of detection tags that facilitate the read-out possibilities including various types of imaging methods and mass spectrometry-based target identification. The power of protease ABPP is illustrated by examples for a selected number of proteases. It is expected that some protease probes that have been evaluated in animal models of human disease will find translation into clinical application in the near future.

1 Introduction

Proteases, the enzymes that hydrolyze peptide bonds in proteins or polypeptides, are involved in numerous biological pathways, such as blood coagulation, apoptosis, antigen presentation, and cytokine processing. Protease activity needs to be tightly controlled, as dysregulation can lead to a variety of different diseases, including cancer, cardiovascular diseases, and neurodegeneration. Activity regulation takes place at several different levels (Fig. 1a). Most proteases are expressed as inactive forms (zymogens), and after activation, they are regulated by various posttranslational mechanisms. These include controlled degradation, covalent modifications, calcium binding and inhibition by endogenous inhibitors. An additional layer of regulation takes place at a spatial level: Although the human genome encodes more than 500 proteases, not all of them are expressed in each cell or tissue, and many show a particular subcellular localization. As a result, proteases only come in contact with a small amount of possible substrates (Verdoes and Verhelst 2016).

Proteases can be classified according to their catalytic mechanism. Currently, the protease community recognizes seven different catalytic types: aspartic, glutamic, cysteine, serine and threonine proteases, metalloproteases, and asparagine lyases (Rawlings et al. 2018). Cysteine, serine, and threonine proteases are named after the residue that performs a direct nucleophilic attack on the scissile peptide bond. Metalloproteases, aspartic, and glutamic proteases utilize their active site residues to activate a water molecule, which then attacks the scissile bond (Sanman and Bogyo 2014). Beside classification according to catalytic mechanism, homologous proteases are grouped into families (e.g., C1: the papain family of cysteine proteases), and families with a single evolutionary origin are sorted into clans (superfamilies, e.g., clan CA, to which the C1 family belongs) (Rawlings et al. 2018).

Proteases distinguish substrates from non-substrates in different ways. Most proteases recognize a sequence of amino acids directly adjacent to the scissile bond

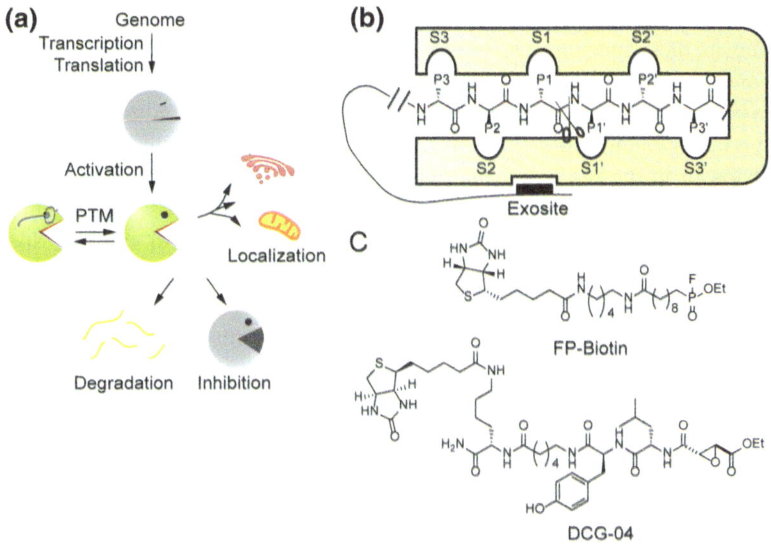

Fig. 1 Proteases and ABPs. **a** The tight activity regulation of proteases: most proteases are translated as inactive forms or zymogens that need an activation step, usually proteolytic removal of a propeptide that blocks the active site. Once active, the protease may be regulated by posttranslational modifications, by controlled degradation or binding to endogenous inhibitors. Restricting the localization of proteases to a subcellular compartment or anchoring to a membrane will limit the number of possible substrates that it may encounter. **b** Protease nomenclature according to Schechter and Berger. Protease pockets are termed S1…Sn N-terminally to the scissile bond (indicated by the scissors) and S1'…Sn' C-terminally to it. The pockets recognize the side chains of the corresponding amino acid residues termed P1…Pn and P1'…Pn'. Exosites may also help binding substrates. **c** The general serine hydrolase ABP FP-Biotin and the general papain-family ABP DCG-04

(Fig. 1b). The residues N-terminal to the scissile bond are termed P1, P2, …Pn, and fit in recognition pockets on the protease called S1, S2, …Sn. Residues and the corresponding recognition pockets on the C-terminal side of the scissile bond are named similarly, but their name receives an apostrophe: P1', S1', etc., (pronounced: P1 prime, S1 prime) (Schechter and Berger 1967). While most proteases cleave proteins somewhere in the middle of a polypeptide chain (endoproteases), some recognize the N- or C-terminus of substrates and cleave off one or two amino acids (exoproteases). Exosites located further away from the cleavage site may also contribute to the recognition.

Naturally, only active proteases can turn over substrates into products and thereby induce a downstream biological effect. Due to the strict regulation of protease activity, activity-based protein profiling (ABPP) is a well-suited method for the study of proteases. The activity-based probes (ABPs) for application in protease ABPP usually contain three elements: (1) an electrophilic group (also termed warhead) that covalently reacts in a mechanism-based manner with the nucleophilic active site amino acid residue. These ABPs are inherently limited to serine, cysteine, and threonine proteases, since other classes use an activated water molecule in their mechanism, which would result in hydrolysis of the warhead. There are two alternatives to overcome this, which we will not further discuss in this review: quinone-imine methide-forming substrates, which are substrates containing a latent electrophile. The electrophile is formed upon cleavage, but is prone to diffusion from the active site and labeling off-targets (Hu et al. 2011). Reversible inhibitors with high affinity can be equipped with a photocrosslinker, which then labels the protease upon irradiation (Saghatelian et al. 2004). Strictly speaking, these are affinity-based probes. (2) An ABP also contains a tag that is used for detection of the covalent enzyme-probe complex by a variety of different methods. (3) In most protease probes, the two previous elements are fused together by a peptide linker, which can induce selectivity for a desired class of proteases.

Many protease ABPs have been derived from covalent inhibitors. Although tagged inhibitors have been reported before the advent of ABPP (Abuelyaman et al. 1994, 1997; Bogyo et al. 1997), the application to profiling enzyme activities was not fully exploited until landmark papers by the labs of Cravatt and Bogyo, published in quick succession around the turn of the millennium. Cravatt developed a general probe for serine hydrolases (to which serine proteases also belong) consisting of a fluorophosphonate attached to a biotin by an alkyl linker (FP-Biotin, based on the serine protease inhibitor DFP, see Fig. 1c) (Liu et al. 1999). Bogyo utilized the same concept for clan CA cysteine proteases (DCG-04, based on the cysteine protease inhibitor E-64, see Fig. 1c) (Bogyo et al. 2000; Greenbaum et al. 2000). Since then ABPP has been used for many different applications, including comparative profiling of protease activity in cancer, screening for protease inhibitors, and functional proteomics studies.

In this chapter, we focus on recent developments in protease ABPP, but will also mention some important concepts from older papers. In Sect. 2, we discuss methods to achieve ABPP with a highly selective read-out for single protease species. In Sect. 3, technical and chemical aspects of detection are described. Finally, Sect. 4 illustrates the application of protease ABPP with examples for a selected number of proteases.

2 Influencing Selectivity

The choice for the use of a general or a selective ABP depends on the specific application. General probes, such as FP-Rh (Liu et al. 1999) or DCG-04 (Greenbaum et al. 2000), have found widespread application in comparative profiling of hydrolase or protease targets. In the past, they have been used, for example, to compare enzyme activities that relate to invasiveness in breast cancer cell lines (Jessani et al. 2002) or in different stages of pancreatic cancer (Joyce et al. 2004). More recent examples include profiling of off-targets of the fatty acid amide hydrolase (FAAH) inhibitor BIA 10-2474 that led to a fatality in a recent phase I clinical trial (Van Esbroeck et al. 2017). General ABPs can also be used for inhibitor screening, in combination with purified enzymes (Bachovchin et al. 2009, 2014; Wolf et al. 2013) or even whole proteomes. Overall, these studies illustrate the value of ABPs with broad selectivity.

In some applications, highly selective probes are desired. This is especially the case if the experiment does not allow separation of the target of interest from other labeled targets (as can be done in gel-based profiling experiments). A prime example is imaging: highly selective ABPs allow the visualization of a single protease species in a cell or in vivo. Compared with the design of broad-spectrum ABPs, the development of ABPP methods for the detection of a single protease is a very challenging task. In the next sections, we will discuss different strategies that utilize synthetic chemistry, biochemistry, and molecular biology approaches, respectively.

2.1 Non-natural Amino Acids

Most ABPs for proteases rely on a peptide linker to induce selectivity toward the target proteases. Usually, the chemistry of the warhead (i.e., targeting serine, cysteine, or threonine proteases) combined with several amino acid residues results in probes with reasonable selectivity. However, some proteases display overlapping substrate specificities, especially within the same family. A clear example is formed by the caspases, which all have an absolute requirement for an aspartate residue in the P1 position, and overlapping requirements in the P2–P4 position. In such cases, it can be extremely challenging to design highly selective ABPs for an individual family member, and detailed knowledge of the substrate specificity is required.

Substrate specificity profiling of proteases can be done in several ways. Phage displays offer the possibility to test a vast amount of peptide sequences for cleavage by a protease (Matthews and Wells 1993). N-terminomics methods, such as TAILS (Kleifeld et al. 2010), COFRADIC (Gevaert et al. 2003) and ChaFRADIC (Venne et al. 2013), identify cleavage sites in natural substrates using proteomics experiments, whereas other proteomics methods, e.g., proteomic identification of protease cleavage sites (PICS) (Schilling and Overall 2008) and direct in-gel profiling of protease specificity (DIPPS) (Vidmar et al. 2017) utilize proteome-derived peptides or denatured proteins to determine consensus sequences of protease cleavage sites.

One way to enhance the selectivity of ABPs is the incorporation of non-natural amino acids in the peptide linker. Because there are many more non-natural amino acids than natural ones, they offer a variety of properties for optimization of the interaction with the protease subsites. However, all of the above-named techniques utilize biological material and, by their nature, do not include non-natural amino acids. Therefore, substrates with non-natural amino acids need to be chemically synthesized. To keep the number of synthesized substrates as low as possible, the usage of positional scanning libraries (PS-SCLs), originally developed for screening protein–ligand interactions (Dooley and Houghten 1993), has been introduced in protease specificity analysis (Thornberry et al. 1997). Within these libraries, each subsite is evaluated independently of the others: sublibraries are created in which the

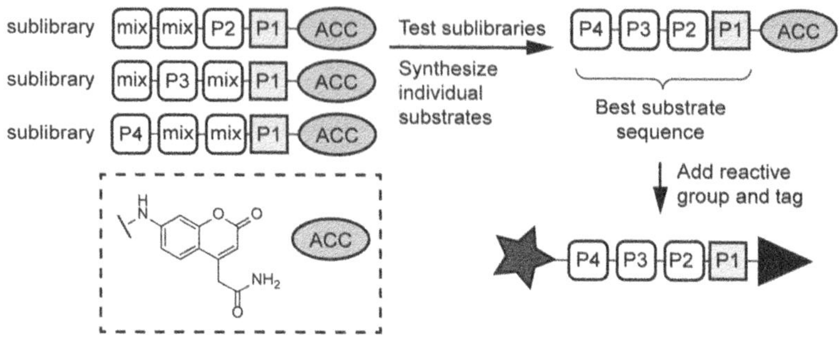

Fig. 2 Substrate specificity determination by HyCoSuL. Fluorogenic peptides are created in sublibraries. For a P2 sublibrary, each well in a microtiter plate will contain a different amino acid in the P2 position, while having a mixture of natural amino acids in the other (P3 and P4) positions. Exposure of this sublibrary to a protease will reveal which amino acids are preferred in the P2 position. This will be repeated for the P3 and the P4 positions. Next, several individual substrates with preferred P2, P3, and P4 positions will be synthesized and tested to determine the best substrate sequence, which may then be utilized to design and synthesize an ABP

amino acid residues in a specific position (e.g., P2, P3, or P4) are varied and the other positions contain an equimolar mixture of natural amino acids (Fig. 2). With these sublibraries, the preferred amino acid in the specific position can be determined. Originally, PS-SCLs have been developed to explore natural amino acids. The Drag laboratory extended this to non-natural amino acids in their so-called hybrid combinatorial substrate library (HyCoSuL) approach that may include more than 100 non-natural amino acids in addition to the proteinogenic amino acids (Poreba, Salvesen, and Drag, 2017). By making use of solid-phase peptide synthesis (SPPS), tetrapeptide sublibraries with a C-terminal 7-aminocoumarin derivative can be rapidly synthesized and tested. Optimal substrate sequences can eventually be utilized to synthesize selective ABPs by adding a warhead and a tag (Fig. 2).

HyCoSuL has been used to develop ABPs for a variety of proteases, including the individual members of the neutrophil serine proteases (NSPs, see also Sect. 3.1) and the Zika Virus NS2B-NS3 protease, which is required for viral replication and regarded as a possible drug target (Rut et al. 2017). In these examples, non-natural amino acids were key in providing higher selectivity. Strikingly, for human neutrophil elastase, all amino acids in the P1–P4 position were replaced by non-natural amino acids, resulting in highly selective targeting over other, related proteases (Kasperkiewicz et al. 2014).

In order to obtain selective ABPs for the executioner caspases-3 and -7, Wolan and co-workers first developed peptidic inhibitors with a C-terminal aldehyde. They substituted the P2, P3, and P4 positions in the known DEVD sequence with 30 natural and non-natural amino acids, and also added an extra P5 position. The individual inhibitors were tested for the IC_{50} ratio between caspase-7 and caspase-3. Eventually, this led to ABPs with an acyloxymethyl ketone (AOMK) warhead and three non-natural amino acids in the linker region, displaying an almost 50-fold selectivity for caspase-3 over caspase-7. However, selectivity over other caspases was lower (Vickers et al. 2013), illustrating the difficulty of the development of highly selective ABPs for individual caspases. In the meantime, the Drag laboratory has used HyCoSuL and made highly selective substrates with non-natural amino acids for caspase-8, -9, and -10 (Poreba et al. 2014). It is expected that these may lead to selective ABPs for individual caspases in the near future.

2.2 *Application of Antibodies*

Classically, antibodies are used to visualize protein abundance with very high specificity, for example, in western blot, enzyme-linked immunosorbent assay (ELISA), or immunohistochemistry. In the case that highly selective ABPs for a specific protease of interest are unavailable, a combination of a general ABP and an antibody can be utilized. Herein, the ABP serves as a read-out of activity, whereas the antibody ensures high specificity.

An early example has been reported by Sieber and Cravatt, who utilized immobilized antibodies on a microarray slide to capture proteases labeled by the general serine hydrolase probe FP-Rh (Fig. 3a). This detection was up to 30 times more sensitive than previously used SDS-PAGE analysis (Sieber et al. 2004). A related and more recent example includes an activity-based ELISA (termed ELASA; Fig. 3b), which was utilized in order to solve the cross-reactivity of a MALT1-targeted ABP with cathepsin B (Eitelhuber et al. 2015). In the ELASA, the covalent complex of the biotinylated ABP with the target protease is enriched on

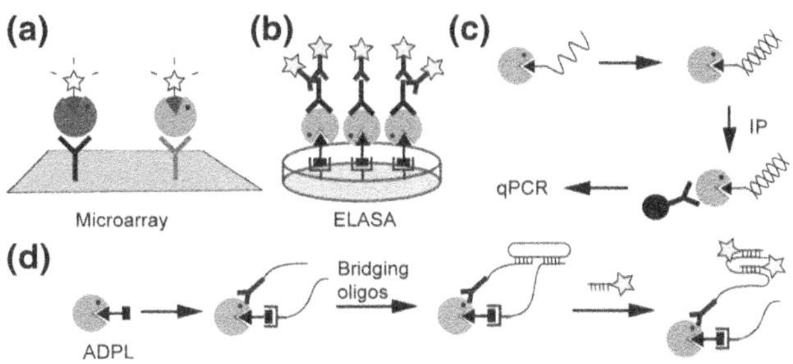

Fig. 3 ABPs combined with antibodies. **a** Different antibodies immobilized on a protein microarray can catch different proteases that have been labeled by the same general ABP. **b** In an ELASA, a streptavidin-coated 96-well plate (of which one well is depicted here) captures proteases that have reacted with a biotinylated ABP. A primary antibody against the target of interest ensures selectivity, whereas a fluorophore- or peroxidase-conjugated secondary antibody gives sensitive read-out. **c** A protease labeled with a DNA-oligo-ABP conjugate is hybridized with amplifiable DNA prior to immunoprecipitation, which ensures selectivity. Finally, qPCR or DNA sequencing is used as a read-out. **d** In ADPL, an enzyme is labeled by a biotinylated ABP inside a cell. After fixation, both biotin and the protease are recognized by separate, DNA-linked antibodies. Only the enzyme that is recognized by both antibodies then undergoes hybridization by a bridging oligo, which is then amplified and hybridized with fluorophore-conjugated oligos

streptavidin-coated 96-well plates. Next, the MALT1 target was selectively visualized by subsequent incubation with a MALT1-antibody. This assay was also more sensitive than a gel-based read-out, due to the usage of an HRP-conjugated secondary antibody, which leads to signal amplification.

A different, but related strategy was recently reported by Krusemark and co-workers (Kim et al. 2017). Here, a serine hydrolase pan-reactive fluorophosphonate was linked to a DNA-oligo for quantification of labeling by qPCR. A selective read-out of a single serine hydrolase is ensured by an immunoprecipitation reaction prior to qPCR (Fig. 3c). This workflow was illustrated by using two serine proteases involved in blood coagulation: thrombin and factor Xa. Although the method has only been demonstrated for purified proteases and not in whole proteomes, it is potentially powerful, since it offers the possibility for multiplexing by using different DNA-oligo sequences and deep sequencing (Kim et al. 2017).

A limitation of the above-described methods is the utilization of homogenized samples (either purified enzymes or lysates of cells). Hence, all spatial information is lost. The group of Moellering recently reported a new strategy that uses ABPs and antibodies that are both labeled with oligonucleotides, allowing activity-dependent proximity ligation (ADPL; Fig. 3d) (Li et al. 2017). ADPL allows imaging of active enzymes with high spatial resolution. The methodology was demonstrated for four specific serine hydrolases in combination with FP-Biotin, and for cathepsin B in combination with a biotinylated AOMK cathepsin probe. Overall, this method overcomes the need for highly selective ABPs for imaging purposes.

In some specific cases, an antibody itself may be utilized as an ABP. LeBeau et al. reported that the human monoclonal antibody A11 was specific for active matriptase (LeBeau et al. 2013). Radiolabeling the antibody with DOTA-chelated ^{111}In or coupling it to a near-infrared (NIR) fluorophore produced two probes which were used to track active matriptase in endothelial cancers through SPECT (single photon emission computed tomography) imaging and fluorescence microscopy. Application of the probes showed active matriptase in human colon cancer cell lines, whereas in healthy colon cancer cell lines all matriptase was bound to its endogenous inhibitor (see also Sect. 4.4).

2.3 *Modification of Target Proteases*

As discussed above, it can be very challenging to develop highly selective ABPs because of similar structures of recognition pockets within a protease family. For example, the different members of the caspase family have overlapping substrate specificities, as all have an absolute requirement for a P1 aspartate residue and a preference for glutamate in the S3 pocket.

In order to study individual members the caspase family, Bogyo and co-workers engineered caspase-1 to contain a non-catalytic cysteine residue in close proximity to the catalytic site (Xiao et al. 2013). This latent cysteine nucleophile, in a

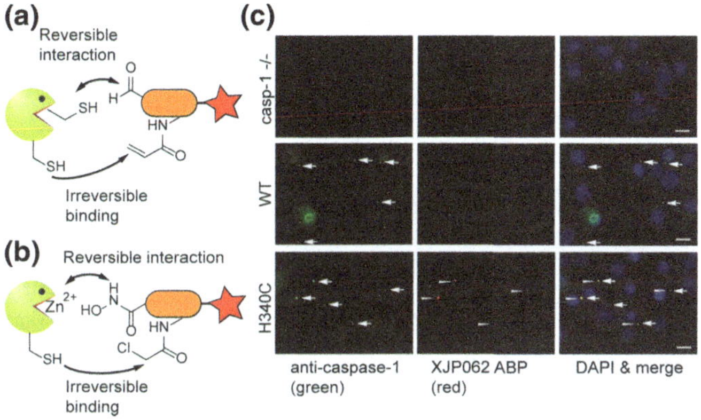

Fig. 4 Selective covalent modification by engineering proteases. **a** The cysteine protease caspase-1 forms a reversible covalent bond with an aldehyde on a probe, bringing an acrylamide in close proximity to the engineered, non-catalytic cysteine, thereby orchestrating covalent bond formation. **b** A hydroxamate group on a probe will chelate to the zinc ion in the target MMPs. In this case, a chloroacetamide reacts with a non-catalytic cysteine. **c** Specific imaging of cysteine-engineered caspase-1. BMM caspase-1$^{-/-}$ (top panels) with engineered wild-type (middle) or H340C mutant caspase-1 (bottom) were infected with *S. typhimurium* and labeled with ABP XJP062 for 1 h. Note that the WT caspase-1 is only detected with the antibody staining, but not with the ABP. Arrows indicate caspase-1 positive inflammasome foci, whereas arrowheads show ABP-labeled inflammasome foci. Reprinted (adapted) with permission from (Xiao et al. 2013). Copyright 2013 American Chemical Society

non-conserved region of the protease, may then react with a strategically placed electrophile on a chemical probe. To this end, probe XJP062 was designed with two different reactive groups: An aldehyde electrophile that reversibly binds to the active site cysteine and steers the probe into the active site, and an acrylamide electrophile, which forms a covalent bond with the non-catalytic cysteine (Fig. 4a). Consequently, only the engineered protease is irreversibly inhibited. The same strategy has been applied to two matrix metalloproteinases (MMPs). Here, the chemical probe contained a hydroxamate group, which coordinates to the active site zinc ion, and an α-chloroacetamide electrophile that covalently binds to the engineered non-catalytic cysteine (Fig. 4b) (Morell et al. 2013). It must be noted that this approach requires genetic engineering. However, with recent developments on various gene editing systems such as CRISPR/Cas9, endogenous proteases can be easily engineered in cell lines and model organisms.

Bogyo and co-workers illustrated the applicability of their approach by selectively targeting caspase-1 in a model of pyroptosis. Bone marrow macrophages (BMMs) undergo pyroptosis upon *S. typhimurium* infection, whereas caspase-1$^{-/-}$ BMMs undergo cell death via apoptosis. Reintroduction of the engineered caspase-1 with the cysteine mutation in caspase-1$^{-/-}$ BMMs restored cell death via pyroptosis and by using probe XJP062 it was possible to specifically image caspase-1 inside these cells. *S typhimurium* infection generally leads to recruitment of active caspase-1 in a typical single focus together with other inflammasome components. These foci were labeled by XJP062 in the engineered and infected cells only (Fig. 3c), further confirming the selectivity of labeling (Xiao et al. 2013).

3 Detection Tags

In comparison with other tools to detect active proteases, such as chromogenic or fluorogenic substrates, a fundamental difference of ABPs is the covalent linkage that these probes make with the protease target. It enables visualization, isolation, and identification by means of the tagging moiety on the ABP.

The choice for a reporter tag depends on the desired detection technique (Fig. 5). For imaging experiments, a fluorescent or radioactive tag can be used. For target identification experiments, an affinity handle (most often biotin) is the method of choice. It must be noted that the tag also influences the physicochemical properties of the probe, such as cell permeability, and even may affect target binding. Therefore, many current probes incorporate an azide, alkyne, or trans-cyclo-octene "mini-tag", which can be functionalized with the reporter of choice in a subsequent step using bioorthogonal chemistry. Because the final read-out still comprises fluorescence, mass spectrometry, or related techniques, we will not further discuss

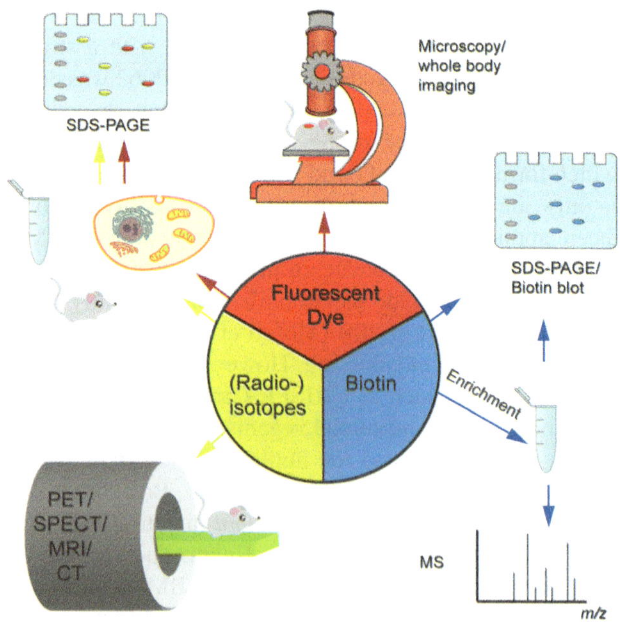

Fig. 5 Overview of three different types of tags and their associated detection techniques

bioorthogonal labeling in ABPP here, but refer to other reviews (Willems et al. 2011; Yang and Liu 2015).

In the following paragraphs, three different types of tags are discussed in depth: fluorescent dyes, (radioactive) isotopes, and biotin (affinity isolation followed by mass spectrometry).

3.1 Fluorophores

Fluorescence detection has been utilized since the early days of ABPP (Greenbaum et al. 2002; Jessani et al. 2002) and remains one of the most popular detection methods because of its versatility: It can be applied in both gel-based experiments with fluorescence scanning and in vivo imaging experiments, such as microscopy or whole body imaging (Fig. 5).

A wide variety of fluorophores has been used in ABPP. Fluorescein- and rhodamine-type dyes are inexpensive, commercially available dyes, hence, a popular choice for gel-based experiments. However, they suffer from rapid photobleaching and are therefore less suitable for imaging experiments that require extended excitation times. In comparison, BODIPY- and cyanine-dyes have better photostability. They are very bright due to high quantum yields and high absorption coefficients. These dyes are also relatively hydrophobic, which will not impede the ABPs from crossing the membrane.

In general, dyes with excitation wavelengths in the visible spectrum are not appropriate for whole body fluorescent imaging in animal models, as imaging will suffer from tissue autofluorescence and poor tissue penetration of visible light. Fluorophores with excitation wavelengths within the near-infrared spectrum (NIR; >700 nm), such as Cy5, are better suited for these applications. With these, a tissue penetration of 5–10 mm is observed and autofluorescence from the surrounding tissue can be neglected.

The availability of highly selective ABPs and fluorophores of different colors has enabled parallel detection of specific proteases (Kasperkiewicz et al. 2017; DeBruin et al. 2016). The Overkleeft laboratory has demonstrated this possibility by conjugating three differently colored fluorescent dyes (blue, green and red) to ABPs for the β1, β2, and β5 catalytic subunits of the proteasome. This probe cocktail was used to develop specific inhibitors for the six different subunits of the constitutive and immunoproteasome (DeBruin et al. 2016). The Drag laboratory developed Cy3-, Cy5-, and Cy7-labeled ABPs for each main member of the neutrophil serine proteases (NSPs) (Kasperkiewicz et al. 2017). More details about ABPs for the proteasome and NSPs can be found in Sects. 4.4 and 4.1, respectively.

Another interesting example illustrating the use of multiple fluorophores involves probes that target penicillin-binding proteins (PBPs). Although PBPs are not proteases, but membrane-bound transpeptidases that crosslink peptidoglycan in the cell walls of bacteria, their mechanism reflects that of serine proteases. Using two β-lactone probes with fluorophores in different colors, each directed against different PBPs, the authors visualized the differences in localization of PBP2x and PBP2b during late stages of cell division in *Streptococcus pneumonia* (Sharifzadeh et al. 2017).

Although fluorophore-conjugated ABPs have been successfully used in imaging experiments, the background fluorescence of unreacted probes requires extensive washing steps and prevents application in real-time imaging. To solve this problem, fluorescently quenched-activity-based probes (qABPs) have been developed. In addition to a detection tag, qABPs contain a quenching moiety that is expelled from the probe upon reaction. Obviously, this poses a restriction to the utilized warhead: it must contain a leaving group that can be connected to a quencher. Acyloxy methyl ketones (AOMKs; Fig. 6a) fulfill this requirement. The Bogyo laboratory was the first to validate this concept in real-time in situ and in vivo experiments using a cathepsin-targeting qABP with an AOMK warhead (Blum et al. 2005, 2007). In a more recent example, Ben-Nun et al. used the qABP approach with a photosensitizer as a fluorophore. Upon reaction with a cathepsin protease, this qABP not only becomes fluorescent, but also starts to induce formation of reactive oxygen species by means of the photosensitizer, eventually leading to cell death (Ben-Nun et al. 2015). Overall, this strategy combines tumor visualization with a dual tumor treatment by cathepsin inhibition and photodynamic therapy. For serine proteases, warheads consisting of mixed alkyl aryl phosphonate esters have been developed as qABPs. Although no imaging experiments employing these probes are published to date, it proves that the qABPs strategy is applicable beyond cysteine proteases (Serim et al. 2015).

To overcome the restriction of a leaving group in qABPs altogether, Lee and co-workers explored the use of the photoinduced electron transfer effect (PeT): Upon an increase in overall electron density on the fluorophore after reaction with

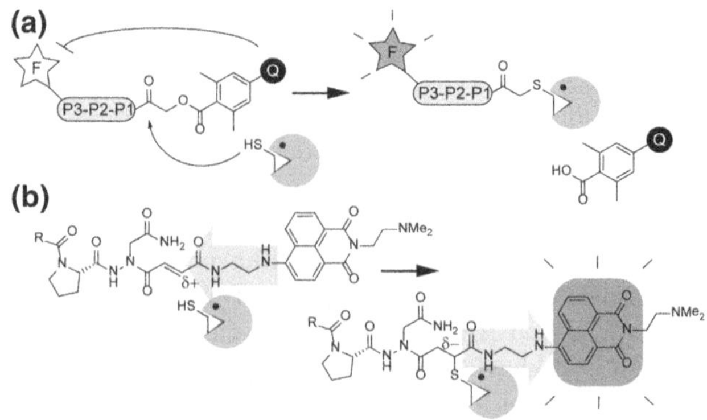

Fig. 6 Two different strategies for qABPs. **a** A warhead with a leaving group enables the incorporation of a fluorophore and a quencher. Upon reaction with the target protease, the leaving group with quencher is expelled and the probe-protease complex becomes fluorescent. **b** The photoinduced electron transfer effect: An increase in electron density after reaction with the target protease leads to an increase in fluorescence

the target protease, the PeT effect leads to an increase in fluorescence. The validity of this strategy was illustrated with a fumarate warhead connected to a 1,8-naphtalimide-based dye and a peptide scaffold that is recognized by asparaginyl endopeptidase (AEP), also known as legumain. This probe showed a twofold increase in fluorescence after reaction with the protease and was successfully used in live cell imaging of AEP. Although there is still room for improvement of signal to background ratio, this probe provides an interesting scaffold that could lead to a new generation of qABPs (Hong et al. 2017).

3.2 *Isotopes*

There are two types of isotopes that can be incorporated in ABPs: stable and radioactive isotopes.

3.2.1 Stable Isotopes

Stable isotopes may be exploited to enhance detection of ABP targets by mass spectrometry (MS). MS detection will be discussed in Sect. 3.3 (biotin tags).

Stable isotopes may also be exploited for imaging purposes, e.g., by using magnetic resonance imaging (MRI). MRI itself is a common technique in hospital settings, and ABPs containing an MRI-active tag could potentially be used in in vivo experiments and diagnostic medicine. To date, several MRI/NMR-active substrate-based protease probes have been developed (Yue et al. 2014; Yuan et al. 2015; Carril 2017), but this strategy has not yet been applied in ABPP.

A recent paper by the Blum laboratory reported ABPs with stable iodine (I-127) for detection by computed tomography (CT) imaging. CT uses X-rays to create detailed three-dimensional images with good spatial resolution, but it requires elements with high atomic weight to be present. Therefore, it was crucial to conjugate the cathepsin-targeting ABP to a dendrimer with up to 48 iodine atoms (**1**, Fig. 7). This strategy provided enough density of iodine to produce contrast in CT images of xenograft tumor model in mice (Gaikwad et al. 2018).

3.2.2 Radioactive Isotopes

Radioisotopes have served as sensitive tools for the detection of various biomolecules for decades, and have been frequently utilized in ABPP. Originally, ^{125}I was a popular choice in ABPP due to the availability of well-defined iodination protocols and the long half-life of 59.4 days (Greenbaum et al. 2000; Falgueyret et al. 2004; Méthot et al. 2004), ensuring that probes can be stored and used for months after their synthesis. An additional benefit is the small size of radioisotopes, resulting in minimal effects of the tag on the properties of the probe. Radioisotopes

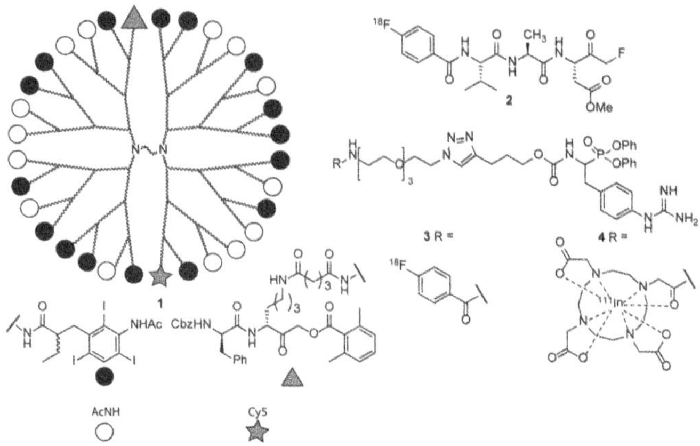

Fig. 7 Examples of ABPs labeled with stable (**1**) or radioactive (**2**, **3**, **4**) isotopes

also provide a highly sensitive signal with extremely low endogenous background. However, the radioactive waste associated with these probes and the development of sensitive gel-based protocols using fluorophores has caused a steep decrease in the usage of radioactive ABPs. One exception is formed by the field of nuclear medicine. Positron and gamma-ray emitting isotopes, which are broadly applied in positron emission tomography (PET) or SPECT, respectively, allow imaging of organs and tissues that are deeply buried under the skin—in contrast to detection by fluorescence. The short half-life of the utilized isotopes has advantages and disadvantages: It makes sure that patients are exposed to radioactivity for only short periods of time, but it also requires that the reagents are synthesized in situ before every experiment and prevents storage for extended periods of time.

PET compatible isotopes, such as ^{18}F and ^{64}Cu, have been incorporated in ABPs and used in in vivo experiments (Ren et al. 2011; Hight et al. 2014; Ides et al. 2014). The caspase inhibitor VAD-FMK, albeit not very selective due to the high reactivity of the FMK reactive group (Berger et al. 2006), was equipped with an ^{18}F tag by Hight and co-workers (**2**, Fig. 7) (Hight et al. 2014). The probe was able to visualize and quantify apoptosis in drug-treated xenograft mouse models of cancer and showed a correlation of apoptosis with a reduction in tumor size.

Ides et al. developed ^{18}F-conjugated phosphonate ABP **3** (Fig. 7) against the serine protease urokinase-type plasminogen activator (uPa). It was utilized in imaging of xenografts of MDA-MB-231 cells, which display high expression of uPA. However, the probe could not reliably visualize uPA on primary and metastatic tumor lesions, because of slow blood clearance (Ides et al. 2014). Therefore, the research group changed detection strategy and designed an ^{111}In-bearing uPA ABP that can be used for in vivo SPECT imaging (**4**, Fig. 7). Indium-111 has a longer half-life compared with ^{18}F (2.8 days vs. 109.7 min), allowing a longer clearance time of unbound probe to improve the tumor/blood signal ratio. In contrast to the ^{18}F probe, the ^{111}In-ABP was able to detect tumors on micro-SPECT images at 95 h post injection. However, the moderate uptake in tumor tissue and the high uptake in non-invaded lymph nodes present obstacles that still need to be overcome in order to allow clinical translation in oncological applications (Vangestel et al. 2016).

3.3 Biotin and Detection by MS

MS is a widely used technique in proteomics for identification of proteins in complex samples. It has also been applied in ABPP in order to identify the targets of ABPs. Biotin-conjugated ABPs have been instrumental for this purpose, because of the efficient enrichment of biotinylated proteins on immobilized avidin resins. The basic workflow for the identification is depicted in Fig. 8a. As early as two decades ago, the first ABPP papers achieved MS identification of some ABP targets by using peptide mass fingerprinting (Liu et al. 1999; Greenbaum et al. 2000). Nowadays, tandem MS experiments are part of the common bottom-up proteomics workflow. These experiments do not only reveal the targets of ABPs, but can also support conclusions about the activity of putative proteases. In 2010, for example, Florea et al. subjected mouse thymus tissue homogenates to a biotinylated epoxomycin derivative for labeling of the then recently discovered thymoproteasome subunit β5t (Florea et al. 2010). Analysis of the fragments from the tryptic digests revealed the labeled active site peptide, thereby confirming its catalytic reactivity.

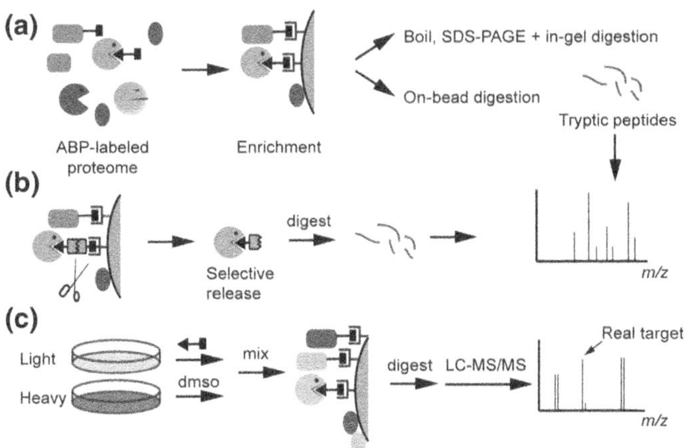

Fig. 8 MS-based target identification. **a** Biotin-streptavidin mediated enrichment can be followed by release, gel-based separation, tryptic digestion and LC-MS/MS. Alternatively, a direct digestion on bead can yield the necessary tryptic peptides for target analysis. Note that endogenously biotinylated proteins and non-specifically enriched proteins (dark oval) may contaminate the sample and lead to false-positive identifications. **b** A cleavable linker can be utilized to selectively release probe targets prior (or after) tryptic digestion. **c** Stable isotope labeling – either in the biological sample with SILAC as depicted here, or at a later stage—can be used to distinguish probe targets from background identifications

The high sensitivity of modern mass spectrometers enables the identification of minute quantities of proteins. Endogenously biotinylated proteins as well as low amounts of non-specifically bound proteins may therefore create "false-positive" identifications in the identification of ABP targets. Although non-protease targets can in principle be "filtered out" when utilizing protease ABPs, this potentially leads to ignoring "real" off-targets. To facilitate a workflow with lower false positives, cleavable linkers have been introduced in ABPP (Fonovic et al. 2007; Verhelst et al. 2007). Strategically placed between the reactive electrophile and the biotin tag, they allow selective elution of probe targets (Fig. 8b). This is effectively illustrated by a diol linker, incorporated in the cathepsin-targeting DCG-04, which led to an almost 90% reduction in false-positive identifications (Yang et al. 2013). Nowadays, various cleavable linkers are available for chemical proteomics applications (Yang et al. 2017).

Another strategy to distinguish genuine ABP targets from background protein identifications involves the use of stable isotope labeling. Stable isotopes can be introduced in samples at different points of the workflow, but at some point, samples are mixed and eventually analyzed by LC-MS/MS. The ratios of isotopic labels in the MS^1 spectra then reflect the relative abundance of the tryptic peptide species and the proteins they were derived from. When applied to an ABP-labeled and a DMSO control sample, this will help to distinguish background proteins from genuine ABP targets (Fig. 8c). Although not many studies that combine protease ABPs and isotope labeling have appeared, we will here highlight two types of workflows: ABPP-SILAC and ABPP combined with dimethyl labeling.

Adibekian et al. used stable isotope labeling in cell culture (SILAC; Ong et al. 2002) combined with FP-biotin labeling to discover serine hydrolases that were targeted by 1,2,3-triazole urea inhibitors (Adibekian et al. 2011). In short, cells cultured in "light" medium were exposed to inhibitor while cells in "heavy" medium were treated with vehicle as a control. After cell lysis and exposure to FP-biotin, samples were mixed and subjected to the general workflow of enrichment, digest, and LC-MS/MS (Fig. 8c). This method did not only reveal which serine hydrolase targets were inhibited, but also how selective these inhibitors were. Li et al. described an ABPP workflow with dimethyl labeling – sometimes referred to as the "poor man's SILAC," since the reagents are much less expensive (Li et al. 2013). In this workflow, the introduction of the isotopes and the mixing of the samples take place at a later stage than in SILAC, which can possibly lead to more variability.

3.4 Dual Detection

Some reported ABPs contain two different tags to allow more than one type of detection. Ideally, two complementary detection techniques are combined to compensate for each other's shortcomings.

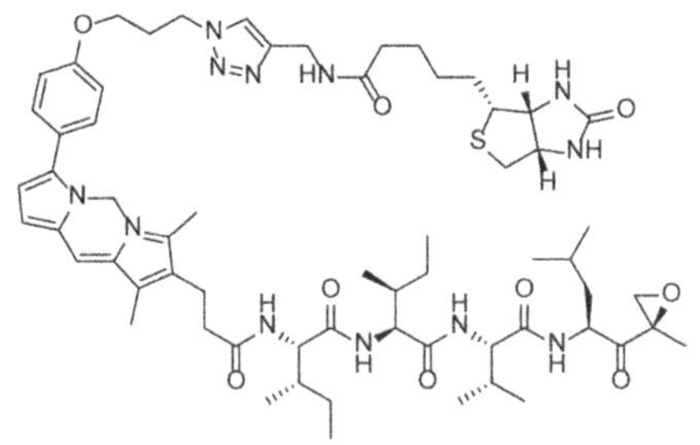

5

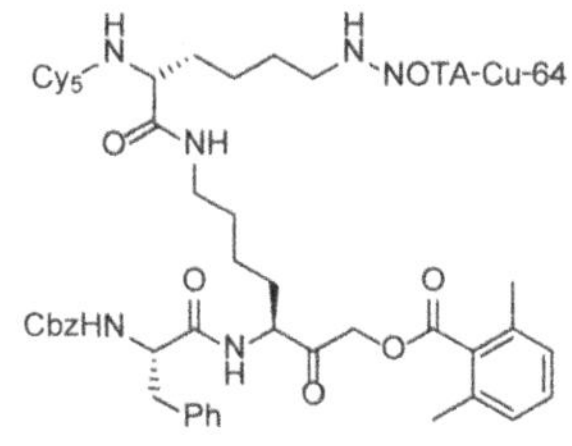

6

Fig. 9 Examples of dual detection ABPs

A convenient choice is the combination of a fluorescent tag and a biotin: It ensures sensitive detection of low abundant probe targets, even after enrichment, and simultaneously permits identification of the probe targets. The biotin and BODIPY-tagged epoxomycin-derived proteasome probe MVB072 (**5**, Fig. 9) has been used by Van der Hoorn and co-workers to detect the inhibition of plant proteasome subunits by the non-ribosomal cyclic peptide Syringolin A, which is made by the bacterial pathogen *Pseudomonas syringae pv syringae* (Kolodziejek et al. 2011). The same laboratory used this probe to show that salt stress in tomato plants leads to alteration of the catalytic β2 and β5 catalytic subunits (Kovács et al. 2017).

Another attractive combination of detection tags is a fluorophore and a radioisotope. Whereas the radioisotope compensates for the poor tissue penetration of light, the fluorophore ensures gel-based detection long after the signal of the radioisotope has faded away. Withana et al. designed a cathepsin-targeting ABP containing both a Cy5 fluorophore and a ^{64}Cu isotope, which is a positron emitter (**6**, Fig. 9) (Withana et al. 2016). This dual optical/PET probe was used in a murine model of atherosclerosis. The method did not only facilitate PET and fluorescence imaging of carotid atherosclerotic plaques, but also biochemical analysis by SDS-PAGE to confirm target engagement.

4 Examples of ABPP on Specific Proteases

4.1 Neutrophil Serine Proteases

Neutrophils are among the first immune cells to arrive at a site of inflammation in order to fight pathogens and noxious substances. They do so by using neutrophil serine proteases (NSPs), which include four members: neutrophil elastase (NE), cathepsin G, proteinase 3 (PR3), and the recently discovered (Perera et al. 2012) neutrophil serine protease 4 (NSP4).

NSPs are primarily stored in azurophilic granules. Upon activation of the neutrophil, they are released into the phagosome, the extracellular space (degranulation), or as part of neutrophil extracellular traps (NETs). NSPs are thought to play a role in several chronic inflammatory diseases, including atherosclerosis, cancer, Alzheimer's disease, and chronic obstructive pulmonary disease (Hoenderdos and Condliffe 2013; Coffelt et al. 2016; Soehnlein et al. 2017). As a result, they represent potential future targets for disease diagnosis and drug development. Hence, selective ABPs against these proteases would form attractive reagents for the study of NSPs and potentially as diagnostic tools.

In some work prior to 2012, ABPs for NE and cathepsin G have been developed (Kam et al. 1993; Zou et al. 2012; Haedke et al. 2013); however, these were not very selective and showed cross-reactivity with other proteases. Since 2014 the Drag laboratory has performed a series of studies aimed at the development of highly selective ABPs for each individual member of the NSPs. To this end, Drag and co-workers first used HyCoSuL approaches (see Sect. 2.1) in order to find the optimal recognition sequence for each NSP and then synthesized ABPs with high selectivity against the other family members (Kasperkiewicz et al. 2014, 2015, 2017). A crystal structure of NE with a NE-specific ABP revealed the details of interaction (Fig. 10a). Especially an interaction of the P3 AA with the backbone of Gly219 in the S3 pocket unique to NE and an *exo*-pocket reached by the unnatural AA in P4 seemed to be of importance for specificity (Lechtenberg et al. 2015). The NE-specific probe was used in fluorescent microscopy to visualize NETs. Antibody

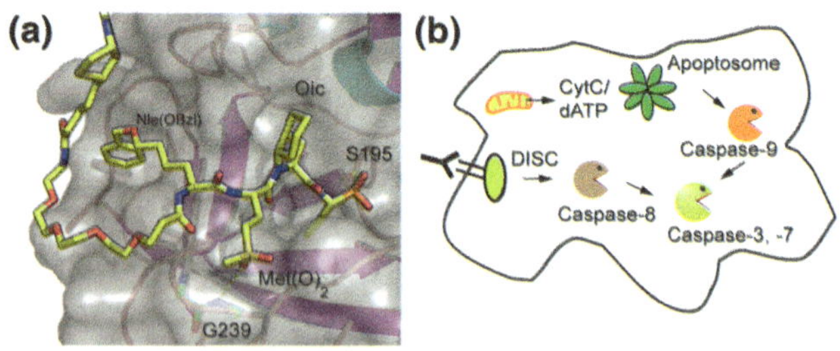

Fig. 10 **a** Crystal structure of ABP PK101 and neutrophil elastase showing key interactions. PK101 is bound covalently via the phosphonate group to the active site serine residue. Abu in the P1 position and Oic in the P2 position occupy the pockets in a similar manner as their natural counterparts Val and Pro. The additional cyclohexane ring of Oic, however, allows for more hydrophobic contacts with Leu99 and His57. Most important for specificity to NE over the closely related NSPs are the interactions of $Met(O)_2$ and Nle(O-Bzl) with the residues in the S3 and S4 pockets, respectively. $Met(O)_2$ forms a hydrogen bond with the backbone amide of Gly219 which is strictly dependent on the unique conformation adopted by Gly218 and Gly219 in NE. Additionally, the O-Bzl group in P4 reaches into a hydrophobic exo-pocket, adding binding strength. This exo-pocket is blocked in all other human serine proteases except in PR3 where it is, however, much shallower. **b** Intrinsic and extrinsic pathways of apoptosis activation. DISC formation will lead to autoactivation of caspase-8, which in turn cleaves caspases-3 and -7, thereby activating them. The intrinsic pathway leads to release of stimuli from the mitochondria, which induce formation of the apoptosome that activates caspase-9. Caspase-9 then cleaves executioner caspases-3 and -7

staining of NE and propidium iodide staining of DNA showed that most NE in the NETs is not active, although this may vary in different individuals (Kasperkiewicz et al. 2014).

Using the HyCoSuL approach, the Drag laboratory also managed to develop selective probes for each of the remaining three NSPs (Kasperkiewicz et al. 2015, 2017). They used their probes, each equipped with a different fluorescent dye, to visualize active NSPs in the azurophilic granules. Interestingly, the majority of azurophilic granules do not show co-localization of the different NSPs suggesting they could be packed into different granules during myelopoiesis (Kasperkiewicz et al. 2017).

4.2 Caspases

Caspases are a family of cysteine proteases with important roles in apoptosis and inflammation. Initiation of apoptosis can happen via two distinct pathways (Boatright and Salvesen 2003; Pop and Salvesen 2009): the extrinsic pathway, which activates initiator caspase-8 and -10, or the intrinsic pathway, which activates initiator caspases-9 and -2 (Fig. 10b). The extrinsic pathway is triggered by external stimuli via the death receptors, leading to formation of the death-inducing signaling complex (DISC), which recruits procaspase-8 for autoactivation. The intrinsic pathway is activated upon cellular damage by, e.g., drugs or radiation. Selective permeabilization of mitochondria releases cytochrome c into the cytosol, causing the formation of the apoptosome complex, which leads to recruitment and activation of procaspase-9. At this point, the two pathways merge by activation of executioner caspase-3 and -7, eventually leading to cell death.

Caspases have an absolute requirement for aspartate in the P1 position (hence their name, for cysteine aspartate-specific protease). Many members have overlapping substrate specificities, which makes the generation of highly specific ABPs very challenging. Early ABPs, such as fluoromethylketone-based FLICA (fluorochrome-labeled inhibitors of caspases) (Bedner et al. 2000) and AOMKs by the Bogyo laboratory (Kato et al. 2005; Berger et al. 2006) were not selective for individual members, but have been useful for labeling active caspases in lysates and living cells. Using caspase ABPs, Berger et al. found that an early full-length intermediate of caspase-7 is catalytically active and is further processed by executioner caspases (Berger et al. 2006). This illustrates the power of ABPs to dissect protease activation pathways.

In an attempt to develop caspase selective probes for optical imaging of apoptosis, Edgington et al. introduced a proline residue at the P2 position in order to decrease cathepsin B cross-reactivity. This probe was used to noninvasively image apoptosis in xenograft mouse models of cancer (Edgington et al. 2009). Cross-reactivity with the clan CD protease legumain was still observed, and the use of non-natural amino acids in the P3 position to overcome this led to probes that were poorly permeable to cells. Recently, Blum and co-workers were able to further optimize qABPs for caspases by decreasing the cross-reactivity with legumain (Shaulov-Rotem et al. 2016). This probe revealed the unexpected presence of active caspase-3 in the endoplasmatic reticulum of apoptotic cells.

4.3 Malt1

MALT1 is a cytosolic cysteine protease and the only human member of the paracaspase family. As part of the CARMA1-BLC10-MALT1(CBM)-complex, it leads to activation of the transcription of NF-κB target genes (Hachmann et al. 2012). Unlike most proteases, MALT1 activation does not require an autoproteolytic cleavage reaction, but relies on dimerization and additional stabilization through substrate/inhibitor binding (Hachmann and Salvesen 2016). Its proteolytic activity is essential for survival of activated B cell subtypes in diffuse large B cell lymphoma (ABC-DLBCL), a common type of non-Hodgkin's lymphoma (Hailfinger et al. 2009), making it a promising therapeutic target. In addition, experiments in MALT-1 deficient mice show that they are protected from experimental autoimmune encephalomyelitis (EAE), suggesting that MALT1 inhibition might be used as a treatment for various central nervous system (CNS) demyelinating diseases, like multiple sclerosis (Mc Guire et al. 2013).

In contrast to the classical caspases, MALT1 has a preference for arginine in the P1 position (Hachmann et al. 2012). The first reported MALT1 inhibitor was a fluoromethyl ketone attached to the tetrapeptide sequence Cbz-VRPR (P4–P1), which is based on the MALT1 substrate BCL-10 (Yu et al. 2011). The inhibitor and further substrate specificity information led to the development of several ABPs. These showed cross-reactivity with cathepsin B, which was overcome by an activity-based ELISA (see Sect. 2.3) (Eitelhuber et al. 2015).

The group of Overkleeft synthesized a probe based on the covalent MALT1 inhibitor MI-2 (Fig. 11a). MI-2 contains a chloromethyl amide moiety, which was thought to react with the active site cysteine (Fontan et al. 2012). Surprisingly

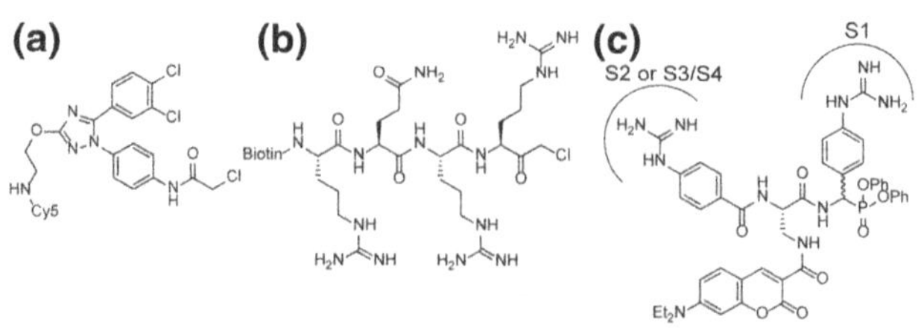

Fig. 11 Selected structures of ABPs for MALT1 and matriptase. **a** MI-2 based probe for MALT1. The chloroacetamide reacts covalently with a nucleophilic residue outside the active site. **b** A chloromethyl ketone ABP for matriptase. **c** A diphenyl phosphonate ABP with two benzguanines. For matriptase, these are thought to interact with the S1 and S2 pockets, for matriptase-2 with the S1 and S3/S4 pockets

however, the MI-2 based ABPs labeled both wild-type and C464A active site mutant of MALT1, suggesting that this inhibitor reacts at another site in the MALT1 protein (Xin et al. 2016). Hence, the mechanism of action of the MI-2 inhibitor needs to be revisited.

4.4 The Proteasome

Apart from general protein turnover, the proteasome plays an important role in antigen presentation and other cellular processes by degradation of, e.g., cell cycle regulators, transcription factors, and enzymes. For this reason, the proteasome is a drug target and multiple proteasome-targeting drugs are on the market for treatment of multiple myeloma. To explore the biological role and biomedical potential of the proteasome, there is a need for proteasome-selective ABPs.

The barrel-shaped assembly of the proteasome holds three different catalytic subunits: β1 (caspase-like), β2 (trypsin-like) and β5 (chymotrypsin-like). Specialized catalytic subunits with slightly different substrate specificities exist in the immunoproteasome (β1i, β2i, β5i) and the thymoproteasome (β1i, β2i, β5t), which occur in specialized immune cells (Tanaka and Kasahara 1998) and cortical thymic epithelial cells (Murata et al. 2007), respectively.

Broad-spectrum probes for the catalytic β-subunits of the constitutive proteasome date back to as early as 1997 (Bogyo et al. 1997). As expected, these also label the immunoproteasome (Kessler et al. 2001) and thymoproteasome subunits (Florea et al. 2010). Highly selective ABPs for the separate subunits have only recently been reported by a series of studies from the Overkleeft laboratory. Of particular interest is a set of ABPs that is selective for each different catalytic subunit β1, β2 or β5 (DeBruin et al. 2016). A set of subunit-selective inhibitors (Britton et al. 2009; Screen et al. 2010; Geurink et al. 2013) were equipped with fluorophores of different colors, enabling easy visualization of each subunit by gel-based ABPP. This led to the development of inhibitors that are selective for each of the six different catalytic subunits in the constitutive or immunoproteasome (DeBruin et al. 2016). When these were again equipped with different fluorophores, FRET experiments after native gel electrophoresis showed that not only constitutive proteasomes (β1c, β2c, β5c) and immunoproteasomes (β1i, β2i, β5i) exist, but also proteasomes with a mix of the active site subunits (de Bruin et al. 2016).

Subunit-specific ABPs have also been applied in plants. A recent study revealed that in the tobacco plant *N. bentamiana* multiple isoforms of proteasome subunits occur, and therefore the coexistence of different types of proteasomes (Misas-Villamil et al. 2017). Moreover, infection with the pathogen *P. syringae pv tomato* led to differential activities of the β1 and β5 subunits.

4.5 *Matriptase*

The matriptase subfamily consists of serine proteases anchored to the membrane by a transmembrane helix near the N-terminus (Bugge, Antalis and Wu, 2009). There are four different members: matriptase, matriptase-2, matriptase-3, and polyserase-1. The protease matriptase, after which the subfamily is named, is expressed as an inactive zymogen, which needs a proteolytic cleavage to become active. After cleavage, the protease domain remains attached to the membrane anchor by a disulfide bond. Under normal physiological conditions, matriptase activity is regulated by an excessive amount of its endogenous inhibitor HGF activator inhibitor-1 (HAI-1) (Oberst et al. 2003). HAI-1 also plays an important role in the expression, intracellular transportation, and activation of matriptase (Oberst et al. 2003, 2005).

Matriptase is linked to several human diseases. Dysregulation of matriptase is associated with various epithelial cancers (Miller and List 2013). Recently, it was also found that matriptase is able to process the amyloid precursor protein (Lanchec et al. 2017), thereby reducing Aβ40 production, one of the toxic peptide species that is responsible for senile plaque formation in the brains of Alzheimer patients. Moreover, matriptase promotes replication of Influenza A viruses (Menou et al. 2017), further illustrating its various roles in human pathology. Its family member matriptase-2 is key in human iron homeostasis and is mainly expressed in hepatocytes (Silvestri et al. 2008; Bugge et al. 2009; Wang et al. 2014). These biomedical roles have spurred an interest in probes for the matriptase family.

Godiksen et al. reported biotin-RQRR-chloromethyl ketone (Fig. 11b) as a probe suitable for detection of active matriptase. This ABP, based on the preferred P_4-P_1 sequence of matriptase (Béliveau et al. 2009), was selective for active matriptase over the matriptase zymogen and HAI-1-bound matriptase. In general, chloromethyl ketones are more reactive against cysteine proteases (Powers et al. 2002) and full selectivity of this probe will need to be explored. The Gütschow lab utilized the more serine protease selective diphenyl phosphonate warhead to design ABPs for matriptase and matriptase-2 (Häußler et al. 2016, 2017). As recognition element, two benzguanidines were used as arginine mimetics (Fig. 11c), which are putatively recognized by the S1 and S2 pocket of matriptase and the S1 and S3/S4 pocket of matriptase-2. Although the ABPs showed no selectivity between matriptase and matriptase-2, the limited tissue distribution of matriptase-2 may allow future visualization of matriptase.

5 Conclusions & Future Directions

Since the early 2000s, protease ABPP has made tremendous progress, as evidenced by a large body of studies reporting new ABPs and new applications in biology.

Broad-spectrum ABPs can be utilized for various purposes, such as protease profiling in different disease states and protease inhibitor screening. In combination with antibodies, broad-spectrum probes can even be used for the detection of specific, active proteases, as discussed in Sect. 2.2.

Besides the generation of general ABPs for protease clans and families, it is now also possible to develop highly selective probes that can discriminate between members of the same protease family. Synthetic chemistry efforts that explore novel reactive groups and recognition elements have been fundamental for these advances.

However, challenges in protease ABPP remain. Although a handful of proteases can now be detected by highly selective ABPs, over 500 proteases in the human genome remain to be selectively targeted. We expect that techniques to interrogate protease substrate specificity, especially in combination with non-natural amino acids, such as HyCoSul (Poreba et al. 2017), will be instrumental in future development of selective probes.

Although ABPs haven't yet undergone translation into clinical settings, promising preclinical animal studies have been reported. Topical administration of a cathepsin-targeting qABP with NIR fluorescence in a glioblastoma mouse model rapidly labeled tumor cells, especially the tumor margins (Cutter et al. 2012). This could be of particular benefit during operative procedures to ensure that all tumor cells are removed. Another study topically applied qABPs for detection of intestinal polyps in mice (Segal et al. 2015). As some polyps can be difficult to detect by conventional colonoscopy, protease-directed molecular imaging by using ABPs may be a valuable clinical tool. Further progression in protease ABPP by a combination of chemical and biological techniques will likely lead to clinical application of protease ABPs in the near future.

Acknowledgements The work in the Verhelst Laboratory is generously supported by the FWO, Stichting Alzheimer Onderzoek, KU Leuven, the Deutsche Forschungsgemeinschaft (DFG), the Ministerium für Innovation, Wissenschaft und Forschung des Landes Nordrhein-Westfalen, the Senatsverwaltung für Technologie und Forschung des Landes Berlin, and the Bundesministerium für Bildung und Forschung.

References

Abuelyaman AS et al (1994) Fluorescent derivatives of diphenyl [1-(N-Peptidylamino)alkyl] phosphonate esters: synthesis and use in the inhibition and cellular localization of serine proteases. Bioconjug Chem 5(5):400–405. https://doi.org/10.1021/bc00029a004

Abuelyaman AS et al (1997) Synthesis and kinetic studies of diphenyl 1-(N-peptidylamino) alkanephosphonate esters and their biotinylated derivatives as inhibitors of serine proteases and probes for lymphocyte granzymes. Arch Biochem Biophys 344(2):271–280. https://doi.org/10.1006/abbi.1997.0231

Adibekian A et al (2011) Click-generated triazole ureas as ultrapotent in vivo-active serine hydrolase inhibitors. Nat Chem Biol 7(7):469–478. https://doi.org/10.1038/nchembio.579

Bachovchin DA et al (2009) Identification of selective inhibitors of uncharacterized enzymes by high-throughput screening with fluorescent activity-based probes. Nat Biotechnol 27(4):387–394. https://doi.org/10.1038/nbt.1531

Bachovchin DA et al (2014) A high-throughput, multiplexed assay for superfamily-wide profiling of enzyme activity. Nat Chem Biol 10(8):656–663. https://doi.org/10.1038/nchembio.1578
Bedner E et al (2000) Activation of caspases measured in situ by binding of fluorochrome-labeled inhibitors of caspases (FLICA): correlation with DNA fragmentation. Exp Cell Res 259 (1):308–313. https://doi.org/10.1006/excr.2000.4955
Béliveau F, Désilets A, Leduc R (2009) Probing the substrate specificities of matriptase, matriptase-2, hepsin and DESC1 with internally quenched fluorescent peptides. The FEBS Journal 276(8):2213–2226. https://doi.org/10.1111/j.1742-4658.2009.06950.x
Ben-Nun Y et al (2015) Photodynamic quenched cathepsin activity based probes for cancer detection and macrophage targeted therapy. Theranostics 5(8):847–862. https://doi.org/10.7150/thno.10854
Berger AB et al (2006a) Identification of early intermediates of caspase activation using selective inhibitors and activity-based probes. Mol Cell 23(4):509–521. https://doi.org/10.1016/j.molcel.2006.06.021
Berger AB, Sexton KB, Bogyo M (2006b) Commonly used caspase inhibitors designed based on substrate specificity profiles lack selectivity. Cell Res 16(12):961–963. https://doi.org/10.1038/sj.cr.7310112
Blum G et al (2005) Dynamic imaging of protease activity with fluorescently quenched activity-based probes. Nat Chem Biol 1(4):203–209. https://doi.org/10.1038/nchembio728
Blum G et al (2007) Noninvasive optical imaging of cysteine protease activity using fluorescently quenched activity-based probes. Nat Chem Biol 3(10):668–677. https://doi.org/10.1038/nchembio.2007.26
Boatright KM, Salvesen GS (2003) Mechanisms of caspase activation. Curr Opin Cell Biol 15 (6):725–731. https://doi.org/10.1016/j.ceb.2003.10.009
Bogyo M et al (1997) Covalent modification of the active site threonine of proteasomal β subunits and the Escherichia coli homolog HslV by a new class of inhibitors. Proc Natl Acad Sci USA 94(13):6629–6634. https://doi.org/10.1073/pnas.94.13.6629
Bogyo M et al (2000) Selective targeting of lysosomal cysteine proteases with radiolabeled electrophilic substrate analogs. Chem Biol 7(1):27–38. https://doi.org/10.1016/S1074-5521(00)00061-2
Britton M et al (2009) Selective inhibitor of proteasome's caspase-like sites sensitizes cells to specific inhibition of chymotrypsin-like sites. Chem Biol 16(12):1278–1289. https://doi.org/10.1016/j.chembiol.2009.11.015
de Bruin G et al (2016) Proteasome subunit selective activity-based probes report on proteasome core particle composition in a native polyacrylamide gel electrophoresis fluorescence-resonance energy transfer assay. J Am Chem Soc 138(31):9874–9880. https://doi.org/10.1021/jacs.6b04207
Bugge TH, Antalis TM, Wu Q (2009) Type II transmembrane serine proteases. J Biol Chem 284 (35):23177–23181. https://doi.org/10.1074/jbc.R109.021006
Carril M (2017) Activatable probes for diagnosis and biomarker detection by MRI. J Mater Chem B 5(5):4332–4347. https://doi.org/10.1039/C7TB00093F
Coffelt SB, Wellenstein MD, de Visser KE (2016) Neutrophils in cancer : neutral no more. Nature Publishing Group 16(7):431–446. https://doi.org/10.1038/nrc.2016.52
Cutter JL et al (2012) Topical application of activity-based probes for visualization of brain tumor tissue. PLoS ONE 7(3):e33060. https://doi.org/10.1371/journal.pone.0033060
DeBruin G et al (2016) A set of activity-based probes to visualize human (Immuno)proteasome activities. Angew Chem—Int Ed 55(13):4199–4203. https://doi.org/10.1002/anie.201509092
Dooley CT, Houghten RA (1993) The use of positional scanning synthetic peptide combinatorial libraries for the rapid determination of opioid receptor ligands. Life Sci 52(18):1509–1517. https://doi.org/10.1016/0024-3205(93)90113-H
Edgington LE et al (2009) Noninvasive optical imaging of apoptosis by caspase-targeted activity-based probes. Nat Med 15(8):967–U177. https://doi.org/10.1038/nm.1938

Eitelhuber AC et al (2015) Activity-based probes for detection of active MALT1 paracaspase in immune cells and lymphomas. Chem Biol (Elsevier Ltd) 22(1):129–138. https://doi.org/10.1016/j.chembiol.2014.10.021

Van Esbroeck ACM et al (2017) Activity-based protein profiling reveals off-target proteins of the FAAH inhibitor BIA 10-2474. Science 356(6342):1084–1087. https://doi.org/10.1126/science.aaf7497

Falgueyret JP et al (2004) An activity-based probe for the determination of cysteine cathepsin protease activities in whole cells. Anal Biochem 335(2):218–227. https://doi.org/10.1016/j.ab.2004.09.005

Florea BI et al (2010) Activity-based profiling reveals reactivity of the murine thymoproteasome-specific subunit b 5t. Chem Biol (Elsevier Ltd) 17(8):795–801. https://doi.org/10.1016/j.chembiol.2010.05.027

Fonovic M et al (2007) Proteomics evaluation of chemically cleavable activity-based probes. Mol Cell Proteomics 6(10):1761–1770. https://doi.org/10.1074/mcp.M700124-MCP200

Fontan L et al (2012) MALT1 small molecule inhibitors specifically suppress ABC-DLBCL In Vitro and In Vivo. Cancer Cell (Elsevier Inc.,) 22(6):812–824. https://doi.org/10.1016/j.ccr.2012.11.003

Gaikwad HK et al (2018) Molecular imaging of cancer using X-ray computed tomography with protease targeted iodinated activity-based probes. Nano Lett 18(3):1582–1591. https://doi.org/10.1021/acs.nanolett.7b03813

Geurink PP et al (2013) Incorporation of non-natural amino acids improves cell permeability and potency of specific inhibitors of proteasome trypsin-like sites. J Med Chem 56(3):1262–1275. https://doi.org/10.1021/jm3016987

Gevaert K et al (2003) Exploring proteomes and analyzing protein processing by mass spectrometric identification of sorted N-terminal peptides. Nat Biotechnol 21(5):566–569

Greenbaum D et al (2000) Epoxide electrophiles as activity-dependent cysteine protease profiling and discovery tools. Chem Biol 7(8):569–581. https://doi.org/10.1016/S1074-5521(00)00014-4

Greenbaum D et al (2002) Chemical approaches for functionally probing the proteome. Mol Cell Proteomics 1(1):60–68. https://doi.org/10.1074/mcp.T100003-MCP200

Hachmann J et al (2012) Mechanism and specificity of the human paracaspase MALT1. Biochem J 443(1):287–295. https://doi.org/10.1042/BJ20120035

Hachmann J, Salvesen GS (2016) The paracaspase MALT1. Biochimie 122:324–338. https://doi.org/10.1016/j.biochi.2015.09.018

Haedke U et al (2013) Tuning probe selectivity for chemical proteomics applications. Curr Opin Chem Biol 17(1):102–109. https://doi.org/10.1016/j.cbpa.2012.11.024

Hailfinger S et al (2009) Essential role of MALT1 protease activity in activated B cell-like diffuse large B-cell lymphoma. Proc Natl Acad Sci 106(47):19946–19951. https://doi.org/10.1073/pnas.0907511106

Häußler D et al (2016) Phosphono bisbenzguanidines as irreversible dipeptidomimetic inhibitors and activity-based probes of matriptase-2. Chem Eur J 22(25):8525–8535. https://doi.org/10.1002/chem.201600206

Häußler D et al (2017) A fluorescent-labeled phosphono bisbenzguanidine as an activity-based probe for matriptase. Chem Eur J 23(22):5205–5209. https://doi.org/10.1002/chem.201700319

Hight MR et al (2014) A peptide-based positron emission tomography probe for in vivo detection of caspase activity in apoptotic cells. Clin Cancer Res 20(8):2126–2135. https://doi.org/10.1158/1078-0432.CCR-13-2444

Hoenderdos K, Condliffe A (2013) The neutrophil in chronic obstructive pulmonary disease: too little, too late or too much, too soon? Am J Respir Cell Mol Biol 48(5):531–539. https://doi.org/10.1165/rcmb.2012-0492TR

Hong J-A et al (2017) Development of a smart activity-based probe to detect subcellular activity of asparaginyl endopeptidase in living cells. Org Biomol Chem (Royal Society of Chemistry) 1:8018–8022. https://doi.org/10.1039/c7ob01467h

Hu M et al (2011) Multicolor, one- and two-photon imaging of enzymatic activities in live cells with fluorescently quenched activity-based probes (qABPs). J Am Chem Soc 133(31):12009–12020

Ides J et al (2014) Synthesis and in vivo preclinical evaluation of an18F labeled uPA inhibitor as a potential PET imaging agent. Nucl Med Biol (Elsevier Inc.,) 41(6):477–487. https://doi.org/10.1016/j.nucmedbio.2014.03.019

Jessani N et al (2002) Enzyme activity profiles of the secreted and membrane proteome that depict cancer cell invasiveness. Proc Natl Acad Sci USA 99(16):10335–10340

Joyce JA et al (2004) Cathepsin cysteine proteases are effectors of invasive growth and angiogenesis during multistage tumorigenesis. Cancer Cell 5(5):443–453. https://doi.org/10.1016/S1535-6108(04)00111-4

Kam CM et al (1993) Biotinylated isocoumarins, new inhibitors and reagents for detection, localization, and isolation of serine proteases. Bioconjug Chem 4(6):560–567. https://doi.org/10.1021/bc00024a021

Kasperkiewicz P et al (2014) Design of ultrasensitive probes for human neutrophil elastase through hybrid combinatorial substrate library profiling. Proc Natl Acad Sci USA 111(7):2518–2523. https://doi.org/10.1073/pnas.1318548111

Kasperkiewicz P et al (2015) Design of a selective substrate and activity based probe for human neutrophil serine protease 4. PloS one 1–12. https://doi.org/10.1371/journal.pone.0132818

Kasperkiewicz P, Altman Y, D'Angelo M et al (2017) Toolbox of fluorescent probes for parallel imaging reveals uneven location of serine proteases in neutrophils. J Am Chem Soc 139 (29):10115–10125. https://doi.org/10.1021/jacs.7b04394

Kato D et al (2005) Activity-based probes that target diverse cysteine protease families. Nat Chem Biol 1(1):33–38. https://doi.org/10.1038/nchembio707

Kessler BM et al (2001) Extended peptide-based inhibitors efficiently target the proteasome and reveal overlapping specificities of the catalytic beta-subunits. Chem Biol 8(9):913–929. https://doi.org/10.1016/S1074-5521(01)00069-2

Kim D, Jetson RR, Krusemark CJ (2017) A DNA-assisted immunoassay for enzyme activity via a DNA-linked, activity-based probe. Chemical communications (Cambridge, England), 53(68), pp. 9474–9477. https://doi.org/10.1039/c7cc05236g

Kleifeld O et al (2010) Isotopic labeling of terminal amines in complex samples identifies protein N-termini and protease cleavage products. Nat Biotechnol 28(3):281–288. https://doi.org/10.1038/nbt.1611

Kolodziejek I et al (2011) Proteasome activity imaging and profiling characterizes bacterial effector syringolin A. Plant Physiol 155(1):477–489. https://doi.org/10.1104/pp.110.163733

Kovács J et al (2017) Proteasome activity profiling uncovers alteration of catalytic β2 and β5 subunits of the stress-induced proteasome during salinity stress in tomato roots. Front Plant Sci 8(107). https://doi.org/10.3389/fpls.2017.00107

Lanchec E et al (2017) The type II transmembrane serine protease matriptase cleaves the amyloid precursor protein and reduces its processing to beta-amyloid peptide. J Biol Chem 292:20669–20682. https://doi.org/10.1074/jbc.M117.792911

LeBeau AM et al (2013) Imaging a functional tumorigenic biomarker in the transformed epithelium. Proc Natl Acad Sci 110(1):93–98. https://doi.org/10.1073/pnas.1218694110

Lechtenberg BC et al (2015) The elastase-PK101 structure: Mechanism of an ultrasensitive activity-based probe revealed. ACS Chem Biol 10(4):945–951. https://doi.org/10.1021/cb500909n

Li G et al (2017) An activity-dependent proximity ligation platform for spatially resolved quantification of active enzymes in single cells. Nat Commun 8(1). https://doi.org/10.1038/s41467-017-01854-0

Li N et al (2013) Relative quantification of proteasome activity by activity-based protein profiling and LC-MS/MS. Nat Protoc (Nature Publishing Group) 8(6):1155–1168. https://doi.org/10.1038/nprot.2013.065

Liu Y, Patricelli MP, Cravatt BF (1999) Activity-based protein profiling: the serine hydrolases. Proc Natl Acad Sci USA 96(26):14694–14699. https://doi.org/10.1073/pnas.96.26.14694

Matthews DJ, Wells JA (1993) Substrate phage: selection of protease substrates by monovalent phage display. Science (New York, N.Y.) 260(5111):1113–1117. https://doi.org/10.1126/science.8493554

Mc Guire C et al (2013) Paracaspase MALT1 deficiency protects mice from autoimmune-mediated demyelination. J Immunol 190(6):2896–2903. https://doi.org/10.4049/jimmunol.1201351

Menou A et al (2017) Human airway trypsin-like protease, a serine protease involved in respiratory diseases. Am J Lung Cell Mol Physiol 312(107):657–668. https://doi.org/10.1152/ajplung.00509.2016

Méthot N et al (2004) A caspase active site probe reveals high fractional inhibition needed to block DNA fragmentation. J Biol Chem 279(27):27905–27914. https://doi.org/10.1074/jbc.M400247200

Miller GS, List K (2013) The matriptase-prostasin proteolytic cascade in epithelial development and pathology. Cell Tissue Res 351:245–253. https://doi.org/10.1007/s00441-012-1348-1

Misas-Villamil JC et al (2017) Subunit-selective proteasome activity profiling uncovers uncoupled proteasome subunit activities during bacterial infections. Plant J 90(2):418–430. https://doi.org/10.1111/tpj.13494

Morell M et al (2013) Coupling protein engineering with probe design to inhibit and image matrix metalloproteinases with controlled specificity. J Am Chem Soc 135(24):9139–9148. https://doi.org/10.1021/ja403523p

Murata S et al (2007) Regulation of CD8+ T cell development by thymus-specific proteasomes. Science 316(5829):1349–1353. https://doi.org/10.1126/science.1141915

Oberst MD et al (2003) The activation of matriptase requires its noncatalytic domains, serine protease domain, and its cognate inhibitor. J Biol Chem 278(29):26773–26779. https://doi.org/10.1074/jbc.M304282200

Oberst MD et al (2005) HAI-1 regulates activation and expression of matriptase, a membrane-bound serine protease. Am J Physiol—Cell Physiol 289:462–470. https://doi.org/10.1152/ajpcell.00076.2005

Ong S-E et al (2002) Stable isotope labeling by amino acids in cell culture, SILAC, as a simple and accurate approach to expression proteomics. Mol Cell Proteomics 1(5):376–386. https://doi.org/10.1074/mcp.M200025-MCP200

Perera NC et al (2012) NSP4, an elastase-related protease in human neutrophils with arginine specificity. Proc Natl Acad Sci 109(12):6229–6234. https://doi.org/10.1073/pnas.1200470109

Pop C, Salvesen GS (2009) Human caspases: activation, specificity, and regulation. J Biol Chem 284(33):21777–21781. https://doi.org/10.1074/jbc.R800084200

Poreba M et al (2014) Unnatural amino acids increase sensitivity and provide for the design of highly selective caspase substrates. Cell Death Differ 21(9):1482–1492

Poreba M, Salvesen GS, Drag M (2017) Synthesis of a HyCoSuL peptide substrate library to dissect protease substrate specificity. Nat. Protoc (Nature Publishing Group) 12(10):2189–2214. https://doi.org/10.1038/nprot.2017.091

Powers JC et al (2002) Irreversible inhibitors of serine, cysteine, and threonine proteases. Chem Rev 102(12):4639–4750. https://doi.org/10.1021/cr010182v

Rawlings ND, Barrett AJ, Finn R (2018) Twenty years of the MEROPS database of proteolytic enzymes, their substrates and inhibitors. Nucleic Acids Res 44(February):343–350. https://doi.org/10.1093/nar/gkv1118

Ren G et al (2011) Non-invasive imaging of cysteine cathepsin activity in solid tumors using a (64)Cu-labeled activity-based probe. PLoS One 6(11)

Rut W et al (2017) Extended substrate specificity and first potent irreversible inhibitor/activity-based probe design for Zika virus NS2B-NS3 protease. Antiviral Res 139:88–94. https://doi.org/10.1016/j.antiviral.2016.12.018

Saghatelian A et al (2004) Activity-based probes for the proteomic profiling of metalloproteases. PNAS 101(27):10000–10005. https://doi.org/10.1073/pnas.0402784101

Sanman LE, Bogyo M (2014) Activity-based profiling of proteases. Annu Rev Biochem 83:249–273. https://doi.org/10.1146/annurev-biochem-060713-035352

Schechter I, Berger A (1967) On the size of the active site in proteases. I. Papain. Biochem Biophys Res Commun 27(2):157–162. https://doi.org/10.1016/j.bbrc.2012.08.015

Schilling O, Overall CM (2008) Proteome-derived, database-searchable peptide libraries for identifying protease cleavage sites. Nat Biotechnol 26(6):685–694. https://doi.org/10.1038/nbt1408

Screen M et al (2010) Nature of pharmacophore influences active site specificity of proteasome inhibitors. The J Biol Chem. Am Soc Biochem Mol Biol 285(51):40125–34. https://doi.org/10.1074/jbc.m110.160606

Segal E et al (2015) Detection of intestinal cancer by local, topical application of a quenched fluorescence probe for cysteine cathepsins. Chem Biol 22(1):148–158. https://doi.org/10.1016/j.chembiol.2014.11.008

Serim S, Baer P, Verhelst SHL (2015) Mixed alkyl aryl phosphonate esters as quenched fluorescent activity-based probes for serine proteases. Org Biomol Chem. Roy Soc Chem 13(8):2293–9. https://doi.org/10.1039/c4ob02444c

Sharifzadeh S et al (2017) Novel electrophilic scaffold for imaging of essential penicillin-binding proteins in *Streptococcus pneumoniae*. ACS Chem Biol 12:2849–2857. https://doi.org/10.1021/acschembio.7b00614

Shaulov-Rotem Y et al (2016) A novel quenched fluorescent activity-based probe reveals caspase-3 activity in the endoplasmic reticulum during apoptosis. Chem. Sci. Roy Soc Chems 7(2):1322–1337. https://doi.org/10.1039/C5SC03207E

Sieber SA et al (2004) Microarray platform for profiling enzyme activities in complex proteomes. J Am Chem Soc 126(48):15640–15641. https://doi.org/10.1021/ja044286+

Silvestri L et al (2008) the serine protease matriptase-2 (TMPRSS6) inhibits hepcidin activation by cleaving membrane hemojuvelin. Cell Metab 8(6):502–511. https://doi.org/10.1016/j.cmet.2008.09.012

Soehnlein O et al (2017) Neutrophils as protagonists and targets in chronic inflammation. Nat Publishing Group 17(4):248–261. https://doi.org/10.1038/nri.2017.10

Tanaka K, Kasahara M (1998) The MHC class I ligand-generating system: roles of immunoproteasomes and the interferon-gamma-inducible proteasome activator PA28. Immunol Rev 163:161–176. https://doi.org/10.1111/j.1600-065X.1998.tb01195.x

Thornberry NA et al (1997) A combinatorial approach defines specificities of members of the caspase family and granzyme B: functional relationships established for key mediators of apoptosis. J Biol Chem 272(29):17907–17911. https://doi.org/10.1074/jbc.272.29.17907

Vangestel C et al (2016) Preclinical evaluation of [(111) In]MICA-401, an activity-based probe for SPECT imaging of in vivo uPA activity. Contrast Media Mol Imaging 11:448–458. https://doi.org/10.1002/cmmi.1706

Venne AS et al (2013) Novel highly sensitive, specific, and straightforward strategy for comprehensive N-terminal proteomics reveals unknown substrates of the mitochondrial peptidase Icp55. J Proteome Res 12(9):3823–3830. https://doi.org/10.1021/pr400435d

Verdoes M, Verhelst SHL (2016) Detection of protease activity in cells and animals Biochim Biophys Acta—Proteins and Proteomics (Elsevier B.V.) 1864(1):130–142. https://doi.org/10.1016/j.bbapap.2015.04.029

Verhelst SHL, Fonovic M, Bogyo M (2007) A mild chemically cleavable linker system for functional proteomic applications. Angew Chem Int Ed Engl 46(8):1284–1286. https://doi.org/10.1002/anie.200603811

Vickers CJ, González-Páez GE, Wolan DW (2013) Selective detection of caspase-3 versus caspase-7 using activity-based probes with key unnatural amino acids. ACS Chem Biol 8(7):1558–1566. https://doi.org/10.1021/cb400209w

Vidmar R et al (2017) Protease cleavage site fingerprinting by label-free in-gel degradomics reveals pH-dependent specificity switch of legumain. The EMBO J 36(16):2455–2465. https://doi.org/10.15252/embj.201796750

Wang C, Meynard D, Lin HY (2014) The role of TMPRSS6/matriptase-2 in iron regulation and anemia. Front in Pharmacol 5(114):1–6. https://doi.org/10.3389/fphar.2014.00114

Willems LI et al (2011) Bioorthogonal chemistry: applications in activity-based protein profiling. Acc Chem Res 44(9):718–729. https://doi.org/10.1021/ar200125k

Withana NP et al (2016) Dual-modality activity-based probes as molecular imaging agents for vascular inflammation. J Nucl Med 57(10):1583–1590. https://doi.org/10.2967/jnumed.115.171553

Wolf EV et al (2013) A new class of rhomboid protease inhibitors discovered by activity-based fluorescence polarization. PLoS ONE 8(8). https://doi.org/10.1371/journal.pone.0072307

Xiao J et al (2013) A coupled protein and probe engineering approach for selective inhibition and activity-based probe labeling of the caspases. J Am Chem Soc 135(24):9130–9138. https://doi.org/10.1021/ja403521u

Xin BT et al (2016) Development of new Malt1 inhibitors and probes. Bioorg Med Chem (Elsevier Ltd) 24(15):3312–3329. https://doi.org/10.1016/j.bmc.2016.03.035

Yang P, Liu K (2015) Activity-based protein profiling: recent advances in probe development and applications. ChemBioChem 16(5):712–724. https://doi.org/10.1002/cbic.201402582

Yang Y et al (2013) a simple and effective cleavable linker for chemical proteomics applications. Mol Cell Proteomics 12(1):237–244. https://doi.org/10.1074/mcp.M112.021014

Yang Y, Fonovic M, Verhelst SH (2017) Cleavable linkers in chemical proteomics applications. Methods Mol Biol 1491:185–203. https://doi.org/10.1007/978-1-4939-6439-0_14

Yu JW et al (2011) Crystal structure of the mucosa-associated lymphoid tissue lymphoma translocation 1 (MALT1) paracaspase region. Proc Natl Acad Sci 108(52):21004–21009. https://doi.org/10.1073/pnas.1111708108

Yuan Y et al (2015) Controlled intracellular self-assembly and disassembly of 19F nanoparticles for MR imaging of caspase 3/7 in zebrafish. ACS Nano 9(1):761–768. https://doi.org/10.1021/nn5062657

Yue X et al (2014) Novel 19 F activatable probe for the detection of matrix metalloprotease-2 Activity by MRI/MRS. Mol Pharm 11(11):4208–4217. https://doi.org/10.1021/mp500443x

Zou F et al (2012) Application of a novel highly sensitive activity-based probe for detection of cathepsin G. Anal Biochem (Elsevier Inc.,) 421(2):667–672. https://doi.org/10.1016/j.ab.2011.11.016

Opportunities for Lipid-Based Probes in the Field of Immunology

Sebastiaan T. A. Koenders, Berend Gagestein and Mario van der Stelt

Contents

Abstract Lipids perform a wide range of functions inside the cell, ranging from structural building block of membranes and energy storage to cell signaling. The mode of action of many signaling lipids has remained elusive due to their low abundance, high lipophilicity, and inherent instability. Various chemical biology approaches, such as photoaffinity or activity-based protein profiling methods, have been employed to shed light on the biological role of lipids and the lipid–protein interaction profile. In this review, we will summarize the recent developments in the field of chemical probes to study lipid biology, especially in immunology, and indicate potential avenues for future research.

S. T. A. Koenders · B. Gagestein · M. van der Stelt (✉)
Leiden University, Leiden, The Netherlands
e-mail: m.van.der.stelt@chem.leidenumiv.nl

Current Topics in Microbiology and Immunology (2019) 420:283–320
DOI 10.1007/82_2018_127

Published Online: 22 September 2018

1 Introduction

Lipids are defined as hydrophobic biomolecules that dissolve in organic solvents, but not in water. They perform a wide range of functions inside the cell, ranging from structural building block of membranes and energy storage to cell signaling. The discovery of the link between aspirin and prostaglandin synthesis showed that lipids can modulate the immune system and that the enzymes involved in their metabolism constitute interesting drug targets (Vane 1971). Since then, many associations have been made between the immune system and signaling lipids in the field of endocannabinoids, resolvins, steroid hormones, and vitamins A and D (Mora et al. 2008; Pandey et al. 2009; Marshall-Gradisnik et al. 2009; Serhan and Petasis 2011). However, due to the low abundance, high lipophilicity, and inherent instability of many lipid signaling molecules, their lipid–protein interaction profile and mode of action have remained largely elusive.

In recent years, several technical advances in mass spectrometry and innovative chemical biology strategies have been developed to shed light on these lipids and their protein interaction landscapes to study their biological function. Since its inception, the field of lipidomics, which is the analysis of lipids and their interacting partners within a biological system, has made great strides forward (Wenk 2005). Standardization of protocols, increased availability of deuterated lipids, and the high mass accuracy and resolution of modern mass spectrometers have made it possible to measure many lipids in complex biological samples (Wenk 2010). In an effort to systematically classify the rapidly expanding database of characterized lipids, the Lipid Metabolites and Pathways Strategy (LIPID MAPS) consortium has come up with a more concise definition of a lipid: a small hydrophobic or amphipathic molecule that is formed at least partially by the condensation of ketoacyl thioesters and/or isoprene units (Fahy et al. 2011). Based on these two building blocks eight major lipid classes are defined: glycerophospholipids, glycerolipids, fatty acyls, sphingolipids, sterol lipids, prenol lipids, saccharolipids, and polyketides (Fig. 1) (Fahy et al. 2009).

The variability in chemical characteristics between the various lipid groups makes it a challenge to measure all lipid species in one experiment. Different sample preparation methods and measuring techniques are required to quantify different classes of lipids (Yang and Han 2016). Adding to the challenge is the wide range of alterations that lipids can undergo to form still non-identified lipid derivatives. By combining the fields of chemistry and biology, some of these challenges can be overcome.

The field of chemical biology has developed two main approaches to study lipids: a) novel chemically modified lipids to study their biological role and b) chemical tools to visualize and modulate the proteins involved in lipid metabolism. To track lipids in a biological system, they can be functionalized with alkyne tags (Gaebler et al. 2013; Robichaud et al. 2016). Using ligation chemistry, these functionalized lipids can be visualized and identified (Hein and Fokin 2010). This method has been used for the visualization of lipids in membranes, modification of

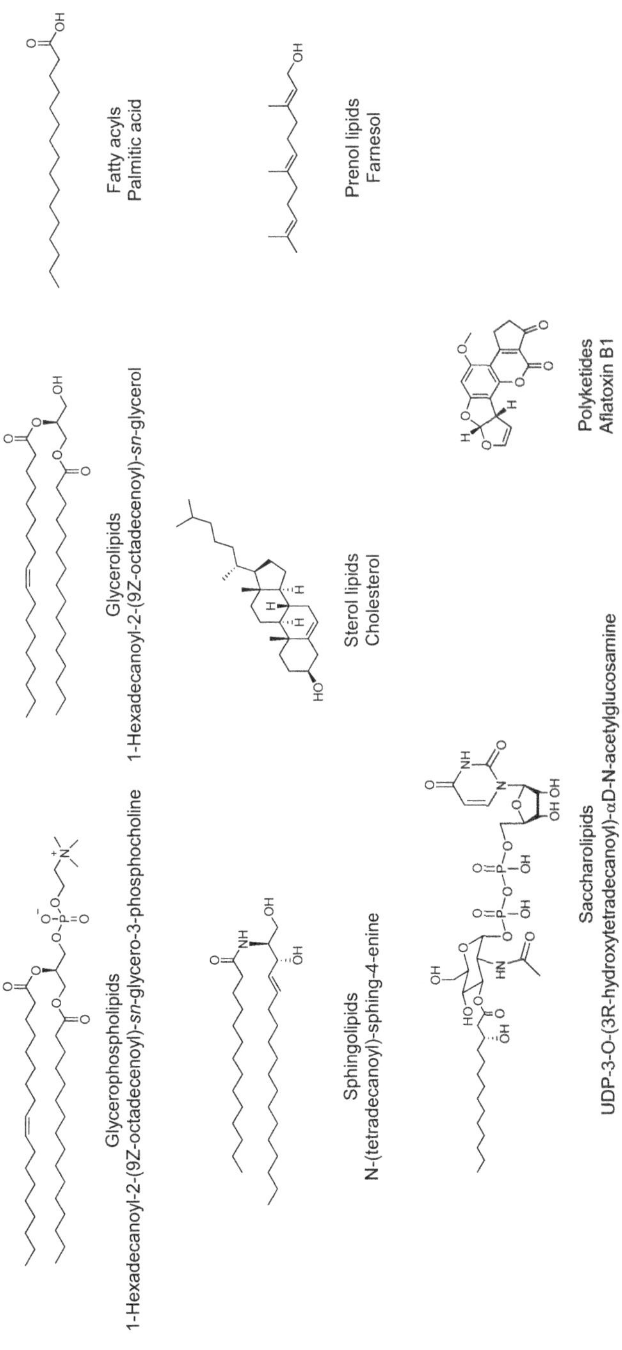

Fig. 1 Representative structures of the eight lipid categories according to LIPID MAPS (Fahy et al. 2009, 2011)

proteins by lipids, and their metabolism (Hofmann et al. 2014; Tate et al. 2015; Gaebler et al. 2016). Although the alkyne functionalization enables affinity purification, this will only reveal protein interaction partners that are covalently bound to the lipid.

To visualize lipid-interacting partners, such as binding proteins or metabolizing proteins, photoaffinity or activity-based labeling can be used. These methods require bifunctionalized lipids and probes. These molecules contain a photoreactive group or an electrophilic warhead and an alkyne or azide, which are employed in affinity-based protein profiling (A*f*BPP) and activity-based protein profiling (ABPP) (Cravatt et al. 2008). Both methods enabled the visualization and identification of the protein–lipid interactions (Wright and Sieber 2016).

In this review, we will summarize the recent developments concerning lipid-based probes. The chemical tools are divided into two classes: affinity- and activity-based probes. The affinity-based probes are discussed using the LIPID MAPS classification system, while the activity-based probes are grouped based on their enzyme class. Finally, we present some opportunities for future research.

2 Lipid Photoaffinity Probes

Classical techniques to study protein–lipid interactions include microscale thermophoresis, isothermal titration calorimetry, and surface plasmon resonance. Recently, new techniques have emerged to study these types of interactions in a native environment, such as live-cell imaging with fluorescent proteins or affinity purification lipidomics. Most of these methods, however, rely on a predetermined protein-of-interest and its modification. Compared to these techniques, photoaffinity lipids offer new possibilities regarding throughput and biocompatibility.

Photoactivatable lipids are synthetic derivatives that retain the overall structure and interactions of the parent lipid, but contain a photoactivatable moiety that may form a covalent bond with its interacting partner. This covalent bond essentially 'freezes' the interaction and allows for affinity purification. Of interest, this interaction can be captured in complex native environments (Haberkant et al. 2013). A number of photoreactive moieties are available, with differing synthetic accessibility, reactivity, efficiency, and structural impact on a probe. Three photoreactive groups are routinely used: diazirines, benzophenones, and aryl azides (Fig. 2). Their properties and reactivity have been previously discussed in excellent reviews (Bush et al. 2013; Sakurai et al. 2014; Kleiner et al. 2017).

The choice of photoactivatable group depends on its application. However, a shift from benzophenones to diazirines has been observed in the last decade. Increased synthetic efforts, the desire for a smaller modification, and the application in a relative water-free inner membrane have contributed to this trend. In this chapter, lipid photoaffinity probes for fatty acyls, glycero(phospho)lipids, sphingolipids, and sterols will be discussed.

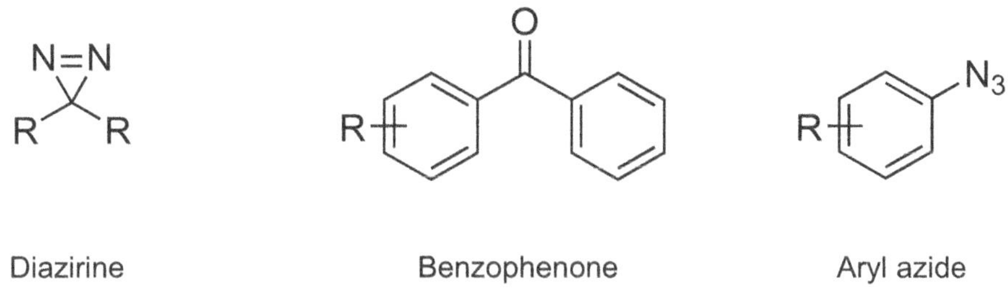

Fig. 2 Structures of three commonly used photoreactive groups

2.1 Glycero(Phospho)Lipids

One of the most well-studied lipid classes is the glycero(phospho)lipids. Their amphipathic nature is essential for the formation of lipid bilayers, and they are universally present in cellular membranes. Glycerolipids are composed of a glycerol backbone that has been mono-, di-, or trisubstituted with a fatty acyl. In most glycerophospholipids, the *sn*-3-position is esterified with a phosphate group, which in turn can be substituted with different head groups. The most common substitutions are a choline, ethanolamine, glycerol, serine, or inositol, giving rise to phosphatidylcholine (PC), phosphatidylethanolamine (PE), phosphatidylglycerol (PG), phosphatidylserine (PS), and phosphatidylinositol (PI), respectively. The length and degree of unsaturation of the fatty acyl groups determine the physical properties of the lipid, and its substitution pattern on the phosphate group dictates its protein interactions partners on the membrane surface (Yeagle 2016). Different types of lipids and their compartmentalization between cellular structures do not only determine the physical properties of the biomembrane, but also regulate protein activity, localization, or signaling events via lipid–protein interactions. Such interactions can be discovered via application of photoaffinity lipid probes.

Phosphatidylserine (PS) lipids are primarily found on intracellular membrane surfaces. When this asymmetry is disturbed during apoptosis, PS presentation is a trigger for phagocytosis. Several PS analogues **1–5** containing a benzophenone and alkyne click-handle were synthesized and shown to label PS-interacting proteins in PS-lipid-mediated engulfment, such as prothrombin-1 (Fig. 3) (Bandyopadhyay and Bong 2011).

To study membrane-associated proteins that are under-represented by traditional proteomics, lipid probes based on phosphatidylcholine were made (Gubbens et al. 2009). The goal was to detect proteins interacting with the phospholipid head groups in inner mitochondrial membranes. To this end, the photoreactive group was installed on the solvent-exposed head group of the lipid. The aryl azide probe **6** and benzophenone probe **7** were used to isolate proteins from *Saccharomyces cerevisiae* mitochondria (Fig. 3). The identified targets consisted of known membrane interacting proteins, like Gut2p and Cox2p, but also new proteins. Of note, the authors

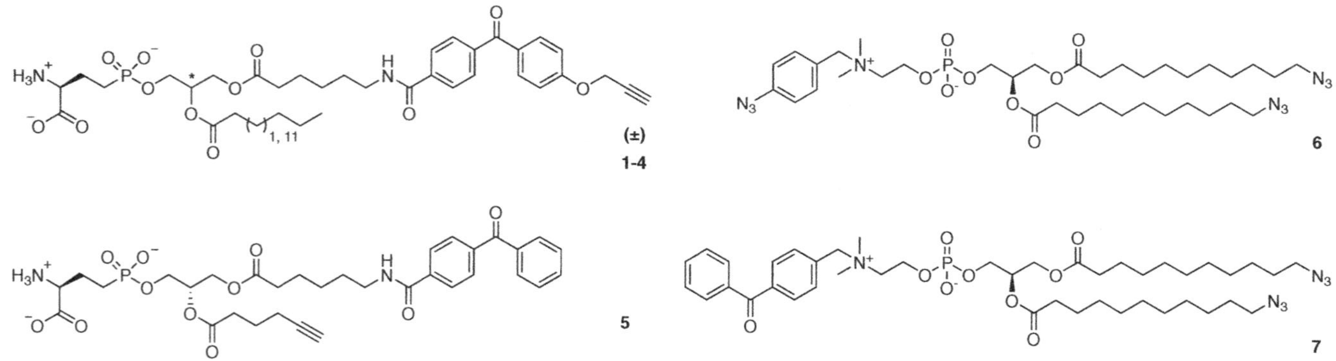

Fig. 3 Structures of photoaffinity probes based on phosphatidylserine **1–5** and phosphatidylcholine **6–7**

concluded that aryl azide **7** is the preferred probe due to high background labeling and labeling of highly abundant proteins with benzophenone **6** (Gubbens et al. 2009).

Phosphatidylinositol (PI) polyphosphates are a family of signaling lipids that act as ligands for protein-membrane association. To discover the proteins that associate with these lipids, two chemical probes based on the head group of phosphatidylinositol 3,4,5-triphosphate have been synthesized (Rowland et al. 2011). Probes **8** and **9** were tested on the purified pleckstrin homology domain of protein kinase B, a known phosphatidylinositol phosphate binder (Fig. 4). The shorter **8** gave better signal, which can be attributed to its shorter linker length which positioned the benzophenone closer to the protein. Probe **8** was applied to MDA-MB-435 cell extracts, and 265 proteins were identified as phosphatidylinositol 3,4,5-triphosphate-binding proteins (Rowland et al. 2011).

To study the membrane-binding domain of PON1, a high-density lipoprotein (HDL)-associated protein, phospholipid probe **10** was made with the photoactivatable group on the head of the lipid (Fig. 4). After covalent photocrosslinking, digestion and analysis by MS several residues localized closely together were found on the surface of PON1, indicating the HDL-binding domain (Gu et al. 2016).

To study the ability of cardiolipin to form a complex with the mitochondrial protein cytochrome *c*, several photoaffinity probes **11–13** were synthesized (Fig. 4). These probes induced similar or higher cytochrome *c* peroxidase activity compared to endogenous tetraoleoyl cardiolipin, indicating that the central hydroxyl group is not necessary for activation. No follow-up study applying the labeling functionality has appeared (Abe et al. 2015).

Several photoactivatable analogs of phosphatidylcholines are commercially available. However, it is argued that lipids supplied to a living cell are differently distributed than lipids synthesized within the cell, which distorts the interaction profile reported by the probe (Peng et al. 2014). Moreover, palmitoleic and myristoleic acid probes that contain both a diazirine and an alkyne can be incorporated in various glycerolipids and used as post-translational modifications (PTMs), thereby further complicating the interpretation of the results of a photoaffinity experiment. The solution was to incubate cells with an alkyne-tagged choline **14** and diazirine-containing fatty acid **15** (Fig. 5). In this manner, the photoaffinity experiment only captured the proteins that interacted with the in situ synthesized phosphatidylcholine. Different combinations of alkyne- and diazirine-containing components of PC were tested, but the combination of **14** and **15** gave the best results (Wang et al. 2017). This strategy yielded a PC-protein interaction map in living cells. Two hundred and eleven high-confidence PC-interacting proteins, mostly present in the cytosol, ER, mitochondria, and nucleus were identified. The authors concluded that the double incorporation strategy offered significant improvements for global mapping of genuine protein–lipid interactions and indicated that the strategy is expandable to different lipid classes, such as phosphatidylinositols (Wang et al. 2017).

Dilinolenoyl monogalactosyldiacylglycerol (MGDG) is an anti-inflammatory signaling lipid (Ulivi et al. 2011). Three photoactivatable probes **16–18** were synthesized to elucidate its mode of action (Fig. 5). The linolenoyl groups were

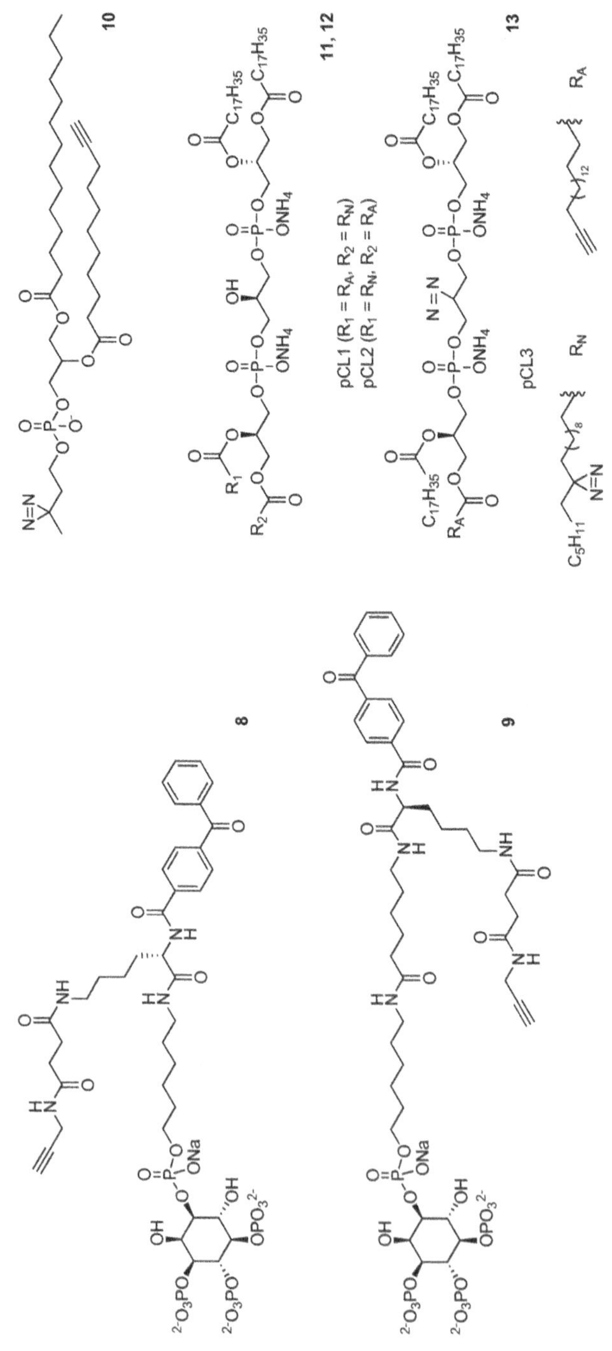

Fig. 4 Structures of photoaffinity probes based on phosphatidylinositol 3,4,5-triphosphate **8–9**, phospholipid **10** and cardiolipin **11–13**

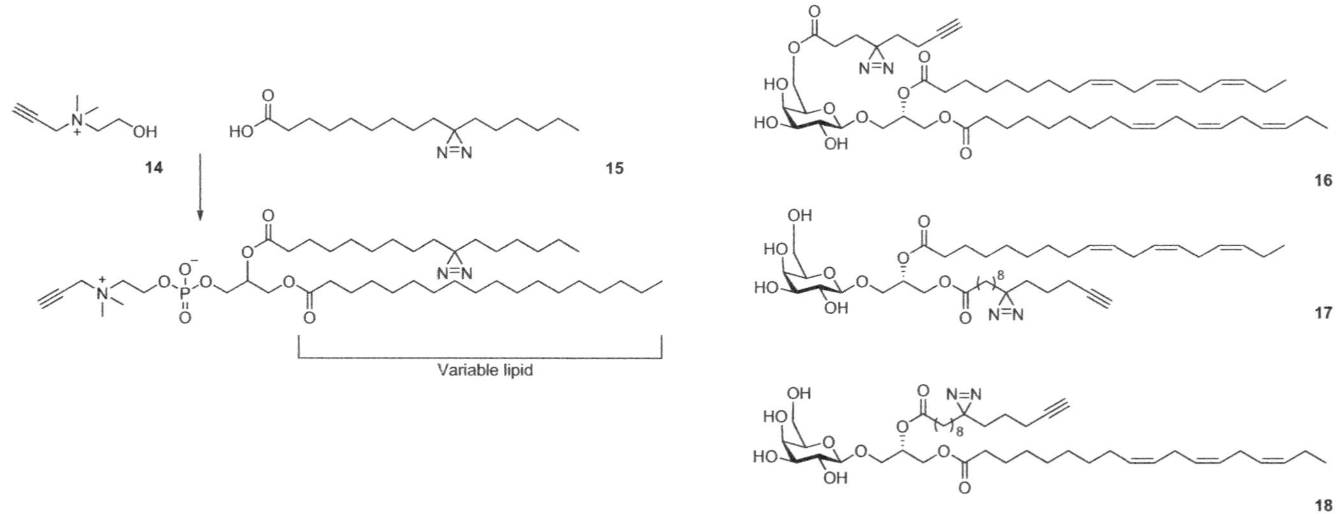

Fig. 5 Structures of photoaffinity probes based on phosphatidylcholine **14+15** and dilinolenoyl MGDG **16–18**

replaced with a similar photoaffinity-click fatty acid (pacFA), or a minimalist linker was tagged to the sugar moiety (Li et al. 2013). **16** Did not show any activity in an inflammatory assay, but **17** and **18** had comparable activity to MGDG. This indicated that an unmodified galactosyl moiety is essential for MGDG activity. Moreover, when incubated, UV-exposed and washed, **17** lost all anti-inflammatory activity, whereas **18** was still active. This implied that **18** was capable of both binding and labeling an unknown target. With structure **16** as negative control, TLR4 was identified as a probe target of **18** in human chondrocytes. Subsequent orthogonal experiments demonstrated MGDG acted as TLR-4 antagonist (Liu et al. 2016).

Recently, caged photoaffinity probes including **19–21** were developed to study biologically active signaling lipids with temporal, spatial, and subcellular resolution (Figs. 6, 7 and 8). These caged probes contain a fluorescent coumarin group, which is released upon irradiation with a certain wavelength light that does not activate the diazirine moiety. Microscopy experiments indicated that all lipids were indiscriminately localized to internal membranes and the cytoplasm. Uncaging, photocrosslinking, and ligation with a different fluorophore showed that each lipid localized to distinct cellular components. Controlled release of DAG via uncaging was also shown to be feasible. Elevated diacylglycerol (DAG) levels are known to trigger translocation of C1-domain-containing proteins to the plasma membrane (Nadler et al. 2013). Immediate translocation of C1-linked green fluorescent protein (GFP) was observed when **19** was uncaged in HeLa cells (Höglinger et al. 2017). DAG turnover was subsequently quantified on a population-wide and single-cell level. The authors suggested that standard biochemical experiments to measure DAG metabolism were inherently flawed, since they only measure combined lipid transport and metabolism. Moreover, striking differences were found between DAG turnover on a cell-to-cell level, indicating that heterogeneity might be an underrated complication of lipid signaling (Höglinger et al. 2017). In addition, a pull-down experiment with probe **19–21** was performed with HeLa cell proteome affording 130 **19**-specific binding proteins. As the probe contains arachidonic acid, the hits were compared to the targets of arachidonic acid-based probes **22** and **23** (Fig. 7). Remarkably, only 17 proteins of the 130 hits were overlapping, showing that probe **19** has mostly DAG-specific interactions. Thus, the two activatable functionalities allowed to investigate different aspects of the lipids with the same probe (Höglinger et al. 2017).

2.2 *Fatty Acids*

Fatty acids are constituents of other lipid classes, such as glycerol(phospho)lipid, glycolipids, and ceramides, but also have their own signaling roles. In addition, they can be incorporated by proteins as a post-translational modification embedding the proteins into the lipid membrane. The length of fatty acid varies, just as its degree of unsaturation. An unsaturated lipid has one or more double bonds, usually in a cis-configuration and with one methylene group in between. Furthermore, oxidative enzymes, such as lipoxygenases, cyclooxygenases, and cytochrome

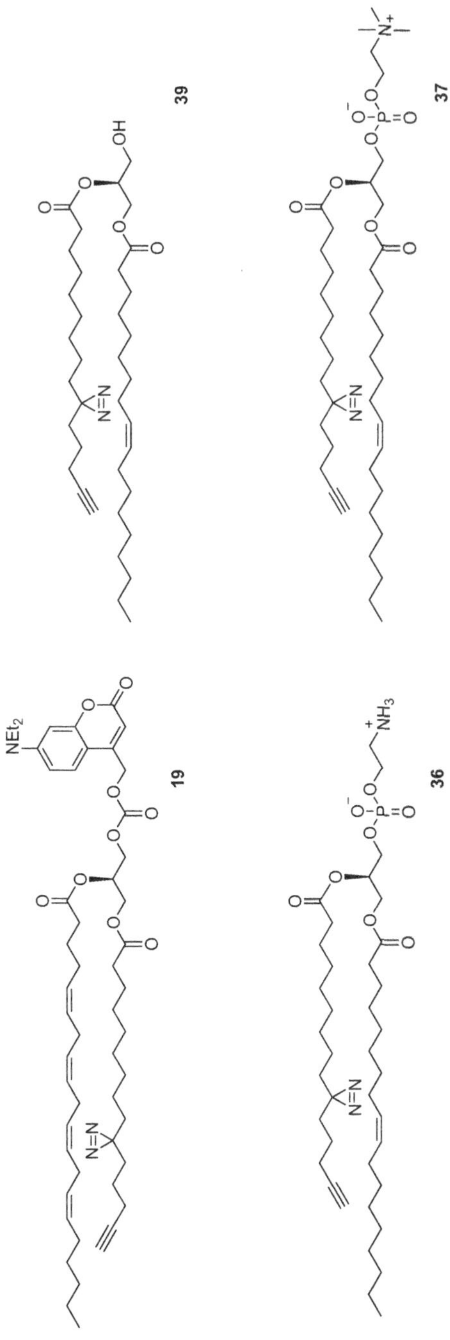

Fig. 6 Structures of photoaffinity probes based on diacylglycerol **19** and **39**, phosphatidylethanolamine **36**, and phosphatidylcholine **37**

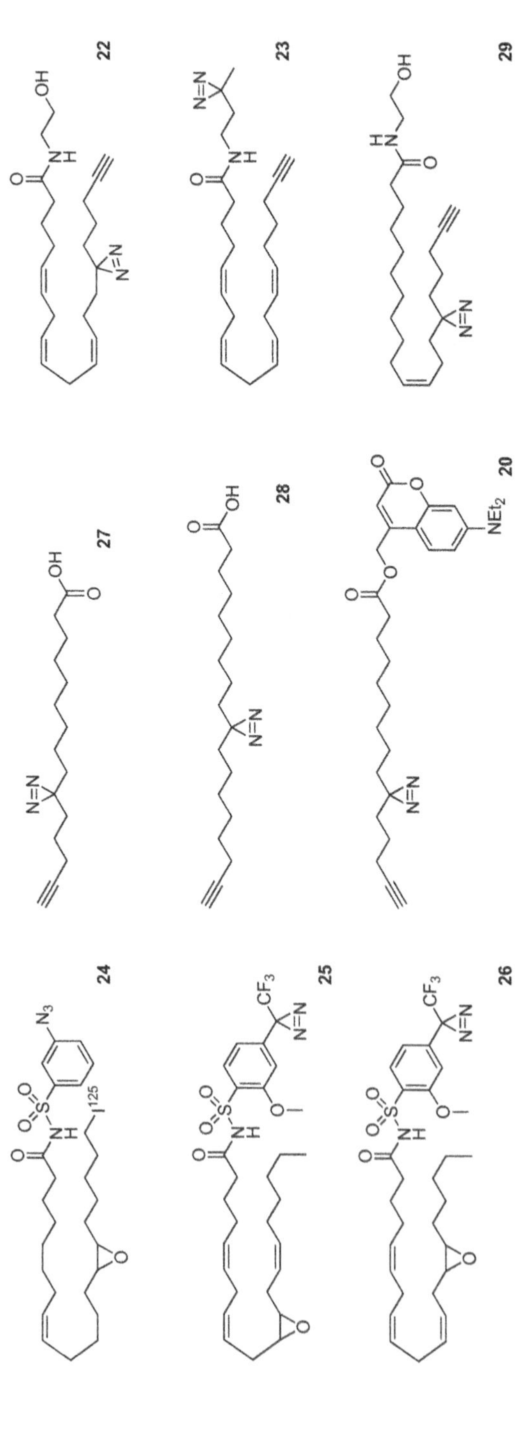

Fig. 7 Structures of photoaffinity probes based on 14,15-epoxyeicosatrienoic acid **24–26**, bifunctional lipid probes **27–28**, caged lipid probe **20**, and anandamide-based probes **22**, **23**, and **29**

P450 s, metabolize polyunsaturated fatty acids (i.e., arachidonic acid) into bioactive signaling molecules. For example, endothelium-derived epoxyeicosatrienoic acids (EETs) are lipid signaling molecules with various biological activities. In search of a high-affinity G protein-coupled EET receptor for 14,15-epoxyeicosatrienoic acid, epoxide-containing lipid **24** equipped with a photoactivatable aryl azide, and radioactive iodide as reporter group was synthesized (Fig. 7). The probe showed EET agonist activity and labeled a 47-kDa band which could be outcompeted with several EET agonists and antagonists (Chen et al. 2011).

The structurally related probes **25** and **26** have been used to study the binding mode of EETs in the soluble epoxide hydrolase (sEH) enzyme (Fig. 7). The stereoselectivity of the epoxide dictated the binding mode, but this has not been confirmed by testing the individual enantiomers of the probes (Lee et al. 2017).

Bifunctional probe **27** was synthesized to study in vivo protein–lipid interactions (Fig. 7) (Haberkant et al. 2013). A palmitic acid mimic was incorporated into different lipid classes and proteins. Proteomic analysis resulted in the identification of 214 lipid-interacting proteins. Ligation to a fluorophore visualized the probe-bound proteins in nematode larvae. Moreover, the probe was metabolically incorporated in 185 additional proteins, thereby highlighting the versatility of bifunctional probes in studying protein–lipid interactions (Haberkant et al. 2013).

Others also exploited this type of metabolic incorporation. For example, probe **28** was incorporated in S-palmitoylated membrane proteins (Fig. 7). The photoaffinity group allowed the capture protein–protein interactions of IFITM3, a protein with antiviral properties. This method was validated by studying VAPA—a known interaction partner of IFITM3—after which a pull-down experiment afforded 12 novel interaction partners of IFITM3 (Peng and Hang 2015).

The endocannabinoid anandamide is a signaling lipid involved in neurotransmission. To map the anandamide-binding proteins, photoaffinity probes **22** and **23** and two control probes based on oleoylethanolamide and palmitoylethanolamide were synthesized (Fig. 7). More than thousand interacting proteins were identified, including NUCB1, NENF, and VAT1. These probes were subsequently employed to discover ligands for the lipid-binding pockets of said targets using competitive fluorescence polarization assays (Niphakis et al. 2015). In addition, lipid probes **22**, **23,** and **29** were utilized to determine the selectivity profile of small molecules occupying lipid-binding pockets (Fig. 7). This strategy was referred to as lipid–protein interaction profiling (LiPIP) (Lum et al. 2017).

2.3 *Sphingolipids*

Sphingolipids are lipids that have sphingosine (or a derivative) as scaffold. Similar to glycerolipids, most of the sphingolipids exist in the form of a phosphate ester and a fatty acyl amide bond. Sphingomyelin, one of the most common sphingolipids, serves mainly a structural purpose, but its derivatives are increasingly recognized as important signaling molecules.

The last decade has witnessed a continuous progression in the development of chemical probes to identify sphingolipid-binding proteins. For example, Haberkant et al. synthesized probes **30** and **31**, which were rapidly incorporated into sphingolipids (Fig. 8). Caveolin-1 and nicastrin were identified as sphingolipid-interacting proteins (Haberkant et al. 2008). In 2010, photoactivatable sphingosine **32** was made and was co-incubated together with radioactive [^{3}H]choline in fibroblasts from healthy subjects or patients with Niemann-Pick A disease (Fig. 8) (Aureli et al. 2010). The storage of sphingomyelin was found to be disregulated in fibroblasts of the patients. In 2015, the photoaffinity-click sphingosine (pacSph) **33** was developed, which led to the identification of 186 pacSph-enriched proteins (Haberkant et al. 2016). As a control, fatty acid probe **27** was used. Although pacSph **27** and **33** have a different structure and are metabolized via separate pathways, substantial overlap between the interacting proteins was found (Figs. 7 and 8). It was suggested that different photoaffinity-click lipids might not be suitable as control compounds. Four potential issues should be taken into consideration: (i) Both lipid probes could be incorporated into phosphatidyl cholines (PCs), (ii) proteins may have two or more lipid-binding sites, (iii) a single lipid-binding site may be able to recognize multiple lipids, and (iv) alterations made on the lipids may alter their physiochemical properties (Haberkant et al. 2016). Nevertheless, pacSph **27** probe seems to be a more versatile tool to discover new sphingosine-binding proteins compared to **30** and **31**. Finally, functional uncaging of pacSph **21** afforded 64 pacSph-specific binding proteins, of which 14 were also found using uncaged pacSph (Höglinger et al. 2017).

Ceramide, a sphingosine containing a fatty acyl amide, is a signaling molecule with pro-apoptotic activity. In search of ceramide-binding proteins, a number of photoaffinity probes with pacFA as basis have been synthesized. The targets of ceramide probe (pacCer) **34** were compared to the interaction partners of glucosylceramide (pacGlcCer) **35** in cytosolic fractions of various cell lines. CERT, a protein with a StAR-related lipid-transfer domain was chosen as model protein to study structure–activity relationships of the lipid-binding pocket using probes **34**–**39** (Figs. 6 and 9) (Bockelmann et al. 2018).

2.4 Sterols

Sterols are lipids with rigid, fused rings with one or more hydroxyl groups, which gives them amphiphilic properties. Cholesterol is the most abundant sterol found in mammalian cells. Cholesterol alters the fluidity of the membrane and is a constituent of lipid rafts, which are liquid-ordered regions of the plasma membrane high in cholesterol and glycosphingolipids, sequestering specific proteins (Pike 2003). The lipid environment alters the biological properties of the embedded proteins, and the noncovalent interaction of sterols to proteins has had a surge of interest in recent years.

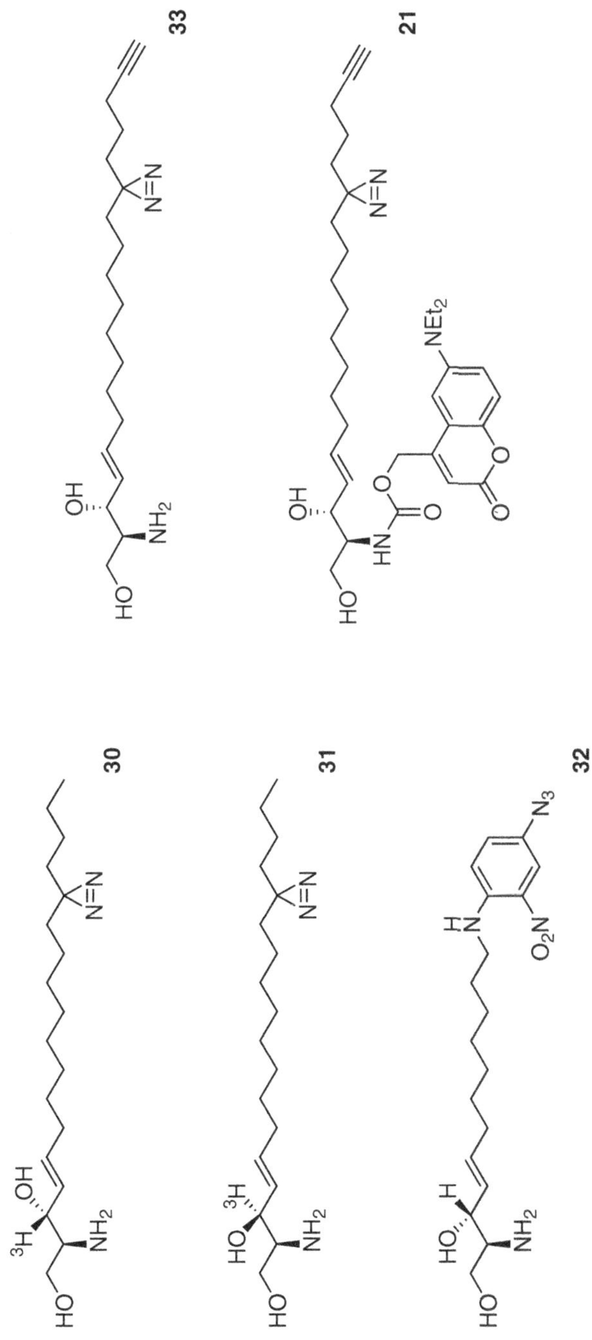

Fig. 8 Structures of photoaffinity probes based on sphingosine **30**, **31**, **32**, **33**, and **21**

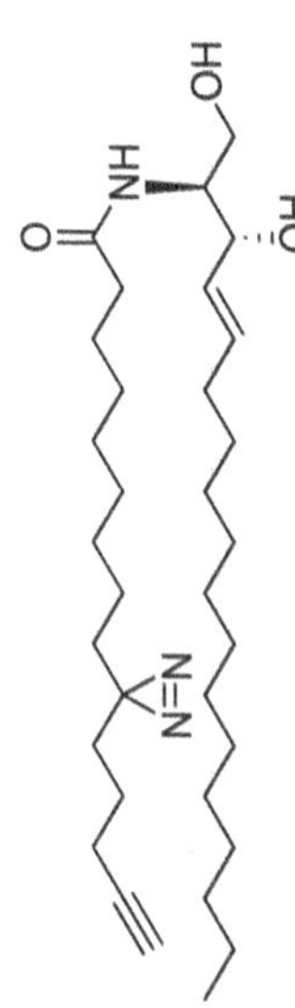

34

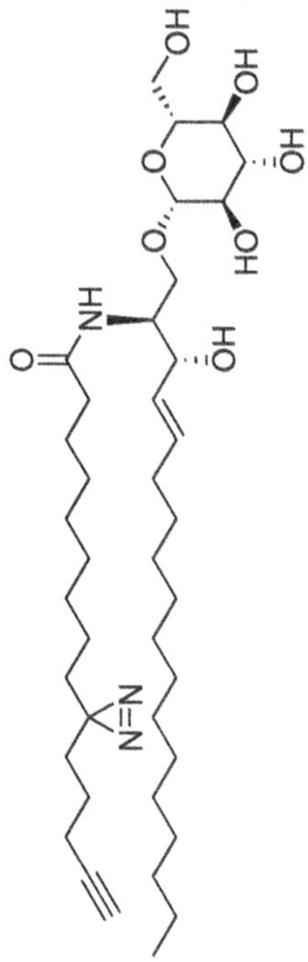

35

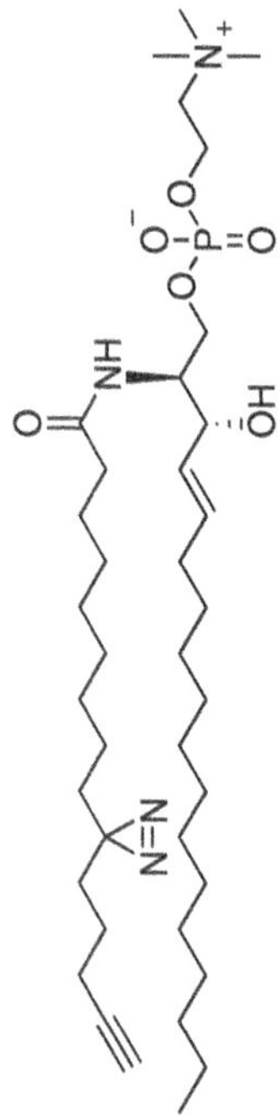

38

Fig. 9 Structures of photoaffinity based probes based on ceramide **34**, glucosylceramide **35**, and sphingomyelin **38**

Various types of cholesterol probes have been synthesized. Structures **40–42** are diastereomers and showed similar labeling patterns on gel (Fig. 10) (Hulce et al. 2013). In a pull-down experiment with trans-sterol probe **40** about 850 proteins were enriched using no UV and no probe as controls. Nearly 700 proteins showed a preference over a palmitoylethanolamide-based probe.

To map cholesterol binding sites in VDAC1, **43** and **44** were synthesized (Fig. 10). Purified recombinantly expressed mouse VDAC1 was used in a top-down and bottom-up proteomics analysis to map the binding pocket. The binding site was found to include Thr83 and Glu73 (Budelier et al. 2017).

To study the transfer of cholesterol between NPC1 and NPC2, the cholesterol derivative **45** has been synthesized to stabilize the protein transition state during the handoff (Fig. 10). Supported by previously reported cross-linked bile acids and modeling studies, the probe was supposed to stabilize protein dimer complexes. No follow-up studies have been reported to date (Byrd et al. 2015).

Bile acids are sterols that aid in dietary lipid digestion, but also act as signaling molecules that regulate lipid and glucose metabolism (Zhou and Hylemon 2014). Three probes based on the general structure of bile acids have been synthesized with the diazirine and alkyne positioned on different parts of the scaffold (Zhuang et al. 2017). Probes **46–48** incubated at 50 μM and competition with a twofold excess of competitor afforded 331 proteins that were labeled by all three structures, which provided evidence to assign them as bile acid interacting proteins (Fig. 11). Six known and unknown bile acid binding proteins were validated by overexpression, labeling with or without a competitor and immunoblotting to show probe- and UV-specific enrichment of these proteins. In view of their structural similarity to cholesterol probes **40–42,** the bile acid binding proteins were compared to the cholesterol targets. 146 proteins were shared by both lipid classes.

The development of betulinic acid-based probes has been reported, including photoactivatable probes **49** and **50** (Fig. 11) (Guo et al. 2017). Probe **49** was armed with a 2-aryl-5-carboxytetrazole, a recently developed photoactivatable linker with high crosslinking efficiency (Herner et al. 2016). A pull-down experiment performed with both probes afforded 150 proteins, which were subsequently triaged using control experiments. This afforded 9 and 13 unique proteins for structure **49** and **50**, respectively. The lack of overlap between the two probes was rationalized by the difference in location and reactivity of the photoactivatable group. Of note, most of the **50**-bound hits coincided with the results of Zhuang et al. 2017, indicating the reliability of these hits (Guo et al. 2017).

Two probes based on the sterol oleanolic acid have been synthesized (Zhang et al. 2012). To test functional similarity to the parent structure, they were tested in a RMPGa inhibition assay, where **51** showed a twofold and **52** a fivefold reduction of potency compared to oleanolic acid (Fig. 11). Probe **51** labeled two bands UV-dependently in soluble proteomes prepared from HepG2 cells which could be outcompeted with oleanolic acid (Zhang et al. 2012).

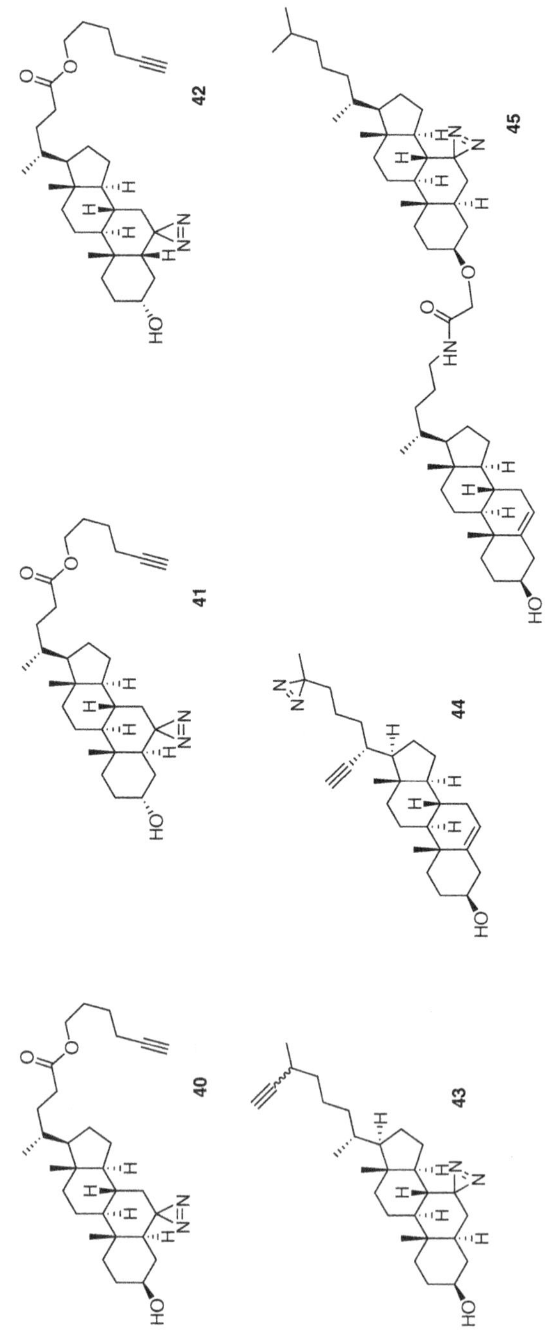

Fig. 10 Structures of photoaffinity probes based on cholesterol **40–45**

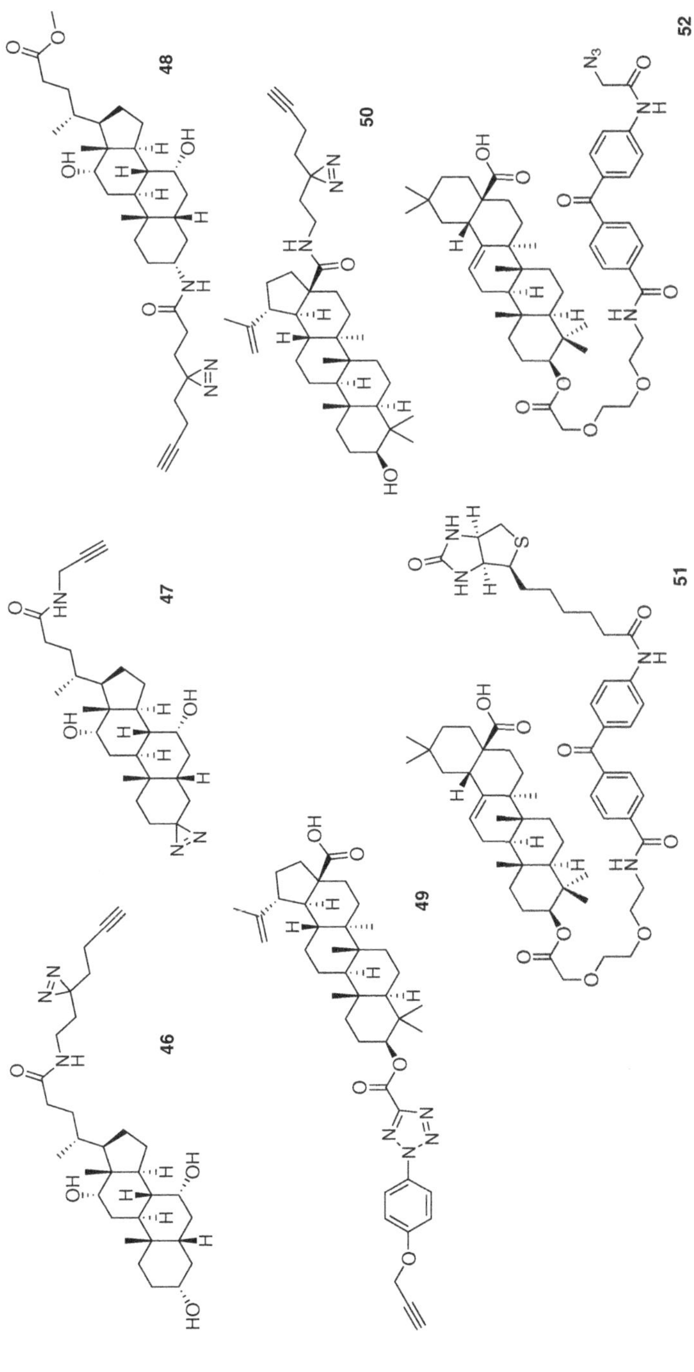

Fig. 11 Structures of photoaffinity probes based on bile acid **46–48**, on betulinic acid **49–50**, and on oleanolic acid **51–52**

2.5 Promiscuous Lipid-Binding Proteins

Chemical proteomic datasets contain an enormous amount of data, making it a challenge to distinguish real specific interacting partners from background proteins for a chemical probe. This necessitates the careful design of the experiment with negative controls. Due to a bias toward highly abundant proteins and potential co-purification of other proteins with probe targets, negative controls do not always cover the whole spectrum of background proteins. To combat this problem, twelve laboratories have combined the data of >300 negative control experiments to establish a database of common background proteins. This contaminant repository for affinity purification (CRAPome) is a useful tool to identify common background proteins (Mellacheruvu et al. 2013). In case of affinity-based protein profiling, one also needs to keep in mind the specific background labeling proteins associated with each individual photoreactive group (Kleiner et al. 2017). The CRAPome database can be used for analyzing (photoaffinity) proteomic datasets. However, it is confined to the background of the purification method (CRAPome) or photoreactive group (Kleiner et al. 2017). When conducting chemical proteomics with lipid probes, one also has to account for the nonspecific interactions due to the lipophilic character of the probes. A different set of background proteins could therefore be envisioned based on hydrophobic interactions with lipid probes.

To compile a database of promiscuous lipid-binding proteins, we have combined, compared, and ranked all the enriched target proteins of probes **8**, **14**, **15**, **19**, **21**, **22**, **23**, **27**, **29**, **33**, **40,** and **47**. For each probe, the criteria of the authors for assigning probe targets were used. Where possible, the proteins identified in multiple cell lines were used. This resulted in a list of 1367 distinct proteins of which 176 targets were found in ≥ 4 experiments (from a total of 11 experiments) (Fig. 12a). Of note, only 13 of these proteins were identified as background proteins

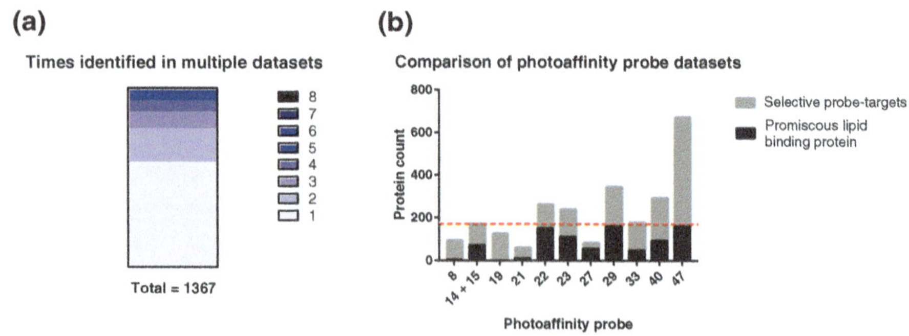

Fig. 12 **a** Overview of how many times a protein was targeted by different probes. In total, 1367 proteins were identified of which 176 (13%) were found in at least 4 of the 11 experiments. **b** Overview of the number of probe targets and promiscuous lipid-binding proteins identified by each probe. The red line is set at 176 and visualizes the maximum amount of promiscuous lipid-binding proteins

by the CRAPome (using >20% of total entries as cutoff criteria) (Mellacheruvu et al. 2013). An overview of the 176 most promiscuous lipid-binding proteins can be found in the supplementary. It was found that (caged) photoaffinity probes **8**, **19,** and **21** were highly selective (Fig. 12b) with the smallest number of common targets. The other probes interacted with targets also detected by different probes to some extent (Fig. 12b). Without diligent controls, such as competition with the endogenous ligand, caution would therefore be advised before assigning these targets as probe-specific proteins.

3 Lipid Activity-Based Probes

Activity-based protein profiling (ABPP) uses a chemical probe to covalently label and identify an enzyme or class of enzymes in a biological sample. Activity-based probes (ABPs) consist of a reactive group or warhead and a reporter tag. ABPs can further be divided into one-step and two-step probes based on the presence of a reporter tag during labeling or ligation of the tag after the labeling event using click chemistry, respectively. The scaffold of the probe combined with the reactive group determines the affinity and selectivity of the ABP.

The advantage of using ABPP over transcriptomics or whole-cell proteomics is the ability to quantify the amount of active proteins, whereas alternative methods do not discriminate between active or inactive protein forms. If a protein is inactive due to a post-translational modification (PTM) or blocked by a non-allosteric inhibitor, it will not react with the ABP, because the reactive group interacts with the active site of the enzyme.

Most of the groundwork for ABPP has been done in the field of serine hydrolases. This class of enzymes consists of more than 200 enzymes of which half are serine proteases and other half metabolic enzymes with an active site serine (Long and Cravatt 2011). The group of metabolic serine hydrolases can be divided into (thio)esterases, lipases, peptidases and thus can hydrolyze a wide variety of bonds (Simon and Cravatt 2010). Lipases hydrolyze triglycerides feeding liberated fatty acids back into the β-oxidative pathway. Next to their role in the energy household of cells, lipases also play a role in cellular signaling as some of their products, such as 2-Arachidonoylglycerol (2-AG) and anandamide are endogenous ligands of the cannabinoid receptors. The amidase FAAH hydrolyzes a variety of endocannabinoid lipid amides, thereby inactivating these lipid messengers. It is because of their wide variety in role and function that this protein class has gotten so much attention in the last years.

The first ABP for this class of enzymes was fluorophosphonate (FP)-biotin which is a one-step probe with a FP-warhead (Liu et al. 1999). Together with the FP-TAMRA probe **53**, they are routinely used in this field for the mapping of activity profiles of serine hydrolases in different animals or tissues or for the determination of selectivity profiles of inhibitors (Baggelaar et al. 2017; Zweerink et al. 2017; Van Esbroeck et al. 2017; Lentz et al. 2018; Rooden et al. 2018).

By tweaking the scaffold of traditional FP probes, subclass-selective probe **54** for serine phospholipases was made (Tully and Cravatt 2010). This demonstrates that the scaffold and reactivity of the warhead are key for affinity and selectivity of a probe (Fig. 13).

The FP probes enabled measurement of the activity of a wide range of serine hydrolases, but do not completely cover the entire family. Several complementary ABPs have therefore been developed. For example, MB064 (**55**) is based on the scaffold of tetrahydrolipstatin (THL), a promiscuous lipase inhibitor that also labeled DAGL-α, which is not targeted by FP-TAMRA (Baggelaar et al. 2013). Compared to FP-TAMRA (**53**), MB064 (**55**) showed a more limited labeling pattern, thereby complementing the serine hydrolase toolbox (Baggelaar et al. 2017). Another example is JW576 (**56**), an ABP selective for KIAA1363, which can also be used as an imaging biomarker (Fig. 13) (Chang et al. 2012).

Another group of selective serine hydrolase probes is the triazole urea probes DH379 (**57**) and HT-01 (**58**) (Hsu et al. 2012; Ogasawara et al. 2016). These probes show selectivity for DAGL-α and DAGL-β. HT-01 has been used in the field of immunology to study the regulatory role of DAGL-β in the inflammatory response of macrophages (Hsu et al. 2012). The last example in this group is the FAAH-selective probe PF-04457845yne (**59**) (Fig. 13) (Ahn et al. 2011). This probe was synthesized to study the off-targets of covalent FAAH inhibitor PF-04457845 directly. This ABP proved to be selective for FAAH in mouse brain and liver tissue, thus illustrating how ABPP can be used to profile the off-targets of covalent drug candidates.

More recently, two probes (**60** and **61**) for a related class of enzymes, the lysosomal cysteine hydrolases, have been published (Fig. 14) (Ouairy et al. 2015; Romeo et al. 2015). This group of hydrolases is also involved in the hydrolysis of lipids, but performs its activity via an active site cysteine instead of a serine. Both probes are activity-based and react with the active site cysteine. However, they use different warheads for this purpose. The carmofur-based probe **60** uses a 5-fluorouracil group and ARN14686 (**61**) uses a β-lactam. Probe **60** labeled acid ceramidase (AC) and ARN14686 (**61**) was shown to label both AC and N-acylethanolamine acid amidase (NAAA). Changing the ligation tag of **61** led to the development of norbornene-ABP **62** and an BODIPY-ABP **63** based on a N, O-substituted threonine-β-lactam (Petracca et al. 2017). Both probes react with NAAA and could be used to label NAAA directly (Fig. 14).

Another interesting group of ABPs is the probes **64** and **65** for beta-glucosidases, such as GBA that hydrolyze glucosylceramide (Fig. 14). ABP **64** was shown to be selective for GBA1 (Witte et al. 2010), whereas ABP **65** acted as a broad-spectrum probe for GBA1, GBA2, GBA3, and LPH (Kallemeijn et al. 2012). These probes are currently used to study Gaucher disease in which deficiency of GBA1 leads to accumulation of its substrate.

RPE65 or retinoid isomerohydrolase is essential in the visual cycle and converts all-*trans*-retinyl esters to 11-*cis*-retinol (Cai et al. 2009). LRAT transfers acyl groups from lecithin to all-*trans*-retinol yielding retinyl esters, which provides the substrate for RPE65 and is therefore also part of the visual cycle (Jahng et al. 2003b). ABPP has also made its entry into the field of the lipid vitamins.

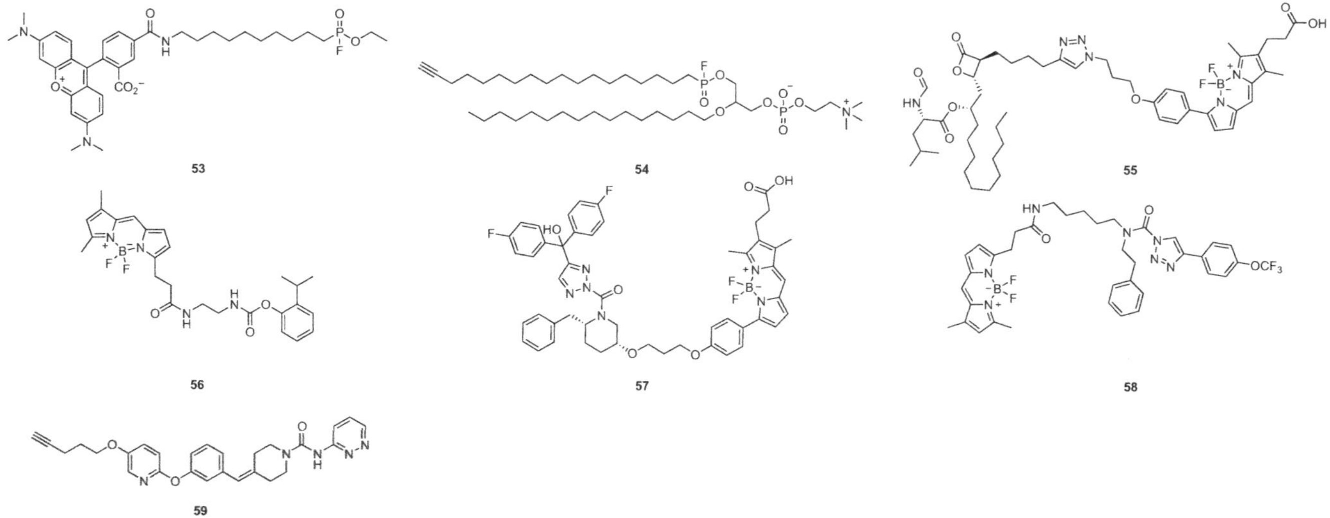

Fig. 13 Structures of activity-based probes for serine hydrolases. Broad-spectrum serine hydrolase FP probes **53–54**, THL-based probe MB064 **55**, KIAA1363 selective probe JW576 **56**, DAGL-α and DAGL-β selective probes DH379 **57** and HT-01 **58**, FAAH-selective probe PF-04457845yne **59**

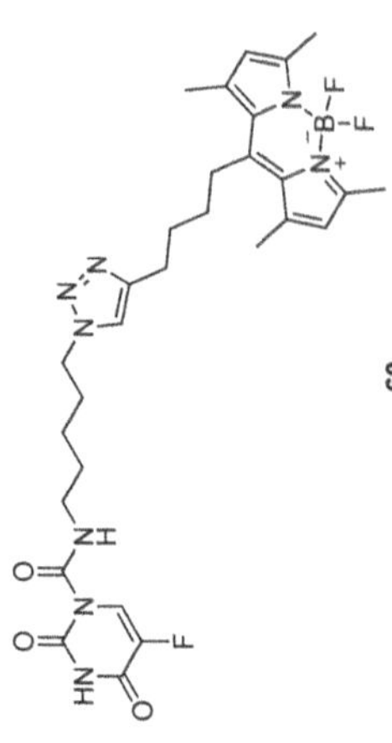

60

61

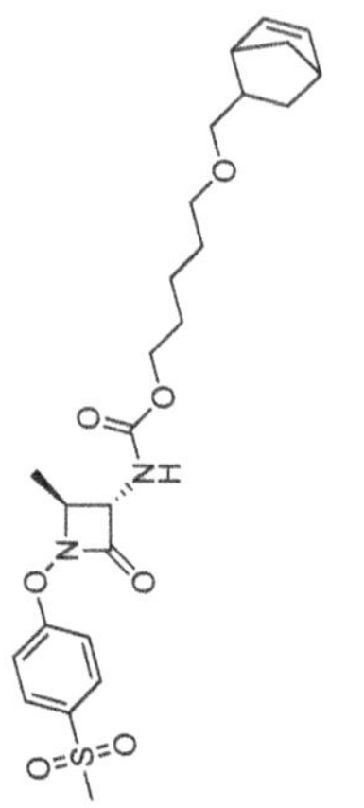

62

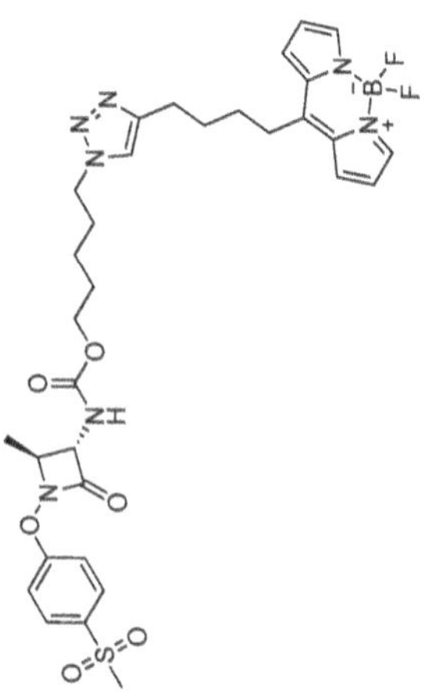

63

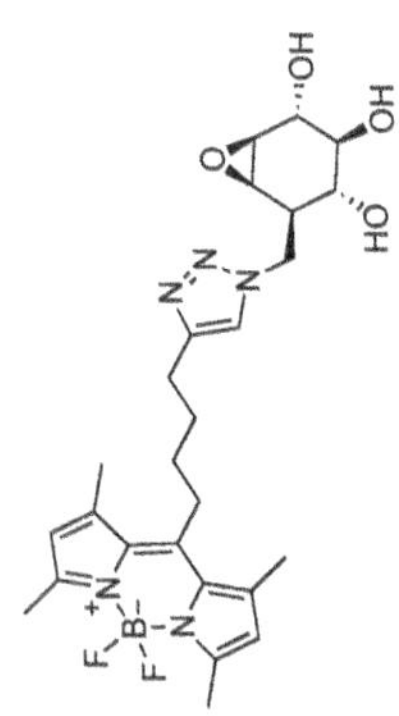

64

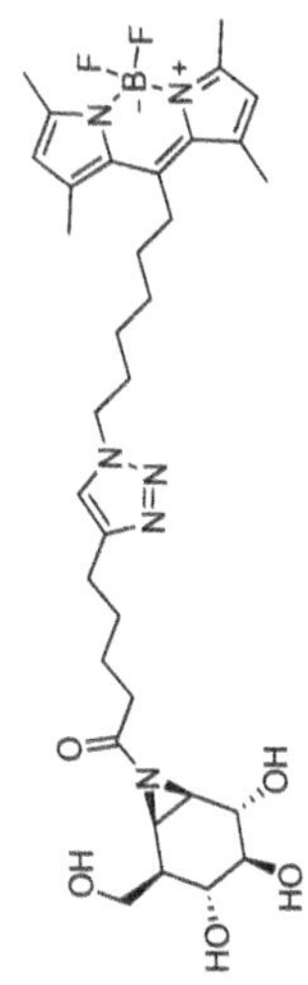

65

Fig. 14 Structures of activity-based probes for cysteine hydrolases **60–63** and beta-glucosidases **64–65**

An one-step probe all-*trans*-retinyl chloroacetate (RCA) (**66**) mimicking retinyl acetate was synthesized in 2002 containing a chloroacetate warhead and a cleavable biotin linker (Fig. 15) (Nesnas et al. 2002). This ABP **66** was shown to label RPE65 and lecithin retinol acyltransferase (LRAT) in retinal pigment epithelial membrane (Jahng et al. 2003a). Although this ABP **66** has proven to be a great tool for visualizing interactions of retinoids with enzymes involved in the visual cycle, it has so far found no application outside of the field.

Another lipid soluble vitamin, vitamin D_3, has a chemical ^{14}C labeled probe **67** visualized by phosphorimaging (Fig. 15). Though not activity-based, this probe **67** was able to covalently label the vitamin D-binding protein (DBP) by reacting with a bromoacetate to the tyrosine of the binding protein (Swamy et al. 2000a). A similar affinity probe based on the 1,25-dihydroxyvitamin D_3 equipped with a bromoacetate was shown to label the vitamin D nuclear receptor (VDR) (Swamy et al. 2000b).

Lipoxygenases oxygenate polyunsaturated fatty acids (PUFAs) and are essential metabolic enzymes in the formation of resolvins. N144 (**68**) is a recently developed ABP for 15-lipoxygenase-1 (15-LOX-1) (Eleftheriadis et al. 2016). N144 (**68**) mimics linoleic acid using a bis(alkyne)core to interact with the active site of the enzyme via its radical mechanism (Fig. 15). It also possesses a terminal alkene, which can be utilized as a handle in an oxidative Heck reaction with biotinylated phenylboronic acid to ligate a biotin reporter tag (Ourailidou et al. 2014).

Although not activity-based, arylfluorosulfate probes **69** have recently been shown to covalently label intracellular lipid-binding proteins (Fig. 15) (Chen et al. 2016). These probes have been shown to react with a reactive tyrosine inside the binding pocket of cellular retinoic acid binding proteins (CRABPs) and fatty acid binding proteins (FABPs). These arylfluorosulfates enable the visualization of these lipid trafficking proteins and their ligands without the need of probes with a photoaffinity group.

4 Opportunities

So far the majority of lipid chemical probes consist of photoaffinity probes. A general trend can be observed when surveying the published structures for lipid A*f*BPP from the last seven to eight years. Before that, many photoaffinity lipids consisted of benzophenone-modified lipids, lacking a reporter group. When they did contain a reporter group, it was usually a fluorophore or radioactive isotope and was exclusively used for the study on purified proteins or cell lysates (Xia and Peng 2013). These probes could be used to locate binding sites on known binders or show binding proteins on a gel. However, identifying novel proteins with these reporters (or lack thereof) is challenging.

In recent years, the radioactive reporters, bulky fluorophores and linker-attached biotin groups are being replaced with an alkyne, ameliorating the challenging synthesis of probes based on biomolecules. Besides its minimal impact on the

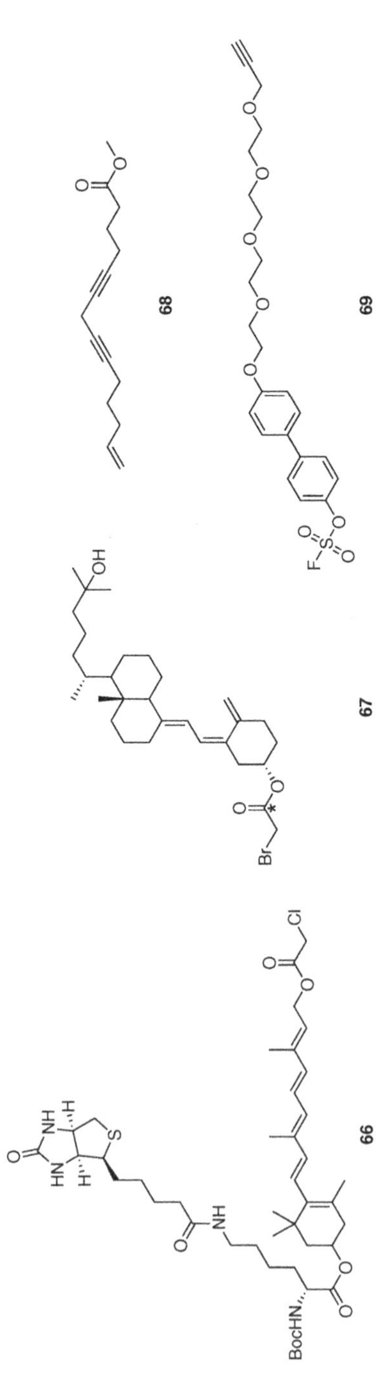

Fig. 15 Structures of activity-based probes based on retinyl acetate **66**, vitamin D **67** (*indicates ^{14}C), linoleic acid **68**, and arylfluorosulfonate probe **69**

structure, the alkyne group also offers increased flexibility in the experimental setup. With an increasing amount of click chemistry possibilities and a large library of commercially available azides, the same alkyne photoaffinity probe can be used for detailed binding studies and global A*f*BPP.

An interesting combination of a global and detailed study is described for lipid probes **22** and **23**, in which a probe target and its binding site were discovered in living cells without any protein purification or overexpression, using the isoTOP-ABPP platform and a isotopically labeled cleavable TEV linker. This double enrichment method, first probe-bound protein enrichment followed by probe-bound peptide enrichment, allows the discovery of probe targets and binding sites in a complex mixture (Weerapana et al. 2007; Niphakis et al. 2015). Techniques like isoTOP-ABPP becoming more feasible will be a major driving force in the applicability of lipid probes.

Protein–protein interactions can be mediated through lipid PTMs. Introduction of both an alkyne and diazirine seems to moderately affect the capability of a lipid to be recognized by its interacting proteins and incorporated as a PTM. This strategy has been used by several groups to discover lipidated proteins and their interaction partners (Peng and Hang 2015; Wright et al. 2015). As lipidation is a common PTM, this approach should be applicable to lipids other than simple fatty acids, most notably the class of prenyl lipids.

Challenges for lipid probes include their metabolism and high lipophilicity. Membrane lipid composition is tightly regulated, and any imbalance is quickly rectified, complicating the identification of the exact lipid species by the tagged protein (Haberkant et al. 2016). Making metabolically inactive, 'caged' derivatives are suggested to ameliorate this problem, allowing the lipid probe to distribute before releasing the bioactive lipid in a controlled manner (Höglinger et al. 2017). Moreover, the metabolism can also be exploited, as demonstrated by Aureli et al. and Wang et al. by incubating the cells with a reporter group and photoactivatable group installed on different parts of a bioactive molecule. Only the assembled combination of the two parts will function as a probe (Aureli et al. 2010; Wang et al. 2017). Of note, different probes based on the same lipid have overlapping, but distinct labeling profiles. To increase the reliability of a photoaffinity project, multiple active and preferably even inactive variants should, therefore, be made (Arrowsmith et al. 2015). Moreover, it suggested that probes should be outcompeted by their parent lipid to increase the amount of specific binders in a probe target dataset. However, a-specific lipophilic interactions often prevail and cannot be outcompeted (Hulce et al. 2013; Zhuang et al. 2017; Bockelmann et al. 2018).

A-specific binding is also an inherent issue with photoactivatable groups, which lead Sieber and colleagues to make an inventory of common off-targets bound by aryl azides, benzophenones, and most thoroughly diazirines. Next to the CRAPome, this database should be consulted when performing photoaffinity pull-down experiments (Mellacheruvu et al. 2013; Kleiner et al. 2017). Moreover, the data in Chapter 2.5 from analysis performed on the targets of over ten different lipid photoaffinity probes can also be used to recognize promiscuous lipid binders.

As noted above, activity-based lipid probes are relatively under-represented, thereby providing great opportunities in the field of chemical biology. Particularly in the field of immunology, where lipid signaling molecules are increasingly recognized as important regulators of the immune response, there is ample room for novel activity-based lipid probes.

4.1 Endocannabinoids

The serine hydrolase ABPs discussed in Sect. 3 have been widely used to study the endocannabinoid system (ECS) and the effect of its lipid messengers on the central nervous system (CNS). The main receptors for these lipids are the cannabinoid receptor type 1 and type 2 (CB_1 and CB_2 receptor respectively). While CB_1 is mainly expressed within the central nervous system, CB_2 is mainly expressed in immune cells and tissue (Cabral and Griffin-Thomas 2009). Since the discovery of the CB_2 receptor, a lot of effort has gone into determining its expression level in different immune cell types (Turcotte et al. 2016). This work is done either by looking at the mRNA levels of the receptor or by western blotting with specific antibodies. Both of these methods, however, have drawbacks. While mRNA levels do intuitively correlate with protein levels, it gives no information about the actual activity of the receptor. Protein degradation, PTMs, and protein–protein interactions are important factors in regulating the amount of active enzyme. The presence of protein mRNA does therefore not predictably correlate with the amount and activity of an enzyme. While correlating with the amount of protein, antibodies on the other hand can have cross-reactivity problems (Weller 2016). Without the necessary quality controls and with different groups using different antibodies, it is a difficult task to determine which results are significant and reproducible. Recently, it has been shown that photoaffinity probes can also be used as an alternative to antibodies (Soethoudt et al. 2018).

With endocannabinoids being implicated in more and more immune diseases, the field of immunology is a clear application area for existing and future lipid chemical probes (Basu and Dittel 2011; Pacher and Mechoulam 2011; Chiurchiù et al. 2015, 2018). These ABPs would enable the visualization of levels of active enzyme in immune cells as has been shown with the HT-01 probe **58** for DAGL-β in macrophages (Hsu et al. 2012). Coupled with cell sorting techniques, the effect of different kinds of stimulation on the ECS can be analyzed. This technique could also be used to make a comparison between tissue/cells in healthy and disease states.

4.2 Steroid Hormones

The steroid hormones can be divided into two categories: sex steroids and corticosteroids. They are all derived from cholesterol and are fat-soluble.

These hormones act as lipid messengers and interact with nuclear steroid receptors influencing gene expression. More recently, they have also been implicated to act via different, faster pathways localized in the cytoplasm (Norman et al. 2004). Due to the sexual dimorphism in the immune system, sex steroids have long been implicated to affect the immune system. Testosterone seems to suppress the immune system leading to a lowered immune response to infections and vaccines in men (Furman 2015; Trigunaite and Dimo 2015). Females seem to have a more robust immune response, but are more susceptible to autoimmune diseases possible due to the modulatory effects of estrogen (Cunningham and Gilkeson 2011; Waldmann et al. 2016). Most research in this field has focused on administering steroid hormones, removing hormone-producing organs or protein knockouts (Homo-Delarche et al. 1991; Lai et al. 2012). We therefore envision potential for steroid hormones based lipid probes. For example, no photoaffinity probes based on the scaffold of its steroid hormone derivatives and no activity-based probes have been published so far. These probes would enable investigation of steroid hormone interactions while keeping the biological system in its natural state. They could also reveal previously unknown hormone–protein interactions.

4.3 Lipid Soluble Vitamins

The lipid soluble vitamins, vitamin A and D, exercise their effect via the same mechanism as the steroid hormones by binding to their nuclear receptors, the retinoic acid receptor (RAR) and vitamin D receptor (VDR), respectively (Carlberg 1999). Vitamin D, synthesized in the skin under influence of light, is a secosteroid and therefore also structurally resembles the steroid hormones. Both lipid vitamins have been implicated to affect the immune system (Mora et al. 2008). A radioactive photoaffinity probe and a radioactive affinity-based probe for Vitamin D have been synthesized and applied (Ray et al. 1991; Swamy et al. 2000a). These probes may be replaced by novel ABPs with a diazirine as a less bulky photoactive group and addition of a click-handle to enable proteomics studies. An activity-based probe to study the metabolism of the active metabolite of vitamin D throughout the immune system would be a new addition to this toolbox.

In the case of vitamin A, direct photoaffinity labeling with radioactive retinoic acid was enabled by its inherent photoreactive characteristics (Bernstein et al. 1995; Chen and Radominska-Pandya 2000). A radioactive photoaffinity probe based on retinoic acid has been synthesized, but A*f*BPs has not been performed due to lack of a ligation handles for application in chemical proteomics studies (Shimazawa et al. 1991). The activity-based probe for retinyl ester processing enzymes has not yet been tested outside of the retinal pigment epithelial membrane (Jahng et al. 2003a). The application of this ABP in the field of immunology would be interesting. Furthermore, activity-based probes based on retinol, retinal and the immunologically active metabolite, retinoic acid, would be valuable additions to this field.

The other lipid soluble vitamins, such as vitamin E and K, have not been studied using ABPs. Vitamin E has been associated with T-cell differentiation and decreased cellular immunity in aging (Moriguchi 1998; Moriguchi and Kaneyasu 2003). Next to its antioxidant activity, it has also been implicated as a potential orphan nuclear receptor ligand (Soontjens et al. 1996; Carlberg 1999). The synthesis of a vitamin E derivative containing a diazirine and an alkyne click-handle would therefore be an important chemical tool to unravel its protein interactions and mechanism of action. Such a photoaffinity probe could help to find the proposed nuclear receptor or discover other cellular pathways involved in Vitamin K biology.

5 Conclusions

Proteomics using chemical probes is an invaluable strategy to study the biology of lipid messengers. A*f*BPP using photoaffinity probes provides insight into the target interaction landscape of lipid messengers with previously unknown proteins, such as transporters and receptors, while ABPP using activity-based probes identifies the enzymes that control the metabolism of these important messengers in health and disease. We believe that the combination of A*f*BPP and ABPP is a powerful approach to obtain a global and detailed view of the biological processes mediated by lipid signaling molecules. The chemical probes, however, do not cover many lipid classes yet. A potential reason could be that most long-chain, polyunsaturated lipids and their metabolites are challenging to synthesize and have an inherent instability. Although chemists are still working on the total synthesis of these low abundant biologically active lipids and their probes (Ogawa et al. 2017; Rodriguez and Spur 2017), there is still a need for further elucidation of important cellular processes performed by lipids, especially in the field of immunology. So we call on chemists, biologists, and immunologists to combine their expertise to tackle these topics and become chemical immunologists.

Supplementary

List of proteins identified as promiscuous lipid-binding proteins. The count shows the amount of experiments in which the protein was identified as a probe target. Proteins colored red are the proteins flagged by the CRAPome database.

Accession	Protein	Count	Accession	Protein	Count	Accession	Protein	Count	Accession	Protein	Count
Q96A33	CCD47_HUMAN	8	Q96LJ7	DHRS1_HUMAN	5	P11021	BIP_HUMAN	4	Q14108	SCRB2_HUMAN	4
P43307	SSRA_HUMAN	8	Q9Y394	DHRS7_HUMAN	5	Q9UBD9	CLCF1_HUMAN	4	Q9UGP8	SEC63_HUMAN	4
Q9BV23	ABHD6_HUMAN	7	Q9UBX3	DIC_HUMAN	5	Q6UW02	CP20A_HUMAN	4	Q15005	SPCS2_HUMAN	4
P51648	AL3A2_HUMAN	7	O75477	ERLN1_HUMAN	5	O15121	DEGS1_HUMAN	4	Q9NQZ5	STAR7_HUMAN	4
P51572	BAP31_HUMAN	7	O00461	GOLI4_HUMAN	5	Q96KC8	DNJC1_HUMAN	4	Q9UJZ1	STML2_HUMAN	4
Q16850	CP51A_HUMAN	7	Q92643	GPI8_HUMAN	5	Q8N766	EMC1_HUMAN	4	P46977	STT3A_HUMAN	4
P50402	EMD_HUMAN	7	Q3SXM5	HSDL1_HUMAN	5	P0DO92	ENOL_HUMAN	4	Q8TCJ2	STT3B_HUMAN	4
Q9NRY5	F1142_HUMAN	7	P07099	HYEP_HUMAN	5	P24390	ERD21_HUMAN	4	Q9UH99	SUN2_HUMAN	4
O75844	FACE1_HUMAN	7	Q06136	KDSR_HUMAN	5	Q14534	ERG1_HUMAN	4	Q9NZ01	TECR_HUMAN	4
Q96AG4	LRC59_HUMAN	7	Q86SY8	KTAS1_HUMAN	5	Q9BSJ8	ESYT1_HUMAN	4	Q8WUY1	THEM6_HUMAN	4
Q86UE4	LYRIC_HUMAN	7	O95202	LETM1_HUMAN	5	A0FGR8	ESYT2_HUMAN	4	O60830	TI17B_HUMAN	4
O00264	PGRC1_HUMAN	7	Q9NZJ7	MTCH1_HUMAN	5	Q96A26	F162A_HUMAN	4	Q9HC07	TM165_HUMAN	4
Q9Y512	SAM50_HUMAN	7	Q969V3	NCLN_HUMAN	5	Q8WVX9	FACR1_HUMAN	4	Q13445	TMED1_HUMAN	4
Q9UBV2	SE1L1_HUMAN	7	Q15738	NSDHL_HUMAN	5	P37268	FDFT_HUMAN	4	Q9Y3B3	TMED7_HUMAN	4
Q99623	PHB2_HUMAN	7	Q02818	NUCB1_HUMAN	5	Q9P035	HACD3_HUMAN	4	O96008	TOM40_HUMAN	4
P53365	ARFP2_HUMAN	6	Q9Y5Y5	PEX16_HUMAN	5	Q9NRV9	HEBP1_HUMAN	4	Q9H4I3	TRABD_HUMAN	4
Q9HD20	AT131_HUMAN	6	Q96S52	PIGS_HUMAN	5	P22830	HEMH_HUMAN	4	P53007	TXTP_HUMAN	4
Q96G23	CERS2_HUMAN	6	P50897	PPT1_HUMAN	5	O60725	ICMT_HUMAN	4	O95292	VAPB_HUMAN	4
Q9NZ45	CISD1_HUMAN	6	P61026	RAB10_HUMAN	5	Q8TCB0	IFI44_HUMAN	4	Q96GC9	VMP1_HUMAN	4
Q07065	CKAP4_HUMAN	6	P04844	RPN2_HUMAN	5	Q70UQ0	IKIP_HUMAN	4	Q93050	VPP1_HUMAN	4
O96005	CLPT1_HUMAN	6	O95197	RTN3_HUMAN	5	Q8N5M9	JAGN1_HUMAN	4	Q5BJH7	YIF1B_HUMAN	4
Q53GQ0	DHB12_HUMAN	6	Q9NQC3	RTN4_HUMAN	5	Q9BZL6	KPCD2_HUMAN	4	Q8TAD4	ZNT5_HUMAN	4
Q15392	DHC24_HUMAN	6	Q8NBX0	SCPDL_HUMAN	5	Q14739	LBR_HUMAN	4	P05141	ADT2_HUMAN	4
P40939	ECHA_HUMAN	6	Q8IWL2	SFTA1_HUMAN	5	Q02978	M2OM_HUMAN	4	P25705	ATPA_HUMAN	4
Q8TCT9	HM13_HUMAN	6	O95470	SGPL1_HUMAN	5	Q6P1A2	MBOA5_HUMAN	4	P06576	ATPB_HUMAN	4
P30519	HMOX2_HUMAN	6	O15260	SURF4_HUMAN	5	Q4ZIN3	MBRL_HUMAN	4	P14625	ENPL_HUMAN	4
P42166	LAP2A_HUMAN	6	O14925	TIM23_HUMAN	5	O43772	MCAT_HUMAN	4	P20700	LMNB1_HUMAN	4
Q9Y6C9	MTCH2_HUMAN	6	O43615	TIM44_HUMAN	5	Q9H2D1	MFTC_HUMAN	4	P07237	PDIA1_HUMAN	4
Q9Y6Q9	NCOA3_HUMAN	6	Q9NX00	TM160_HUMAN	5	Q16891	MIC60_HUMAN	4	P08670	VIME_HUMAN	4
Q9NX40	OCAD1_HUMAN	6	Q6NUQ4	TM214_HUMAN	5	Q10713	MPPA_HUMAN	4			
Q9UHG3	PCYOX_HUMAN	6	Q15363	TMED2_HUMAN	5	Q9BTX1	NDC1_HUMAN	4			
O15173	PGRC2_HUMAN	6	P49755	TMEDA_HUMAN	5	Q9Y639	NPTN_HUMAN	4			
P35232	PHB_HUMAN	6	P57088	TMM33_HUMAN	5	P80303	NUCB2_HUMAN	4			
P51571	SSRD_HUMAN	6	Q9BTV4	TMM43_HUMAN	5	Q8NFH5	NUP53_HUMAN	4			
Q9Y4P3	TBL2_HUMAN	6	Q9H3N1	TMX1_HUMAN	5	O60313	OPA1_HUMAN	4			
P21796	VDAC1_HUMAN	6	Q9Y320	TMX2_HUMAN	5	Q9H7Z7	PGES2_HUMAN	4			
Q8N0U8	VKORL_HUMAN	6	Q9NS69	TOM22_HUMAN	5	Q9H490	PIGU_HUMAN	4			
P16615	AT2A2_HUMAN	6	O14656	TOR1A_HUMAN	5	Q8TEM1	PO210_HUMAN	4			
P27824	CALX_HUMAN	6	P45880	VDAC2_HUMAN	5	P18031	PTN1_HUMAN	4			
P04843	RPN1_HUMAN	6	O95831	AIFM1_HUMAN	5	P62820	RAB1A_HUMAN	4			
P28288	ABCD3_HUMAN	5	Q3ZCQ8	TIM50_HUMAN	5	Q9H0U4	RAB1B_HUMAN	4			
Q9BRR6	ADPGK_HUMAN	5	Q9NRG9	AAAS_HUMAN	4	P62491	RB11A_HUMAN	4			
P24539	AT5F1_HUMAN	5	Q8WTS1	ABHD5_HUMAN	4	Q8TC12	RDH11_HUMAN	4			
Q8WY22	BRI3B_HUMAN	5	O95870	ABHGA_HUMAN	4	Q9NTJ5	SAC1_HUMAN	4			
P07339	CATD_HUMAN	5	O95573	ACSL3_HUMAN	4	Q9NR31	SAR1A_HUMAN	4			
O75746	CMC1_HUMAN	5	Q8NHH9	ATLA2_HUMAN	4	Q9Y6B6	SAR1B_HUMAN	4			
P23786	CPT2_HUMAN	5	Q6DD88	ATLA3_HUMAN	4	P67812	SC11A_HUMAN	4			
Q8WVC6	DCAKD_HUMAN	5	Q9UHQ4	BAP29_HUMAN	4	O43819	SCO2_HUMAN	4			
Q9BUN8	DERL1_HUMAN	5	P35613	BASI_HUMAN	4	Q8WTV0	SCRB1_HUMAN	4			

References

Abe M, Nakano M, Kosaka A, Miyoshi H (2015) Syntheses of photoreactive cardiolipins for a photoaffinity labeling study. Tetrahedron Lett 56:2258–2261. https://doi.org/10.1016/j.tetlet.2015.03.056

Ahn K, Smith SE, Liimatta MB et al (2011) Mechanistic and pharmacological characterization of PF-04457845: a highly potent and selective fatty acid amide hydrolase inhibitor that reduces inflammatory and noninflammatory pain. J Pharmacol Exp Ther 338:114–124. https://doi.org/10.1124/jpet.111.180257

Arrowsmith CH, Audia JE, Austin C et al (2015) The promise and peril of chemical probes. Nat Chem Biol 11:536–541

Aureli M, Prioni S, Mauri L et al (2010) Photoactivable sphingosine as a tool to study membrane microenvironments in cultured cells. J Lipid Res 51:798–808. https://doi.org/10.1194/jlr.M001974

Baggelaar MP, Janssen FJ, van Esbroeck ACM et al (2013) Development of an activity-based probe and in silico design reveal highly selective inhibitors for diacylglycerol lipase-α in brain. Angew Chemie Int Ed 52:12081–12085. https://doi.org/10.1002/anie.201306295

Baggelaar MP, Van Esbroeck ACM, Van Rooden EJ et al (2017) Chemical proteomics maps brain region specific activity of endocannabinoid hydrolases. ACS Chem Biol 12:852–861. https://doi.org/10.1021/acschembio.6b01052

Bandyopadhyay S, Bong D (2011) Synthesis of trifunctional phosphatidylserine probes for identification of lipid-binding proteins. European J Org Chem 751–758. https://doi.org/10.1002/ejoc.201001264

Basu S, Dittel BN (2011) Unraveling the complexities of cannabinoid receptor 2 (CB2) immune regulation in health and disease. Immunol Res 51:26–38. https://doi.org/10.1007/s12026-011-8210-5

Bernstein PS, Choi S-Y, Ho Y-C, Rando RR (1995) Photoaffinity labeling of retinoic acid-binding proteins. Biochemistry 92:654–658

Bockelmann S, Mina JGM, Korneev S, et al (2018) A search for ceramide binding proteins using bifunctional lipid analogs yields CERT-related protein StarD7. J Lipid Res. https://doi.org/10.1194/jlr.m082354 (jlr.M082354)

Budelier MM, Cheng WWL, Bergdoll L et al (2017) Photoaffinity labeling with cholesterol analogues precisely maps a cholesterol-binding site in voltage-dependent anion channel-1. J Biol Chem 292:9294–9304. https://doi.org/10.1074/jbc.M116.773069

Bush JT, Walport LJ, McGouran JF et al (2013) The Ugi four-component reaction enables expedient synthesis and comparison of photoaffinity probes. Chem Sci 4:4115. https://doi.org/10.1039/c3sc51708j

Byrd KM, Arieno MD, Kennelly ME et al (2015) Design and synthesis of a crosslinker for studying intracellular steroid trafficking pathways. Bioorganic Med Chem 23:3843–3851. https://doi.org/10.1016/j.bmc.2015.03.053

Cabral GA, Griffin-Thomas L (2009) Emerging role of the cannabinoid receptor CB2 in immune regulation: therapeutic prospects for neuroinflammation. Expert Rev Mol Med 11:e3. https://doi.org/10.1017/S1462399409000957

Cai X, Conley SM, Naash MI (2009) RPE65: Role in the visual cycle, human retinal disease, and gene therapy. Ophthalmic Genet 30:57–62

Carlberg C (1999) Lipid soluble vitamins in gene regulation. In: BioFactors, IOS Press, Oxford, pp 91–97

Chang JW, Moellering RE, Cravatt BF (2012) An activity-based imaging probe for the integral membrane hydrolase KIAA1363. Angew Chemie Int Ed 51:966–970. https://doi.org/10.1002/anie.201107236

Chen G, Radominska-Pandya A (2000) Direct photoaffinity labeling of cellular retinoic acid-binding protein I (CRABP-I) with all-trans-retinoic acid: identification of amino acids in the ligand binding site. Biochemistry 39:12568–12574. https://doi.org/10.1021/bi000321n

Chen Y, Falck JR, Manthati VL et al (2011) 20-Iodo-14,15-epoxyeicosa-8(Z)-enoyl-3-azidophenylsulfonamide: photoaffinity labeling of a 14,15-epoxyeicosatrienoic acid receptor. Biochemistry 50:3840–3848. https://doi.org/10.1021/bi102070w

Chen W, Dong J, Plate L et al (2016) Arylfluorosulfates inactivate intracellular lipid binding protein(s) through chemoselective SuFEx reaction with a binding site Tyr residue. J Am Chem Soc 138:7353–7364. https://doi.org/10.1021/jacs.6b02960

Chiurchiù V, Battistini L, Maccarrone M (2015) Endocannabinoid signalling in innate and adaptive immunity. Immunology 144:352–364. https://doi.org/10.1111/imm.12441

Chiurchiù V, van der Stelt M, Centonze D, Maccarrone M (2018) The endocannabinoid system and its therapeutic exploitation in multiple sclerosis: clues for other neuroinflammatory diseases. Prog Neurobiol 160:82–100. https://doi.org/10.1016/j.pneurobio.2017.10.007

Cravatt BF, Wright AT, Kozarich JW (2008) Activity-based protein profiling: from enzyme chemistry to proteomic chemistry. Annu Rev Biochem 77:383–414. https://doi.org/10.1146/annurev.biochem.75.101304.124125

Cunningham M, Gilkeson G (2011) Estrogen receptors in immunity and autoimmunity. Clin Rev Allergy Immunol 40:66–73. https://doi.org/10.1007/s12016-010-8203-5
Eleftheriadis N, Thee SA, Zwinderman MRH et al (2016) Activity-based probes for 15-lipoxygenase-1. Angew Chemie Int Ed 55:12300–12305. https://doi.org/10.1002/anie.201606876
Fahy E, Subramaniam S, Murphy RC et al (2009) Update of the LIPID MAPS comprehensive classification system for lipids. J Lipid Res 50(Suppl):S9–S14. https://doi.org/10.1194/jlr.R800095-JLR200
Fahy E, Cotter D, Sud M, Subramaniam S (2011) Lipid classification, structures and tools. Biochim Biophys Acta 1811:637–647. https://doi.org/10.1016/j.bbalip.2011.06.009
Furman D (2015) Sexual dimorphism in immunity: improving our understanding of vaccine immune responses in men. Expert Rev Vaccines 14:461–471
Gaebler A, Milan R, Straub L et al (2013) Alkyne lipids as substrates for click chemistry-based in vitro enzymatic assays. J Lipid Res 54:2282–2290. https://doi.org/10.1194/jlr.D038653
Gaebler A, Penno A, Kuerschner L, Thiele C (2016) A highly sensitive protocol for microscopy of alkyne lipids and fluorescently tagged or immunostained proteins. J Lipid Res 57:1934–1947. https://doi.org/10.1194/jlr.D070565
Gu X, Huang Y, Levison BS et al (2016) Identification of critical paraoxonase 1 residues involved in high density lipoprotein interaction. J Biol Chem 291:1890–1904. https://doi.org/10.1074/jbc.M115.678334
Gubbens J, Ruijter E, de Fays LEV et al (2009) Photocrosslinking and click chemistry enable the specific detection of proteins interacting with phospholipids at the membrane interface. Chem Biol 16:3–14. https://doi.org/10.1016/j.chembiol.2008.11.009
Guo H, Xu J, Hao P et al (2017) Competitive affinity-based proteome profiling and imaging to reveal potential cellular targets of betulinic acid. Chem Commun 53:9620–9623. https://doi.org/10.1039/C7CC04190J
Haberkant P, Schmitt O, Contreras F-X et al (2008) Protein-sphingolipid interactions within cellular membranes. J Lipid Res 49:251–262. https://doi.org/10.1194/jlr.D700023-JLR200
Haberkant P, Raijmakers R, Wildwater M et al (2013) In vivo profiling and visualization of cellular protein-lipid interactions using bifunctional fatty acids. Angew Chemie Int Ed 52:4033–4038. https://doi.org/10.1002/anie.201210178
Haberkant P, Stein F, Höglinger D et al (2016) Bifunctional sphingosine for cell-based analysis of protein-sphingolipid interactions. ACS Chem Biol 11:222–230. https://doi.org/10.1021/acschembio.5b00810
Hein JE, Fokin VV (2010) Copper-catalyzed azide-alkyne cycloaddition (CuAAC) and beyond: new reactivity of copper(I) acetylides. Chem Soc Rev 39:1302–1315
Herner A, Marjanovic J, Lewandowski TM et al (2016) 2-Aryl-5-carboxytetrazole as a new photoaffinity label for drug target identification. J Am Chem Soc 138:14609–14615. https://doi.org/10.1021/jacs.6b06645
Hofmann K, Thiele C, Schött H-F et al (2014) A novel alkyne cholesterol to trace cellular cholesterol metabolism and localization. J Lipid Res 55:583–591. https://doi.org/10.1194/jlr.D044727
Höglinger D, Nadler A, Haberkant P et al (2017) Trifunctional lipid probes for comprehensive studies of single lipid species in living cells. Proc Natl Acad Sci 114:1566–1571. https://doi.org/10.1073/pnas.1611096114
Homo-Delarche F, Fitzpatrick F, Christeff N et al (1991) Sex steroids, glucocorticoids, stress and autoimmunity. J Steroid Biochem Mol Biol 40:619–637. https://doi.org/10.1016/0960-0760(91)90285-D
Hsu KL, Tsuboi K, Adibekian A et al (2012) DAGLβ inhibition perturbs a lipid network involved in macrophage inflammatory responses. Nat Chem Biol. https://doi.org/10.1038/nchembio.1105
Hulce JJ, Cognetta AB, Niphakis MJ et al (2013) Proteome-wide mapping of cholesterol-interacting proteins in mammalian cells. Nat Methods 10:259–264. https://doi.org/10.1038/nmeth.2368

Jahng WJ, David C, Nesnas N et al (2003a) A cleavable affinity biotinylating agent reveals a retinoid binding role for RPE65. Biochemistry 42:6159–6168. https://doi.org/10.1021/bi034002i

Jahng WJ, Xue L, Rando RR (2003b) Lecithin retinol acyltransferase is a founder member of a novel family of enzymes. Biochemistry 42:12805–12812. https://doi.org/10.1021/bi035370p

Kallemeijn WW, Li KY, Witte MD et al (2012) Novel activity-based probes for broad-spectrum profiling of retaining β-exoglucosidases in situ and in vivo. Angew Chemie Int Ed 51:12529–12533. https://doi.org/10.1002/anie.201207771

Kleiner P, Heydenreuter W, Stahl M et al (2017) A whole proteome inventory of background photocrosslinker binding. Angew Chemie Int Ed 56:1396–1401. https://doi.org/10.1002/anie.201605993

Lai JJ, Lai KP, Zeng W et al (2012) Androgen receptor influences on body defense system via modulation of innate and adaptive immune systems: lessons from conditional AR knockout mice. Am J Pathol 181:1504–1512

Lee KSS, Henriksen NM, Ng CJ et al (2017) Probing the orientation of inhibitor and epoxy-eicosatrienoic acid binding in the active site of soluble epoxide hydrolase. Arch Biochem Biophys 613:1–11. https://doi.org/10.1016/j.abb.2016.10.017

Lentz CS, Sheldon JR, Crawford LA et al (2018) Identification of a S. aureus virulence factor by activity-based protein profiling (ABPP) article. Nat Chem Biol 14:609–617. https://doi.org/10.1038/s41589-018-0060-1

Li Z, Hao P, Li L et al (2013) Design and synthesis of minimalist terminal alkyne-containing diazirine photo-crosslinkers and their incorporation into kinase inhibitors for cell- and tissue-based proteome profiling. Angew Chemie Int Ed 52:8551–8556. https://doi.org/10.1002/anie.201300683

Liu Y, Patricelli MP, Cravatt BF (1999) Activity-based protein profiling: the serine hydrolases. Proc Natl Acad Sci 96:14694–14699. https://doi.org/10.1073/pnas.96.26.14694

Liu X, Dong T, Zhou Y et al (2016) Exploring the binding proteins of glycolipids with bifunctional chemical probes. Angew Chemie Int Ed 55:14330–14334. https://doi.org/10.1002/anie.201608827

Long JZ, Cravatt BF (2011) The metabolic serine hydrolases and their functions in mammalian physiology and disease. Chem Rev 111:6022–6063

Lum KM, Sato Y, Beyer BA et al (2017) Mapping protein targets of bioactive small molecules using lipid-based chemical proteomics. ACS Chem Biol 12:2671–2681. https://doi.org/10.1021/acschembio.7b00581

Marshall-Gradisnik S, Green R, Brenu E, Weatherby R (2009) Anabolic androgenic steroids effects on the immune system: a review. Open Life Sci 4:19–33. https://doi.org/10.2478/s11535-008-0058-x

Mellacheruvu D, Wright Z, Couzens AL et al (2013) The CRAPome: a contaminant repository for affinity purification-mass spectrometry data. Nat Methods 10:730–736. https://doi.org/10.1038/nmeth.2557

Mora JR, Iwata M, Von Andrian UH (2008) Vitamin effects on the immune system. Nat Rev Immunol 8:685–698. https://doi.org/10.1038/nri2378.Vitamin

Moriguchi S (1998) The role of vitamin E in T-cell differentiation and the decrease of cellular immunity with aging. BioFactors 7:77–86. https://doi.org/10.1002/biof.5520070111

Moriguchi S, Kaneyasu M (2003) Role of Vitamin E in immune system. Ser Rev JClin Biochem 34:97–109

Nadler A, Reither G, Feng S et al (2013) The fatty acid composition of diacylglycerols determines local signaling patterns. Angew Chemie Int Ed 52:6330–6334. https://doi.org/10.1002/anie.201301716

Nesnas N, Rando RR, Nakanishi K (2002) Synthesis of biotinylated retinoids for cross-linking and isolation of retinol binding proteins. Tetrahedron 58:6577–6584. https://doi.org/10.1016/S0040-4020(02)00667-1

Niphakis MJ, Lum KM, Cognetta AB et al (2015) A global map of lipid-binding proteins and their ligandability in cells. Cell 161:1668–1680. https://doi.org/10.1016/j.cell.2015.05.045

Norman AW, Mizwicki MT, Norman DPG (2004) Steroid-hormone rapid actions, membrane receptors and a conformational ensemble model. Nat Rev Drug Discov. https://doi.org/10.1038/nrd1283

Ogasawara D, Deng H, Viader A et al (2016) Rapid and profound rewiring of brain lipid signaling networks by acute diacylglycerol lipase inhibition. Proc Natl Acad Sci 113:26–33. https://doi.org/10.1073/pnas.1522364112

Ogawa N, Sugiyama T, Morita M et al (2017) Total synthesis of resolvin D5. J Org Chem 82:2032–2039. https://doi.org/10.1021/acs.joc.6b02870

Ouairy CMJ, Ferraz MJ, Boot RG et al (2015) Development of an acid ceramidase activity-based probe. Chem Commun 51:6161–6163. https://doi.org/10.1039/C5CC00356C

Ourailidou ME, van der Meer J-Y, Baas B-J et al (2014) Aqueous oxidative heck reaction as a protein-labeling strategy. ChemBioChem 15:209–212. https://doi.org/10.1002/cbic.201300714

Pacher P, Mechoulam R (2011) Is lipid signaling through cannabinoid 2 receptors part of a protective system? Prog Lipid Res 50:193–211. https://doi.org/10.1016/j.plipres.2011.01.001

Pandey R, Mousawy K, Nagarkatti M, Nagarkatti P (2009) Endocannabinoids and immune regulation. Pharmacol Res 60:85–92. https://doi.org/10.1016/j.phrs.2009.03.019

Peng T, Hang HC (2015) Bifunctional fatty acid chemical reporter for analyzing S-palmitoylated membrane protein-protein interactions in mammalian cells. J Am Chem Soc 137:556–559. https://doi.org/10.1021/ja502109n

Peng T, Yuan X, Hang HC (2014) Turning the spotlight on protein-lipid interactions in cells. Curr Opin Chem Biol 21:144–153. https://doi.org/10.1016/j.cbpa.2014.07.015

Petracca R, Romeo E, Baggelaar MP et al (2017) Novel activity-based probes for N-acylethanolamine acid amidase. Chem Commun 53:11810–11813. https://doi.org/10.1039/C7CC06838G

Pike LJ (2003) Lipid rafts. J Lipid Res 44:655–667. https://doi.org/10.1194/jlr.R200021-JLR200

Ray R, Bouillon R, Van Baelen H, Holick MF (1991) Synthesis of 25-hydroxyvitamin D3 3β-3′-[N-(4-azido-2-nitrophenyl)amino]propyl ether, a second-generation photoaffinity analogue of 25-hydroxyvitamin D3: photoaffinity labeling of rat serum vitamin D binding protein. Biochemistry 30:4809–4813. https://doi.org/10.1021/bi00233a024

Robichaud PP, Poirier SJ, Boudreau LH et al (2016) On the cellular metabolism of the click chemistry probe 19-alkyne arachidonic acid. J Lipid Res 57:1821–1830. https://doi.org/10.1194/jlr.M067637

Rodriguez AR, Spur BW (2017) First total synthesis of pro-resolving and tissue-regenerative resolvin sulfido-conjugates. Tetrahedron Lett 58:1662–1668. https://doi.org/10.1016/J.TETLET.2017.03.041

Romeo E, Ponzano S, Armirotti A et al (2015) Activity-based probe for *N*-acylethanolamine acid amidase. ACS Chem Biol 10:2057–2064. https://doi.org/10.1021/acschembio.5b00197

Rowland MM, Bostic HE, Gong D et al (2011) Phosphatidylinositol 3,4,5-trisphosphate activity probes for the labeling and proteomic characterization of protein binding partners. Biochemistry 50:11143–11161. https://doi.org/10.1021/bi201636s

Sakurai K, Ozawa S, Yamada R et al (2014) Comparison of the reactivity of carbohydrate photoaffinity probes with different photoreactive groups. ChemBioChem 15:1399–1403. https://doi.org/10.1002/cbic.201402051

Serhan CN, Petasis NA (2011) Resolvins and protectins in inflammation resolution. Chem Rev 111:5922–5943. https://doi.org/10.1021/cr100396c

Shimazawa R, Sanda R, Mizoguchi H et al (1991) Fluorescent and photoaffinity labeling probes for retinoic acid receptors. Biochem Biophys Res Commun 179:259–265. https://doi.org/10.1016/0006-291X(91)91363-H

Simon GM, Cravatt BF (2010) Activity-based proteomics of enzyme superfamilies: serine hydrolases as a case study. J Biol Chem 285:11051–11055

Soethoudt M, Stolze SC, Westphal MV et al (2018) Selective photoaffinity probe that enables assessment of cannabinoid CB2Receptor expression and ligand engagement in human cells. J Am Chem Soc 140:6067–6075. https://doi.org/10.1021/jacs.7b11281

Soontjens CD, Rafter JJ, Gustafsson JÅ (1996) Ligands for orphan receptors? J Endocrinol 150:241–257

Swamy N, Addo J, Vskokovic MR, Ray R (2000a) Probing the vitamin D sterol-binding pocket of human vitamin D-binding protein with bromoacetate affinity labeling reagents containing the affinity probe at C-3, C-6, C-11, and C-19 positions of parent vitamin D sterols. Arch Biochem Biophys 373:471–478. https://doi.org/10.1006/ABBI.1999.1537

Swamy N, Xu W, Paz N et al (2000b) Molecular modeling, affinity labeling, and site-directed mutagenesis define the key points of interaction between the ligand-binding domain of the vitamin D nuclear receptor and 1α,25-dihydroxyvitamin D3. Biochemistry 39:12162–12171. https://doi.org/10.1021/bi0002131

Tate EW, Kalesh KA, Lanyon-Hogg T et al (2015) Global profiling of protein lipidation using chemical proteomic technologies. Curr Opin Chem Biol 24:48–57. https://doi.org/10.1016/j.cbpa.2014.10.016

Trigunaite A, Dimo J (2015) Suppressive effects of androgens on the immune system. Cell Immunol 294:87–94. https://doi.org/10.1016/J.CELLIMM.2015.02.004

Tully SE, Cravatt BF (2010) Activity-based probes that target functional subclasses of phospholipases in proteomes. J Am Chem Soc 132:3264–3265. https://doi.org/10.1021/ja1000505

Turcotte C, Blanchet MR, Laviolette M, Flamand N (2016) The CB2 receptor and its role as a regulator of inflammation. Cell Mol Life Sci 73:4449–4470

Ulivi V, Lenti M, Gentili C et al (2011) Anti-inflammatory activity of monogalactosyldiacylglycerol in human articular cartilage in vitro: Activation of an anti-inflammatory cyclooxygenase-2 (COX-2) pathway. Arthritis Res Ther 13:R92. https://doi.org/10.1186/ar3367

Van Esbroeck ACM, Janssen APA, Cognetta AB et al (2017) Activity-based protein profiling reveals off-target proteins of the FAAH inhibitor BIA 10-2474. Science (80–)356:1084–1087. https://doi.org/10.1126/science.aaf7497

van Rooden EJ, van Esbroeck ACM, Baggelaar MP et al (2018) Chemical proteomics analysis of endocannabinoid hydrolase activity in Niemann-Pick Type C mouse brain. Manuscr Submitt 12. https://doi.org/10.3389/fnins.2018.00440

Vane JR (1971) Inhibition of prostaglandin synthesis as a mechanism of action for aspirin-like drugs. Nat New Biol 231:232–235. https://doi.org/10.1038/newbio231232a0

Waldmann H, Ahmed SA, Khan D (2016) The immune System is a natural target for estrogen action: opposing effects of estrogen in two prototypical autoimmune diseases. Front Immunol 6:6353389–6353635. https://doi.org/10.3389/fimmu.2015.00635

Wang D, Du S, Cazenave-Gassiot A et al (2017) Global mapping of protein-lipid interactions by using modified choline-containing phospholipids metabolically synthesized in live cells. Angew Chemie Int Ed 56:5829–5833. https://doi.org/10.1002/anie.201702509

Weerapana E, Speers AE, Cravatt BF (2007) Tandem orthogonal proteolysis-activity-based protein profiling (TOP-ABPP)—a general method for mapping sites of probe modification in proteomes. Nat Protoc 2:1414–1425. https://doi.org/10.1038/nprot.2007.194

Weller MG (2016) Quality issues of research antibodies. Anal Chem Insights 2016:21–27. https://doi.org/10.4137/Aci.s31614

Wenk MR (2005) The emerging field of lipidomics. Nat Rev Drug Discov 4:594–610. https://doi.org/10.1038/nrd1776

Wenk MR (2010) Lipidomics: new tools and applications. Cell 143:888–895. https://doi.org/10.1016/j.cell.2010.11.033

Witte MD, Kallemeijn WW, Aten J et al (2010) Ultrasensitive in situ visualization of active glucocerebrosidase molecules. Nat Chem Biol 6:907–913. https://doi.org/10.1038/nchembio.466

Wright MH, Paape D, Storck EM et al (2015) Global analysis of protein N-myristoylation and exploration of N-myristoyltransferase as a drug target in the neglected human pathogen leishmania donovani. Chem Biol 22:342–354. https://doi.org/10.1016/j.chembiol.2015.01.003

Wright MH, Sieber SA (2016) Chemical proteomics approaches for identifying the cellular targets of natural products. Nat Prod Rep 33:681–708. https://doi.org/10.1039/C6NP00001K

Xia Y, Peng L (2013) Photoactivatable lipid probes for studying biomembranes by photoaffinity labeling. Chem Rev 113:7880–7929. https://doi.org/10.1021/cr300419p

Yang K, Han X (2016) Lipidomics: techniques, applications, and outcomes related to biomedical sciences. Trends Biochem Sci 41:954–969

Yeagle PL (2016) The membranes of cells. Academic Press

Zhang L, Zhang Y, Dong J et al (2012) Design and synthesis of novel photoaffinity probes for study of the target proteins of oleanolic acid. Bioorganic Med Chem Lett 22:1036–1039. https://doi.org/10.1016/j.bmcl.2011.11.123

Zhou H, Hylemon PB (2014) Bile acids are nutrient signaling hormones. Steroids 86:62–68

Zhuang S, Li Q, Cai L et al (2017) Chemoproteomic profiling of bile acid interacting proteins. ACS Cent Sci 3:501–509. https://doi.org/10.1021/acscentsci.7b00134

Zweerink S, Kallnik V, Ninck S et al (2017) Activity-based protein profiling as a robust method for enzyme identification and screening in extremophilic Archaea. Nat Commun 8:15352. https://doi.org/10.1038/ncomms15352

Activity-Based Protein Profiling of Non-ribosomal Peptide Synthetases

Fumihiro Ishikawa, Genzoh Tanabe and Hideaki Kakeya

Contents

Abstract Non-ribosomal peptide (NRP) natural products are one of the most promising resources for drug discovery and development because of their wide-ranging of therapeutic potential, and their behavior as virulence factors and signaling molecules. The NRPs are biosynthesized independently of the ribosome

F. Ishikawa (✉) · G. Tanabe
Faculty of Pharmacy, Kindai University, 3-4-1 Kowakae, Higashi-Osaka, Osaka 577-8502, Japan
e-mail: ishikawa@phar.kindai.ac.jp

H. Kakeya (✉)
Graduate School of Pharmaceutical Sciences, Kyoto University, Sakyo, Kyoto 606-8501, Japan
e-mail: scseigyo-hisyo@pharm.kyoto-u.ac.jp

Current Topics in Microbiology and Immunology (2019) 420:321–350
DOI 10.1007/82_2018_133

Published Online: 04 September 2018

by enzyme assembly lines known as the non-ribosomal peptide synthetase (NRPS) machinery. Genetic, biochemical, and bioinformatics analyses have provided a detailed understanding of the mechanism of NRPS catalysis. However, proteomic techniques for natural product biosynthesis remain a developing field. New strategies are needed to investigate the proteomes of diverse producer organisms and directly analyze the endogenous NRPS machinery. Advanced platforms should verify protein expression, protein folding, and activities and also enable the profiling of the NRPS machinery in biological samples from wild-type, heterologous, and engineered bacterial systems. Here, we focus on activity-based protein profiling strategies that have been recently developed for studies aimed at visualizing and monitoring the NRPS machinery and also for rapid labeling, identification, and biochemical analysis of NRPS enzyme family members as required for proteomic chemistry in natural product sciences.

1 Introduction

1.1 Structural Diversity of Non-ribosomal Peptide Natural Products

Non-ribosomal peptide (NRP) natural products display a broad range of important medicinal agents such as the antibiotic vancomycin, the anticancer bleomycin, and the immunosuppressant cyclosporine, as well as virulence factors and signaling molecules (Walsh 2004). These peptide-based natural products are synthesized by enzyme assembly lines that are collectively known as the non-ribosomal peptide synthetase (NRPS) machinery (Fischbach and Walsh 2006). NRPS biosynthetic enzymes utilize a range of amino acid building blocks (to date, more than 500 different amino acids) to construct the structurally and biologically diverse NRP natural products (Walsh et al. 2013).

1.2 Amino Acid Selection, Activation, and Incorporation of the Non-ribosomal Peptide Machinery

In contrast to the extensive structural diversity of the NRP natural products, their biosynthetic strategies are highly conserved. Typically, the NRPS machinery can be categorized into linear-, iterative-, and nonlinear-type NRPS enzymes (Fig. 1a) (Mootz et al. 2002). The majority of the NRPS machinery is linear, in which the individual catalytic components are organized into modules that function in the incorporation of cognate amino acid building blocks into the NRP natural products.

The standard NRPS module comprises three catalytic domains: adenylation (A), a peptidyl carrier protein (PCP, also known as the thiolation domain, T), and condensation (C) (Hur et al. 2012). The proteinogenic and non-proteinogenic amino acid building blocks are selected and activated by the A-domains. Accordingly, the A-domains of the NRPS machinery function as gatekeepers for the incorporation of cognate amino acids into the NRP natural products. Genetics, bioinformatics, and biochemical analyses have revealed the enzymology of the NRPS machinery. An 8–10 kDa PCP domain acts as a central cargo for amino acid substrates and the growing peptide chains. To convert the inactive *apo*-PCP to active *holo*-PCP, PCP domains must be posttranslationally modified with a 4′-phosphopantetheine (Ppant) group. 4′-Phosphopantetheinyl transferase (PPTase) catalyzes the attachment of the Ppant group to a conserved Ser of the PCP domain using coenzyme A (CoA) as a substrate (Fig. 1b). In the first step, the A-domain catalyzes the formation of aminoacyl adenylate (aminoacyl-AMP) using adenosine triphosphate (ATP) and Mg^{2+} (Fig. 1c). In the second step of NRP biosynthesis, the adenylated amino acid undergoes a nucleophilic attack of thiol functionality of the Ppant arm of a downstream PCP domain, yielding an aminoacyl thioester tethered to the PCP domain (aminoacyl-S-PCP) (Fig. 1c). Finally, a C-domain catalyzes the formation of an amide linkage between two aminoacyl-*S*-PCP species in both upstream and downstream modules (Fig. 1d). In addition to the A, PCP, and C-domains, tailoring enzymes provide a variety of chemical modifications for the diversification of NRPs. These chemical modifications are incorporated into the growing peptidyl intermediates during NRP biosynthesis using the following tailoring enzyme domains: epimerase (E; Stachelhaus and Walsh 2000; Samel et al. 2014; Chen et al. 2016), methyl transferase (MT; Hornbogen et al. 2007; Brieke et al. 2016), cyclization (Cy; Marshall et al. 2001), oxidation (Ox; Schneider et al. 2003), thioesterase (TE; Kohli et al. 2002; Hoyer et al. 2007), and reductase (R; Keating et al. 2001). We refer the reader to recent review articles (Hur et al. 2012; Gulik 2016; Walsh 2016) for a detailed understanding of the NRPS machinery.

1.3 Engineering the Non-ribosomal Peptide Synthetase Machinery

The utility and advantage of the NRPS machinery carry the potential for the production of non-native peptide products with desired properties. The strict substrate recognition properties of A-domains control the incorporation of the 20 proteinogenic amino acids and the unusual non-proteinogenic amino acids into the final products. The X-ray crystal structure of the Phe-activating domain of GrsA (domain structure: A-PCP-E) involved in the gramicidin S biosynthesis was first reported in a complex with AMP and *L*-Phe (Conti et al. 1997). Subsequent structural investigations of GrsA NRPS have greatly accelerated in silico and structure–function

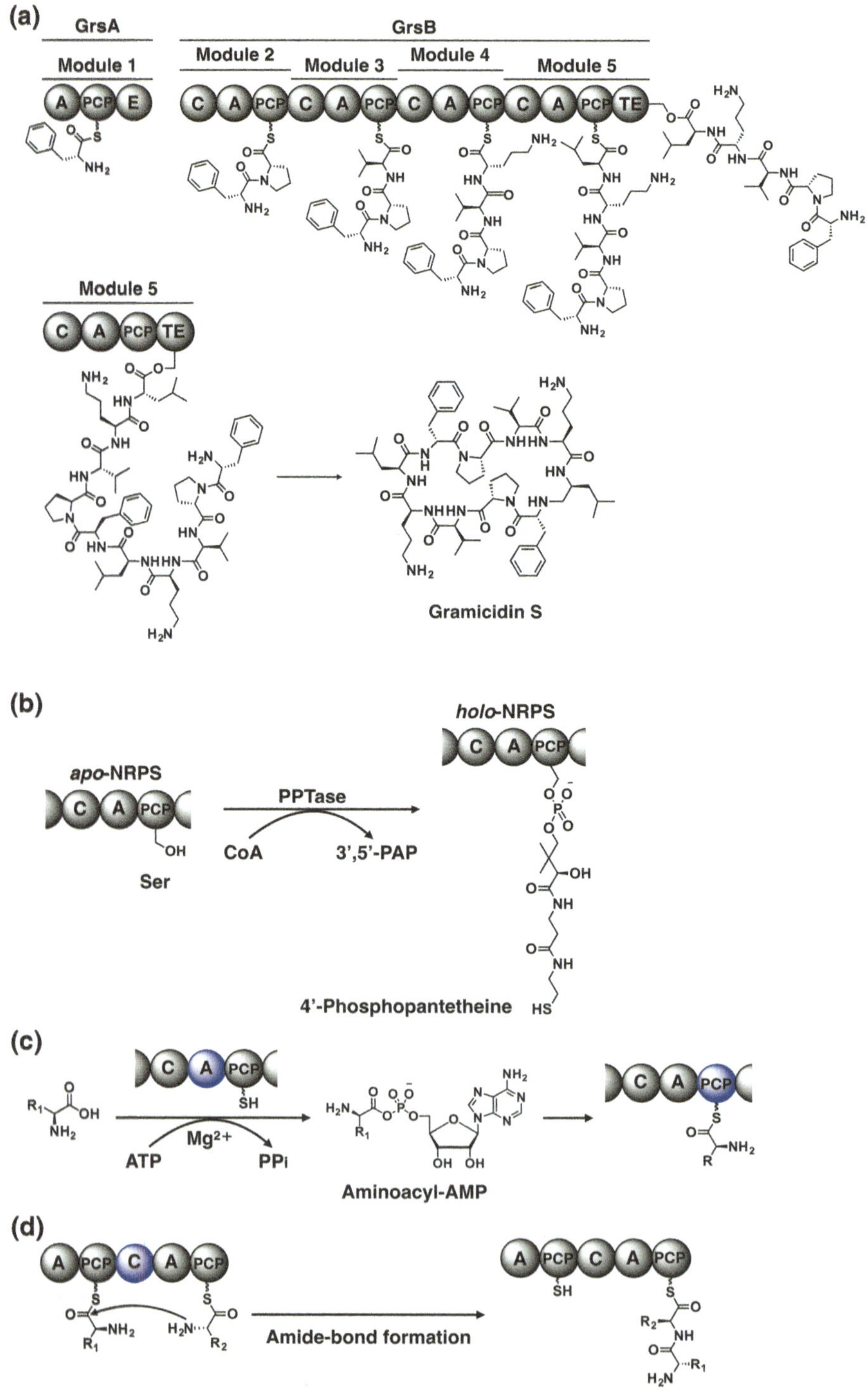
(a)
GrsA
GrsB
Module 1
Module 2
Module 3
Module 4
Module 5
Module 5
Gramicidin S
(b)
apo-NRPS
holo-NRPS
PPTase
CoA
3',5'-PAP
Ser
4'-Phosphopantetheine
(c)
Mg2+
ATP
PPi
Aminoacyl-AMP
(d)
Amide-bond formation

◂**Fig. 1** Non-ribosomal peptide synthetase (NRPS) machinery.**a** The NRPS assembly line of gramicidin S biosynthesis. **b** Posttranscriptional phosphopantetheinylation catalyzed by 4′-phosphopantetheinyl transferase (PPTase) enzymes. **c** Loading amino acids onto NRPS machinery. **d** Amide bond formation is catalyzed by condensation (C) domains. Modules are comprised of adenylation (A), peptidyl carrier protein (PCP), condensation (C), epimerase (E), and thioesterase (TE) domains. CoA: Coenzyme A; 3′,5′-PAP: adenosine 3′,5′-diphosphate; AMP: adenosine monophosphate

mutagenesis studies that have identified the core residues responsible for substrate specificities of A-domains of the NRPS machinery (Stachelhaus et al. 1999; Challis et al. 2000). On the basis of a profound structural and biochemical understanding of A-domain enzymology, NRPS enzyme engineering has manipulated these NRPS code residues to generate active-site chemistries with desired substrate specificities (Evans et al. 2011; Thirlway et al. 2012; Zhang et al. 2013). The NRPS specificity code enables us to predict the substrate specificities of uncharacterized A-domains extracted from the primary sequence information (Röttig et al. 2011). The straightforward biochemical logic of the A-domains has made them attractive targets for repurposing and/or engineering in the NRPS machinery to rationally produce non-native products. The NRPS biosynthetic machinery could provide templates for biosynthetic systems that could be exploited to develop novel bioactive peptides. Engineering on the NRPS biosynthetic level is not a versatile strategy yet. Nonetheless, many peptide-based non-native products have been designed and biosynthesized by precursor-directed biosynthesis (Moran et al. 2009), mutasynthesis (Thirlway et al. 2012), active-site manipulation of A-domains (Kries et al. 2014), directed evolution of A-domains (Evans et al. 2011; Zhang et al. 2013; Niquille et al. 2018), swapping domains and modules (Calcott et al. 2014), subdomain swapping (Kries et al. 2015), and de novo NRPSs (Bozhüyük et al. 2018). The search continues for novel strategies and assay platforms, preferably by developing non-traditional methods that accelerate the redesign and optimization of the NRPS machinery. NRPS manipulation is beyond the scope of this chapter, and we refer the reader to recent review articles (Baltz 2014; Kries 2016; Winn et al. 2016).

1.4 Non-ribosomal Peptide Synthetases as a New Antibiotic Strategy

Clinically, significant antibiotics faced the challenge of the development of antibiotic-resistant strains. The development of antimicrobials with novel modes of action has lagged behind the development of resistance. Therefore, new antibiotic strategies are needed. Natural products frequently play essential roles in in vivo bacterial processes involved in nutrient acquisition, quorum sensing, biofilm formation, infection, and virulence. One promising approach for the development of novel antibiotics is targeting bacterial nutrient acquisition processes (Clatworthy et al. 2007). Iron is an important cofactor required for all bacteria as well as humans. Iron is a constituent of a number of enzymes that catalyze essential chemical transformations in vivo for cell viability. The iron(III) observed in aqueous and aerobic environments can be insoluble and inaccessible, which results in the low concentration of free iron(III) ion available to pathogenic bacteria in the human host. In response, bacteria have developed elaborate systems to acquire iron from the host, involving heme acquisition, transferrin/lactoferrin receptors, and siderophore mediation (Skaar 2010). A primary iron acquisition pathway involves the use of small-molecule iron chelators known as siderophores. In the iron-limiting conditions, pathogens biosynthesize, secrete, or steal iron from the host proteins, and then take up the ferric-siderophore complexes into the bacteria, which are required for cellular processes that are critical to the growth and virulence of pathogens. Furthermore, siderophore biosynthetic machinery proteins have no human homologues (Quadri 2007). The disruption of siderophore biosynthetic pathways represents an attractive antimicrobial strategy for the development of antibiotics that have novel modes of action (Cisar and Tan 2008; Lamb 2015). A variety of siderophores are biosynthesized by the NRPS machinery in gram-positive and gram-negative pathogens including mycobactins produced by *Mycobacterium tuberculosis* (Mbt) (Quadri et al. 1998a), petrobactin produced by *Bacillus anthracis* (Lee et al. 2007), pyochelin and pyoverdine produced by *Pseudomonas aeruginosa* (Visca et al. 2007), yersiniabactin produced by *Yersinia pestis* (Miller DA et al. 2002), vibriobactin produced by *Viblio cholerae* (Keating et al. 2000), acinetobactin produced by *Acinetobacter baumannii* (Mihara et al. 2004), and enterobactin produced by *Escherichia coli* (Walsh et al. 1990). Methods that allow the dynamic assessment of NRPS enzymatic activities in complex biological systems are required to accurately understand human bacterial pathogenesis. Activity-based protein profiling (ABPP) strategies in natural product biosynthesis will be important for studies aimed at visualizing, monitoring, and tracking NRPS enzyme activities in vitro (proteomes) and in vivo as a viable drug target, and also for studies aimed at understanding the transcriptional controls, expression patterns, functional states of proteins, and infection processes in pathogenic microbiology.

1.5 Proteomics Strategies for Natural Product Biosynthesis

Despite the significant progress in genetic, biochemical, structural, bioinformatics studies over the past three decades, the knowledge of natural product biosynthesis at the proteomic level remains limited. Analogous to the studies of the human proteome, proteomic analyses of natural product producing microorganisms have provided useful insights into biological processes that cannot be addressed by genetic manipulations (Reyes et al. 2010). Schley et al. (2006) have firstly demonstrated a whole proteome approach based on liquid chromatography-tandem mass spectrometry (LC-MS/MS) to elucidate the expression profiles of the NRPS and polyketide synthase (PKS) machinery in proteomes from *Myxococcus xanthus*.

This approach has identified many of the peptide fragments from the NRPS and PKS machinery that are responsible for the biosynthesis of myxalamid, DKxanthene, myxovirescin, and myxochromid. It is noteworthy that peptides from six orphan NRPS and PKS gene clusters have been identified, with known natural products, demonstrating the potential of proteomics in discovering natural products and their biosynthetic pathways. However, further proteomic studies were hampered by the low expression levels of NRPS and/or PKS biosynthetic machinery. To overcome these obstacles, researchers have developed new proteomic methodologies for the study of natural product biosynthesis. These techniques include proteomic interrogation of secondary metabolism (PrISM) (Bumpus et al. 2009) and orthogonal active site identification system (OASIS) (Meier et al. 2009). The PrISM strategy permits the accurate detection of peptides from the NRPS and/or PKS machinery using a combination of size-based fractionation by SDS-PAGE and the Ppant ejection assay by Fourier transform-ion cyclotron resonance (FT-ICR) LC-MS/MS. FT-ICR MS methods have been applied in the in vitro studies of NRPS and PKS machinery to directly observe 4′-Ppant bound substrates, intermediates, and products (Bumpus and Kelleher 2008). Furthermore, tandem MS methods are used to eject the Ppant functionality from the intact protein and peptide forming two ejection ions, which results in the rapid assignment of the covalently bound substrates, intermediates, and products and the presence of CP domains (Dorrestein et al. 2006a, b; Meluzzi et al. 2008). This technique has been extended to detect and identify NRPS and PKS machinery in proteomes. Accordingly, tryptic digests of gel-fractionated high molecular weight proteins are analyzed by FT-ICR LC-MS/MS. Combined with database searching, the PrISM method has enabled the discovery of a new NRP and its biosynthetic gene cluster, which represents the first example of an orphan NRPS pathway analysis by an MS-based approach. The OASIS method utilizes chemical proteomic probes to label, enrich, and identify the particular NRPS and/or PKS machinery in an activity-based manner in complex proteomes. Briefly, natural product producer proteomes are tagged with domain-specific probes. The tagged NRPS and/or PKS machinery are then enriched using affinity purification methods. Samples of the enriched NRPS and/or PKS machinery are evaluated by multidimensional protein identification technology (MudPIT) MS analysis. The collection of multiple datasets using ABPP probes allows selective identification of NRPS and/or PKS machinery, which contain multiple active sites on a single protein. Finally, the cloned genes are identified by reverse genetics. The concept of the OASIS method would be useful for the discovery of expressed machinery proteins and genes and their small-molecule products, and also for studies aimed at visualizing, tracking, and imaging biosynthetic enzyme activities as required for proteomics in natural product science. We refer the reader to a review article (Meier and Burkart 2010) for a detailed understanding of the PrISM and OASIS methods.

Genome sequencing projects have demonstrated the wide distribution of NRPS biosynthetic gene clusters in a broad range of organisms (Bode and Müller 2005). In addition, recent advances in gene manipulation methodologies are making it possible to characterize and manipulate the NRPS machinery via the heterologous expression systems of biosynthetic gene clusters in host organisms including *E. coli*, *Streptomyces* strains, yeast, and *Aspergillus* strains (Zhang et al. 2008). However, some pathways of the NRPS machinery can be resistant to conventional assessments, because of large molecular weight (100–1600 kDa) and general intractability of the producer bacteria to laboratory gene manipulation and protein expression methodologies (Wakimoto et al. 2014). Moreover, the discovery, isolation, and evaluation of NRPS biosynthetic gene clusters from the complex ensembles of symbiont microorganisms have remained challenging because most genetic approaches are optimized for pure and isolated strains (Paul et al. 2007). Furthermore, natural product biosynthetic pathways responsible for siderophore production in pathogens are an emerging antibiotic strategy. ABPP strategies should provide highly complementary to genetic strategies, which will provide insights into the activities, transcriptional regulation, posttranslational processes, infection, and virulence and which will also accelerate the visualization, monitoring, identification, and functional assessments of NRPS machinery in biological samples.

2 Activity-Based Protein Profiling of Non-ribosomal Peptide Synthetases

2.1 Benzophenone-Based Photo-affinity Labeling Probes for Aryl Acid-Adenylating Enzymes

Before describing specific examples of benzophenone (BP)-based photo-affinity probes and their biological applications, we briefly mention the general design features for ABPP probes and chemical properties of the BP group. An ABPP probe contains a reactive group, a spacer or binding group, and a reporter tag. Reactive groups are categorized as two general classes: (i) electrophilic groups (fluorophosphonate, epoxyketone, and vinyl sulfone) that capture conserved active-site nucleophiles (Ser, Thr, and Cys) (Sanman and Bogyo 2014) and (ii) photoreactive groups (BP, diazirine, and azide) that cross-link nearby residues in enzyme active sites by ultraviolet (UV) light (Saario et al. 2012). Reporter tags can be used for enzyme visualization and enrichment including fluorophores (rhodamine), biotin, and clickable functionalities azide and alkyne [Huisgen's copper(I)-catalyzed azide-alkyne cycloaddition (CuAAC) (Kolb and Sharpless 2003)]. We refer the reader to excellent review articles (Cravatt et al. 2008; Sanman and Bogyo 2014) that explain the basic technology of ABPP.

Photoactivation of BPs enables the generation of a triplet diradical on the carbonyl and the insertion into proximal C–H bonds (Dormán et al. 2016). A major advantage is that BPs are activated by UV light in the range of 350–360 nm, which reduces the damage of biomacromolecules (Dormán et al. 2016). If the activated BPs are not able to cross-link with proximal molecular species, the generated diradical can relax back to the ground state to reproduce BP group (Dormán et al. 2016). In terms of advantage, the chemical property enables the high-efficient cross-linking of targets of interest with increasing UV photolysis time. One disadvantage is that this property may lead to an increasing nonspecific cross-linking in proteomic environments. Another disadvantage is that BP groups are bulkier than the photoreactive groups of diazirine and azide, which may result in the disruption of interactions between ligands (natural product, drug, and synthetic molecule) and target biomacromolecules (DNA, RNA, and protein). Accordingly, a highly specific ligand and its modification sites will be important for ABPP applications. To convert reversible inhibitors to ABPP probes, a photoreactive BP group and propargyl functionality can be embedded in pharmacophores, which allows the covalent labeling of proximal molecular species following UV irradiation and the installation of a variety of reporter tags for the characterization of molecular targets by CuAAC, the most used "click" reaction.

Mbt can enable the biosynthesis of the mycobactin siderophore by the assembly of a hybrid NRPS-PKS megasynthetase (Quadri et al. 1998a). The production of mycobactins plays an essential role in cell viability and virulence of Mbt. Because of this significant function in the human host, there is considerable interest in drug development with novel modes of action (De Voss et al. 2000; Tim et al. 2003). The MbtA aryl acid-adenylating enzyme catalyzes the adenylation reaction of salicylic acid (Sal) substrate to form Sal-AMP in the active site, followed by the loading of Sal onto the cognate aryl carrier protein (ArCP) domain of MbtB, which is the initiating step in the construction of mycobactins. Various groups have reported the development of an 5′-*O*-*N*-(salicyl)sulfamoyladenosine (Sal-AMS) compound, which is a highly specific inhibitor of MbtA in vitro and in vivo, where the highly reactive acylphosphate linkage is substituted by a non-hydrolyzable acylsulfamate (Fig. 2a) (Ferreras et al. 2005; Miethke et al. 2006; Somu et al. 2006). While Sal-AMS displayed anti-tuberculosis activity in iron-deficient conditions with a minimum inhibitory concentration (MIC) of 0.39 μM, it still exhibited inhibitory activity in iron-rich conditions, with a MIC of 1.56 μM (Ferreras et al. 2005). These results strongly suggest the presence of off-target binding proteins. On the basis of medicinal chemistry, biochemical, and structural studies of MbtA (Neres et al. 2008), another study described the development of a BP-based photo-affinity labeling probe for aryl acid-adenylating enzymes to profile the protein targets of Sal-AMS in proteomic context (Duckworth et al. 2012). A BP group with a terminal alkyne functionality was incorporated into the C-2 position of the adenosine ring of Sal-AMS to generate Sal-AMS-based activity-based probe (ABP) (Fig. 2a). Sal-AMS ABP retained tight-binding properties toward the MbtA enzyme in vitro ($K_i^{app} = 0.94$ nM) and in vivo in iron-deficient conditions (MIC = 3–6 μM). After in vitro labeling experiments using recombinant MbtA, ABPP experiments were conducted using mycobacterial cell extracts. *M. smegmatis* proteomes were treated with Sal-AMS ABP in the absence or presence of Sal-AMS, irradiated, and subjected to CuAAC with a biotin azide group. Biotin-tagged proteins in proteomes were enriched using streptavidin beads, digested by trypsin, and analyzed by LC-MS/MS. The presence of endogenous MbtA was evident. In the

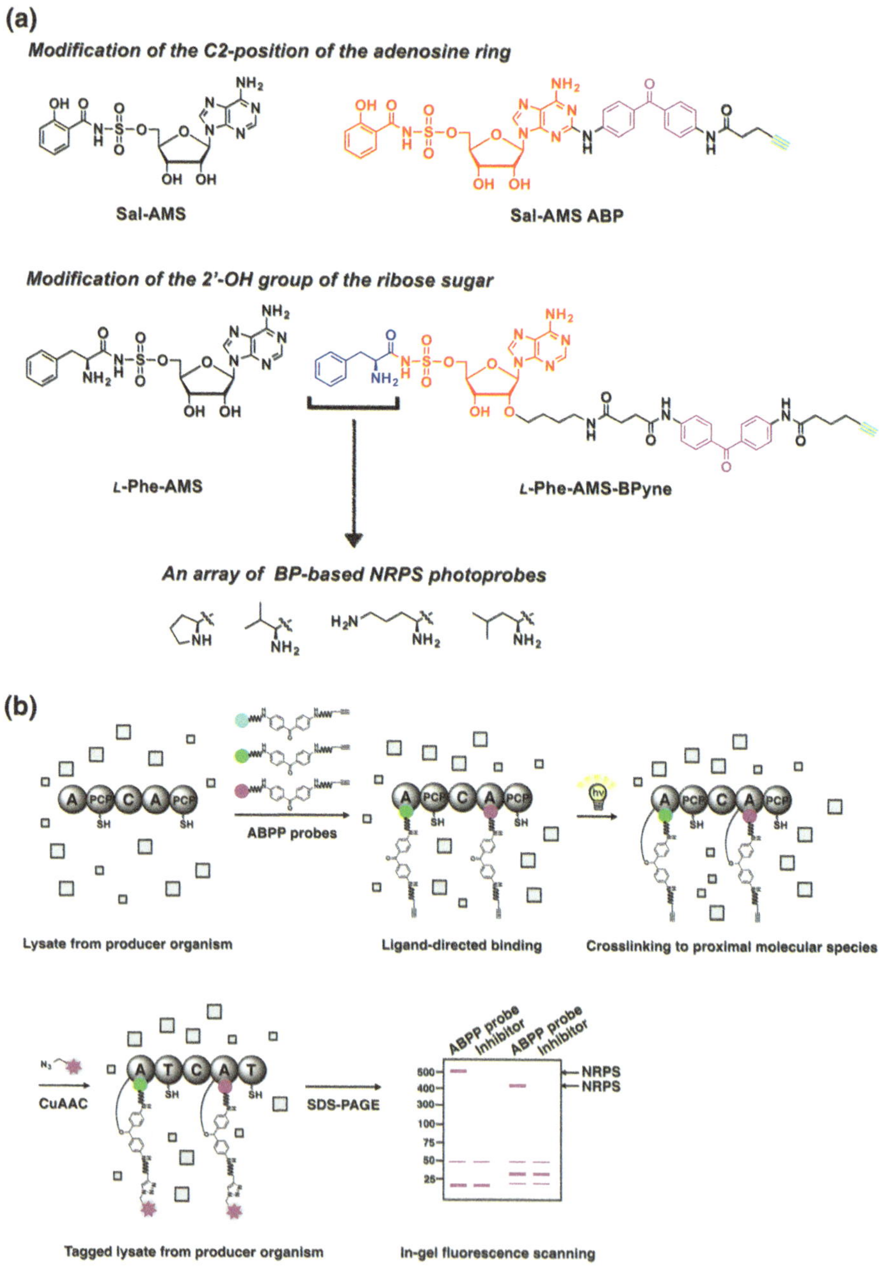

future, this ABPP strategy will elucidate the off targets of Sal-AMS, provide clues to develop more selective inhibitors of the MbtA enzyme, and deepen the understanding of *Mycobacterium* virulence.

◂**Fig. 2** Activity-based protein profiling (ABPP) of non-ribosomal peptide synthetases (NRPSs). **a** Benzophenone (BP)-based ABPP probes for the adenylation (A) domains in NRPS machinery (modifications of the C2 position of the adenosine ring and 2′-OH group of ribose sugar). **b** Methods for gel-based ABPP of NRPS machineries. The rectangles display nonspecific proteins. In gel-based ABPP, proteomes from producer organisms were treated with BP-based NRPS photoprobes and exposed to UV light (365 nm) to conduct cross-linking. Probes were conjugated rhodamine azide under Huisgen's copper(I)-catalyzed azide-alkyne cycloaddition (CuAAC) conditions. Proteomes were analyzed by SDS-PAGE and in-gel fluorescence imaging

2.2 *Benzophenone-Based Photo-affinity Labeling Probes for Adenylation Domains in Non-ribosomal Peptide Synthetases*

An ideal ABPP strategy for the NRPS machinery can selectively detect this enzyme class in an activity-based manner. We chose A-domains from many catalytic domains housed on NRPS enzyme family members. The A-domain is a gatekeeping domain to incorporate an amino acid substrate into peptide-based natural products and is located on each NRPS module. Therefore, the A-domains are a reasonable target to design ABPP probes, develop ABPP strategies, and conduct functional analysis of the NRPS machinery in wild-type, heterologous, and engineered systems. Using the strict substrate recognition properties of A-domains, we expected that an ABPP strategy would accelerate the detection of NRPS activities in complex biological systems, provide detailed analysis of the NRPS enzymology, and deepen the understanding of NRPS-related biochemical pathways in vivo.

We initially demonstrated our ABPP strategy in natural product biosynthesis using gramicidin S NRPSs (GrsA and GrsB synthetases) in the producer proteomes. Gramicidin S is a macrocyclic non-ribosomal peptide antibiotic with the structure cyclo (-*D*-Phe$_1$-*L*-Pro$_2$-*L*-Val$_3$-*L*-Orn$_4$-*L*-Leu$_5$-)$_2$. GrsA (module 1) is a single protein (127 kDa) containing the *L*-Phe-activating A-domain (A$_{Phe}$), PCP domain, and E domain (Fig. 1A) and converts *L*-Phe into *L*-Phe-AMP intermediate (Krätzschmar et al. 1989). In contrast, GrsB is a multifunctional mega-enzyme (508 kDa) comprised of four NRPS modules: C-A$_{Pro}$-PCP (module 2), C-A$_{Val}$-PCP (module 3), C-A$_{Orn}$-PCP (module 4), and C-A$_{Leu}$-PCP-TE (module 5) (Fig. 1a) (Krätzschmar et al. 1989). Each NRPS module (2–5) is responsible for the incorporation of the amino acid substrates, *L*-Pro, *L*-Val, *L*-Orn, and *L*-Leu into the gramicidin S, respectively (Fig. 1a).

The isolation, structure, and biological activities of the antibiotics ascamycin (Isono et al. 1984) indicated a selective inhibitor of alanyl-tRNA synthetase (Ueda et al. 1991). The aminoacylation reactions catalyzed by aminoacyl-tRNA synthetases correspond identically to those of the A-domains of the NRPS machinery. On the basis of ascamycin structure, an 5′-*O*-*N*-(aminoacyl)sulfamoyladenosine (aminoacyl-AMS) inhibitor was designed and synthesized for the *L*-Phe-activating A-domain excised from GrsA (Finking et al. 2003). *L*-Phe-AMS exhibited strong binding to the cognate A-domain, with K_i values of 61 nM (Finking et al. 2003).

A-domain-directed probes were designed based on the aminoacyl-AMS scaffold to retain low nM inhibition properties. We and others have described the development of aminoacyl-AMS with a pegylated biotin linker at the 2′-OH group of the ribose sugar (Finking et al. 2003; Ishikawa and Kakeya 2014; Ishikawa et al. 2015a). The modification at the 2′-OH position of the ribose sugar minimized the binding affinity of the A-domains within the tested NRPS machinery (Finking et al. 2003; Ishikawa and Kakeya 2014; Ishikawa et al. 2015a). On the basis of these results, we attached a functionalized linker including a photoreactive BP and alkyne functionality at this position. A-domain-directed ABPP probes possess three elements: (i) an aminoacyl-AMS ligand that allows ligand-directed binding at the cognate A-domain, (ii) a photoreactive BP that induces cross-linking to nearby molecular species, and (iii) an alkyne functionality that facilitates CuAAC with reporter tags for the rapid and sensitive visualization and affinity purification of labeled proteins (Fig. 2a). The resulting A-domain-directed probe was termed aminoacyl-AMS-BPyne.

Our biochemical studies revealed that BP-based NRPS photoprobes, *L*-Phe-, *L*-Pro-, and *L*-Val-AMS-BPynes, retained tight-binding properties ($K_i^{\text{app.}}$ = 1.43–327 nM) toward the recombinant *L*-Phe-, *L*-Pro-, and *L*-Val-activating A-domains (Konno et al. 2015; Ishikawa et al. 2015b). Using a suite of BP-based NRPS photoprobes *L*-Phe-, *L*-Pro-, *L*-Val-, *L*-Orn, and *L*-Leu-AMS-BPynes, we demonstrated the applicability of probes by applying them to the gramicidin S producers' proteomes (*Aneurinibacillus miglanus* ATCC 9999 and DSM 5759). BP-based NRPS photoprobes were capable of highly selective binding, cross-linking, and labeling the cognate A-domains by ligand-directing targeting in native proteomic environments, even if multiple A-domains were housed on a single megasynthetase (Fig. 2b) (Konno et al. 2015; Ishikawa et al. 2015b). In sequenced producers, the BP-based NRPS photoprobes were used to track the expression dynamics and to optimize bacterial culture conditions (Ishikawa et al. 2015b). In non-sequenced bacteria, this ABPP strategy for the NRPS machinery could facilitate the discovery of peptide-based natural products and their biosynthetic pathways (genes and proteins) using a combination of recent genome sequencing and metagenomics techniques. Furthermore, we developed a highly sensitive detection technique for multidomain NRPS machineries using a cocktail of BP-based NRPS photoprobes (i.e., quadruple labeling of GrsB), affording a new analytical option for the high-resolution detection and functional analysis of low-abundance active NRPS machinery in proteomes, providing further the evidence of an unidentified inactive form of GrsB synthetase in a non-producing strain *A. migulanus* DSM 2895 (Ishikawa et al. 2015c). The ABPP probes described here enable the selective labeling, visualization, and identification of the NRPS machinery in native proteomes from producer organisms. The ABPP platform affords the strategies to discover, quantify, and analyze NRPS-related pathways that are limited to genetic approaches, opening a wide variety of chemical biology studies in natural product sciences.

2.3 Functional Profiling of Adenylation Domains in Complex Proteomes

Because of the multidomain structural property of the NRPS machinery and the general intractability of methods that analyze A-domain functions in complex proteomic samples, advanced strategies that can detect the enzymatic activities of A-domains in NRPSs of proteomic samples from wild-type, heterologous, and engineered systems are required. Indeed, all conventional methods to assess the activities of the A-domains in NRPS machinery rely on pure recombinant enzyme systems (Dorrestein et al. 2006a, b; Otten et al. 2007; McQuade et al. 2009; Phelan et al. 2009; Wilson and Aldrich 2010). Simple, rapid, and reliable proteomic platforms should facilitate the evaluation of enzymatic activities of A-domains in the NRPS machinery on a proteome-wide scale. We and others have described sulfamoyloxy-linked aminoacyl-AMP analogs that are capable of selectively binding A-domains with substrate specificity that corresponds to a conjugated amino acid (Finking et al. 2003; Ishikawa et al. 2015b; Konno et al. 2015). Accordingly, the aminoacyl-AMP analogs discriminate the differential active-site chemistries of the A-domains of the NRPS machinery. On the basis of these results, we have recently described a functional profiling technique for the prediction of A-domain functions in the NRPS machinery by the competitive ABPP mode (Fig. 3) (Kasai et al. 2015). In this technique, a set of sulfamoyloxy-linked aminoacyl-AMP analogs plays a key role in a direct readout of enzyme substrate specificities and the active-site architectures of A-domains in the NRPS machineries. To this end, we prepared sulfamoyloxy-linked aminoacyl-AMP analogs with 19 proteinogenic amino acids (except for *L*-Cys) and *L*-Orn as ligands (Kasai et al. 2015). For the proof-of-principle studies, we tested whether the inhibitor sensitivity profiles of A-domains on the endogenous NRPS machinery corresponded to enzyme substrate specificities using a recombinant NRPS module (GrsA).

The BP-based NRPS photoprobe *L*-Phe-AMS-BPyne was incubated with *A. migulanus* ATCC 9999 proteomes in the absence or presence of the 20 sulfamoyloxy-linked aminoacyl-AMP analogs. In-gel fluorescence scanning showed that *L*-Phe-AMS, *L*-Leu-AMS, *L*-Met-AMS, *L*-Trp-AMS, and *L*-His-AMS inhibited the A-domain of endogenous GrsA. To examine the correlation between the IC_{50} values and substrate specificities, we determined the Michaelis constants (K_m values) of the corresponding amino acid substrates (*L*-Phe, *L*-Leu, *L*-Met, *L*-Trp, and *L*-His). Dose–response competitive ABPP studies of the A-domain of endogenous GrsA by *L*-Phe-AMS, *L*-Leu-AMS, *L*-Met-AMS, *L*-Trp-AMS, and *L*-His-AMS provided IC_{50} values of 0.38 ± 0.14 μM, 23.0 ± 0.12 μM, 27.2 ± 0.22 μM, 9.9 ± 0.13 μM, and 126 ± 0.18 μM, respectively. The Michaelis constants (K_m) toward the amino acid substrates *L*-Phe, *L*-Leu, *L*-Met, *L*-Trp, and *L*-His measured using recombinant GrsA were 0.0248 ± 0.0023 mM, 2.85 ± 0.57 mM, 18.5 ± 3.3 mM, 1.49 ± 0.19 mM, and 17.5 ± 3.0 mM, respectively. In contrast, *L*-Thr-AMS and *L*-Lys-AMS did not show inhibition

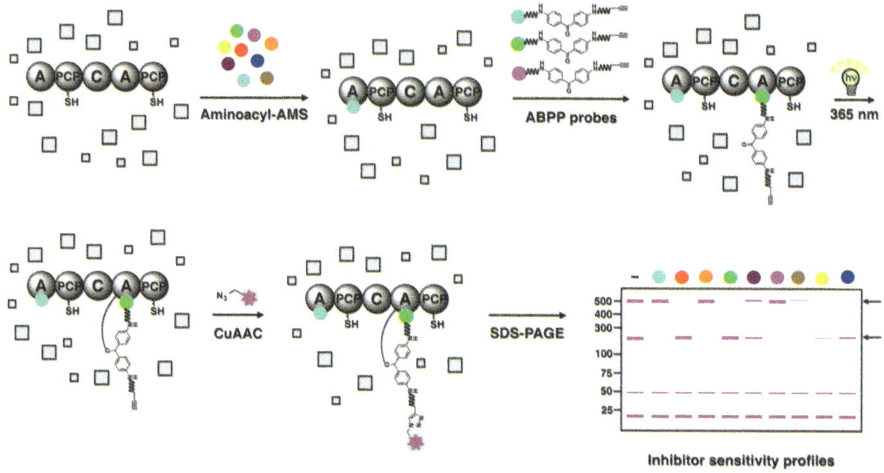

Fig. 3 Profiling of adenylation (A) domain functions of the non-ribosomal peptide synthetase (NRPS) machinery by competitive activity-based protein profiling (ABPP). The substrate candidates and active-site architectures of the A-domains of NRPS machinery can be imaged by pre-incubation of a proteome with a set of sulfamoyloxy-linked aminoacyl-AMP analogs followed by probe treatment. Enzyme substrate candidates are estimated by a reduction in probe labeling intensities (dose–response competitive ABPP experiments). Modules are comprised of peptidyl carrier protein (PCP), adenylation (A), and condensation (C) domains. The rectangles display nonspecific proteins

properties toward the A-domains of endogenous GrsA. Significantly, the K_m value of the A-domain of recombinant GrsA with the corresponding amino acid substrate *L*-Thr was 246 mM, and *L*-Lys resulted in no detectable enzymatic activity. It is noteworthy that the competitive ABPP assay can be used to directly gain substrate candidates and image the active-site architectures of A-domains of endogenous NRPS machinery in complex biological systems. Furthermore, we have demonstrated the applicability of this competitive ABPP toward the four A-domains of endogenous NRPS GrsB megasynthetase in the proteomes (Kasai et al. 2015).

The competitive ABPP mode offers several advantages over conventional techniques. First, the analysis of A-domains in the NRPS machinery is conducted directly in wild-type biological samples, alleviating the need for experimental steps involving isolating, cloning, and sequencing genes and protein expression and purification. Second, a set of BP-based photoprobes and sulfamoyloxy-linked aminoacyl-AMP analogs specifically targets the A-domains with substrate specificity identical to an attached amino acid moiety even if multiple A-domains are housed on multidomain NRPS machinery. Third, in-gel fluorescence scanning provides experimental simplicity, rapidness, and robustness. The competitive ABPP technique should greatly accelerate systematic enzymatic analysis of A-domains in NRPS machinery using a combination of a large set of NRPS gene cluster information and bioinformatics tools (Stachelhaus et al. 1999; Challis et al. 2000; Röttig et al. 2011).

The competitive ABPP technique should be applied in cases where gene cloning or the expression of the NRPS machinery is not successful, and also in producer organisms that are difficult to use the conventional genetic technique. Furthermore, the competitive ABPP strategy could be used to characterize orphan NRPS machinery in native proteomes. It is important to note that functional analysis and engineering of NRPS machinery have been performed using the heterologous expression systems of yeast, fungus, *Streptomyces* strains, and *Escherichia coli* (Zhang et al. 2008). Therefore, the competitive ABPP technique can be further used as a powerful diagnostic tool to investigate substrate candidates and active-site chemistries of the engineered A-domains in NRPSs. Using a combination of non-proteinogenic amino acid sulfamoyl adenosines with functionalized amino acids (i.e., propargyl-*L*-Gly and azide-*L*-Ala), the competitive ABPP technique enables us to construct inhibitor sensitivity profiles toward non-proteinogenic amino acids, which should greatly facilitate the discovery and generation of A-domains with broad substrate specificities and incorporation of non-proteinogenic amino acids into non-ribosomal peptide natural products.

3 Activity-Based Probes for Carrier Proteins in Non-ribosomal Peptide Synthetases

Posttranslational protein modifications (PTMs) regulate protein function in most eukaryotes and prokaryotes and have a key role in a wide range of cellular functions. An in-depth understanding of protein modifications is important to elucidate the in vivo activity of proteins. However, traditional profiling methods can be difficult to permit a comprehensive understanding of biochemical processes at the cellular level (Chuh et al. 2016). Phosphopantetheinylation, a PTM that is crucial in primary and secondary metabolic pathways, is germane natural product biosynthesis (Mercer and Burkart 2007). A carrier protein (CP), such as the acyl carrier protein (ACP) domain in PKS and fatty acid synthase (FAS), the T-domain in NRPS, and ArCP found in siderophore NRPS, must be posttranslationally installed with a Ppant cofactor. The Ppant group from CoA is attached to a conserved Ser residue of inactive *apo*-CPs by PPTase enzymes to form active *holo*-CPs (Beld et al. 2014). Bacteria possess the AcpS-type and Sfp-type PPTases, based on amino acid sequences, primary structures, and their target synthases (Lambalot et al. 1996). The AcpS–PPTase acts on type II FAS ACP (AcpP) (Lambalot and Walsh 1995). Surfactin PPTase (Sfp) is generally responsible for installing Ppant on the CP domains of type I PKS and NRPS machinery (Lambalot et al. 1996; Quadri et al. 1998b). The phosphopantetheinylated CP domains are responsible for carrying substrates and intermediates from the active site to active site in FAS, PKS, and the NRPS machinery to produce primary and secondary metabolites (e.g., lipid A, lipoteichoic acid, and siderophores) that are required for bacteria to survive in environmental and infection circumstances. Accordingly, the phosphopantetheinylation of CP domains is involved in bacterial cell viability and virulence and is also used in the biosynthesis of many pharmaceuticals. The functional expression of these biosynthetic gene clusters depends on complex regulation processes in vivo, involving transcription of related genes to mRNAs, translation of mRNAs to proteins, and modifications of proteins. Overexpression of PPTase genes in *Actinomycetes* awakes cryptic and silenced biosynthetic pathways through the phosphopantetheinyl modification-level regulation of CP domains (Zhang et al. 2017). Furthermore, blocking of the PPTase (AcpS) in *Streptomyces coelicolor* improves actinorhodin production by switching on the upregulation of Sfp (Foley et al. 2009). Taken together, these results emphasize that an accurate assessment of the functional states of proteins in complex proteomes and inside bacterial cells contributes to our understanding of phosphopantetheinylation dynamics in physiological and pathological processes, providing an advance in microbiology.

3.1 Designing Activity-Based Probes for Peptidyl Carrier Proteins

The A-domains in the NRPS machinery catalyze the adenylation reactions of cognate amino acid substrates to form aminoacyl-AMP intermediates in the active sites at the expense of ATP. The A-domains then interact with downstream PCP domains and transfer the amino acid building blocks to the Ppant prosthetic group of the PCP domains. Qiao and co-workers (2007) cleverly designed 5′-(vinylsulfonylamino)deoxyadenosine (AVSN), which covalently attaches to the terminal thiol functionality of the Ppant group on the ArCP domain through the Michael addition reaction. Briefly, the mechanism-based inactivators are non-covalently binding A-domains that covalently trapping the Ppant cofactors of partner PCP domains through highly specific protein–protein interactions. On the basis of these results, we have incorporated an alkyne reporter tag into the scaffold, allowing the PCP domain-tethered ABPs to be visualized and identified (Fig. 4a, b). We recently described a set of photo-crosslinking ABPs for the study of A-domains coupled to aminoacyl-AMS appended to a BP cross-linker and an alkyne tag at the 2′-OH group of the adenosine skeleton. The incorporation of the clickable BP functionality had no effect on the binding properties of a variety of A-domains within the NRPS machinery tested (Ishikawa et al. 2015b; Konno et al. 2015). To this end, we converted the AVSN scaffold to an ABP by installing an alkyne tag at the 2′-OH group of the adenosine skeleton and *L*-Val as a ligand (*L*-Val-AVSN-yne) (Kasai et al. 2016).

After demonstrating the selective labeling of cognate A-PCP motif in recombinant NRPS enzymes by *L*-Val-AVSN-yne, we investigated its use in proteomic samples (Kasai et al. 2016). *B. subtilis* ATCC 21332 biosynthesizes a lipopeptide antibiotic and biosurfactant surfactin (Arima et al. 1968). Three multifunctional NRPS enzymes, SrfAA, SrfAB, and SrfAC, involve in the production of this

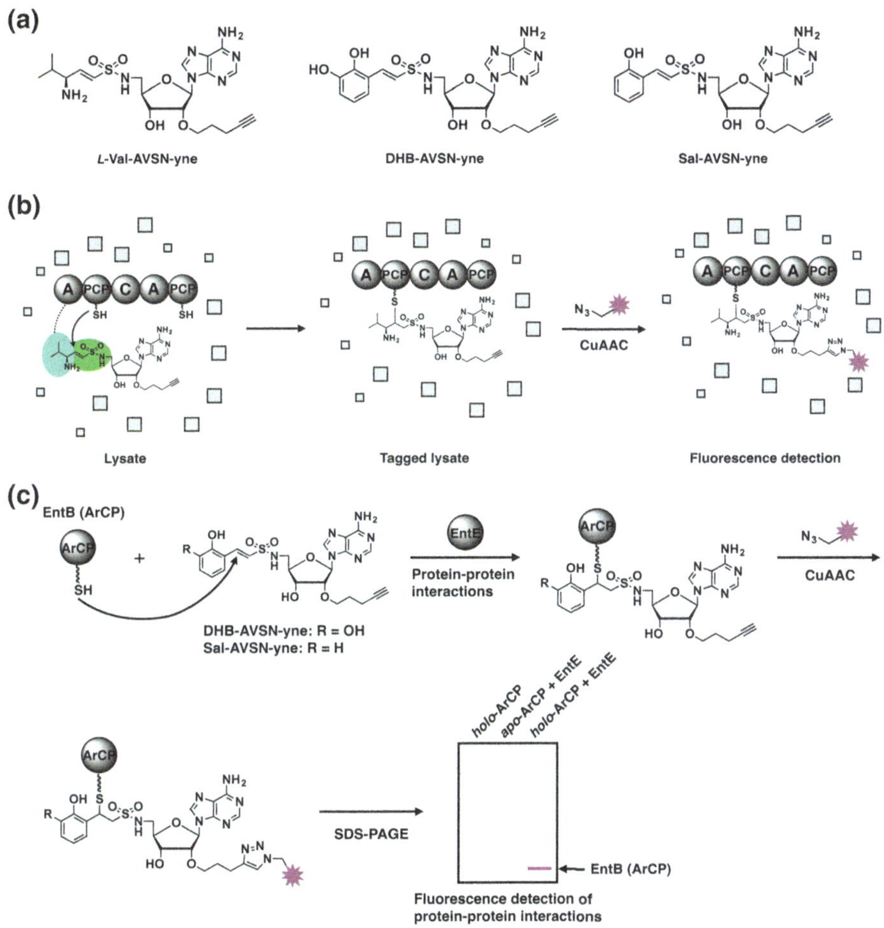

Fig. 4 Activity-based protein profiling (ABPP) of carrier protein (CP) domains in the non-ribosomal peptide synthetase (NRPS) machinery. **a** Structures of ABPP probes. **b** Chemical proteomics for the analysis of native CP motifs in NRPS machinery. The Val-activating domain of the NRPS machinery selectively binds *L*-Val-AVSN-yne by ligand (*L*-Val)-directed targeting, which then facilitates the selective capture of the phosphopantetheine (Ppant) group of a cognate CP through the Michael addition reaction to the vinylsulfonamide reactive group. The tagged lysate treated with rhodamine azide under Huisgen's copper(I)-catalyzed azide-alkyne cycloaddition (CuAAC) conditions and separated by SDS-PAGE. NRPS modules are comprised of peptidyl carrier protein (PCP), adenylation (A), and condensation (C) domains. The rectangles display nonspecific proteins. (c) Fluorescence detection of protein–protein interactions between adenylation (A) and aryl carrier protein (ArCP) domains in NRPS machinery. EntE A-domain recognizes DHB-AVSN-yne as a substrate analog and is capable of transferring it to the Ppant arm of the ArCP domain of EntB to form a covalent bond. The selective protein–protein interactions of EntE-EntB (ArCP) pair were visualized by CuAAC using rhodamine azide

lipopeptide. SrfAA and SrfAB are large multidomain synthetases with calculated molecular weights of 402 and 401 kDa, respectively (Cosmina et al. 1993). SrfAA and SrfAB consist of three NRPS modules, C_1-A_1 (*L*-Glu)-PCP_1-C_2-A_2 (*L*-Leu)-PCP_2-C_3-A_3 (*L*-Leu)-PCP_3-E_3 and C_4-A_4 (*L*-Val)-PCP_4-C_5-A_5 (*L*-Asp)-PCP_5-C_6-A_6 (*L*-Leu)-PCP_6-E_6. Each A-domain activates amino acid substrates (their cognate substrates are shown in parentheses) to construct the lipopeptide scaffold. In contrast, SrfAC is a single-module NRPS containing the domain structure C_7-A_7 (*L*-Leu)-PCP_7-TE_7 and is responsible for the incorporation of *L*-Leu during the surfactin biosynthesis. Accordingly, surfactin synthetases (SrfAA, SrfAB, and SrfAC) possess seven PCP domains of the 24 catalytic components, which are modified with the Ppant cofactor. Selective labeling of one of the seven PCP domains (e.g., PCP_4 domain) is therefore extremely difficult. The ligand amino acid moiety provides absolute specificity to an ABP by binding the Val-activating A-domain (A_4-domain) as the substrate surrogate (Fig. 4b). *B. subtilis* ATCC 21332 lysate was treated with *L*-Val-AVSN-yne and subsequently incubated with Rh-azide under the CuAAC conditions. We observed the selective labeling of SrfAB synthetase, which was the only fluorescent protein in the cellular lysate. Furthermore, the fluorescence labeling by *L*-Val-AVSN-yne was abrogated in the presence of *L*-Val-AMS, which is a tight-binding, reversible inhibitor of the Val-activating A-domain (Ishikawa et al. 2015b). These results indicate that the ligand *L*-Val of the probe is directed to the active site of Val-activating domain (A_4-domain) of SrfAB synthetase and then drives the Michael addition reaction with the thiol of the Ppant cofactor on the PCP_4 domain through the mechanism-based enzymatic reaction. This labeling strategy should accelerate the understanding of activity, transcriptional regulation, and modification processes of NRPS machinery and provide avenues for studies of the phosphopantetheinylation dynamics important for both primary and secondary metabolic pathways.

3.2 A Chemoproteomics Platform for Phosphopantetheine Transferase Activity

Significant antibiotic resistance has emerged toward virtually all clinically relevant antibiotics, which is a growing global health problem. We now are in a race to discover and develop new classes of antibiotics to combat the growing emergence of antibiotic-resistant strains. The discovery and development of new antimicrobials with distinct mechanisms of action are urgently required. New strategies and targets have been revealed by the effort over the last two decades by investigation of how bacteria cause infection, host damage, and disease.

Many natural products frequently function as virulence factors for infection (Clatworthy et al. 2007). They are not essential for bacterial viability but are required to cause host damage and disease. PPTase enzymes catalyze the attachment of a 4′-phosphopantetheinyl moiety of CoA to a conserved Ser residue of CP domains in FAS, PKS, and NRPS biosynthetic enzymes, which convert the inactive *apo*-CP to the active *holo*-CP (Beld et al. 2014). Therefore, PPTase enzymes represent promising new antibiotic targets because their inhibition should antagonize the production of primary and secondary metabolites involving virulence factors essential to many pathogenic bacteria. A majority of bacteria contain two structurally distinct PPTase enzymes, AcpS (acyl carrier protein synthase) in *E. coli* and Sfp from *B. subtilis*. Mbt has the Sfp-type PPTase PptT, which modifies various type I PKS enzymes for the production of mycolic acids and lipid virulence factors (Quadri et al. 1998a). PptT activated NRPS biosynthetic enzymes, MbtB and MbtE in vitro biochemical experiments (Quadri et al. 1998a). MbtB and MbtE enzymes are involved in the biosynthesis of the siderophore mycobactin that functions as an Mbt virulence factor. PptT plays a central role in the production of secondary metabolites that are essential for in vivo viability (Chalut et al. 2006). Thus, small molecules that inhibit PptT represent a promising new class of antibiotics.

A high-throughput screening campaign was done to search for small-molecule inhibitors toward Sfp-PPTase (Foley et al. 2014). The authors discovered 2-pyridinyl-*N*-(4-aryl)-piperazine-1-carbothioamides and subsequently conducted structure–activity relationship study, which resulted in the development of 4-(3-chloro-5-(trifluoromethyl)pyridine-2-yl)-*N*-(4-methoxypyridin-2-yl)piperidine-1-carbothioamide (ML267). Biological activity studies indicated that ML267 possesses dual activity toward the bacterial Sfp-PPTase and AcpS-PPTase, displaying IC_{50} values of 290 nM and 8.1 μM, respectively (Foley et al. 2014). Furthermore, no cross-reactivity was observed toward the human ortholog (Foley et al. 2014). Moreover, ML267 has been shown to reduce the production of a Sfp-PPTase-dependent surfactin and the viability of bacterial cells when *B. subtilis* OKB105 was treated with this small-molecule inhibitor. However, while ML267 has displayed the potential activity in vivo, very little is known about the changes in protein expression, activity, and PTMs in living bacterial cells.

We established a chemoproteomics platform to explore the PPTase activity at the cellular level and applied it to gain a deeper understanding of the in vivo activity of ML267 (Konno et al. 2017). Using the ABPP probe *L*-Val-AVSN-yne, we have demonstrated the mechanism-based labeling of the PCP_4 domain of SrfAB synthetase in proteomes (Kasai et al. 2016). Furthermore, we have demonstrated that the active-site-directed proteomic probe *L*-Leu-AMS-BPyne allows selective labeling of the Leu-activating domains of NRPSs in proteomic samples (Ishikawa et al. 2015b). In order to investigate the phosphopantetheinylated PCP_4 domain of SrfAB synthetase, proteomes from cultures in the absence and presence of ML267 were incubated with *L*-Val-AVSN-yne. To track the expression and activities of surfactin synthetases (SrfAA, SrfAB, and SrfAC), lysates were treated with ML267 or dimethylsulfoxide (DMSO) was treated with *L*-Leu-AMS-BPyne. After in-gel fluorescence imaging, the signals of SrfAB labeled by *L*-Val-AVSN-yne were quantified and expressed relative to SrfAB intensities labeled by *L*-Leu-AMS-BPyne.

When surfactin production was periodically measured, the production levels of surfactin by *B. subtilis* ATCC 21332 were attenuated in the presence of ML267.

NRPS activity-profiling experiments of the proteomes using *L*-Leu-AMS-BPyne provided the different patterns of enzyme activities of SrfAA, SrfAB, and SrfAC synthetases treated with ML267, relative to the lysates treated with DMSO. Furthermore, assessment of the functional states of PCP domains of NRPSs in the proteomes using *L*-Val-AVSN-yne revealed a labeled endogenous SrfAB synthetase using the proteome from DMSO treatment, whereas the labeling of proteomic samples from cultures in the presence of ML267 was a weak fluorescent signal. Quantification analysis of phosphopantetheinylated SrfAB (PCP_4 domain) bands showed a 45% reduction by the addition of ML267 relative to the DMSO control culture, validating a lack of phosphopantetheinyl modifications (down-regulation of the Sfp-PPTase activity) and the inhibitory activity of ML267 in vivo, for which no previous survey to changes in enzyme expression and activity in bacterial cells has been demonstrated.

3.3 Visualizing the Protein–Protein Interactions of Aryl Acid-Adenylating Enzymes

Biochemical, structural, and activity investigations have revealed the dynamic, transient, and specific roles of ACP domains in FASs and PKSs (Parris et al. 2000; Płoskoń et al. 2010; Whicher et al. 2014; Ishikawa et al. 2016). The ACP domains found in FASs and PKSs are highly flexible proteins that must specifically interact with all enzyme domains responsible for loading carbon units, condensing these units, decorating the condensation products, and releasing the final product (Crosby and Crump 2012). Each step requires highly specific protein–protein interactions to channel the biosynthetic intermediates to the partner enzymes. However, the transient ACP–partner enzyme interactions impose major obstacles to the study of ACP activity and protein–protein interactions at a molecular level (Jenni et al. 2006; Maier et al. 2006). To solve these problems, several chemistry-based techniques have been developed over the last decade. A set of synthetic probes was developed to investigate ACP activity and trap transient protein–protein interactions with ketosynthase (KS), thioesterase (TE), and dehydratase (DH) in FAS and PKS biosynthetic machineries, including β-chloroacryl amide, α-bromopalmitic amide, and sulfonyl 3-alkyne amide-based pantetheine analogs, respectively (Worthington et al. 2006; Blatti et al. 2012; Ishikawa et al. 2013) (Fig. 5). These chemical probes enable the irreversible capture of the active-site residues of partner enzymes in an activity-based manner, resulting in the covalent cross-linking of protein partners in functional states. A pantetheineamide probe containing a tight-binding reversible inhibitor triclosan was used to study the interface between AcpP (ACP) and FabI (enoyl reductase, ER) in the *E. coli* FAS (Tallorin et al. 2016). Using a coupled chemoenzymatic protocol described previously (Worthington and Burkart 2006), Naguyen et al. (2014) described the active-site covalent cross-linking of AcpP to FabA (DH) in the *E. coli* FAS and reported the crystal structure of the cross-linked

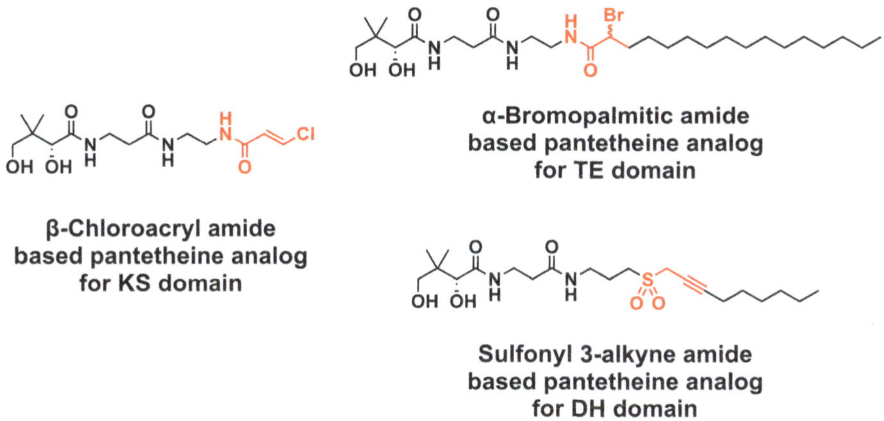

Fig. 5 Synthetic probes for trapping protein–protein interactions between acyl carrier protein (ACP) and ketosynthase (KS)/thioesterase (TE)/dehydratase (DH) domains in fatty acid and polyketide synthases

AcpP-FabA complex, representing the first snapshots of ACP interactions (Naguyen et al. 2014). An in vitro cross-linking strategy was used to improve ACP compatibility with non-cognate product template (PT) domains in PKS enzymes (Barajas et al. 2016).

Liu and Bruner (2007) have synthesized α-chloroacetyl-amino-CoA to covalently capture a catalytic Ser residue in the active site of TE domain, which enabled to gain snapshots of PCP-TE didomain protein from enterobactin synthetase (EntF) in a functional state. Qiao et al. (2007) developed a suite of mechanism-based probes that embedded a vinylsulfonamide Michael acceptor functionality. The mechanism-based probes are recognized by adenylating enzymes as substrate surrogates and shuttled to CPs through the highly specific protein–protein interactions. The substrate recognition characteristics of the adenylating enzymes allow mechanism-based trapping of the thiol group of the Ppant prosthetic group on CPs. Other studies explored the biochemical and structural challenges of the functional interactions of PCP and ArCP domains with A-domains and aryl acid-adenylating enzymes, which provided views of A-PCP and A-ArCP interactions in the NRPS machinery (Mitchell et al. 2012; Sundlov et al. 2012). Accordingly, these studies revealed structural properties that are important for CP-mediated protein–protein interactions and provided a guideline for engineering efforts. As well as PKS and FAS biosynthetic machinery, the structural basis for catalysis with NRPS machinery will accelerate engineering to generate non-native products. Therefore, the development of rapid and simple assay systems to study protein–protein interactions should enable us to evaluate key surface-interacting residues and also to improve CP compatibility with non-cognate partner complexes.

Type II NRPS enzymes function in concert with FASs, PKSs, and type I NRPSs to generating peptide-based natural products. Whether incorporated into large multifunctional enzymes (type I system) or in discrete enzymes functioning independently (type II system), CP domains must interact with partner enzymes. The activities of PCPs/ArCPs and A-domains in all NRPS-related pathways are particularly important for the incorporation of amino acid and aryl acid substrates into NRPS machinery. Aryl acids represented by Sal, 2′,3′-dihydroxybenzoic acid (DHB), 2-aminobenzoic acid, 3-hydroxypicolinic acid, and 5-methyl orsellinic acid (5-MOA) are an important class of building blocks in non-ribosomal peptide products (Ames and Walsh 2010; Huang et al. 2011; Wang et al. 2014). Aryl acid building blocks involving Sal and DHB are most commonly observed in siderophore molecules (Barry and Challis 2009). Enterobactin is a cyclic catecholic siderophore biosynthesized by *E. coli*. The biosynthetic machinery is comprised of six enzymes encoded by the *entABCDEF* genes (Walsh et al. 1990). The NRPS gene products EntB, EntE, and EntF achieve the biosynthesis of enterobactin (Gehring et al. 1998). EntB and EntE are type II NRPS enzymes that act as an ArCP and aryl acid-adenylating enzyme, respectively. The EntB protein is a bifunctional enzyme consisting of a *C*-terminal ArCP domain and an *N*-terminal isochorismate lyase (ICL) (Gehring et al. 1998). The EntE-adenylating enzyme catalyzes the formation of DHB-AMP intermediate. The adenylated DHB substrate, in turn, undergoes a nucleophilic attack by the thiol group of the Ppant prosthetic group of the ArCP domain of EntB (Rusnak et al. 1989; Gehring et al. 1998). The type I NRPS enzyme EntF consists of four catalytic domains C-A-PCP-TE and is responsible for the activation, loading, and incorporation of serine, forming the three amide and three ester bonds during the biosynthesis of iron-chelating siderophore enterobactin. Finally, the terminal TE domain of EntF catalyzes the hydrolysis of three molecules of DHB-Ser by intermolecular cyclization. EntE, the ArCP domain of EntB, and the C-domain of EntF must therefore interact with each other through intramolecular protein–protein interactions during the enterobactin biosynthesis. Significantly, understanding the interaction surfaces of the ArCP and A-domain at molecular levels may enable us to fine-tune the recognition specificities for use in the combinatorial biosynthesis of aryl acid-containing non-ribosomal peptides. Herein, we have developed a new gel-based fluorescence method coupled with ABPs to visualize and access the adenylation activities and functional protein–protein interactions of aryl acid-adenylating enzymes (Fig. 4c) (Ishikawa et al. 2017).

A clickable mechanism-based probe, *L*-Val adenosine vinylsulfoneamide alkyne (*L*-Val-AVSN-yne), has been used to selectively visualize posttranslationally modified PCP domains in NRPSs in native proteomic samples described above (Kasai et al. 2016). The strict substrate recognition properties of A-domains enable covalent capture of the terminal thiol of a Ppant functionality through a mechanism-based Michael addition. On the basis of these results, we incorporated DHB and Sal components into the AVSN-yne scaffold to construct an in-gel fluorescence assay system and further visualize adenylation activities and protein–protein interactions by using the type II ArCP (EntB) and aryl acid-adenylating enzymes (wild-type EntE and modified EntE) (Figs. 4a, c). Briefly, ABPs (DHB- and Sal-AVSN-yne) were incubated with *apo*- and *holo*-ArCP domain of EntB in the absence or presence of EntE. These samples were then treated with Rh-azide under CuAAC conditions and were simply analyzed by in-gel fluorescence scanning. The EntE-adenylating enzyme recognized DHB-AVSN-yne as a substrate surrogate and could shuttle it to the ArCP domain of EntB, which resulted in trapping the Ppant functionality of ArCP domain. The transient and dynamic recognition processes could be visualized by the fluorescent labeling of the ArCP protein. Notably, the DHB-AVSN-yne did not show any reactivity toward the inactive *apo*-form. Furthermore, the labeling of *holo*-ArCP domain of EntB in the absence of EntE showed markedly low levels of fluorescence signals. Also, Sal-AVSN-yne was not capable of binding the active site of EntE as a substrate, which is in agreement with the substrate specificity of EntE enzyme (Ishikawa et al. 2015a). Accordingly, the EntE-adenylating enzyme was able to accept DHB-AVSN-yne as a specific substrate and transport it to the downstream ArCP domain of EntB through a highly specific protein–protein interaction. Non-ribosomal pathways rely on the fidelity of protein–protein interactions with cognate A-PCP pairs to incorporate amino acid and aryl acid building blocks into peptide products (Jaremko et al. 2017). This assay system should be immediately useful for identifying key interacting regions and residues manipulating protein–protein interactions between non-native pairs toward the combinatorial biosynthesis of new NRP products.

4 Concluding Remarks

The NRPS machinery plays a key role in the production of many pharmaceuticals, as well as virulence factors. Because of the large molecular weight with a range between 100 and 1600 kDa, the routine cloning and heterologous expression of whole NRPS genes remain challenging. Advanced strategies that access endogenous NRPS machinery in complex biological systems provide highly complementary genetic approaches and accelerate enzymology and chemical biology studies. In addition, the natural product biosynthetic machinery involving virulence factors represents an attractive antimicrobial target for the development of novel antibiotics. It is noteworthy that the assessment of these activity changes in infection processes and in in vivo biochemical pathways they affect has lagged behind because of the incomplete understanding of the proteomic levels. New methods to not only rapidly verify the activity parameters, functional prediction, and assessment, but also facilitate the detection, validation, and imaging of NRPS machinery in complex biological systems are needed. Moreover, the understanding of protein–protein interactions between PCP and ArCP domains and partner enzymes at molecular levels is prerequisite for the successful engineering of NRPS machinery. Accordingly, simple and rapid assay systems are required to investigate CP-mediated protein–protein interactions. Our ongoing studies toward the application of ABPP strategies provide new insights into NRPS enzymology, in-depth understanding of NRPS-related biochemical pathways in vivo, and validating and imaging of NRPS activities within the native producer bacteria.

Acknowledgements This work was supported by the Japan Society for the Promotion of Science (JSPS) through a Grant-in Aid for Scientific Research on Innovative Areas [17H05438 (F.I.) and 17H06401 (H.K.)] and the Antiaging Project for Private Universities.

Conflict of Interest
The authors declare no competing financial interest.

References

Ames BD, Walsh CT (2010) Anthranilate-activating modules from fungal nonribosomal peptide assembly lines. Biochemistry 49:3351–3365

Arima K, Kakinuma A, Tamura G (1968) Surfactin, a crystalline peptidelipid surfactant produced by Bacillus subtilis: isolation, characterization and its inhibition of fibrin clot formation. Biochem Biophys Res Commun 31:488–494

Baltz RH (2014) Combinatorial biosynthesis of cyclic lipopeptide antibiotics: a model for synthetic biology to accelerate the evolution of secondary metabolite biosynthetic pathways. ACS Synth Biol 3:748–758

Barajas JF, Finzel K, Valentic TR et al (2016) Structural and biochemical analysis of protein–protein interactions between the acyl-carrier protein and product template domain. Angew Chem Int Ed 55:13005–13009

Barry SM, Challis GL (2009) Recent advances in siderophore biosynthesis. Curr Opin Chem Biol 13:205–215

Beld J, Sonnenschein EC, Vickery CR et al (2014) The phosphopantetheinyl transferases: catalysis of a post-translational modification crucial for life. Nat Prod Rep 31:61–108

Blatti JL, Beld J, Behnke CA et al (2012) Manipulating fatty acid biosynthesis in microalgae for biofuel through protein-protein interactions. PLoS ONE 7:e42949

Bode HB, Müller R (2005) The impact of bacterial genomics on natural product research. Angew Chem Int Ed 44:6828–6846

Bozhüyük KAJ, Fleischhacker F, Linck A et al (2018) De novo design and engineering of non-ribosomal peptide synthetases. Nat Chem. https://doi.org/10.1038/NCHEM.2890

Brieke C, Yim G, Peschke M et al (2016) Catalytic promiscuity of glycopeptide *N*-methyltransferases enables bio-orthogonal labelling of biosynthetic intermediates. Chem Commun 52:13679–13682

Bumpus SB, Kelleher NL (2008) Accessing natural product biosynthetic processes by mass spectrometry. Curr Opin Chem Biol 12:475–482

Bumpus SB, Evans BS, Thomas PM et al (2009) A proteomics approach to discovering natural products and their biosynthetic pathways. Nat Biotechnol 27:951–956

Calcott MJ, Owen JG, Lamont IL et al (2014) Biosynthesis of novel pyoverdines by domain substitution in a nonribosomal peptide synthetase of *Pseudomonas aeruginosa*. Appl Environ Microbiol 80:5723–5731

Challis GL, Ravel J, Townsend CA (2000) Predictive, structure-based model of amino acid recognition by nonribosomal peptide synthetase adenylation domains. Chem Biol 7:211–224

Chalut C, Botella L, Sousa-D'Auris C et al (2006) The nonredundant roles of two 4'-phosphopantetheinyl transferases in vital processes of *Mycobacteria*. Proc Natl Acad Sci U S A 103:8511–8516

Chen WH, Li K, Sandhya N et al (2016) Interdomain and intermodule organization in epimerization domain containing nonribosomal peptide synthetases. ACS Chem Biol 11:2293–2303

Chuh KN, Batt AR, Pratt MR (2016) Chemical methods for encoding and decoding of posttranslational modifications. Cell Chem Biol 23:86–107

Cisar JS, Tan DS (2008) Small molecule inhibition of microbial natural product biosynthesis—an emerging antibiotic strategy. Chem SocRev 37:1320–1329

Clatworthy AE, Pierson E, Hung DT (2007) Targeting virulence: a new paradigm for antimicrobial therapy. Nat Chem Biol 3:541–548

Conti E, Stachelhaus T, Marahiel MA et al (1997) Structural basis for the activation of phenylalanine in the non-ribosomal biosynthesis of gramicidin S. EMBO J 16: 4174–4183

Cosmina P, Rodriguez F, de Ferra F et al (1993) Sequence and analysis of the genetic locus responsible for surfactin synthesis in *Bacillus subtilis*. Mol Microbiol 8:821–831

Cravatt BF, Wright AT, Kozarich JW (2008) Activity-based protein profiling: from enzyme chemistry to proteomic chemistry. Annu Rev Biochem 77:383–414

Crosby J, Crump MP (2012) The structural role of the carrier protein—active controller or passive carrier. Nat Prod Rep 29:1111–1137

De Voss JJ, Rutter K, Schroeder BG et al (2000) The salicylate-derived mycobactin siderophores of *Mycobacterium tuberculosis* are essential for growth in macrophages. Proc Natl Acad Sci U S A 97:1252–1257

Dormán G, Nakamura H, Pulsipher A et al (2016) The life of Pi star: exploring the exciting and forbidden worlds of the benzophenone photophore. Chem Rev 116:15284–15398

Dorrestein PC, Blackhall J, Straight PD et al (2006a) Activity screening of carrier domains within nonribosomal peptide synthetases using complex substrate mixtures and large molecule mass spectrometry. Biochemistry 45:1537–1546

Dorrestein PC, Bumpus SB, Calderone CT et al (2006b) Facile detection of acyl and peptidyl intermediates on thiotemplate carrier domains via phosphopantetheinyl elimination reactions during tandem mass spectrometry. Biochemistry 45:12756–12766

Duckworth BP, Wilson DJ, Nelson KM et al (2012) Development of a selective activity-based probe for adenylating enzymes: profiling MbtA involved in siderophore biosynthesis from mycobacterium tuberculosis. ACS Chem Biol 7:1653–1658

Evans BS, Chen Y, Metcalf WW et al (2011) Directed evolution of the nonribosomal peptide synthetase AdmK generates new andrimid derivatives in vivo. Chem Biol 18:601–607

Ferreras JA, Ryu JS, Di Lello F et al (2005) Small-molecule inhibition of siderophore biosynthesis in *Mycobacterium tuberculosis* and *Yersinia petis*. Nat Chem Biol 1:29–32

Finking R, Neumüller A, Solsbacher J et al (2003) Aminoacyl adenylate substrate analogues for the inhibition of adenylation domains of nonribosomal peptide synthetases. ChemBioChem 4:903–906

Fischbach MA, Walsh CT (2006) Assembly-line enzymology for polyketide and nonribosomal peptide antibiotics: logic, machinery, and mechanisms. Chem Rev 106:3468–3496

Foley TL, Young BS, Burkart MD (2009) Phosphopantetheinyl transferase inhibition and secondary metabolism. FEBS J 276:7134–7145

Foley TL, Rai G, Yasgar A et al (2014) 4-(3-Chloro-5-(trifluoromethyl)pyridin-2-yl)-*N*-(4-methoxypyridin-2-yl) piperazine-1-carbothioamide (ML267), a potent inhibitor of bacterial phosphopantetheinyl transferase that attenuates secondary metabolism and thwarts bacterial growth. J Med Chem 57:1063–1078

Gehring AM, Mori I, Walsh CT (1998) Reconstitution and characterization of the Escherichia coli enterobactin synthetase from EntB, EntE, and EntF. Biochemistry 37:2648–2659

Gulick AM (2016) Structural insights into the necessary conformational changes of modular nonribosomal peptide synthetases. Curr Opin Chem Biol 35:89–96

Hornbogen T, Riechers SP, Prinz B et al (2007) Functional characterization of the recombinant *N*-methyltransferase domain from the multienzyme enniatin synthetase. ChemBioChem 8:1048–1054

Hoyer KM, Mahlert C, Marahiel MA (2007) The iterative gramicidin S thioesterase catalyzes peptide ligation and cyclization. Chem Biol 14:13–22

Huang T, Wang Y, Yin J et al (2011) Identification and characterization of the pyridomycin biosynthetic gene cluster of *Streptomyces pyridomyceticus* NRRL B-2517. J Biol Chem 286:20648–20657

Hur GH, Vickery CR, Burkart MD (2012) Exploration of catalytic domains in non-ribosomal peptide synthetase enzymology. Nat Prod Rep 10:1074–1098

Isono K, Uramoto M, Kusakabe H et al (1984) Ascamycin and dealanylascamycin, nucleoside antibiotics from *Streptomyces* sp. J Antibiot 37:670–672

Ishikawa F, Haushalter RW, John Lee D et al (2013) Sulfonyl 3-alkynyl pantetheinamides as mechanism-based cross-linkers of acyl carrier protein dehydratase. J Am Chem Soc 135:8846–8849

Ishikawa F, Kakeya H (2014) Specific enrichment of nonribosomal peptide synthetase module by an affinity probe for adenylation domains. Bioorg Med Chem Lett 24:865–869

Ishikawa F, Miyamoto K, Konno S et al (2015a) Accurate detection of adenylation domain functions in nonribosomal peptide synthetases by an enzyme-linked immunosorbent assay system using active site-directed probes for adenylation domains. ACS Chem Biol 10:2816–2826

Ishikawa F, Konno S, Suzuki T et al (2015b) Profiling nonribosomal peptide synthetase activities using chemical proteomic probes for adenylation domains. ACS Chem Biol 10:1989–1997

Ishikawa F, Suzuki T, Dohmae N et al (2015c) A multiple-labeling strategy for nonribosomal peptide synthetases using active-site-directed proteomic probes for adenylation domains. ChemBioChem 16:2590–2594

Ishikawa F, Sugimoto H, Kakeya H (2016) In vitro investigation of crosstalk between fatty Acid and polyketide synthases in the andrimid biosynthetic assembly line. ChemBioChem 17:2137–2142

Ishikawa F, Kasai S, Kakeya H et al (2017) Visualizing the adenylation activities and protein–protein interactions of aryl acid adenylating enzymes. ChemBioChem 18:2199–2204

Jaremko MJ, John Lee D, Patel A et al (2017) Manipulating protein–protein interactions in nonribosomal peptide synthetase type II peptidyl carrier proteins. Biochemistry 56:5269–5273

Jenni S, Leibundgut M, Maier T et al (2006) Architecture of a fungal fatty acid synthase at 5 Å resolution. Science 311:1263–1268

Kasai S, Konno S, Ishikawa F et al (2015) Functional profiling of adenylation domains in nonribosomal peptide synthetases by competitive activity-based protein profiling. Chem Commum 51:15764–15767

Kasai S, Ishikawa F, Suzuki T et al (2016) A chemical proteomic probe for detecting native carrier protein motifs in nonribosomal peptide synthetases. Chem Commum 52:14129–14132

Keating TA, Gary Marshall C, Walsh CT (2000) Reconstitution and characterization of the Vibrio cholerae vibriobactin synthetase from VibB, VibE, VibF, and VibH. Biochemistry 39:15522–15530

Keating TA, Ehmann DE, Kohli RM et al (2001) Chain termination steps in nonribosomal peptide synthetase assembly lines: directed acyl-*S*-enzyme breakdown in antibiotic and siderophore biosynthesis. ChemBioChem 2:99–107

Kohli RM, Walsh CT, Burkart MD (2002) Biomimetic synthesis and optimization of cyclic peptide sntibiotics. Nature 418:658–661

Kolb HC, Sharpless KB (2003) The growing impact of click chemistry on drug discovery. Drug Discov Today 8:1128–1137

Konno S, Ishikawa F, Suzuki T et al (2015) Active site-directed proteomic probes for adenylation domains in nonribosomal peptide synthetases. Chem Commum 51:2262–2265

Konno S, Ishikawa F, Suzuki T et al (2017) A chemoproteomics approach to investigate phosphopantetheine transferase activity at the cellular level. ChemBioChem 18:1855–1862

Krätzschmar J, Krause M, Marahiel MA (1989) Gramicidin S biosynthesis operon containing the structural genes *grsA* and *grsB* has an open reading frame encoding a protein homologous to fatty acid thioesterases. J Bacteriol 171:5422–5429

Kries H, Wachtel R, Pabst A et al (2014) Reprogramming nonribosomal peptide synthetases for "clickable" amino acids. Angew Chem Int Ed 53:10105–10108

Kries H, Niquille DL, Hilvert D (2015) A subdomain swap strategy for reengineering nonribosomal peptides. Chem Biol 22:640–648

Kries H (2016) Biosynthetic engineering of nonribosomal peptide synthetases. J Pept Sci 9:564–570

Lamb AL (2015) Breaking a pathogen's iron will: inhibiting siderophore production as an antimicrobial strategy. Biochim Biophys Acta 1854:1054–1070

Lambalot RH, Walsh CT (1995) Cloning, overproduction, and characterization of the *Escherichia coli* holo-acyl carrier protein synthase. J Biol Chem 270:24658–24661

Lambalot RH, Gehring AM, Flugel RS et al (1996) A new enzyme superfamily—the phosphopantetheinyl transferases. Chem Biol 3:923–936

Lee JY, Janes BK, Passalacqua KD et al (2007) Biosynthetic analysis of the petrobactin siderophore pathway from *Bacillus anthracis*. J Bacteriol 189:1698–1710

Liu Y, Bruner SD (2007) Rational manipulation of carrier-domain geometry in nonribosomal peptide synthetases. ChemBioChem 8:617–621

Maier T, Jenni S, Ban N (2006) Architecture of mammalian fatty acid synthase at 4.5 Å resolution. Science 311:1258–1263

Marshall CG, Burkart MD, Keating TA et al (2001) Heterocycle formation in vibriobactin biosynthesis: alternative substrate utilization and identification of a condensed intermediate. Biochemistry 40:10655–10663

McQuade TJ, Shallop AD, Sheoran A et al (2009) A nonradioactive high-throughput assay for screening and characterization of adenylation domains for nonribosomal peptide combinatorial biosynthesis. Anal Biochem 386:244–250

Meier JL, Niessen S, Hoover HS et al (2009) An orthogonal active site identification system (OASIS) for proteomic profiling of natural product biosynthesis. ACS Chem Biol 11:948–957

Meier JL, Burkart MD (2010) Proteomic analysis of polyketide and nonribosomal peptide biosynthesis. Curr Opin Chem Biol 15:1–9

Meluzzi D, Zheng WH, Hensler M et al (2008) Top-down mass spectrometry on low-resolution instruments: characterization of phosphopantetheinylated carrier domains in polyketide and non-ribosomal biosynthetic pathways. Bioorg Med Chem Lett 18:3107–3111

Mercer AC, Burkart MD (2007) The ubiquitous carrier protein—a window to metabolite biosynthesis. Nat Prod Rep 24:750–773

Miethke M, Bisseret P, Beckering CL et al (2006) Inhibition of aryl acid adenylation domains involved in bacterial siderophore synthesis. FEBS J 273:409–419

Mihara K, Tanabe T, Yamakawa Y et al (2004) Identification and transcriptional organization of a gene cluster involved in biosynthesis and transport of acinetobactin, a siderophore produced by *Acinetobacter baumannii* ATCC 19606T. Microbiology 150:2587–2597

Miller DA, Luo L, Hillson N et al (2002) Yersiniabactin synthetase: a four-protein assembly line producing the nonribosomal peptide/polyketide hybrid siderophore of *Yersinia pestis*. Chem Biol 9:333–344

Mitchell CA, Shi C, Aldrich CC et al (2012) Structure of PA1221, a nonribosomal peptide synthetase containing adenylation and peptidyl carrier protein domains. Biochemistry 51:3252–3263

Mootz HD, Schwarzer D, Marahiel MA (2002) Ways of assembling complex natural products on modular nonribosomal peptide synthetases. ChemBioChem 3:490–504

Moran S, Rai DK, Clark BR et al (2009) Precursor-directed biosynthesis of fluorinated iturin A in *Bacillus* spp. Org Biomol Chem 7:644–646

Neres J, Labello NP, Somu RV et al (2008) Inhibition of siderophore biosynthesis in *Mycobacterium tuberculosis* with nucleoside bisubstrate analogues: Structure–activity relationships of the nucleobase domain of 5′-*O*-[*N*-(salicyl)sulfamoyl]adenosine. J Med Chem 51:5349–5370

Nguyen C, Haushalter RW, John Lee D et al (2014) Trapping the dynamic acyl carrier protein in fatty acid biosynthesis. Nature 505:427–431

Niquille DL, Hansen DA, Mori T et al (2018) Nonribosomal biosynthesis of backbone-modified peptides. Nat Chem 10:282–287. https://doi.org/10.1038/NCHEM.2891

Otten LG, Schaffer ML, Villiers BRM et al (2007) An optimized ATP/PP_i-exchange assay in 96-well format for screening of adenylation domains for applications in combinatorial biosynthesis. Biotechnol J 2:232–240

Parris KD, Lin L, Tam A et al (2000) Crystal structures of substrate binding to *Bacillus subtilis* holo-(acyl carrier protein) synthase reveal a novel trimeric arrangement of molecules resulting in three active sites. Structure 8:883–895

Paul VJ, Arthur KE, Ritson-Williams R et al (2007) Marine biological laboratory chemical defenses: from compounds to communities. Biol Bull 213:226–251

Phelan VV, Du Y, McLean JA et al (2009) Adenylation enzyme characterization using γ-$^{18}O_4$-ATP pyrophosphate exchange. Chem Biol 16:473–478

Płoskoń E, Arthur CJ, Kanari ALP et al (2010) Recognition of intermediate functionality by acyl carrier protein over a complete cycle of fatty acid biosynthesis. Chem Biol 17:776–785

Qiao C, Wilson DJ, Bennett EM et al (2007) A mechanism-based aryl carrier protein/thiolation domain affinity probe. J Am Chem Soc 129:6350–6351

Quadri LEN, Sello J, Keating TA et al (1998a) Identification of a *Mycobacterium tuberculosis* gene cluster encoding the biosynthetic enzymes for assembly of the virulence-conferring siderophore mycobactin. Chem Biol 5:631–645

Quadri LEN, Weinreb PH, Lei M et al (1998b) Characterization of Sfp, a *Bacillus subtilis* phosphopantetheinyl transferase for peptidyl carder protein domains in peptide synthetases. Biochemistry 37:1585–1595

Quadri LEN (2007) Strategic paradigm shifts in the antimicrobial drug discovery process of the 21st century. Infect Disord Drug Targets 7:230–237

Reyes CP, La Clair JJ, Burkart MD (2010) Metabolic probes for imaging endosymbiotic bacteria within toxic dinoflagellates. Chem Commum 46:8151–8153

Röttig M, Medema MH, Blin K et al (2011) NRPSpredictor2-a web server for predicting NRPS adenylation domain specificity. Nucleic Acids Res 39(suppl 2):W362–W367

Rusnak F, Stephen Faraci W, Walsh CT (1989) Subcloning, expression, and purification of the enterobactin biosynthetic enzyme 2,3-dihydroxybenzoate-AMP ligase: demonstration of enzyme-bound (2,3-dihydroxybenzoyl)adenylate product. Biochemistry 28:6827–6835

Saario SM, McKinney MK, Speers AE et al (2012) Clickable,photoreactive inhibitors to probe the active site microenvironment of fatty acid amide hydrolase. Chem Sci 3:77–83

Samel SA, Czodrowski P, Essen LO (2014) Structure of the epimerization domain of tyrocidine synthetase A. Acta Cryst D70:1442–1452

Sanman LE, Bogyo M (2014) Activity-based profiling of proteases. Annu Rev Biochem 83:249–273

Schley C, Altmeyer MO, Swart R et al (2006) Proteome analysis of *Myxococcus xanthus* by off-line two-dimensional chromatographic separation using monolithic poly-(styrene-divinylbenzene) columns combined with ion-trap tandem mass spectrometry. J Proteome Res 5:2760–2768

Schneider TL, Shen B, Walsh CT (2003) Oxidase domains in epothilone and bleomycin biosynthesis: thiazoline to thiazole oxidation during chain elongation. Biochemistry 42:9722–9730

Skaar EP (2010) The battle for iron between bacterial pathogens and their vertebrate hosts. PLoS Pathog 6:e1000949

Somu RV, Boshoff H, Qiao C et al (2006) Rationally-designed nucleoside antibiotics that inhibit siderophore biosynthesis of *Mycobacterium tuberculosis*. J Med Chem 49:31–34

Stachelhaus T, Mootz HD, Marahiel MA (1999) The specificity-conferring code of adenylation domains in nonribosomal peptide synthetases. Chem Biol 6:493–505

Stachelhaus T, Walsh CT (2000) Mutational analysis of the epimerization domain in the initiation module PheATE of gramicidin S synthetase. Biochemistry 39:5775–5787

Sundlov JA, Shi C, Wilson DJ et al (2012) Structural and functional investigation of the intermolecular interaction between NRPS adenylation and carrier protein domains. Chem Biol 19:188–198

Tallorin L, Finzel K, Nguyen QG et al (2016) Trapping of the enoyl-acyl carrier protein reductase–acyl carrier protein interaction. J Am Chem Soc 138:3962–3965

Thirlway J, Lewis R, Nunns L et al (2012) Introduction of a non-natural amino acid into a nonribosomal peptide antibiotics by modification of adenylation domain specificity. Angew Chem Int Ed 51:7181–7184

Tim J, Post FA, Bekker LG et al (2003) Differential expression of iron-, carbon-, and oxygen-responsive mycobacterial genes in the lungs of chronically infected mice and tuberculosis patients. Proc Natl Acad Sci U S A 100:14321–14326

Ueda H, Shoku Y, Hayashi N et al (1991) X-ray crystallographic conformational study of 5′-*O*-[*N*-(L-alanyl)-sulfamoyl]adenosine, a substrate analogue for alanyl-tRNA synthetase. Biochim Biophys Acta 1080:126–134

Visca P, Imperi F, Lamont IL (2007) Pyoverdine siderophores: from biogenesis to biosignificance. Trends Microbiol 15:22–30

Wakimoto T, Egami Y, Nakashima Y et al (2014) Calyculin biogenesis from a pyrophosphate protoxin produced by a sponge symbiont. Nat Chem Biol 10:648–655

Walsh C, Liu J, Rusnak F et al (1990) Molecular studies on enzyme in chorismate metabolism and the enterobactin biosynthetic pathway. Chem Rev 90:1105–1129

Walsh CT (2004) Polyketide and nonribosomal peptide antibiotics: modularity and versatility. Science 303:1805–1810

Walsh CT, O'Brien RV, Khosla C (2013) Nonproteinogenic amino acid building blocks for nonribosomal peptide and hybrid polyketide scaffolds. Angew Chem Int Ed 52:7098–7124

Walsh CT (2016) Insights into the chemical logic and enzymatic machinery of NRPS assembly lines. Nat Prod Rep 33:127–135

Wang M, Beissner M, Zhao H (2014) Aryl-aldehyde formation in fungal polyketides: discovery and characterization of a distinct biosynthetic mechanism. Chem Biol 21:257–263

Whicher JR, Dutta S, Hansen DA et al (2014) Structural rearrangements of a polyketide synthase module during its catalytic cycle. Nature 510:560–564

Wilson DJ, Aldrich CC (2010) A continuous kinetic assay for adenylation enzyme activity and inhibition. Anal Biochem 404:56–63

Winn M, Fyans JK, Zhuo Y et al (2016) Recent advances in engineering nonribosomal peptide assembly lines. Nat Prod Rep 33:317–347

Worthington AS, Rivera H, Torpey JW et al (2006) Mechanism-based protein cross-linking probes to investigate carrier protein-mediated biosynthesis. ACS Chem Biol 1:687–691

Worthington AS, Burkart MD (2006) One-pot chemo-enzymatic synthesis of reporter-modified proteins. Org Biomol Chem 4:44–46

Zhang B, Tian W, Wang S et al (2017) Activation of natural products biosynthetic pathways via a protein modification level regulation. ACS Chem Biol 12:1732–1736

Zhang H, Wang, Y. Pfeifer BA (2008) Bacterial hosts for natural product production. Mol Pharmaceutics 5:212–225

Zhang K, Nelson KM, Bhuripanyo K et al (2013) Engineering the substrate specificity of the DhbE adenylation domain by yeast cell surface display. Chem Biol 20:92–101

Target Identification of Bioactive Covalently Acting Natural Products

Daniel K. Nomura and Thomas J. Maimone

Contents

Abstract There are countless natural products that have been isolated from microbes, plants, and other living organisms that have been shown to possess therapeutic activities such as antimicrobial, anticancer, or anti-inflammatory effects. However, developing these bioactive natural products into drugs has remained challenging in part because of their difficulty in isolation, synthesis, mechanistic understanding, and off-target effects. Among the large pool of bioactive natural products lies classes of compounds that contain potential reactive electrophilic centers that can covalently react with nucleophilic amino acid hotspots on proteins and other biological molecules to modulate their biological action. Covalently

---

D. K. Nomura (✉) · T. J. Maimone (✉)
Department of Chemistry, University of California, 94720 Berkeley, CA, USA
e-mail: dnomura@berkeley.edu

T. J. Maimone
e-mail: maimone@berkeley.edu

Current Topics in Microbiology and Immunology (2019) 420:351–374
DOI 10.1007/82_2018_121

Published Online: 14 August 2018

acting natural products are more amenable to rapid target identification and mapping of specific druggable hotspots within proteins using activity-based protein profiling (ABPP)-based chemoproteomic strategies. In addition, the granular biochemical insights afforded by knowing specific sites of protein modifications of covalently acting natural products enable the pharmacological interrogation of these sites with more synthetically tractable covalently acting small molecules whose structures are more easily tuned. Both discovering binding pockets and targets hit by natural products and exploiting druggable modalities targeted by natural products with simpler molecules may overcome some of the challenges faced with translating natural products into drugs.

1 Introduction

Natural products from microbes, plants, and other living organisms are a rich source of medicinal compounds that have been utilized by humankind for thousands of years. Their further characterization and exploitation of important biological pathways have led to many different types of modern medicines to treat pathologies such as bacterial and fungal infections, inflammation, diabetes, neurodegenerative diseases, and cancer (Koehn and Carter 2005; Kingston 2011; Harvey et al. 2015; Laraia et al. 2018). Between 25 and 50% of marketed drugs are thought to be derived from natural products, and while such compounds populate many therapeutic areas, they have been particularly important to the areas of oncology and infectious diseases (Kingston 2011). However, while many natural products have shown therapeutic potential, developing natural products into drugs has remained challenging and most US pharmaceutical companies have reduced or eliminated their natural product groups. This is in part due to the difficulty of synthesis and execution of global structure–activity relationship (SAR) studies, challenges in large-scale isolation, slower discovery, and development compared to target-based high-throughput screening campaigns involving simpler, often "flat" molecules, hurdles in understanding mode of action or direct targets, and complex polypharmacology (Beutler 2009; Romo and Liu 2016). Despite these challenges, natural products have not only yielded classical active-site inhibitors against druggable targets, but have also been able to uniquely access druggable modalities that would be considered undruggable or difficult to tackle with traditional drug discovery efforts. These include, for example, FK506 or tacrolimus, which reduces peptidylprolyl isomerase activity by binding to FKBP12, creating a new complex FKBP12–FK506 complex, leading to the interaction with and inhibition of calcineurin resulting in inhibition of T cell signaling (Liu et al. 1991). Another example comes from rapamycin or sirolimus, which also binds to FKBP12 to form a FKBP12–rapamycin complex which then interacts with and inhibits mTORC1 signaling (Saxton and Sabatini 2017). Even in the twenty-first century, these molecular functions are difficult to "design" from first principles. Thus, target identification of natural products remains a critically important endeavor not just for translational development of natural products into drugs, but also to gain insight into new modalities that can be targeted for drug discovery and to gain insight into natural strategies for target engagement. Upon identifying particular binding pockets within proteins targeted by natural products, synthetically simpler libraries of small molecules can potentially be developed against these sites to overcome the synthetic and isolation challenges faced with natural products. In this review, we will highlight activity-based protein profiling (ABPP) chemoproteomic strategies for mapping targets of covalently acting natural products and covalent ligand discovery approaches that can be utilized to potentially develop more synthetically tractable compounds that mimic the actions of the more complex natural products. We will also discuss diverse examples of electrophilic natural product scaffolds and their targets.

2 Covalently Acting Natural Products

Hundreds of natural products have been shown, or proposed, to engage a wide array of biological targets in covalent bond formation including proteins, nucleic acids, and even small biological molecules themselves. Equally diverse are the chemical structures of the "reactive warhead" embedded into the parent natural product scaffold–motifs which in some cases would often be difficult to design from first principles (Potashman and Duggan 2009; Baillie 2016; Lagoutte et al. 2017), yet are continually influencing modern covalent drug design (Lagoutte and Winssinger 2017). Several excellent reviews have been written on covalently acting natural products, particularly with respect to their protein targeting abilities (Drahl et al. 2005; Leslie and Hergenrother 2008; Gersch et al. 2012; Wright and Sieber 2016; Pan et al. 2016; Jackson et al. 2017). Here, we briefly highlight various classes of electrophilic species presented by natural products that have been implicated in their mechanism of action.

2.1 *Strained Ring-Containing Electrophiles*

Nature extensively exploits ring strain in the creation of covalent warheads. Reactive carbonyl groups embedded in strained rings, as typified by the β-lactam and β-lactone functional group, are some of the oldest known and most intensely studied covalent pharmacophores found in natural products (Fig. 1a) (Böttcher and Sieber 2012). Indeed, the β-lactam antibiotics, typified by penicillin V (**1**), are perhaps the greatest success story in the use of covalently acting natural products to treat human disease. Their primary target (an active-site serine of bacterial transpeptidases) and other modes of action are discussed in Sect. 3. In the past three decades, β-lactones have garnished significant attention due to their protein targeting ability (Böttcher and Sieber 2012). Naturally occurring lipstatin (**2**) inspired

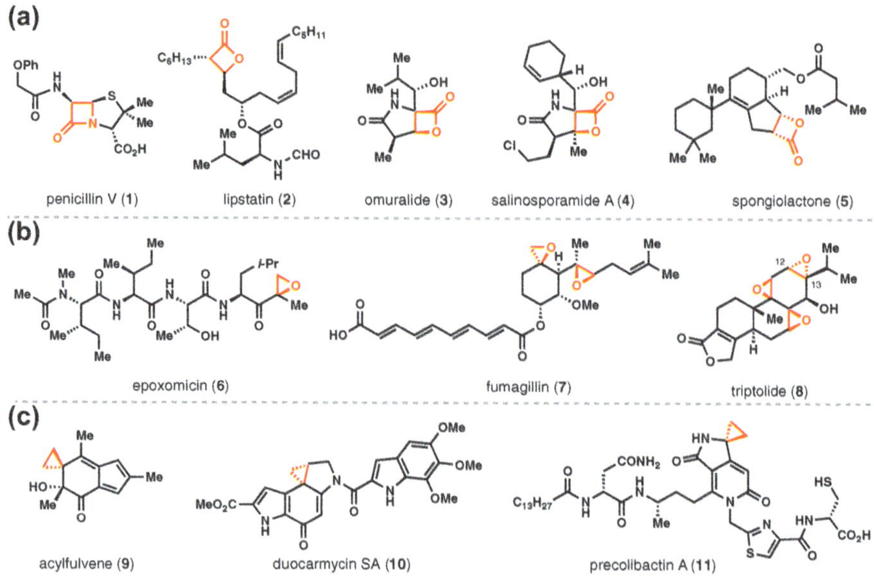

Fig. 1 Covalently acting natural products exploiting strained ring electrophiles. **a** β-lactams and β-lactones, **b** reactive epoxides, and **c** electrophilic cyclopropanes

the antiobesity drug and lipase inhibitor orlistat (Bialecka-Florjanczyk et al. 2018). Omuralide (**3**) and salinosporamide A (**4**) are both potent cytotoxic agents which covalently modify an active-site threonine of the 20S proteasome subunit (Gulder and Moore 2010). This family of natural products is involved in various clinical trials for the treatment of myriad cancers (Potts et al. 2011). Despite the target specificity for **1**–**4**, proteomic profiling studies of spongiolactone (**5**) revealed a complex portrait of targets and thus likely polypharmacological effects dictating its antiproliferative effects (Wright et al. 2017).

The epoxide functional group is another major class of reactive electrophile frequently exploited by natural products (Fig. 1b). The intrinsic affinity of the parent molecule to the target, prior to covalent engagement, is a common theme for promoting reactivity of otherwise unreactive substrates (Swinney 2009; Bauer 2015; Strelow 2017). Thus, many complex epoxide-containing molecules studied are not promiscuous, and in the case of multiple epoxidized molecules, only a single epoxide often reacts as discussed below. The bacterial-derived natural product epoxomicin (**6**) is a potent covalent proteasome inhibitor whose understanding of mechanism of action led to the discovery of the anticancer drug carfilzomib (Kim and Crews 2013). Fumagillin (**7**) targets the methionine aminopeptidase MetAP-2 wherein only the spiroepoxide is essential. Target identification for both **6** and **7** is discussed in Sect. 3. The diterpene triepoxide triptolide (**8**) has long been used in traditional Chinese medicine to treat a host of diseases including inflammation and cancer, and derivatives of **8** have entered clinical trials for the treatment of cancer. Triptolide induces apoptosis and cell growth arrest and is a potent inhibitor of

NF-κB- and NF-AT-mediated transcription (Meng et al. 2014). Many additional targets of **8** (both covalent and non-covalent) have been proposed (Leuenroth and Crews 2005). Triptolide covalently modifies a reactive cysteine on the XPB subunit of the transcription factor TFIIH (Titov et al. 2011). Notably, the covalent locus of reactivity is believed to be the 12, 13-epoxide (He et al. 2015).

Cyclopropanes adjacent to electron-deficient π-systems (i.e., homo-Michael acceptors) have been frequently implicated in the alkylation of biological nucleophiles, particularly DNA bases (Fig. 1c). Not surprisingly, many natural products containing such motifs are potent cytotoxins including both acylfulvene (**9**) and duocarmycin SA (**10**), derivatives of which are being investigated for the treatment of cancer (Tanasova and Sturla 2012). DNA alkylation by **9** impairs transcription by RNA Pol II (Malvezzi et al. 2017), while **10** is a highly sequence-selective covalent alkylator of adenine-N^3 in AT-rich sequences (Boger et al. 1994). The human gut microbe-derived family of colibactins, as typified by precolibactin A (**10**), alkylate and cross-link DNA and have been implicated in human colorectal cancers (Vizcaino and Crawford 2015).

2.2 Unsaturated Ketones, Esters, and Amides

By far, the largest class of covalently acting natural product is those susceptible to hetero-Michael addition reactions–often by the nucleophilic amino acid side chains of proteins (Fig. 2) (Jackson et al. 2017). These addition reactions can be both reversible and irreversible. Notably, hetero-Michael acceptors are prevalent across much of natural product space (peptides, alkaloids, polyketides, and terpenes) and serve as the basis for much of modern covalent drug design (De Cesco et al. 2017). Nature employs both α,β-unsaturated amides (see **12**), esters (see **13**), and ketones (see **14**) to achieve covalent target modification (Fig. 2a). Particularly important from a natural product's perspective is the large class of exocyclic methylene lactones of which over 5000 members are believed to exist (see **15**–**18**, Fig. 2b) (Kitson et al. 2009). Many of these compounds are reactive to thiols, both cysteine side chains and the pool of free intracellular thiols. The anti-inflammatory and anticancer sesquiterpene parthenolide (**15**) has been used for proteomics-based target identification as discussed in Sect. 3. The anticancer pseudoguaianolide helenalin (**16**) is also thiol-reactive (Widen et al. 2017). Nature also constructs various dimeric and higher-order structures featuring multiple electrophilic sites as typified by ainsliadimer A (**17**) and ainsliatrimer A (**18**). Mode of action studies by Lei et al. has found these compounds target IKKα/β and PPARγ (Li et al. 2014; Dong et al. 2015).

An important but less frequently encountered class of hetero-Michael acceptors is the family of quinone methides, which aromatize the following conjugate addition and thus providing an additional driving force for conjugation (Fig. 2c) (Bai et al. 2014). Celastrol (**19**), which possesses a myriad of biological activities including anticancer and anti-inflammatory effects, is cysteine-reactive (Klaić et al.

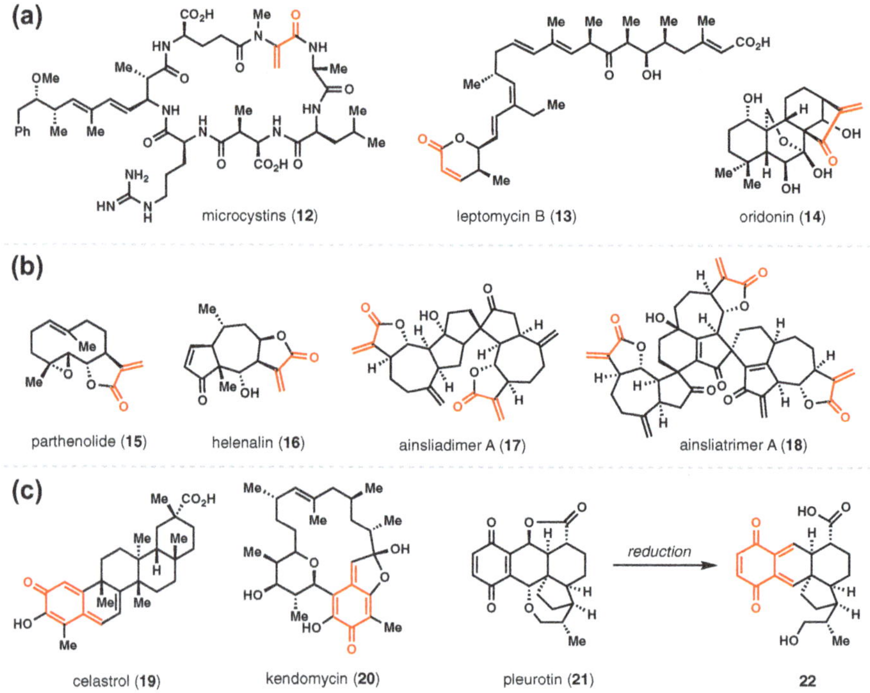

Fig. 2 Natural products susceptible to hetero-Michael addition. **a** α,β-unsaturated amides, esters, and ketones. **b** selected exocyclic methylene lactone **c** electrophilic quinone methides

2012; Zhou et al. 2016a). The antibacterial and cytotoxic natural product kendomycin (**20**) reacts with a proteasome histidine residue (Beck et al. 2014). Finally, it should be noted that few "stable" quinone methides exist; in many cases, the reactive electrophile is generated in situ (Bai et al. 2014). Pleurotin (**21**), for example, has been hypothesized to undergo *in situ* reduction to generate the reactive quinone methide **22** (Moore 1977).

2.3 Aldehydes, Hemiacetals, and Aminals

Nature also utilizes natural products containing electrophilic carbonyl groups at the aldehyde oxidation state, including free aldehydes, aminals, and hemiacetals, for covalent target engagement. The phytotoxic sesterterpene ophiobolin A (**23**) has been shown to condense with a lysine side chain of the calcium-binding protein calmodulin, presumably forming an extended enamine intermediate (Leung et al. 1988). Of note, 1, 4-dicarbonyl-containing natural products, including **23**, have also been implicated in the formation of pyrroles via a Paal–Knorr-type process with intracellular amines (Chidley et al. 2016; Kornienko and La Clair 2017).

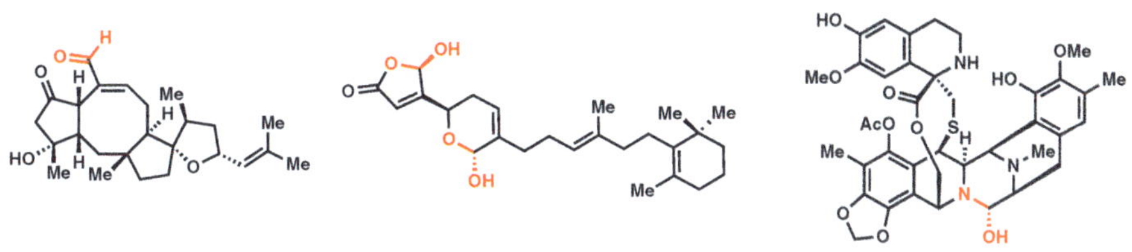

ophiobolin A (**23**) manoalide (**24**) ecteinascidin 743 (**25**)

Fig. 3 Natural products containing reactive functional groups capable of condensation reactions

The *bis*-hemiacetal marine natural product manoalide (**24**) has also been proposed to react with lysines (Pettinger et al. 2017). Ecteinascidin 743 (**25**), an aminal-containing anticancer agent and approved drug (trabectedin), generates an iminium ion which covalently bonds to DNA bases (Aune et al. 2002) (Fig. 3).

In summary, nature has evolved a myriad of chemical warheads to covalently modify a range of biological nucleophiles and many historically successful natural product-based drugs have exploited these features. Given the enormous number of naturally occurring epoxide and α,β-unsaturated carbonyl-containing metabolites, one can assume that we have only scratched the surface with respect to detailed target annotation as well as understanding the complex polypharmacology of bioactive natural products. Moreover, nature continues to impress the organic and medicinal chemist with its complex arsenal and sometimes bizarre mechanisms for generating reactive warheads (Fig. 4) (Kwan and Luesch 2010). One cannot help but stand in awe at processes such as the thiol-mediated conversion of leinamycin

Fig. 4 Natural products containing exotic warheads: Unsaturated nitrones and episulfonium ions

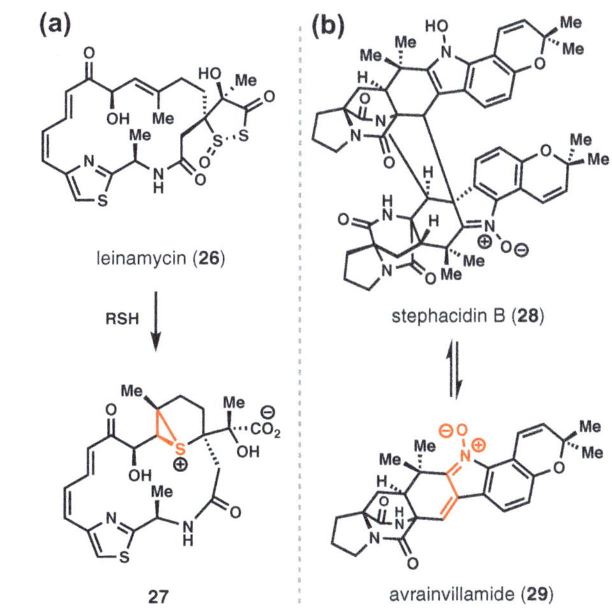

(**26**) to DNA-reactive episulfonium ion **27** or the retro-dimerization of the alkaloid stephacidin B (**28**) into protein-reactive, α,β-unsaturated nitrone-containing avrainvillamide (**29**) (Asai et al. 1996; Myers and Herzon 2003; Wulff et al. 2007). By studying these processes, searching for new ones, and distilling the essential features needed for efficient bond formation, new chemistry concepts relevant to small molecule drug discovery will no doubt be unearthed (Crane and Gademann 2016).

3 Chemoproteomic Approaches to Target Identification of Covalently Acting Natural Products

3.1 Reversible Versus Irreversible Natural Products and Methods for Target Identification

Many strategies have arisen to tackle the challenge of identifying direct protein targets of natural products. These include classical approaches wherein natural products can be immobilized onto solid supports for affinity-based isolation of protein targets (Wright and Sieber 2016). Approaches that enable target identification without chemical modification have also arisen, including cellular thermal shift assay (CETSA) or thermal proteome profiling (TPP) (Chang et al. 2016). Among various approaches, chemoproteomic strategies have been thus far been the dominant approach for natural product target identification, which relies on: (1) direct derivatization of natural products to incorporate photoaffinity cross-linkers (e.g., diazirines, benzophenones), biorthogonal handles (e.g., alkynes or azides), and/or biotin enrichment handles to enable covalent capture and enrichment; or (2) activity-based protein profiling (ABPP)-based competitive chemoproteomic profiling approaches pioneered by Benjamin Cravatt, wherein the parent natural product, usually covalently acting, is competed against a broadly reactive probe—for subsequent identification of the natural product protein targets by mass-spectrometry-based proteomics (Wright and Sieber 2016; Kanoh 2016; Pan et al. 2016). For reversibly acting natural products, these target identification approaches still remain quite challenging and laborious due to low abundance of the targets, high degree of non-specific binding, or synthetic difficulty in natural product derivatization without losing bioactivity. Furthermore, while these types of experiments can provide direct protein targets, identifying the specific site of binding within the protein target often requires additional structural elucidation.

As mentioned, many natural products possess electrophilic fragments capable of covalently reacting with nucleophilic amino acids within proteins to disrupt their biological functions (Drahl et al. 2005; Gersch et al. 2012). As seen in Sect. 2, β-lactones/lactams, exocyclic methylene lactones, spiroepoxides, enones, α,β-unsaturated amides and esters, quinone methides, and aldehydes all take part in protein modification. These reactive groups have been shown to covalently modify mostly serine, cysteine, threonine, and lysine, but also other amino acids such as histidine, arginine, and aspartic and glutamic acid on particular protein targets (Drahl et al. 2005; Gersch et al. 2012). The advantage of covalently acting natural products is their relative ease in target identification compared to reversibly acting molecules. This is in part because direct covalent attachment by the natural product itself enables harsher washing steps of the protein during the enrichment processes to reduce non-specific binding. It also abrogates non-specific reactivity issues that arise from photoaffinity tags and photo-crosslinking. It is also easier to identify the specific site of covalent modification or the druggable hotspots targeted by reactive natural products compared to reversibly acting natural products using competitive ABPP methods subsequently described. As previously mentioned, understanding the specific amino acid targeted by the natural product within a specific protein directly from complex proteomes or living systems provides significant benefit over just the identification of the protein target; the druggable hotspot(s) may represent a unique and functional modality that can be exploited for drug discovery without the laborious and sometimes impossible efforts of having to solve the three-dimensional structure of the natural product bound to the protein by either crystallographic or spectroscopic means. Furthermore, these sites of covalent natural product modification now represent ligandable hotspots that can be interrogated with far simpler and more synthetically tractable covalently acting small-molecule libraries using high-throughput competitive ABPP screening approaches to be subsequently discussed. We will discuss the two primary chemoproteomic methods for target identification of covalently acting natural products.

3.2 Target Identification Using Natural Product-Based Probes

The chemoproteomics approach that has most commonly been used for covalently acting natural products is to derivatize or synthesize the natural product analog incorporating either a biotin enrichment handle or an alkyne to which a fluorophore-azide or biotin-azide handle can be appended by copper-catalyzed azide–alkyne cycloaddition reactions (CuAAC) (Fig. 5a) (Tornøe et al. 2002; Rostovtsev et al. 2002; Heeres and Hergenrother 2011; Gersch et al. 2012; Ursu and Waldmann 2015; Wright and Sieber 2016; Pan et al. 2016). This general strategy can also be coupled with competition studies with the parent natural product and quantitative proteomic approaches with stable isotopic labeling of cells (SILAC) or isotopic tagging of peptides (e.g., TMT labeling) to identify the direct targets of these natural products (Ong et al. 2002; Thompson et al. 2003; McAlister et al. 2014).

With natural product probes bearing the full linker and biotin handles, the utility of these probes is often limited to in vitro experiments since the larger nature of the probe precludes cell penetrance. Nonetheless, this approach has been used repeatedly to identify the targets of many covalently acting natural products. Craig

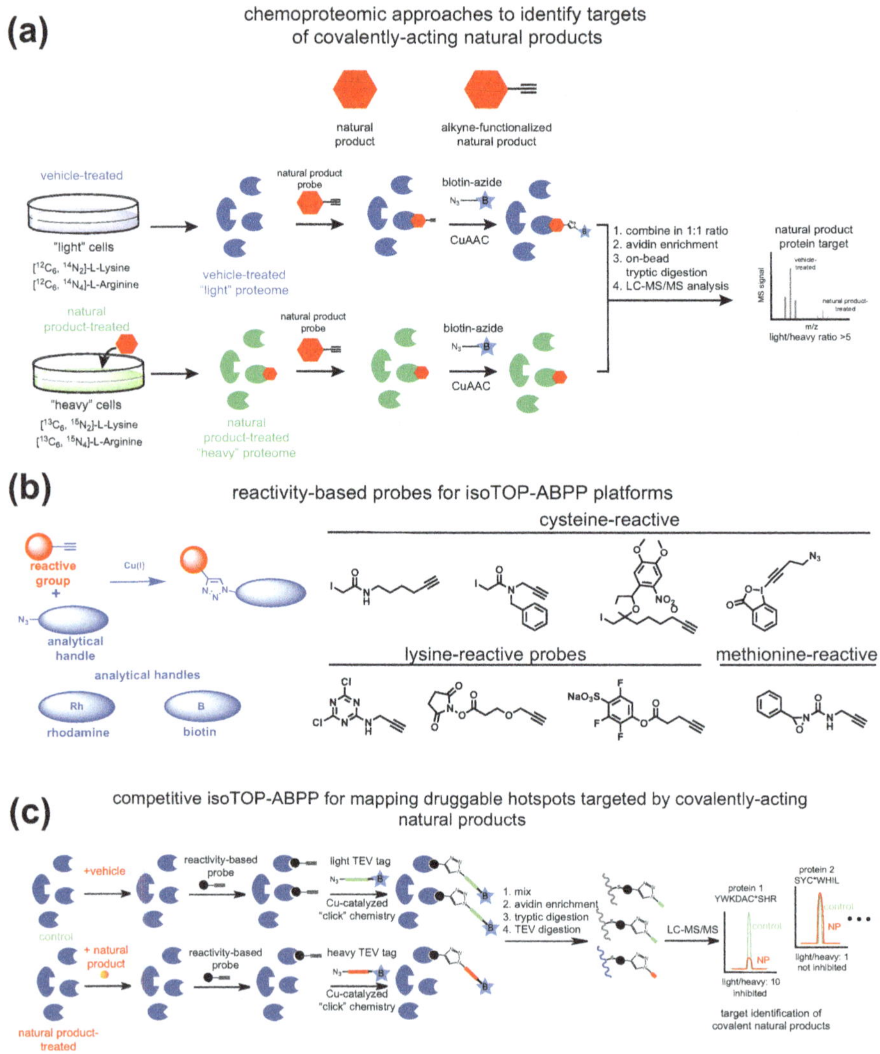

Crews' group has used this approach to gain invaluable insights into the mechanisms of action of multiple covalently acting natural products (Sin et al. 1997; Meng et al. 1999; Kwok et al. 2001). His group synthesized a biotinylated parthenolide analog and showed that this natural product derived from the feverfew plant exerts anti-inflammatory activity through targeting IKKβ and showed through mutagenesis studies that it reacts with cysteine C179 (Kwok et al. 2001). His team also discovered that epoxomicin (**6**), isolated from *Actinomyces,* covalently reacts with the LMP7, X, MECL1, and Z catalytic subunits of the proteasome using a biotinylated epoxomicin derivative (Sin et al. 1999; Meng et al. 1999). Subsequent studies by the Crews Laboratory showed that epoxomicin formed a morpholino ring

◂**Fig. 5** Chemoproteomic approaches to target identification of covalently acting natural products. **a** Natural products can be derivatized with functional biorthogonal alkyne handles to enable chemoproteomic target identification. Cells subjected to SILAC can be treated with vehicle or natural product. Cell proteomes can then be treated with the alkyne-functionalized natural product probe and subjected to CuAAC to append a biotin-azide handle after which probe-labeled isotopically light or heavy proteins can be combined in a 1:1 ratio, avidin-enriched, subjected to on-bead tryptic digestion, and subsequent tryptic peptides can be analyzed by LC-MS/MS. Those proteins showing high light-to-heavy SILAC ratios are targets of the covalently acting natural product probe which are competed by the parent natural product. **b** isoTOP-ABPP uses reactivity-based probes to profile proteome-wide reactive and ligandable hotspots in complex proteomes. Reactivity-based probes consist of a broadly reactive warhead and an alkyne or azide handle which can be appended onto an analytical handle (e.g., fluorophore, biotin, or cleavable biotin handles) by CuAAC. Shown are examples of cysteine, lysine, and methionine reactive probes. (C) In an isoTOP-ABPP experiment, proteomes are treated with vehicle or covalently acting natural product and subsequently labeled with reactivity-based probes. Probe-labeled proteins are then appended to a TEV protease-cleavable biotin-azide tag bearing isotopically light or heavy handles for control or treated proteomes, respectively. Control and treated proteomes are subsequently mixed in a 1:1 ratio, probe-labeled proteins are avidin-enriched, tryptically digested, and probe-modified tryptic peptides are eluted by TEV protease and analyzed by LC-MS/MS. Those probe-modified peptides that show high light-to-heavy ratios indicate the druggable hotspot targeted by the covalently acting natural product

between the amino-terminal threonine and the epoxomicin pharmacophore (Groll et al. 2000) and epoxomicin analogs were eventually developed into the multiple myeloma cancer drug carfilzomib or kyprolis (Kim and Crews 2013). Similarly, his group discovered that the antiangiogenic natural product fumagillin (**7**) was found to covalently react with methionine aminopeptidase, MetAP-2, using a biotinylated derivative of **7** (Sin et al. 1997). Through site-directed mutagenesis studies using a fluorescein-labeled fumagillin analog, Jun Liu's group subsequently identified the site of covalent modification as His231 in MetAP-2 (Griffith et al. 1998). The anticancer natural product withaferin A, a steroidal lactone from the withanolide class of natural products found in *Withania somnifera* and other plant species, was shown to react with C328 of the intermediate filament protein vimentin using a biotinylated derivative of withaferin A (Bargagna-Mohan et al. 2007). Herbert Waldmann's group also synthesized a biotinylated derivative of the tetramic acid natural product melophlin A and found that it targeted dynamin to modulate RAS signaling (Knoth et al. 2009).

Natural product probes bearing a biorthogonal alkyne handle can be used in complex proteomes, live cells, or in live animals, after which fluorescent or enrichment handles can be appended onto probe-labeled proteins for visualization or enrichment and identification of natural product targets. Cravatt et al., who have pioneered these types of chemoproteomic approaches, have utilized this approach successfully to map the on and off targets of countless pharmaceutical drugs, drug candidates, and tool compounds (Ahn et al. 2009, 2011; Bachovchin et al. 2011a; Lanning et al. 2014; Whitby et al. 2017; Niessen et al. 2017). In relation to natural products, Cravatt et al. synthesized a diverse alkyne-functionalized library of spiroepoxides, inspired by the many covalently acting natural products that bear this scaffold including fumagillin and lumanicin D. They then screened this library for antiproliferative effects in breast cancer cells and identified a particular spiroepoxide MJE3 that covalently modified and inhibited the glycolytic enzyme PGAM1 (Evans et al. 2005). Edward Tate's group has also made seminal contributions to chemoproteomic discoveries in natural product target identification. Their group showed, for example, that the reactive cyclic sesquiterpene zerumbone that has been shown to possess anti-inflammatory and anticancer activity reacts with numerous targets in cells involved in cell death, metabolism, cell cycle, and cytoskeletal pathways (Kalesh et al. 2015). Stephen Sieber's group has also repeatedly used these approaches to successfully identify protein targets of covalently acting antibacterial natural products (Staub and Sieber 2008; Böttcher and Sieber 2010; Zeiler et al. 2011; Wirth et al. 2012; Nodwell et al. 2012; Battenberg et al. 2013; Wright and Sieber 2016). To highlight a few examples, the Sieber group functionalized multiple β-lactam antibiotics with alkyne handles for labeling diverse penicillin-binding proteins and subsequently introduced a series of new β-lactam probes that covalently labeled and inhibited a selection of additional bacterial protein targets, including the virulence-associated enzyme ClpP and resistance-associated β-lactamase (Staub and Sieber 2009). In another study, they functionalized a potent nucleoside antibiotic showdomycin with an alkyne handle and showed that it covalently binds to cysteines and inhibits multiple bacterial pathogenesis-associated enzymes, including MurA1 and MurA2 enzymes required for cell wall biosynthesis (Böttcher and Sieber 2010). They also defined multiple antibacterial targets of the dihydro-α-pyrone natural product rugulactone that bears two potentially cysteine-reactive Michael acceptors using an alkyne-functionalized rugulactone derivative, including covalent inhibition of 4-amino-5-hydroxymethyl-2-methylpyrimidine phosphate kinase (ThiD), involved in thiamine biosynthesis (Nodwell et al. 2012).

3.3 *Activity-Based Protein Profiling (ABPP) for Identifying Proteome-Wide Druggable Hotspots Targeted by Covalently Acting Natural Products*

The aforementioned approaches have been very successful at identifying the targets of covalently acting natural products. However, many natural products are incredibly complex to synthesize or may not necessarily have appropriate sites for derivatization. Furthermore, identifying the specific site that has been covalently modified within the target can be challenging and laborious and requires considerable mutagenesis experiments upfront to nail down the specific site of modification. Identifying the specific sites of modification of large reactive natural products by mass spectrometry may also be challenging if the specific modified adduct on the protein is unstable, semi-reversible, or does not ionize well for mass spectrometry detection.

A chemoproteomic platform pioneered by the Cravatt group that greatly facilitates the identification of specific druggable hotspots targeted by covalently acting natural products is ABPP, which uses reactivity-based chemical probes to profile proteome-wide reactive, functional, and ligandable hotspots directly in complex proteomes (Fig. 5b, c). ABPP utilizes active-site-directed probes to profile the activities of entire enzyme classes. ABPP has now evolved to utilize broader reactivity-based probes to map proteome-wide ligandable amino acid hotspots. Pioneered by Benjamin Cravatt, Eranthie Weerapana, and Chu Wang, this modern version of ABPP termed isotopic tandem orthogonal proteolysis–ABPP (isoTOP-ABPP) enables the specific and quantitative mapping of proteome-wide sites of probe modifications (Weerapana et al. 2010). Reactivity-based probes consist of three features which enable this approach: (1) chemical probes that consist of an electrophilic warhead that can react with nucleophilic hotspots on proteins; (2) an alkyne handle for CuAAC conjugation of an enrichment handle for probe-labeled proteins and peptides; and (3) an azide-functionalized TEV protease recognition peptide linker bearing an isotopically light or heavy valine and a biotin group which can be appended onto probe-labeled proteins for subsequent avidin enrichment of probe-labeled proteins, digestion, isolation of probe-labeled tryptic peptides, and TEV release of probe-labeled peptides for subsequent quantitative proteomic analyses comparing isotopically light-to-heavy peptide ratios of probe-modified tryptic peptides. These probes react not only with catalytic sites within enzymes, but also solvent-accessible binding pockets, post-translational modification sites, cysteine oxidation sites, protein–protein interaction sites, and other types of regulatory or functional domains across the proteome. When coupled with isoTOP-ABPP to map reactivity of specific sites of probe modification, this overall approach enables a global method for mapping protein functionality and, more importantly, facilitates the identification of ligandable hotspots within protein targets that may have previously been undruggable (Weerapana et al. 2010).

When used in a competitive manner (competitive isoTOP-ABPP) (Fig. 5c), covalently acting small molecules, such as electrophilic natural products, can be competed against the binding of their corresponding reactivity-based probes to rapidly identify not only the targets, but also the specific sites of modifications, of these molecules (Weerapana et al. 2008; Wang et al. 2014; Backus et al. 2016; Hacker et al. 2017a). This isoTOP-ABPP approach has also evolved to employ chemically or UV-cleavable isotopic linkers, SILAC, multiplexed isotopic labeling strategies, isotopic incorporation into reactivity-based probes, and caged reactivity-based probes for in-cell use (Adibekian et al. 2011; Qian et al. 2013; Abo and Weerapana 2015; Abegg et al. 2015; Abo et al. 2017; Tian et al. 2017). There are multiple advantages of this method compared to the direct incorporation of analytical handles, conjugation to solid supports, or thermal stability assays. First, because the method involves not just the enrichment of proteins labeled with the broad reactivity-based probes, but also further enrichment of the specific probe-modified peptides, the proteome-wide coverage of probe-labeled sites is substantial. Second, this method allows for the use of the parent covalent natural product directly in vitro, in situ, or in vivo and does not require synthesis of derivatives which may result in altered biological activity. Third and most importantly, this approach enables the identification of not only the direct target of the reactive compound, but also the specific amino acid druggable hotspot that is targeted by the natural product. There are also drawbacks to this approach though. First, this method is most amenable for covalently acting compounds as these entities are competed against other covalently acting reactivity-based probes. Second, this approach does not encompass all types of natural product reactivities. The chemical biology field has developed versatile reactivity-based probes that are compatible with proteomic profiling of probe-modified peptides for many amino acids, such as cysteines, lysines, serines, and methionines (Liu et al. 1999; Weerapana et al. 2008, 2010; Adibekian et al. 2011; Shannon et al. 2014; Lin et al. 2017; Ward et al. 2017; Hacker et al. 2017a). While there are many chemoselective bioconjugation reactions that have been reported and are being discovered for other amino acids, it is unclear whether these probes will be compatible with ABPP methods (Shannon and Weerapana 2015; Mix et al. 2016; deGruyter et al. 2017; Martín-Gago et al. 2017). Thus, if the covalently acting natural product (e.g., epoxomicin, fumagillin) reacts outside of the cysteine, lysine, serine, and methionine functional space, this approach currently does not work. Also, currently existing reactivity-based probes do not likely functionalize all possible ligandable sites within the amino acid scope of the probe and thus may still miss natural product targets. Third, there is no direct enrichment of natural product targets, so low abundance targets of natural products may be missed.

Nonetheless, ABPP platforms have been successfully used to identify targets and druggable hotspots targeted by many covalently acting natural products. The isoTOP-ABPP platform (pioneered by the Cravatt, Weerapana, and Wang groups) has been used to map druggable hotspots targeted by a multitude of reactive metabolites, drugs, drug candidates, and tool compounds (Weerapana et al. 2010; Banerjee et al. 2013; Wang et al. 2014; Shannon et al. 2014; Lewallen et al. 2015; Blewett et al. 2016; Backus et al. 2016; Zhou et al. 2016b; Chen et al. 2017, 2018; Hacker et al. 2017a; Bar-Peled et al. 2017). Focusing on natural products, the Cravatt group has also used ABPP platforms and the serine hydrolase-directed fluorophosphonate activity-based probe to profile the targets of β-lactone natural product tetrahydrolipstatin (THL) in mouse brain proteomes and showed that this compound inhibits multiple serine hydrolase enzymes, including Bat5, Abhd12, Pla2g7, and Tpp2 (Hoover et al. 2008). In another study, Porco et al. used ABPP platforms to discover that a novel rocaglate-derived β-lactone inhibits serine hydrolases ABHD10, ACOT1/2, CTSA, and SCPEP1 in prostate cancer cells (Lajkiewicz et al. 2014). Adibekian et al. adapted isoTOP-ABPP platforms using a cysteine-reactive alkynyl benziodoxolone probe and a UV-cleavable linker bearing a biotin enrichment handle to discover the proteome-wide targets of curcumin, an anticancer diarylheptanoid natural product and showed that curcumin covalently modifies several key players of cell signaling and metabolism, including casein kinase I gamma (Abegg et al. 2015). Yang et al. used a similar multiplexed iTRAQ-ABPP strategy to profile the disparate proteome-wide cysteine reactivity of seven exocyclic methylene lactone-containing natural products (Tian et al. 2017).

A recent elegant collaborative study led by Michael Fischbach's group showed that a family of non-ribosomal peptide synthetase gene clusters in gut bacteria encode for pyrazinone and dihydropyrazinone natural products. They showed in collaboration with the Cravatt group using isoTOP-ABPP platforms that the active form of these molecules is the initially released peptide aldehyde, which bears potent protease inhibitory activity and selectively targets the catalytic cysteines of a subset of human cathepsins (Guo et al. 2017). In another study, Nomura and colleagues used isoTOP-ABPP to map the proteome-wide cysteine reactivity of licochalcone A and found that this natural product impairs breast cancer cell viability through targeting C239 of prostaglandin reductase (Roberts et al. 2017).

3.4 Covalent Ligand Discovery Against Druggable Hotspots Targeted by Natural Products

The advantage of identifying the direct targets and druggable hotspots targeted by covalently acting anticancer natural products is that these targets can then be further pharmacologically interrogated with more synthetically accessible covalently acting chemical scaffolds for drug discovery efforts. Upon identifying the protein and druggable hotspot targeted by the covalently acting natural product that is responsible for its biological action, ABPP-based covalent ligand screening can potentially be used to pharmacologically interrogate this site in a target-based manner using moderate- to high-throughput screening approaches utilized by traditional target-based drug discovery efforts (Fig. 6a, b) (Bachovchin et al. 2009, 2010, 2011b). This method contrasts with having to perform medicinal chemistry efforts on natural product scaffolds that can be synthetically challenging, with readouts based on their bioactivity rather than affinity to specific protein targets. Cravatt and Backus showed the power of chemoproteomics-enabled covalent ligand discovery approaches in pharmacologically targeting classically undruggable or therapeutically intractable ligandable hotspots with cysteine-reactive covalent ligands.[47] This work has been followed up by Cravatt, Weerapana, Nomura, and many others showing that cysteine- and lysine-reactive covalent ligands can be used to target classically intractable targets for potential therapy (Shannon et al. 2014; Bateman et al. 2017; Hacker et al. 2017a, b; Anderson et al. 2017; Grossman et al. 2017).

In Cravatt and Backus's study, they identified covalent ligands against >700 cysteines found in both druggable and more intractable proteins, including transcription factors, adaptor/scaffolding proteins, and uncharacterized proteins. Among the many validated examples of covalent ligands targeting characterized and previously unrecognized functional cysteines, one particular unique example they demonstrated was a cysteine-reactive covalent ligand that targeted the catalytic cysteine C360 of caspase 8, only in its inactive zymogen form, but not its active form, wherein they used these compounds to distinguish extrinsic apoptosis pathways in human cell lines versus primary human T cells (Backus et al. 2016). In another study, Nomura and Bateman performed a phenotypic screen with

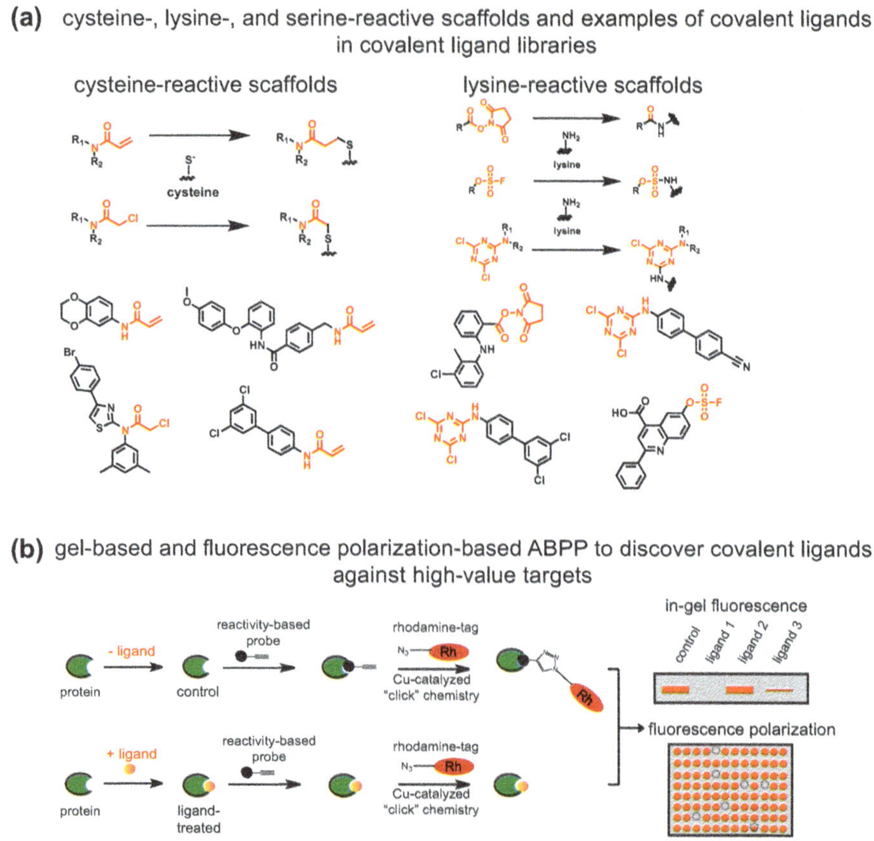

Fig. 6 Covalent ligand discovery against druggable hotspots targeted by covalently acting natural products. Upon identifying druggable hotspots targeted by covalently acting natural products using chemoproteomic platforms, such as isoTOP-ABPP, more synthetically tractable covalent ligand libraries can be competed against the natural product target using higher throughput ABPP approaches. **a** Shown are reactive scaffolds for covalent ligand libraries for cysteines and lysines. **b** Covalent ligands can be screened against targets using gel-based or fluorescence polarization-based ABPP assays in a target-based screening paradigm

cysteine-reactive covalent ligands to identify hits that impaired colorectal cancer cell viability. From this screen, they identified a cysteine-reactive covalent ligand that targeted C1101 of reticulon 4 (RTN4), a protein involved in endoplasmic reticulum (ER) membrane curvature and tubule formation. The authors showed that targeting this C1101 of RTN4 impaired ER tubular formation and nuclear morphology during cell division leading to impaired colorectal cancer pathogenicity (Bateman et al. 2017).

The feasibility of simpler covalent ligands that can react with druggable hotspots targeted by covalently acting natural products was shown by Nomura, Grossman, and Ward recently (Grossman et al. 2017). The authors used isoTOP-ABPP platforms to map the proteome-wide cysteine reactivity of withaferin A and found that the primary target in breast cancer cells was an activating C377 on the protein phosphatase 2A (PP2A) regulatory subunit PPP2R1A, which led to the activation of PP2A activity, dephosphorylation and inactivation of AKT signaling, and impaired breast cancer pathogenicity (Grossman et al. 2017). They performed a parallel phenotypic screen with a simpler cysteine-reactive library and identified a simpler chloroacetamide that targeted this same C377 on PPP2R1A to recapitulate many of the properties observed with withaferin A. Further medicinal chemistry efforts led to the generation of JNS 1-40, a chloroacetamide that was in vitro, in situ, and in vivo selective for C377 on PPP2R1A, which activated PP2A activity, inactivated AKT signaling, and impaired breast cancer tumorigenesis in vivo (Grossman et al. 2017). While the simper covalent ligand identified in this study was fortuitously identified to modify the same site as withaferin A, this study shows that gel-based or fluorescence polarization-based moderate- to high-throughput competitive ABPP assays can likely be performed against natural product targets to rapidly screen for simpler ligands that bind to the same sites and recapitulate the actions of their more complex natural product counterparts (Fig. 6b).

4 Conclusions

In conclusion, there are countless reactive natural products that likely target unique druggable hotspots across a diversity of therapeutic targets. Moreover, many more are likely yet to be discovered, some of which may exploit completely novel reaction chemistry and mechanisms of action. The translational potential of these natural products into drugs has been hampered in part by difficulty of synthesis, isolation, mechanistic understanding, and speed of translation compared to target-based drug discovery efforts. At least with covalently acting natural products, chemoproteomic platforms are becoming increasingly powerful and robust to enable target identification in a reasonable time frame. Covalent ligand discovery against druggable hotspots targeted by reactive natural products using target-based ABPP screening strategies may help to overcome some of the other challenges faced with natural product translation by developing more synthetically tractable small molecule leads that are inspired, not necessarily by the natural product scaffold itself, but rather the target and druggable hotspot.

While this review focused on covalently acting natural products and covalent ligand discovery, these strategies can also be applied to reversibly acting natural products and chemoproteomics-enabled fragment screening using photoaffinity-labeling strategies. Recent studies by the Cravatt group showed how proteome-wide ligandable hotspots can be massively expanded through the use of quantitative chemoproteomic platforms and "fully functionalized" fragment probes bearing a variable small-molecule fragment conjugated to a constant tag bearing an alkyne and photoactivable diazirine group (Parker et al. 2017). These types of strategies may be coupled with traditional photoaffinity-labeling-based natural product target identification approaches to discover more synthetically tractable compounds that target druggable sites targeted by reversibly acting natural products. The future of this research area will no doubt greatly benefit small-molecule drug discovery.

References

Abegg D, Frei R, Cerato L et al (2015) Proteome-Wide Profiling of Targets of Cysteine reactive Small Molecules by using Ethynyl Benziodoxolone Reagents. Angew Chem Int Ed Engl 54:10852–10857. https://doi.org/10.1002/anie.201505641

Abo M, Bak DW, Weerapana E (2017) Optimization of caged electrophiles for improved monitoring of cysteine reactivity in living cells. ChemBioChem 18:81–84. https://doi.org/10.1002/cbic.201600524

Abo M, Weerapana E (2015) A Caged Electrophilic Probe for Global Analysis of Cysteine Reactivity in Living Cells. J Am Chem Soc 137:7087–7090. https://doi.org/10.1021/jacs.5b04350

Adibekian A, Martin BR, Wang C et al (2011) Click-generated triazole ureas as ultrapotent in vivo-active serine hydrolase inhibitors. Nat Chem Biol 7:469–478. https://doi.org/10.1038/nchembio.579

Ahn K, Johnson DS, Mileni M et al (2009) Discovery and characterization of a highly selective FAAH inhibitor that reduces inflammatory pain. Chem Biol 16:411–420. https://doi.org/10.1016/j.chembiol.2009.02.013

Ahn K, Smith SE, Liimatta MB et al (2011) Mechanistic and pharmacological characterization of PF-04457845: a highly potent and selective fatty acid amide hydrolase inhibitor that reduces inflammatory and noninflammatory pain. J Pharmacol Exp Ther 338:114–124. https://doi.org/10.1124/jpet.111.180257

Anderson KE, To M, Olzmann JA, Nomura DK (2017) Chemoproteomics-Enabled Covalent Ligand Screening Reveals a Thioredoxin-Caspase 3 Interaction Disruptor That Impairs Breast Cancer Pathogenicity. ACS Chem Biol 12:2522–2528. https://doi.org/10.1021/acschembio.7b00711

Asai A, Hara M, Kakita S et al (1996) Thiol-Mediated DNA Alkylation by the Novel Antitumor Antibiotic Leinamycin. J Am Chem Soc 118:6802–6803. https://doi.org/10.1021/ja960892w

Aune GJ, Furuta T, Pommier Y (2002) Ecteinascidin 743: a novel anticancer drug with a unique mechanism of action. Anticancer Drugs 13:545–555

Bachovchin DA, Brown SJ, Rosen H, Cravatt BF (2009) Identification of selective inhibitors of uncharacterized enzymes by high-throughput screening with fluorescent activity-based probes. Nat Biotechnol 27:387–394. https://doi.org/10.1038/nbt.1531

Bachovchin DA, Ji T, Li W et al (2010) Superfamily-wide portrait of serine hydrolase inhibition achieved by library-versus-library screening. Proc Natl Acad Sci U S A 107:20941–20946. https://doi.org/10.1073/pnas.1011663107

Bachovchin DA, Mohr JT, Speers AE et al (2011a) Academic cross-fertilization by public screening yields a remarkable class of protein phosphatase methylesterase-1 inhibitors. Proc Natl Acad Sci U S A 108:6811–6816. https://doi.org/10.1073/pnas.1015248108

Bachovchin DA, Mohr JT, Speers AE et al (2011b) Academic cross-fertilization by public screening yields a remarkable class of protein phosphatase methylesterase-1 inhibitors. Proc Natl Acad Sci U S A 108:6811–6816. https://doi.org/10.1073/pnas.1015248108

Backus KM, Correia BE, Lum KM et al (2016) Proteome-wide covalent ligand discovery in native biological systems. Nature 534:570–574. https://doi.org/10.1038/nature18002

Bai W-J, David JG, Feng Z-G et al (2014) The domestication of ortho-quinone methides. Acc Chem Res 47:3655–3664. https://doi.org/10.1021/ar500330x

Baillie TA (2016) Targeted Covalent Inhibitors for Drug Design. Angew Chem Int Ed Engl 55:13408–13421. https://doi.org/10.1002/anie.201601091

Banerjee R, Pace NJ, Brown DR, Weerapana E (2013) 1, 3, 5-Triazine as a modular scaffold for covalent inhibitors with streamlined target identification. J Am Chem Soc 135:2497–2500. https://doi.org/10.1021/ja400427e

Bargagna-Mohan P, Hamza A, Kim Y et al (2007) The tumor inhibitor and antiangiogenic agent withaferin A targets the intermediate filament protein vimentin. Chem Biol 14:623–634. https://doi.org/10.1016/j.chembiol.2007.04.010

Bar-Peled L, Kemper EK, Suciu RM et al (2017) Chemical Proteomics Identifies Druggable Vulnerabilities in a Genetically Defined Cancer. Cell 171(696–709):e23. https://doi.org/10.1016/j.cell.2017.08.051

Bateman LA, Nguyen TB, Roberts AM et al (2017) Chemoproteomics-enabled covalent ligand screen reveals a cysteine hotspot in reticulon 4 that impairs ER morphology and cancer pathogenicity. Chem Commun Camb Engl 53:7234–7237. https://doi.org/10.1039/c7cc01480e

Battenberg OA, Yang Y, Verhelst SHL, Sieber SA (2013) Target profiling of 4-hydroxyderricin in S. aureus reveals seryl-tRNA synthetase binding and inhibition by covalent modification. Mol BioSyst 9:343–351. https://doi.org/10.1039/c2mb25446h

Bauer RA (2015) Covalent inhibitors in drug discovery: from accidental discoveries to avoided liabilities and designed therapies. Drug Discov Today 20:1061–1073. https://doi.org/10.1016/j.drudis.2015.05.005

Beck P, Heinemeyer W, Späth A-L et al (2014) Interactions of the natural product kendomycin and the 20S proteasome. J Mol Biol 426:3108–3117. https://doi.org/10.1016/j.jmb.2014.06.019

Beutler JA (2009) Natural Products as a Foundation for Drug Discovery. Curr Protoc Pharmacol Editor Board SJ Enna Ed–Chief Al 46:9.11.1–9.11.21. https://doi.org/10.1002/0471141755.ph0911s46

Bialecka-Florjanczyk E, Fabiszewska AU, Krzyczkowska J, Kurylowicz A (2018) Synthetic and Natural Lipase Inhibitors. Mini Rev Med Chem 18:672–683. https://doi.org/10.2174/1389557516666160630123356

Blewett MM, Xie J, Zaro BW et al (2016) Chemical proteomic map of dimethyl fumarate-sensitive cysteines in primary human T cells. Sci Signal 9:rs10. https://doi.org/10.1126/scisignal.aaf7694

Boger DL, Johnson DS, Yun W, Tarby CM (1994) Molecular basis for sequence selective DNA alkylation by (+)- and ent-(-)-CC-1065 and related agents: alkylation site models that accommodate the offset AT-rich adenine N3 alkylation selectivity. Bioorg Med Chem 2:115–135

Böttcher T, Sieber SA (2010) Showdomycin as a versatile chemical tool for the detection of pathogenesis-associated enzymes in bacteria. J Am Chem Soc 132:6964–6972. https://doi.org/10.1021/ja909150y

Böttcher T, Sieber SA (2012) β-Lactams and β-lactones as activity-based probes in chemical biology. MedChemComm 3:408–417. https://doi.org/10.1039/c2md00275b

Chang J, Kim Y, Kwon HJ (2016) Advances in identification and validation of protein targets of natural products without chemical modification. Nat Prod Rep 33:719–730. https://doi.org/10.1039/C5NP00107B

Chen Y-C, Backus KM, Merkulova M et al (2017) Covalent Modulators of the Vacuolar ATPase. J Am Chem Soc 139:639–642. https://doi.org/10.1021/jacs.6b12511

Chen Y, Liu Y, Lan T et al (2018) Quantitative Profiling of Protein Carbonylations in Ferroptosis by an Aniline-Derived Probe. J Am Chem Soc 140:4712–4720. https://doi.org/10.1021/jacs.8b01462

Chidley C, Trauger SA, Birsoy K, O'Shea EK (2016) The anticancer natural product ophiobolin A induces cytotoxicity by covalent modification of phosphatidylethanolamine. eLife 5. https://doi.org/10.7554/elife.14601

Crane EA, Gademann K (2016) Capturing Biological Activity in Natural Product Fragments by Chemical Synthesis. Angew Chem Int Ed Engl 55:3882–3902. https://doi.org/10.1002/anie.201505863

De Cesco S, Kurian J, Dufresne C et al (2017) Covalent inhibitors design and discovery. Eur J Med Chem 138:96–114. https://doi.org/10.1016/j.ejmech.2017.06.019
deGruyter JN, Malins LR, Baran PS (2017) Residue-Specific Peptide Modification: a chemist's guide. Biochemistry (Mosc) 56:3863–3873. https://doi.org/10.1021/acs.biochem.7b00536
Dong T, Li C, Wang X et al (2015) Ainsliadimer A selectively inhibits IKKα/β by covalently binding a conserved cysteine. Nat Commun 6:6522. https://doi.org/10.1038/ncomms7522
Drahl C, Cravatt BF, Sorensen EJ (2005) Protein-reactive natural products. Angew Chem Int Ed Engl 44:5788–5809. https://doi.org/10.1002/anie.200500900
Evans MJ, Saghatelian A, Sorensen EJ, Cravatt BF (2005) Target discovery in small-molecule cell-based screens by in situ proteome reactivity profiling. Nat Biotechnol 23:1303–1307. https://doi.org/10.1038/nbt1149
Gersch M, Kreuzer J, Sieber SA (2012) Electrophilic natural products and their biological targets. Nat Prod Rep 29:659–682. https://doi.org/10.1039/c2np20012k
Griffith EC, Su Z, Niwayama S et al (1998) Molecular recognition of angiogenesis inhibitors fumagillin and ovalicin by methionine aminopeptidase 2. Proc Natl Acad Sci U S A 95:15183–15188
Groll M, Kim KB, Kairies N et al (2000) Crystal Structure of Epoxomicin:20S Proteasome Reveals a Molecular Basis for Selectivity of α′, β′-Epoxyketone Proteasome Inhibitors. J Am Chem Soc 122:1237–1238. https://doi.org/10.1021/ja993588m
Grossman EA, Ward CC, Spradlin JN et al (2017) Covalent Ligand Discovery against Druggable Hotspots Targeted by Anti-cancer Natural Products. Cell Chem Biol 24(1368–1376):e4. https://doi.org/10.1016/j.chembiol.2017.08.013
Gulder TAM, Moore BS (2010) Salinosporamide natural products: Potent 20 S proteasome inhibitors as promising cancer chemotherapeutics. Angew Chem Int Ed Engl 49:9346–9367. https://doi.org/10.1002/anie.201000728
Guo C-J, Chang F-Y, Wyche TP et al (2017) Discovery of Reactive Microbiota-Derived Metabolites that Inhibit Host Proteases. Cell 168(517–526):e18. https://doi.org/10.1016/j.cell.2016.12.021
Hacker SM, Backus KM, Lazear MR et al (2017a) Global profiling of lysine reactivity and ligandability in the human proteome. Nat Chem advance online publication. https://doi.org/10.1038/nchem.2826
Hacker SM, Backus KM, Lazear MR et al (2017b) Global profiling of lysine reactivity and ligandability in the human proteome. Nat Chem 9:1181–1190. https://doi.org/10.1038/nchem.2826
Harvey AL, Edrada-Ebel R, Quinn RJ (2015) The re-emergence of natural products for drug discovery in the genomics era. Nat Rev Drug Discov 14:111–129. https://doi.org/10.1038/nrd4510
He Q-L, Titov DV, Li J et al (2015) Covalent modification of a cysteine residue in the XPB subunit of the general transcription factor TFIIH through single epoxide cleavage of the transcription inhibitor triptolide. Angew Chem Int Ed Engl 54:1859–1863. https://doi.org/10.1002/anie.201408817
Heeres JT, Hergenrother PJ (2011) High-throughput screening for modulators of protein-protein interactions: use of photonic crystal biosensors and complementary technologies. Chem Soc Rev 40:4398–4410. https://doi.org/10.1039/b923660k
Hoover HS, Blankman JL, Niessen S, Cravatt BF (2008) Selectivity of inhibitors of endocannabinoid biosynthesis evaluated by activity-based protein profiling. Bioorg Med Chem Lett 18:5838–5841. https://doi.org/10.1016/j.bmcl.2008.06.091
Jackson PA, Widen JC, Harki DA, Brummond KM (2017) Covalent Modifiers: A Chemical Perspective on the Reactivity of α,β-Unsaturated Carbonyls with Thiols via Hetero-Michael Addition Reactions. J Med Chem 60:839–885. https://doi.org/10.1021/acs.jmedchem.6b00788
Kalesh KA, Clulow JA, Tate EW (2015) Target profiling of zerumbone using a novel cell-permeable clickable probe and quantitative chemical proteomics. Chem Commun Camb Engl 51:5497–5500. https://doi.org/10.1039/c4cc09527h

Kanoh N (2016) Photo-cross-linked small-molecule affinity matrix as a tool for target identification of bioactive small molecules. Nat Prod Rep 33:709–718. https://doi.org/10.1039/C5NP00117J
Kim KB, Crews CM (2013) From epoxomicin to carfilzomib: chemistry, biology, and medical outcomes. Nat Prod Rep 30:600–604. https://doi.org/10.1039/c3np20126k
Kingston DGI (2011) Modern Natural Products Drug Discovery and its Relevance to Biodiversity Conservation. J Nat Prod 74:496–511. https://doi.org/10.1021/np100550t
Kitson RRA, Millemaggi A, Taylor RJK (2009) The renaissance of alpha-methylene-gamma-butyrolactones: new synthetic approaches. Angew Chem Int Ed Engl 48:9426–9451. https://doi.org/10.1002/anie.200903108
Klaić L, Morimoto RI, Silverman RB (2012) Celastrol analogues as inducers of the heat shock response. Design and synthesis of affinity probes for the identification of protein targets. ACS Chem Biol 7:928–937. https://doi.org/10.1021/cb200539u
Knoth T, Warburg K, Katzka C et al (2009) The Ras Pathway Modulator Melophlin A Targets Dynamins. Angew Chem Int Ed 48:7240–7245. https://doi.org/10.1002/anie.200902023
Koehn FE, Carter GT (2005) The evolving role of natural products in drug discovery. Nat Rev Drug Discov 4:206–220. https://doi.org/10.1038/nrd1657
Kornienko A, La Clair JJ (2017) Covalent modification of biological targets with natural products through Paal-Knorr pyrrole formation. Nat Prod Rep 34:1051–1060. https://doi.org/10.1039/c7np00024c
Kwan JC, Luesch H (2010) Weapons in disguise–activating mechanisms and protecting group chemistry in nature. Chem Weinh Bergstr Ger 16:13020–13029. https://doi.org/10.1002/chem.201001562
Kwok BH, Koh B, Ndubuisi MI et al (2001) The anti-inflammatory natural product parthenolide from the medicinal herb Feverfew directly binds to and inhibits IkappaB kinase. Chem Biol 8:759–766
Lagoutte R, Patouret R, Winssinger N (2017) Covalent inhibitors: an opportunity for rational target selectivity. Curr Opin Chem Biol 39:54–63. https://doi.org/10.1016/j.cbpa.2017.05.008
Lagoutte R, Winssinger N (2017) Following the Lead from Nature with Covalent Inhibitors. Chimia 71:703–711. https://doi.org/10.2533/chimia.2017.703
Lajkiewicz NJ, Cognetta AB, Niphakis MJ et al (2014) Remodeling natural products: chemistry and serine hydrolase activity of a rocaglate-derived β-lactone. J Am Chem Soc 136:2659–2664. https://doi.org/10.1021/ja412431g
Lanning BR, Whitby LR, Dix MM et al (2014) A road map to evaluate the proteome-wide selectivity of covalent kinase inhibitors. Nat Chem Biol 10:760–767. https://doi.org/10.1038/nchembio.1582
Laraia L, Robke L, Waldmann H (2018) Bioactive compound collections: from design to target identification. Chem 4:705–730. https://doi.org/10.1016/j.chempr.2018.01.012
Leslie BJ, Hergenrother PJ (2008) Identification of the cellular targets of bioactive small organic molecules using affinity reagents. Chem Soc Rev 37:1347–1360. https://doi.org/10.1039/b702942j
Leuenroth SJ, Crews CM (2005) Studies on calcium dependence reveal multiple modes of action for triptolide. Chem Biol 12:1259–1268. https://doi.org/10.1016/j.chembiol.2005.09.009
Leung PC, Graves LM, Tipton CL (1988) Characterization of the interaction of ophiobolin A and calmodulin. Int J Biochem 20:1351–1359
Lewallen DM, Bicker KL, Subramanian V et al (2015) Chemical proteomic platform to identify citrullinated proteins. ACS Chem Biol 10:2520–2528. https://doi.org/10.1021/acschembio.5b00438
Li C, Dong T, Li Q, Lei X (2014) Probing the anticancer mechanism of (-)-ainsliatrimer A through diverted total synthesis and bioorthogonal ligation. Angew Chem Int Ed Engl 53:12111–12115. https://doi.org/10.1002/anie.201407225
Lin S, Yang X, Jia S et al (2017) Redox-based reagents for chemoselective methionine bioconjugation. Science 355:597–602. https://doi.org/10.1126/science.aal3316
Liu J, Farmer JD, Lane WS et al (1991) Calcineurin is a common target of cyclophilin-cyclosporin A and FKBP-FK506 complexes. Cell 66:807–815

Liu Y, Patricelli MP, Cravatt BF (1999) Activity-based protein profiling: the serine hydrolases. Proc Natl Acad Sci U S A 96:14694–14699

Malvezzi S, Farnung L, Aloisi CMN et al (2017) Mechanism of RNA polymerase II stalling by DNA alkylation. Proc Natl Acad Sci U S A 114:12172–12177. https://doi.org/10.1073/pnas.1706592114

Martín-Gago P, Fansa EK, Winzker M et al (2017) Covalent Protein Labeling at Glutamic Acids. Cell Chem Biol 24(589–597):e5. https://doi.org/10.1016/j.chembiol.2017.03.015

McAlister GC, Nusinow DP, Jedrychowski MP et al (2014) MultiNotch MS3 enables accurate, sensitive, and multiplexed detection of differential expression across cancer cell line proteomes. Anal Chem 86:7150–7158. https://doi.org/10.1021/ac502040v

Meng L, Mohan R, Kwok BHB et al (1999) Epoxomicin, a potent and selective proteasome inhibitor, exhibits in vivo antiinflammatory activity. Proc Natl Acad Sci 96:10403–10408. https://doi.org/10.1073/pnas.96.18.10403

Meng C, Zhu H, Song H et al (2014) Targets and molecular mechanisms of triptolide in cancer therapy. Chin J Cancer Res 26:622–626. https://doi.org/10.3978/j.issn.1000-9604.2014.09.01

Mix KA, Aronoff MR, Raines RT (2016) Diazo compounds: versatile tools for chemical biology. ACS Chem Biol 11:3233–3244. https://doi.org/10.1021/acschembio.6b00810

Moore HW (1977) Bioactivation as a model for drug design bioreductive alkylation. Science 197:527–532

Myers AG, Herzon SB (2003) Identification of a novel Michael acceptor group for the reversible addition of oxygen- and sulfur-based nucleophiles. Synthesis and reactivity of the 3-alkylidene-3H-indole 1-oxide function of avrainvillamide. J Am Chem Soc 125:12080–12081. https://doi.org/10.1021/ja0372006

Niessen S, Dix MM, Barbas S et al (2017) Proteome-wide Map of Targets of T790M-EGFR-Directed Covalent Inhibitors. Cell Chem Biol 24(1388–1400):e7. https://doi.org/10.1016/j.chembiol.2017.08.017

Nodwell MB, Menz H, Kirsch SF, Sieber SA (2012) Rugulactone and its analogues exert antibacterial effects through multiple mechanisms including inhibition of thiamine biosynthesis. Chembiochem Eur J Chem Biol 13:1439–1446. https://doi.org/10.1002/cbic.201200265

Ong S-E, Blagoev B, Kratchmarova I et al (2002) Stable isotope labeling by amino acids in cell culture, SILAC, as a simple and accurate approach to expression proteomics. Mol Cell Proteomics MCP 1:376–386

Pan S, Zhang H, Wang C et al (2016) Target identification of natural products and bioactive compounds using affinity-based probes. Nat Prod Rep 33:612–620. https://doi.org/10.1039/C5NP00101C

Parker CG, Galmozzi A, Wang Y et al (2017) Ligand and target discovery by fragment-based screening in human cells. Cell 168(527–541):e29. https://doi.org/10.1016/j.cell.2016.12.029

Pettinger J, Jones K, Cheeseman MD (2017) Lysine-targeting covalent inhibitors. Angew Chem Int Ed Engl 56:15200–15209. https://doi.org/10.1002/anie.201707630

Potashman MH, Duggan ME (2009) Covalent modifiers: an orthogonal approach to drug design. J Med Chem 52:1231–1246. https://doi.org/10.1021/jm8008597

Potts BC, Albitar MX, Anderson KC et al (2011) Marizomib, a proteasome inhibitor for all seasons: preclinical profile and a framework for clinical trials. Curr Cancer Drug Targets 11:254–284

Qian Y, Martell J, Pace NJ et al (2013) An isotopically tagged azobenzene-based cleavable linker for quantitative proteomics. Chembiochem Eur J Chem Biol 14:1410–1414. https://doi.org/10.1002/cbic.201300396

Roberts LS, Yan P, Bateman LA, Nomura DK (2017) Mapping novel metabolic nodes targeted by anti-cancer drugs that impair triple-negative breast cancer pathogenicity. ACS Chem Biol 12:1133–1140. https://doi.org/10.1021/acschembio.6b01159

Romo D, Liu JO (2016) Editorial: Strategies for cellular target identification of natural products. Nat Prod Rep 33:592–594. https://doi.org/10.1039/C6NP90016J

Rostovtsev VV, Green LG, Fokin VV, Sharpless KB (2002) A stepwise huisgen cycloaddition process: copper(I)-catalyzed regioselective "ligation" of azides and terminal alkynes. Angew

Chem Int Ed Engl 41:2596–2599. https://doi.org/10.1002/1521-3773(20020715)41:14%3c2596:AID-ANIE2596%3e3.0.CO;2-4

Saxton RA, Sabatini DM (2017) mTOR Signaling in Growth, Metabolism, and Disease. Cell 168:960–976. https://doi.org/10.1016/j.cell.2017.02.004

Shannon DA, Banerjee R, Webster ER et al (2014) Investigating the proteome reactivity and selectivity of aryl halides. J Am Chem Soc 136:3330–3333. https://doi.org/10.1021/ja4116204

Shannon DA, Weerapana E (2015) Covalent protein modification: the current landscape of residue-specific electrophiles. Curr Opin Chem Biol 24:18–26. https://doi.org/10.1016/j.cbpa.2014.10.021

Sin N, Kim KB, Elofsson M et al (1999) Total synthesis of the potent proteasome inhibitor epoxomicin: a useful tool for understanding proteasome biology. Bioorg Med Chem Lett 9:2283–2288

Sin N, Meng L, Wang MQW et al (1997) The anti-angiogenic agent fumagillin covalently binds and inhibits the methionine aminopeptidase, MetAP-2. Proc Natl Acad Sci U S A 94:6099–6103

Staub I, Sieber SA (2008) Beta-lactams as selective chemical probes for the in vivo labeling of bacterial enzymes involved in cell wall biosynthesis, antibiotic resistance, and virulence. J Am Chem Soc 130:13400–13409. https://doi.org/10.1021/ja803349j

Staub I, Sieber SA (2009) Beta-lactam probes as selective chemical-proteomic tools for the identification and functional characterization of resistance associated enzymes in MRSA. J Am Chem Soc 131:6271–6276. https://doi.org/10.1021/ja901304n

Strelow JM (2017) A Perspective on the Kinetics of Covalent and Irreversible Inhibition. SLAS Discov Adv Life Sci R D 22:3–20. https://doi.org/10.1177/1087057116671509

Swinney DC (2009) The role of binding kinetics in therapeutically useful drug action. Curr Opin Drug Discov Devel 12:31–39

Tanasova M, Sturla SJ (2012) Chemistry and biology of acylfulvenes: sesquiterpene-derived antitumor agents. Chem Rev 112:3578–3610. https://doi.org/10.1021/cr2001367

Thompson A, Schäfer J, Kuhn K et al (2003) Tandem mass tags: a novel quantification strategy for comparative analysis of complex protein mixtures by MS/MS. Anal Chem 75:1895–1904

Tian C, Sun R, Liu K et al (2017) Multiplexed Thiol Reactivity Profiling for Target Discovery of Electrophilic Natural Products. Cell Chem Biol 24(1416–1427):e5. https://doi.org/10.1016/j.chembiol.2017.08.022

Titov DV, Gilman B, He Q-L et al (2011) XPB, a subunit of TFIIH, is a target of the natural product triptolide. Nat Chem Biol 7:182–188. https://doi.org/10.1038/nchembio.522

Tornøe CW, Christensen C, Meldal M (2002) Peptidotriazoles on solid phase: [1, 2, 3]-triazoles by regiospecific copper(i)-catalyzed 1,3-dipolar cycloadditions of terminal alkynes to azides. J Org Chem 67:3057–3064

Ursu A, Waldmann H (2015) Hide and seek: Identification and confirmation of small molecule protein targets. Bioorg Med Chem Lett 25:3079–3086. https://doi.org/10.1016/j.bmcl.2015.06.023

Vizcaino MI, Crawford JM (2015) The colibactin warhead crosslinks DNA. Nat Chem 7:411–417. https://doi.org/10.1038/nchem.2221

Wang C, Weerapana E, Blewett MM, Cravatt BF (2014) A chemoproteomic platform to quantitatively map targets of lipid-derived electrophiles. Nat Methods 11:79–85. https://doi.org/10.1038/nmeth.2759

Ward CC, Kleinman JI, Nomura DK (2017) NHS-Esters as Versatile Reactivity-Based Probes for Mapping Proteome-Wide Ligandable Hotspots. ACS Chem Biol 12:1478–1483. https://doi.org/10.1021/acschembio.7b00125

Weerapana E, Simon GM, Cravatt BF (2008) Disparate proteome reactivity profiles of carbon electrophiles. Nat Chem Biol 4:405–407. https://doi.org/10.1038/nchembio.91

Weerapana E, Wang C, Simon GM et al (2010) Quantitative reactivity profiling predicts functional cysteines in proteomes. Nature 468:790–795. https://doi.org/10.1038/nature09472

Whitby LR, Obach RS, Simon GM et al (2017) Quantitative Chemical Proteomic Profiling of the in Vivo Targets of Reactive Drug Metabolites. ACS Chem Biol 12:2040–2050. https://doi.org/10.1021/acschembio.7b00346

Widen JC, Kempema AM, Villalta PW, Harki DA (2017) Targeting NF-κB p65 with a Helenalin Inspired Bis-electrophile. ACS Chem Biol 12:102–113. https://doi.org/10.1021/acschembio.6b00751

Wirth T, Schmuck K, Tietze LF, Sieber SA (2012) Duocarmycin analogues target aldehyde dehydrogenase 1 in lung cancer cells. Angew Chem Int Ed Engl 51:2874–2877. https://doi.org/10.1002/anie.201106334

Wright MH, Sieber SA (2016) Chemical proteomics approaches for identifying the cellular targets of natural products. Nat Prod Rep 33:681–708. https://doi.org/10.1039/c6np00001k

Wright MH, Tao Y, Drechsel J et al (2017) Quantitative chemoproteomic profiling reveals multiple target interactions of spongiolactone derivatives in leukemia cells. Chem Commun Camb Engl 53:12818–12821. https://doi.org/10.1039/c7cc04990k

Wulff JE, Siegrist R, Myers AG (2007) The natural product avrainvillamide binds to the oncoprotein nucleophosmin. J Am Chem Soc 129:14444–14451. https://doi.org/10.1021/ja075327f

Zeiler E, Braun N, Böttcher T et al (2011) Vibralactone as a tool to study the activity and structure of the ClpP1P2 complex from Listeria monocytogenes. Angew Chem Int Ed Engl 50:11001–11004. https://doi.org/10.1002/anie.201104391

Zhou Y, Li W, Wang M et al (2016a) Competitive profiling of celastrol targets in human cervical cancer HeLa cells via quantitative chemical proteomics. Mol BioSyst 13:83–91. https://doi.org/10.1039/c6mb00691d

Zhou Y, Wynia-Smith SL, Couvertier SM et al (2016b) Chemoproteomic Strategy to Quantitatively Monitor Transnitrosation Uncovers Functionally Relevant S-Nitrosation Sites on Cathepsin D and HADH2. Cell Chem Biol 23:727–737. https://doi.org/10.1016/j.chembiol.2016.05.008

Applications of Reactive Cysteine Profiling

Keriann M. Backus

Contents

Abstract Cysteine thiols are involved in a diverse set of biological transformations, including nucleophilic and redox catalysis, metal coordination and formation of both dynamic and structural disulfides. Often posttranslationally modified, cysteines are also frequently alkylated by electrophilic compounds, including electrophilic metabolites, drugs, and natural products, and are attractive sites for covalent probe and drug development. Quantitative proteomics combined with activity-based protein profiling has been applied to annotate cysteine reactivity, susceptibility to posttranslational modifications, and accessibility to chemical probes, uncovering thousands of functional and small-molecule targetable cysteines across a diverse set

K. M. Backus (✉)
Department of Biological Chemistry, University of California, Los Angeles, 615 Charles E. Young Dr South, BSRB, Room 350A, Los Angeles, CA 90095, USA
e-mail: kbackus@mednet.ucla.edu

Current Topics in Microbiology and Immunology (2019) 420:375–418
DOI 10.1007/82_2018_120

Published Online: 14 August 2018

of proteins, proteome-wide in an unbiased manner. Reactive cysteines have been targeted by high-throughput screening and fragment-based ligand discovery efforts. New cysteine-reactive electrophiles and compound libraries have been synthesized to enable inhibitor discovery broadly and to minimize nonspecific toxicity and off-target activity of compounds. With the recent blockbuster success of several covalent inhibitors, and the development of new chemical proteomic strategies to broadly identify reactive, ligandable and posttranslationally modified cysteines, cysteine profiling is poised to enable the development of new potent and selective chemical probes and even, in some cases, new drugs.

1 Introduction

With its nucleophilicity and sensitivity to oxidative modifications, cysteine is unique among proteinacious amino acids. Cysteines contribute to protein stability, through structural disulfides and are frequently found in enzyme active sites, both acting as catalytic nucleophiles and non-catalytic residues. Cysteine residues play an essential role in protecting cells from reactive oxygen species (ROS) and other cellular oxidants, both as part of the antioxidant tripeptide glutathione and in proteins that regulate the intracellular redox state, including thioredoxin (Holmgren 1985), peroxiredoxin (Wood et al. 2003), protein disulfide isomerases (Hatahet and Ruddock 2009), and glutathione S transferases (Ricci et al. 1995). Cysteines are susceptible to oxidative posttranslational modifications (Cys Ox-PTMs) and are central to redox-dependent signaling (Klomsiri et al. 2011; Holmstrom and Finkel 2014; Yang et al. 2016). Cysteines are involved in metal coordination in many different proteins, including zinc finger domains (Berg 1990; Eom et al. 2016), copper- (Han et al. 1991; Ansbacher et al. 2013) and iron-binding proteins (Kranz et al. 2009). Although cysteine is underrepresented in the human proteome (Go et al. 2015), cysteine residues are highly conserved (Marino and Gladyshev 2010) and upwards of 97% of proteins contain at least one cysteine (Miseta and Csutora 2000). Cysteine residues are among the most frequently mutated disease-causing residues, suggesting a strong link between functional cysteines and disease etiology/progression (Khan and Vihinen 2007). Gain of cysteine mutations is also not uncommon in disease. 12% of oncogenic KRAS mutations, 9% of β-catenin mutations, and a shocking 88% of fibroblast growth factor 3 mutations are acquired cysteine mutations (Visscher et al. 2016; Zhou et al. 2016). A recessive mutation in NFU1 (G208C) impairs iron-sulfur cluster biogenesis (Wachnowsky et al. 2017) and a dominant mutation in the ubiquitin ligase RNF170 (R119C) destabilizes the enzyme (Wright et al. 2015). The identification of other gain of function cysteine mutations linked to disease and small molecules tailored to react with these residues should yield new precision therapeutics.

Given their central roles in many biological processes, there has been a long-standing biochemical interest in the identification of pKa perturbed or "hyperreactive" cysteine residues, as, at least in many cases, elevated cysteine reactivity towards electrophiles has served as a good indication of functional importance. In contrast with many other well-characterized functional residues, reactive cysteine residues do not share a well-defined sequence motif. As such, the de novo identification of such residues has been challenging. These sites have often been identified in recombinant purified proteins by assessing reactivity towards iodoacetamide, acrylamide, and maleimide reagents (Kallis and Holmgren 1980; Gan and Wells 1987). More recently, quantitative proteomic has been combined with activity-based protein profiling (ABPP) to measure relative cysteine reactivities in native biological settings (Weerapana 2010; Martell et al. 2016). Alongside advances in cysteine reactivity profiling, recently new chemical probes and proteomic methods have also enabled the discovery of cysteines sensitive to posttranslational modifications (PTMs), particularly oxidative modifications (Yang et al. 2015; 2016; Gupta et al. 2017) and cysteines that can be targeted by small-molecule chemical probes (Backus et al. 2016; Bar-Peled et al. 2017). In the context of this chapter, reactive cysteines are defined as any cysteines sensitive to chemical probes and hyperreactive cysteines can be broadly defined as those pKa-perturbed thiols that are highly susceptible to alkylation by electrophilic reagents, with saturable labeling at low concentration or short incubations with probes.

Chemical probes—defined here as small molecules that can be applied to selectively manipulate and characterize biological processes—can antagonize or agonize protein function and allow an exquisite level of control over the magnitude of the perturbation (e.g., graded or partial antagonism or agonism). With the seeming prevalence of cysteine involvement in biochemical pathways, the development of small molecules that react irreversibly and selectively with individual cysteine residues is an attractive strategy to inhibit (or activate) proteins with suitable residues. In fact, considerable precedent exists for chemical probes targeting reactive cysteine residues in proteins, including electrophilic natural products and several approved drugs in current clinical use (Drahl et al. 2005; Singh et al. 2011; Gersch et al. 2012; De Cesco et al. 2017) (Fig. 1; Table 1).

Three of the top-selling drugs in the United States, Plavix (P2Y Purinceptor), Prevacid (H^+/K^+ ATPase), and Nexium (H^+/K^+ ATPase), are cysteine-mediated covalent inhibitors (Simon et al. 1997; Kim et al. 2003). Four covalent kinase inhibitors, afatinib, ibrutinib, osimertinib, and neratinib, which all react with conserved non-catalytic, cysteine residues in the kinases' ATP binding sites, have received FDA approval and have become blockbuster drugs for the treatment of various cancers. Afatinib targets Cys797 in the Epidermal Growth Factor Receptor Tyrosine Kinase (EGFR) and is approved for the treatment of non-small cell lung cancers (Carmi et al. 2012; Liu et al. 2013). Osimertinib, which targets the T790M afatinib-resistant mutant EGFR and Neratinib, which inhibits EGFR and HER2 (Zhou et al. 2009; Yver 2016) label Cys797 in EGFR and Cys805 in Her2. Ibrutinib, which targets Cys481 in Bruton's Tyrosine Kinase, is approved to treat cancers (leukemia, Waldenström's macroglobulinemia) and autoimmune disorders (graft disease).

There are numerous examples of electrophilic natural products, including those shown in Table 1. Quite strikingly, several natural products, including

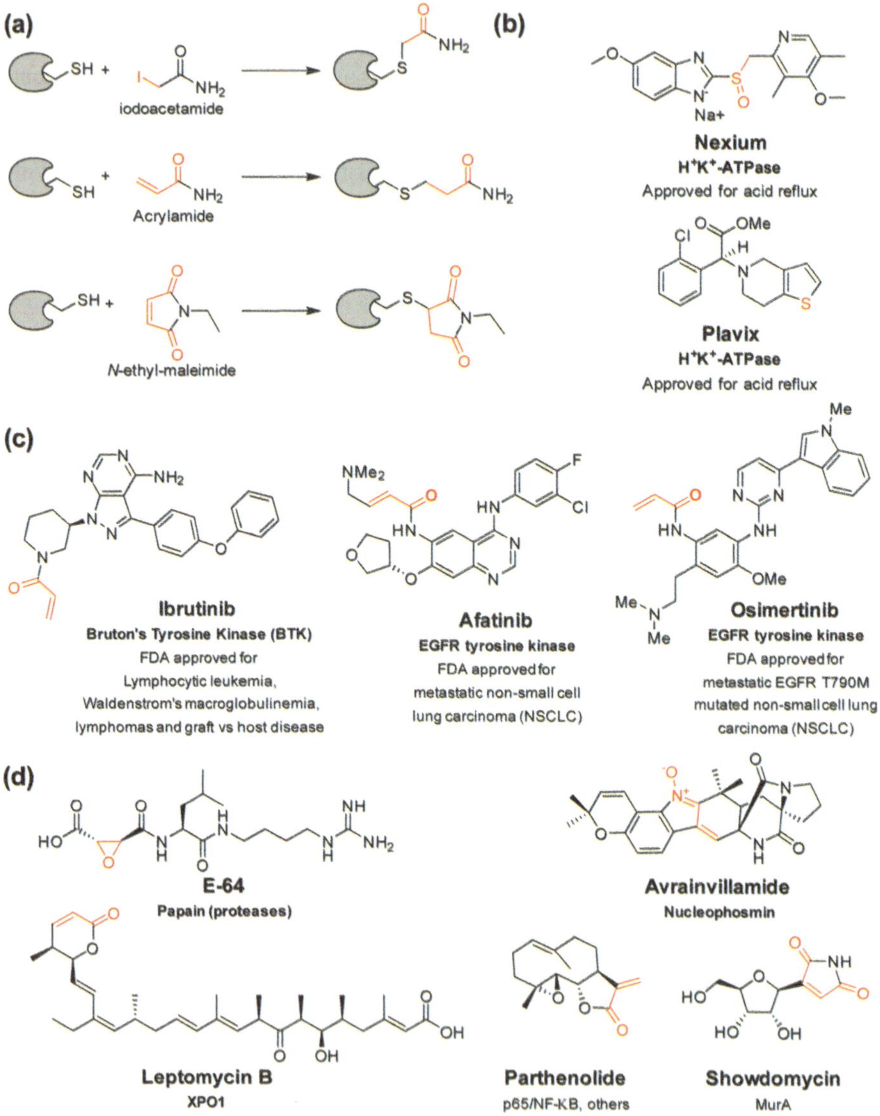

Fig. 1 **a** Labeling of proteinacious cysteines by iodoacetamide, maleimide, and acrylamide electrophiles, **b–d** The structures of representative cysteine-reactive compounds, including, **b** Metabolically activated covalent acid reflux drugs, **c** FDA-approved covalent kinase inhibitors, and **d** Natural products. Electrophiles are highlighted in red

Laptomycin B and Ratjadone have been identified that label Cys529 in Exportin 1 (XPO1 or CRM1) (Jang et al. 2003; Meissner et al. 2004), which is the same cysteine that is targeted by Selinexor (KPT330) and related compounds that are in clinical trials for the treatment of ovarian cancers (Lapalombella et al. 2012;

Table 1 Proteins and cysteines labeled by representative covalent cysteine-reactive compounds

Compound	Target	Labeled residue	Catalytic nucleophile?	References
Afatinib	EGFR	Cys797	No	Carmi et al. (2012)
Ibrutinib	Ibrutinib	Cys481	No	Ponader (2012), #11
Osimertinib	EGFR T790M	Cys797	No	Zhou et al. (2009), Yver (2016)
Neratinib	EGFR, HER2	Cys797 Cys805	No	Wissner et al. (2003), Segovia-Mendoza et al. (2015)
Plavix	P2Y12 subunit of ADP Receptor	Cys17, Cys270	No	Ding et al. (2003)
Prevacid, Nexium	ATP4A	Cys813	No	Besancon et al. (1997)
F-fluorouracil	Thymidylate synthase	Cys195	Yes	Liu et al. (1999)
Selinexor	XPO1 (CRM1)	Cys529	No	Lapalombella et al. (2012)
ARS-1620	KRAS	Gly12Cys	No	Janes et al. (2018)
Hypothemycin	Kinases, including MEK, ERK, PDGFR, VEGFR2, and FLT3	CDXG loop	No	Tanaka et al. (1999)
Leptomycin B	XPO1 (CRM1)	Cys529	No	Jang et al. (2003)
Ratjadone	XPO1 (CRM1)	Cys529	No	Meissner et al. (2004)
E-64	Papain	Cys259	Yes	Barrett et al. (1982)
Microcystin	PP1	Cys273	No	MacKintosh et al. (1995)
Avrainvillamide	Nucleophosmin	Cys275	No	Wulff et al. (2007)
Curcurmin	STAT3	Cys259	No	Hahn et al. (2018)
Andrographolide	p50/NF-κB	Cys62	No	Xia et al. (2004)
Parthenolide	p65/NF-κB, others	Cys38	No	Garcia-Pineres et al. (2001), Ploger et al. (2015)
Fosfomycin	MurA	Cys115	Yes	Eschenburg et al. (2005)
Showdomycin	MurA	Cys115	Yes	Böttcher and Sieber (2010)
Zerumbone	Keap1, others	Multiple	No	Ohnishi et al. (2009, 2013)

Hing et al. 2016), suggesting that electrophilic natural products can, at least in some cases, be used to identify clinically relevant targets. The fungal natural product hypothemycin inhibits a subset of human kinases by labeling a conserved active site cysteine (Tanaka et al. 1999), which provided precedent for many other covalent kinase inhibitors that have been developed to date. Other electrophilic compounds of note include mechanism-based inhibitors (e.g., 5-fluorouracil) and covalent inhibitors of the G12C mutated form of KRAS (Ostrem et al. 2013; Patricelli et al. 2016; Janes et al. 2018), a protein that was previously thought to be "undruggable," showcasing the utility of cysteine-reactive compounds to access tough-to-drug proteins. These examples and many others exemplify how covalent ligands that target cysteine residues offer an attractive strategy for small-molecule probe development that is complementary to more conventional ligand discovery efforts (Erlanson et al. 2000a, b; Jeffery and Bogyo 2003; Erlanson and Hansen 2004; Erlanson et al. 2004; Blum et al. 2007; Cohen et al. 2007; Garske et al. 2011; Sadowsky et al. 2011; Serafimova et al. 2012).

2 Identification of Reactive Cysteines

2.1 Biochemical Characterization of Cysteine Thiol Reactivity

Biochemically, "hyperreactive" cysteines have traditionally been characterized by reactivity with maleimide, iodocetamide, and iodoacetic acid reagents, including calculations of the second-order rate constants for the alkylation reactions and pH titrations or by measuring the rate of a chemical reaction involving the thiol to calculate the pKas of individual cysteine residues (Kallis and Holmgren 1980; Gan and Wells 1987; Grauschopf et al. 1995; Dyson et al. 1997; Jacobi et al. 1997; Hansen et al. 2005). Although the pKa of the average thiol centers around 8.5, thiol pKas has been found to vary widely (from 3.5 to 10) (Nelson and Creighton 1994), depending on the surrounding microenvironment. Thiolate anions are stabilized by positive charges and destabilized by proximal negative charges. For example, electrostatic influence of a nearby lysine residue has been proposed to contribute to the abnormally low pKa of Cys32 in thioredoxin (Kallis and Holmgren 1980) and the presence of a proximal protonated glutamic acid (E18) reduced the pKa of Cys106 in PARK7 (DJ-1) from 8.3 to 5.4 by a stabilizing interaction between the thiolate and neighboring residues (Witt, Lakshminarasimhan et al. 2008). Cysteine location at the N-terminal end of an α-helix (Ncap) was also been found to increase thiol reactivity (Kortemme and Creighton 1995; Iqbalsyah et al. 2006; Weerapana et al. 2010). Limitations of biochemical cysteine reactivity profiling include a requirement for large amounts of purified protein, a lack of throughput, and incompatibility with proteins of unknown function and proteins that are difficult to isolate, including membrane-bound proteins and those found in larger complexes.

2.2 Proteomic Profiling of Cysteine Residues

With the advent of shotgun proteomics (Eng et al. 1994; Shevchenko et al. 1996; Link et al. 1997; Wolters et al. 2001; McDonald and Yates 2002; Washburn 2015), chemical probes were rapidly developed to label and enrich alkylation-sensitive cysteine-containing peptides from complex biological mixtures. Gygi and colleagues developed fully functionalized electrophilic probes, termed isotope-coded affinity tags (ICAT), which feature an electrophile (iodoacetamide or maleimide), an isotopically encoded linker, and a biotin moiety for capture on avidin or streptavidin resin (Fig. 2a) (Gygi et al. 1999). Gygi and coworkers labeled proteins from two distinct samples with either isotopically "light" or "heavy" ICAT reagents and the samples were combined, digested into peptides, enriched on affinity resin, eluted from the resin, and analyzed by liquid chromatography–tandem mass spectrometry (LC-MS/MS). Early ICAT reagents featured deuterium labeled linkers, but in later iterations (Yu et al. 2004; Yi et al. 2005), ^{13}C-labeled linkers were employed to address deuterium–resin interactions, thereby improving the co-elution of light and heavy peptide peaks (Zhang et al. 2001). Building upon the ICAT-approach, Dennehy and colleagues profiled the cytosolic and nuclear protein targets of maleimide and acrylamide probes (Dennehy et al. 2006), revealing that a subset of cysteines label preferentially with each electrophile. Wong et al. used similar probes to profile alkylation sites linked to apoptosis (Wong and Liebler 2008). Cleavable ICAT probes improved the detection of low-abundance peptides (Hansen et al. 2003). ICAT has also successfully identified redox-sensitives cysteine thiols (Wu et al. 2011; Lindemann et al. 2013; Garcia-Santamarina et al. 2014).

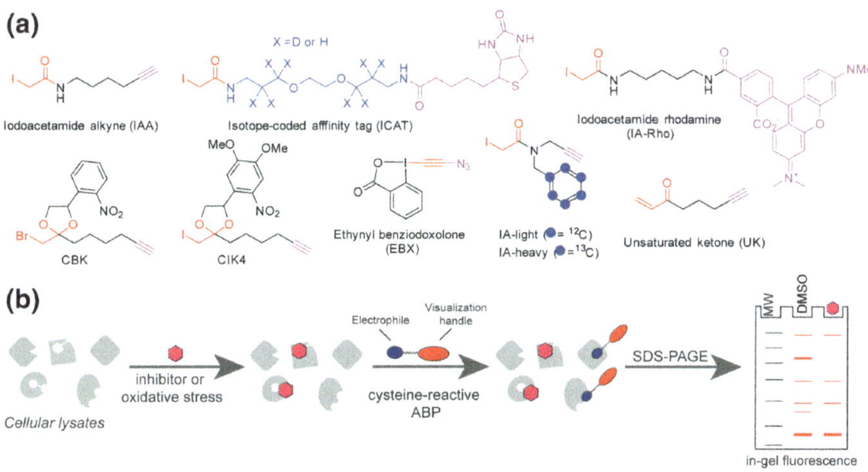

Fig. 2 **a** Cysteine-reactive chemical probes. Electrophiles in red, isotopic labels in blue and enrichment or visualization handles (biotin, alkyne, rhodamine) in magenta. **b** gel-based ABPP at cysteine residues. Cells or cellular lysates are subjected to electrophilic compounds or oxidative stressors, followed by labeling with an activity-based probe (ABP), SDS-PAGE and in-gel fluorescence analysis

In addition to proteomic profiling, in many of these studies, fully functionalized, biotinylated or fluorescent cysteine-reactive probes were also employed to visualize thiol-labeling by western blot or in-gel fluorescence (Baty et al. 2002; Rogers et al. 2006), providing an excellent precedent for more tailored cysteine-based competitive ABPP experiments. In these experiments, lysates were pretreated with an oxidant or alkylating agent and then subjected to labeling by a cysteine-reactive probe, SDS-PAGE and in-gel fluorescence profiling (Fig. 2b). One of the limitations of ICAT and other early cysteine reactivity studies was the that the fully functionalized nature of the probes, containing both electrophile and biotin, precluded profiling of a more diverse set of electrophilic compounds, and largely restricted studies to in lysate experiments, given the poor cellular uptake of biotin-containing compounds and cytotoxicity of iodoacetamide and maleimide reagents (Wong and Liebler 2008). Given the extremely strong interaction between biotin and streptavidin ($K_d \approx 10^{-15}$ M) (Green 1990), a lack of mild conditions to elute labeled peptides from resin was another limitation of these approaches.

2.3 TOP-ABPP and IsoTOP-ABPP

With the introduction of copper-catalyzed azide–alkyne cycloaddition (CuAAC) or "click" chemistry (Kolb et al. 2001) and its application to ABPP and bioorthoganol labeling (Speers and Cravatt 2004) coupled with Tandem Orthogonal Proteolysis-Activity-Based Protein Profiling (TOP-ABPP) (Speers and Cravatt 2005; Weerapana et al. 2007, 2008) enabled the characterization of a more diverse set of electrophilic compounds and the development of more advanced proteomics platforms to quantify cysteine reactivity proteome-wide. To identify the amino acids labeled by a diverse set of electrophilic compounds, Weerapana, Speers, and colleagues synthesized and applied peptide enrichment tags, featuring biotin linked to a Tobacco Etch Virus (TEV) cleavage motif and an azide for "click" chemistry (Speers and Cravatt 2005; Weerapana et al. 2007, 2008), termed "TEV-tags". In these experiments, lysates were treated with chemical probes, containing alkynes and electrophiles and, after incubation, the compounds were then conjugated by "click" chemistry to the TEV-tags, enriched on avidin resin and subjected to sequential on-bead digests with trypsin and TEV protease to release the labeled peptides. TOP-APPP identified the sites of labeling and amino acid specificity of electrophilic compounds, revealing that alpha halides (e.g., chloroacetamide) show remarkable selectivity for cysteine thiols (Weerapana et al. 2008).

2.4 IsoTOP-ABPP Data Collection and Analysis

The synthesis of isotopic labels into the TEV-tags, (^{12}C- and ^{13}C-) TEV-tags to conduct isotopic Tandem Orthogonal Proteolysis-Activity-Based Protein Profiling

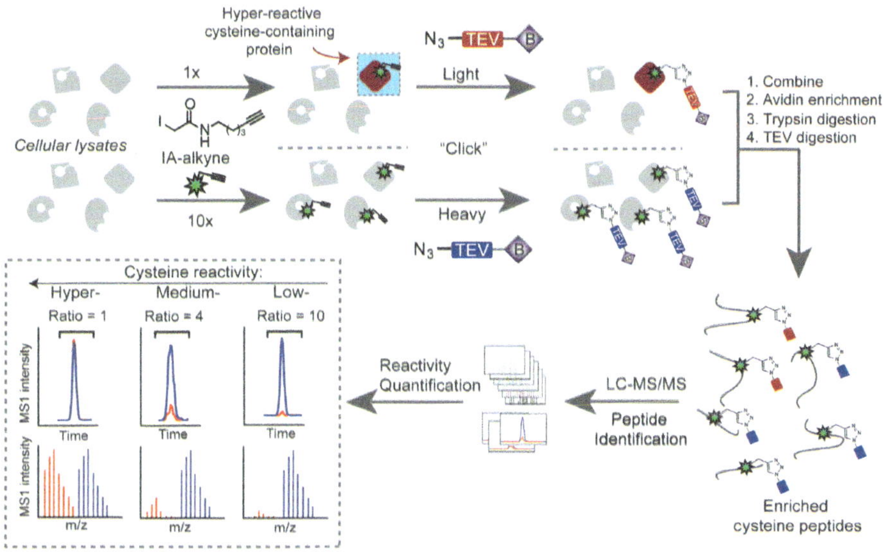

Fig. 3 isoTOP-ABPP to identify hyperreactive cysteine residues. Samples are labeled by cysteine-reactive probes at the indicated concentrations, conjugated to isotopically differentiated TEV-tags by click chemistry, enriched, digested, and analyzed by LC-MS/MS

(isoTOP-ABPP) built upon TOP-ABPP to enable quantitative reactivity profiling at cysteine residues (Weerapana et al. 2010). A typical workflow for isoTOP-ABPP reactivity profiling is shown in Fig. 3. Cellular lysates are treated with a cysteine-reactive chemical probe, such as iodoacetamide-alkyne (IAA), at either two different concentrations of probe or with equal amounts of probe for two time points (e.g., 10× and 1×). Next, the probe labeled samples are conjugated by "click" chemistry to isotopically differentiated, TEV-cleavable peptide tags. The samples are then combined pairwise, reduced with 1,4-dithiothreitol (DTT) or tris (2-carboxyethyl)phosphine (TCEP), and all remaining cysteines alkylated with iodoacetamide, and then the samples are enriched on streptavidin resin and subjected to sequential on-bead digests first with trypsin and then with TEV protease to release the labeled peptides. TEV-eluted peptides are analyzed and identified by LC-MS/MS and the data processed using specialized software, discussed in the next section. Hyperreactive cysteines are identified as those residues that label equally at low and high concentrations or short and long time periods.

During LC-MS/MS analysis of a typical isoTOP-ABPP sample, both full mass spectra (MS1) and tandem MS spectra (MS/MS or MS2) are acquired, which in aggregate provide both the amino acid composition and sequence for each peptide detected. After data acquisition, the experimentally obtained spectra are compared to or "searched," using standard search algorithms [e.g., Mascot (Perkins et al. 1999), SEQUEST (Eng, McCormack et al. 1994), ProLuCid (Xu et al. 2015) or Percolator (The et al. 2016)] against an in silico generated database of MS/MS spectra, originating from the sequences of all known proteins from the organism of interest.

During this search, any known modifications are included, for example, the TEV-tag amino acid adduct, cysteine carbamidomethylation, or methionine oxidation. These modifications can be searched as either dynamic (either present or absent) or static (always present in the sample) modifications. For isoTOP-ABPP, the samples are searched with both a static (always present in the sample) modification for carbamidomethylation of all cysteine thiols as well as a dynamic (either present or absent) modification for the iodoacetamide alkyne-TEV modification of a subset of reactive cysteine thiols. After search and statistical filtering, the samples are quantified by comparing the integrated areas of the MS1 chromatographic peaks for the light- and heavy-labeled samples. For isoTOP-ABPP, MS1-based quantification is advantageous because it is highly quantitative and requires a limited number of fragmented peptide spectra, which reduces scan time and increases coverage. The resulting ratios or R values correspond to the difference in relative peak areas with an R value ≈ 1, corresponding to a light/heavy peak pairs with similar areas and a maximal R value = 10, corresponding to a tenfold difference in labeling between light and heavy samples. Hyperreactive cysteines, as identified by isoTOP-ABPP, have L/H peak ratios close to 1, whereas unreactive cysteines have ratios close to 10. IsoTOP-ABPP reactivity profiling demonstrates key features of all thiol reactivity proteomic profiling platforms, which include a chemical probe that selectively labels thiols, bioorthogonal chemistry to capture labeled peptides, an isotopic label to distinguish between samples, and a biotinylated, cleavable linker for selective enrichment and release of labeled peptides, identifying thousands of cysteines and hundreds of hyperreactive cysteines in single proteomic experiments.

2.5 Modifications and Improvements to Reactivity Profiling

Modified protocols and new probes (see Fig. 2a for representative probes) for sample labeling, enrichment, elution, and quantification have addressed some of the shortcomings of isoTOP-ABPP. The reactivity of cysteines with 2-haloacetic acids has been known since 1933 (Dickens 1933) and, to date, most cysteine reactivity profiling studies have employed haloacetamide-based chemical probes, particularly chloroacetamide and iodoacetamide probes, which react with cysteines via nucleophilic substitution. Numerous cysteines across many different protein classes have been shown to react with chloroacetamide (Barglow and Cravatt 2006; Weerapana et al. 2008; Couvertier and Weerapana 2014; Backus et al. 2016). Iodoacetamide, which is significantly more reactive than chloroacetamide, is still remarkably selective for cysteines with only some off-target reactivity towards lysine residues at high millimolar concentrations (Nielsen et al. 2008). Although the iodoacetamide-alkyne probe used to identify hyperreactive cysteines is broadly reactive and labels most cysteines efficiently, the cytotoxicity of the IA group precludes live-cell labeling. Photocaged probes, including bromomethyl ketone (CBK) (Abo and Weerapana 2015) and iodomethylketone probes (e.g., CIK4) (Abo et al. 2017), which show minimal cytotoxicity, enable cell cysteine profiling and spatial and temporal control over electrophile activation. These probes are particularly suited to the identification of transient modifications that may be sensitive to cellular lysis. In addition to α-halides, many other classes of cysteine-reactive electrophiles have been developed. An unsaturated ketone (UK) probe (Weerapana et al. 2008) exhibited cysteine-selective labeling, but has not to date been used for reactivity profiling. Shannon and colleagues have developed halo-triazines (Shannon et al. 2014) as reagents that allow for modular probe synthesis and selective cysteine labeling. Adibekian and colleagues have demonstrated that hypervalent iodine ethynyl benziodoxolone (EBX) (Frei et al. 2014) are highly selectively for cysteines (97% vs. 91% for IAA) and exhibit increased reactivity when compared with IAA (Abegg et al. 2015). Fluorescent probes, including iodoacetamide rhodamine (IA-Rho) have been developed for in-gel visualization (Fig. 2) and fluorescence polarization screening (Fluopol), discussed below.

New cleavable linkers (Fig. 4) have been synthesized to improve the selective release of labeled peptides. While TEV protease cleavage is versatile and effective, it requires long cleavage times and can lead to contamination of samples by peptides derived from TEV protease autoproteolysis. Also, given that the TEV protease recognition motif (Carrington and Dougherty 1988) [ENLYFQ(G/S)] contains both phenylalanine and a glutamate residues, TEV-tags are incompatible with many sequence-specific proteases, including the endoproteinase Glu-C and chymotrypsin that could enable the identification of additional labeled peptides. Comparable to TEV-tags, isotopically differentiated, dithionite cleavable, azobenzene tags enable reactivity profiling of cysteine residues (Qian et al. 2013) and, given their

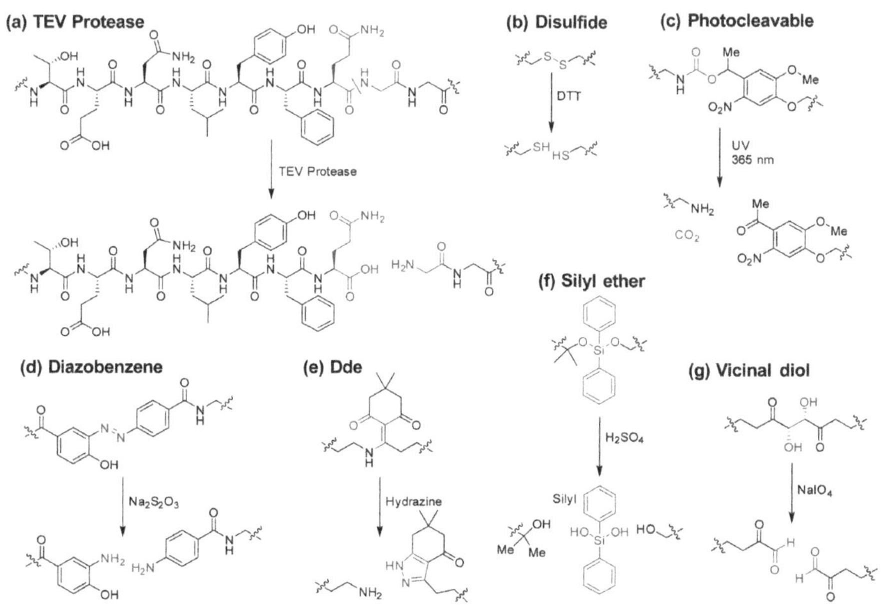

Fig. 4 a–g Cleavable linkers for site of labeling mass spectrometry

non-peptidic structure, should prove compatible with all common proteases. Several other alternative chemically cleavable linkers have been developed, including *o*-nitrobenzyl (cleaved with 365 nm UV light) (Olejnik et al. 1995; Zhou et al. 2002; Wang et al. 2010), a 1-(4,4-dimethyl-2,6-dioxocyclohexylidene)ethyl (Dde, cleaved with hydrazine) (Yang and Verhelst 2013), a vicinal diol (cleaved with periodate) (Maurer et al. 2013; Yang and Verhelst 2013), disulfide (cleaved with DTT) (Gartner et al. 2007), and silyl (cleaved with low pH or fluoride) (Szychowski et al. 2010; Gao et al. 2018).

Two alternative approaches to cysteine profiling that do not require isotopically differentiated enrichment tags have also recently been reported. First, instead of the isotopically differentiated TEV-tags, isotopically labeled (heavy- and light-) iodoacetamide-alkyne probes have been synthesized and applied to profile cysteine residues (Abo et al. 2017). These probes feature a benzyl moiety that has either ^{12}C- (light) or ^{13}C- (heavy) isotopes (Fig. 2a), which build upon and provide a much more affordable alternative to ICAT reagents (Gygi et al. 1999). Isotopically differentiated iodoacetamide probes also support quantitative profiling of reversible cysteine PTMS, which will be discussed in more detail in the next section. Multiplexed thiol reactivity profiling (MTRP), which combines thiol labeling and "isobaric tags for relative and absolute quantization" (iTRAQ) (Ross et al. 2004; Pierce et al. 2008), also circumvents the need for isotopically labeled enrichment tags, through the use of isobaric chemical reporters. MTRP offers the added advantage of high-throughput quantification and comparison of up to eight samples in parallel (Tian et al. 2017), dramatically improving sample throughput and accuracy of quantification. Similar to ITRAQ reagents, tandem mass tags (TMT) (Thompson et al. 2003; Dayon et al. 2008) could also be used in this format and cysteine reactivity profiling should also be compatible with other isotopic labeling strategies, including stable isotope labeling by amino acids in cell culture (SILAC) (Ong and Mann 2007) and dimethyl labeling (DML) (Hsu et al. 2003).

Recent advances in proteomics sample acquisition, particularly Data Independent Acquisition (DIA), including SWATH-MS (Gillet et al. 2012; Rost et al. 2014)—where, in an unbiased manner, all peptides within defined mass-to-charge (*m/z*) windows are sequentially subjected to fragmentation across the full m/z range of the mass spectrometer—or other implementations such as the MS^E approach (Silva et al. 2006; Li et al. 2009; Levin and Bahn 2010), offer the future promise of improved coverage for reactivity profiling experiments. DIA is especially useful for the quantification of low-abundance species, as it is not biased by peptide abundance and does not require predetermined peptide species for identification. However, because multiple peptides are fragmented together in DIA, the resulting MS/MS spectra are complex, and require specialized software for interpretation (Li et al. 2009; Doerr 2014; Röst et al. 2014). As such, methods that combine Data Dependent Analysis (DDA) with DIA may allow for both improved coverage and more straightforward analysis (Geromanos et al. 2009). If applied to cysteine reactivity profiling, a combination of DIA and DDA acquisition would improve coverage and quantification of low-abundance labeled residues. For a detailed review of recent advances and applications of quantitative proteomics, refer to (Bantscheff et al. 2012; Uzozie and Aebersold 2018).

2.6 General Applications of Reactive Cysteine Profiling

IsoTOP-ABPP-based cysteine reactivity profiling identified hundreds of reactive thiols shared across multiple cancer cell lines and demonstrated that hyperreactive cysteines are highly enriched in functional sites (Weerapana, Wang et al. 2010). Included among the many reactive and residues uncovered by Weerapana and coworkers, was a cysteine (Cys161 in FAM96B) essential for iron–sulfur cluster biogenesis and a cysteine (Cys109) in the protein arginine methyl transferase PRMT1 that, when labeled, blocks methyl transferase activity (Weerapana et al. 2010). Cysteine reactivity profiling has uncovered mediators of longevity in *C. elegans* (Martell et al. 2016) and enabled the discovery of reactive residues in subcellular fractions, including the mitochondria (Bak and Weerapana 2015), suggesting that, given the differences in subcellular pH of the different cellular compartments (Llopis et al. 1998), a more comprehensive survey of compartmentalized thiol reactivity would uncover additional functional residues.

With the advent of large datasets of highly reactive cysteines, there has been a recent interest in developing methods to computationally predict cysteine reactivity and to design reactive cysteines in silico. Machine learning has been applied to accurately predict thiol reactivity on a sequence basis (Wang et al. 2018). Reactive cysteines have also been computationally designed (Weerapana et al. 2010) and engineered into proteins and used to generate site-specifically labeled antibody-drug conjugates (Jeffrey et al. 2013; Sussman et al. 2018). Reactive cysteines have also been used as a handle to chemically incorporate the nonnatural amino dehydroalanine, which enables site-specific incorporation of diverse nucleophiles into proteins (Chalker et al. 2011, 2012).

Competitive isoTOP-ABPP, a modified version of reactivity profiling has identified posttranslationally modified cysteines (Wang et al. 2014; Briggs et al. 2016; Bar-Peled et al. 2017), cysteines susceptible to chemical probes (Backus et al. 2016), drug vulnerabilities in cancer (Bar-Peled et al. 2017), the targets of natural products (Guo et al. 2017), and FDA-approved drugs (Blewett et al. 2016), which will be discussed in the following sections. Overall, cysteine reactivity profiling is a powerful and unbiased approach to identify functional cysteines.

3 Identification of Redox-Sensitive Cysteines

The properties that make thiols excellent modulators of biological process, namely their intrinsic nucleophilicity and pKa close to neutral pH, also leave them susceptible to oxidative damage, including oxidation by reactive oxygen (ROS) or nitrogen species (RNS) and labeling by endogenous and exogenous electrophiles (Giles and Jacob 2002; Paulsen and Carroll 2013). Cysteine oxidative posttranslational modifications (Cys Ox-PTMs) include sulfhydration (SSH), S-glutathionylation (SSG), disulfide bonds (RS-SR) (Cumming et al. 2004),

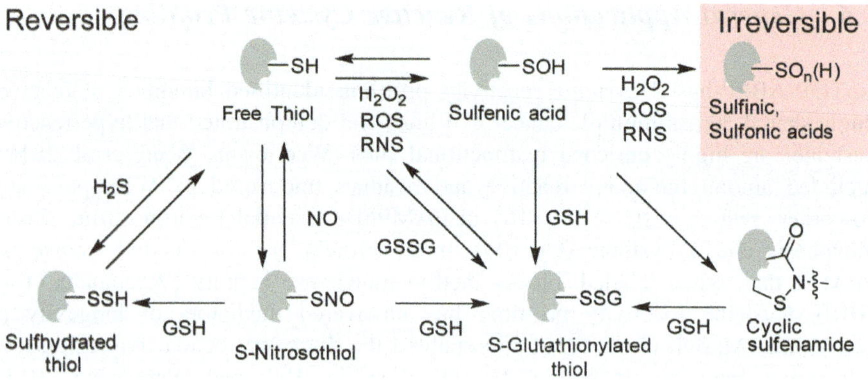

Fig. 5 Reversible and irreversible Cys OX-PTMs. Reactive oxygen and reactive nitrogen stressors (ROS and RNS)

S-nitrosylation (also called S-nitrosation, SNO) (Hess et al. 2005), sulfenylation (SOH), sulfinic acid (SO_2H), and sulfonic acid (SO_3H) (Klatt and Lamas 2000; Paul and Snyder 2015) (Eaton 2006; Poole and Nelson 2008; Lo Conte and Carroll 2013; Paulsen and Carroll 2013). As is shown in Fig. 5, most Cys Ox-PTMs are induced by reactive nitrogen or oxygen species and these modifications are usually dynamic and reversible by intracellular reducing agents, such as glutathione. Cys Ox-PTMs have been implicated in many cancers (Sajadimajd and Khazaei 2017), inflammation (Hoffman et al. 2015), stroke (Andersson et al. 2000; Maron et al. 2013) as well as neurodegenerative diseases (Canet-Aviles et al. 2004; Chung 2006; Blackinton et al. 2009; Fox et al. 2011).

Detection and quantification of Cys OX-PTMs have been limited by the dynamic and reversible nature of these modifications and a lack of chemical probes to directly label modified cysteines. Recently, both *indirect* and *direct* labeling strategies have identified Cys OX-PTMs. Indirect methods fall into two broad categories (1) those that rely on a restoration of labeling by treatment with reducing agents such as dithiothreitol (DTT) (Fig. 6a) or (2) those that detect a loss of reactivity with thiol-modifying agents (Fig. 6b). Direct methods rely on PTM-specific antibodies or tailored chemical probes (Fig. 7) that will react directly modified cysteines. Indirect methods are typically restricted to profiling in lysates or purified proteins whereas direct chemical probe-based methods are compatible within cell profiling. For general reviews of detection of Cys OX-PTMs, refer to (Paulsen and Carroll 2013; Duan et al. 2017). An overview of key methods as they relate to cysteine reactivity is provided below.

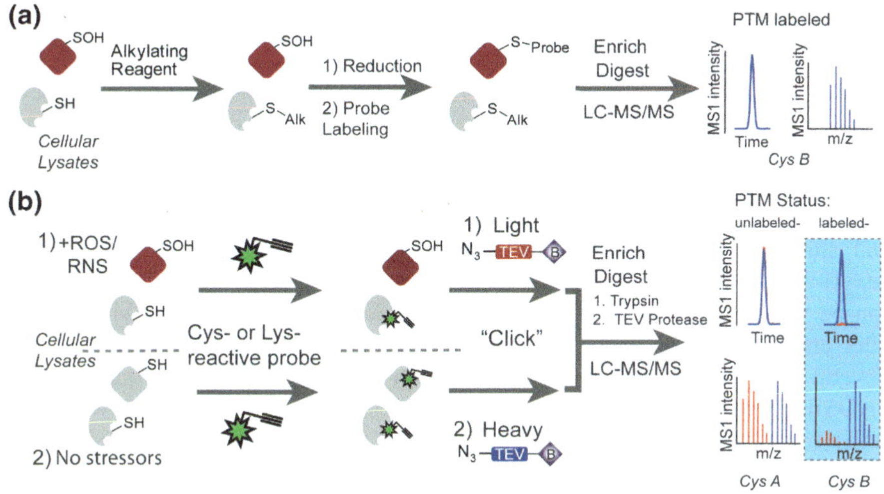

Fig. 6 Proteomic methods to identifying Cys OX-PTMS. **a** In reduction-based methods, such as the biotin switch, all free cysteines are first alkylated and then modified cysteines selectively reduced and labeled with chemical probes and identified by LC-MS/MS. **b** IsoTOP-ABPP to identify the sites of reactive oxygen and reactive nitrogen stressors (ROS and RNS) modifications. Samples are subjected to stressor "1" or control "2" and then are labeled with cysteine-reactive probe. ROS/RNS modifications block probe labeling and this blockade of probe labeling and percentage modification can be quantified by isoTOP-ABPP

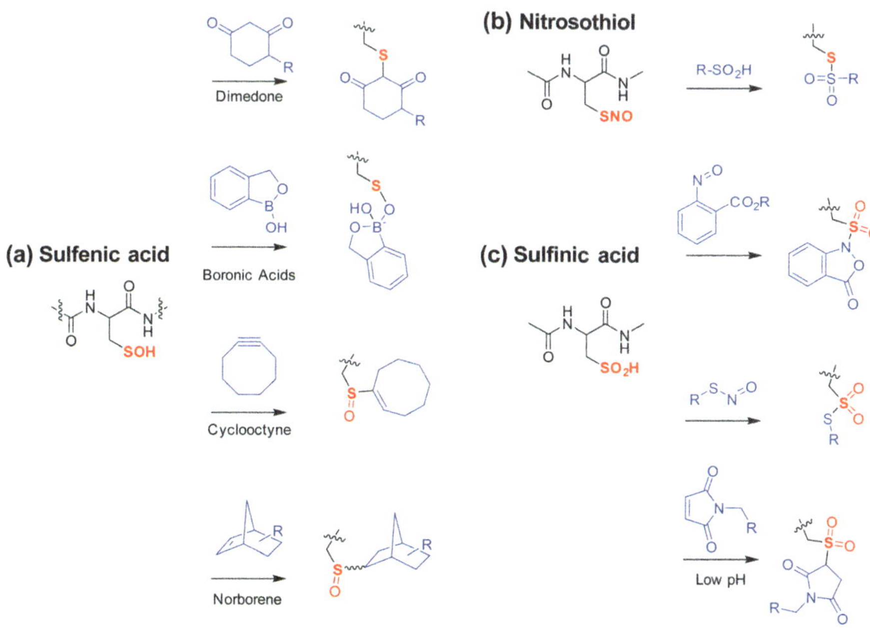

Fig. 7 Chemical probes to detect **a** S-sulfenylation (sulfenic acid), **b** S-nitrosylation (nitrosothiol) and **c** S-sulfinylation (sulfinic acid)

3.1 Indirect Cys OX-PTM Detection

Indirect, reduction-based methods (see Fig. 6a for a general workflow) identify all reversibly modified cysteines, including those that are partial or low-occupancy labeling events. Biotin Switch is one such method. In a Biotin Switch experiment, free cysteines are blocked by N-ethylmaleimide (NEM), methyl methanothiosulfonate (MMTS), or iodoacetamide, and then modified thiols are selectively reduced and simultaneously labeled with a reversible thiol-reactive *N*-[6-(biotinamido) hexyl]-3′-(2′-pyridyldithio)propionamide (biotin-HPDP). The biotinylated proteins can be visualized by western blot and cysteines detected by LC-MS/MS analysis after enrichment, tryptic digest, and elution of labeled peptides from the resin. Several modifications have since been made to the Biotin Switch method, including SNOSID (Hao et al. 2006) and SNO-RAC (Forrester et al. 2009). However, all suffer from the same limitations, including that, unless combined with SILAC or ITRAQ labeling, Biotin Switch and variants rely on spectral counting for quantification and provide only an indirect readout of the specific type of modification.

To quantify cysteine oxidation on the MS1 chromatographic peak level, the OxICAT method was developed by Leichert et al. (2008) and based on the original ICAT method pioneered by Gygi et al. (1999). OxICAT employs sequential labeling with isotopically differentiated ICAT probes. The proteome samples are labeled with a light ICAT reagent under denaturing conditions to cap all free cysteines. Then, the samples are subject to a strong reducing agent, such as tris (2-carboxyethyl)-phosphine hydrochloride (TCEP) and the newly reduced cysteines labeled with the heavy ICAT reagent. ICAT reagents can also be used for pairwise comparisons in a manner similar to the Biotin Switch assay, where free cysteines are first blocked and then oxidized cysteines reduced and labeled with ICAT probes (Garcia-Santamarina et al. 2014). Abo and colleagues recently reported a similar method, which employs less costly ^{13}C-isotopically differentiated iodoacetamide alkyne reagents (Abo et al. 2017). Reduction-based methods identify modified peptides, including low-occupancy labeling events (e.g., thiols that are only partially modified). However, irreversible modifications, such as sulfinic and sulfonic acids, will not be detected.

Indirect, loss of reactivity methods, including ICAT and isoTOP-ABPP, have been used to identify high-occupancy Cys OX-PTMs, including irreversible PTMs. In ICAT, oxidatively stressed samples are labeled with heavy-ICAT probes and control samples are labeled with light-ICAT reagents and then the relative probe labeling is compared by LC-MS/MS. Using ICAT probes, Sethuraman and coworkers found that cysteines with OX-PTMs were resistant to labeling by the ICAT probes and that the fractional occupancy of the OX-PTM could be determined by comparing the relative peak intensities for labeled peptides derived from the heavy- and light-labeled samples, respectively (Sethuraman et al. 2004a, b). ICAT identified cysteines in VDAC and GAPDH that were sensitive to hydrogen peroxide (Sethuraman et al. 2004a, b). In a similar manner, isoTOP-ABPP can also be used to identify Cys OX-PTM sites (Fig. 6b). In these experiments, samples with and without the Cys OX-PTMs are labeled in parallel with iodoacetamide alkyne, conjugated to isotopically differentiated TEV-tags, combined pairwise, and analyzed by LC-MS/MS. Cys OX-PTMs are identified by the decrease in MS1 chromatographic peak signal for peptides containing a modified residue, when compared to the control sample. This approach, termed competitive isoTOP-ABPP, identified cysteines sensitive to oxidative stress in cancers, including cysteines in the nuclear hormone receptor NR0B and in the prolyl hydroxylase EGLN1 (Briggs et al. 2016; Bar-Peled et al. 2017). One obvious limitation of loss of reactivity methods is that they will not identify low-level labeling sites, and as most sites of cysteine modification are only partially labeled (Sethuraman et al. 2004a, b), most Cys OX-PTMs will not be captured by these methods.

3.2 Chemical Probes to Directly Detect Cys OX-PTMs

For the direct detection of PTMs, both antibodies and chemical probes have been employed. Antibodies have been developed that detect protein S-glutathionylation (Prinarakis et al. 2008; Chen et al. 2010), sulfonation (Persson et al. 2004), and nitrosylation (Sun et al. 2001; Matsushita et al. 2003). As antibodies are not compatible with in cell experiments, frequently fail to capture unstable modifications, and exhibit biased detection and enrichment of a subset of proteins, antibody-based Cys OX-PTM profiling methods have largely been supplanted with chemical probes (Fig. 7). First shown to react with nucleophiles in 1974 (Benitez and Allison 1974), protein sulfenic acids have been labeled by a variety of nucleophiles, particularly 1,3-diketones such as dimedone (Paulsen and Carroll 2013). In contrast with other nucleophiles, dimedone-based probes selectively react with sulfenylated cysteines (Paulsen et al. 2011; Yang et al. 2015; Gupta and Carroll 2016) and have been applied to broadly profile S-sulfenylation. Dimedone-based covalent fragments have also been demonstrated to selectively inhibit proteins containing suitable sulfenylation sites. Gupta et al. reported covalent ligands for >1280 S-sulfenylated cysteines (Gupta et al. 2017). Sulfenylation of a conserved non-catalytic cysteine, C797 in EGFR during EGF stimulation, has been found to enhance EGFR kinase activity (Paulsen et al. 2011) and chronic oxidative stress-mediated sulfenylation of EGFR has been shown to render a subpopulation of cancer cells resistant to the FDA-approved drug afatinib (Truong et al. 2016). It is attractive to speculate that combination chemotherapy, including both afatinib and nucleophilic inhibitors targeting 797Cys-SOH would enable the treatment of refractory cancers.

In addition to dimedone, many other nucleophiles and electrophiles directly label Cys OX-PTMs (Fig. 7). Protein sulfenic acids also react with electrophiles, including 7-chloro-4-nitrobenzo-2-oxa-1,3-diazole (NBD-Cl) (Ellis and Poole 1997), boronic acids, and benzoxaborole (Liu and Benkovic 2013). *Trans*-cyclooctene and cyclooctyne probes have also been recently applied to trap sulfenic acids through a cycloaddition to produce a stable sulfoxide adduct (Poole et al. 2014).

Alcock and colleagues found that norborene probes trap sulfenic acids as stable norborane sulfoxides (Alcock et al. 2018). S-sulfinylation has also proven amendable to probe-based labeling, with aryl-nitroso compounds demonstrating rapid and selective labeling of sulfinylated cysteines in the thiol peroxidase GPX3 (Lo Conte et al. 2015). Majmudar and colleagues demonstrated that sulfinate-linked probes enrich S-nitrosated proteins and nitrosothiol probes enrich S-sulfinated proteins (Majmudar et al. 2016). Maleimide probes at low pH have also been shown to react with sulfinated thiols (Kuo et al. 2017). S-nitrosylated residues also reacted with triaryl phosphine reagents to form stable S-alkylphosphonium adducts (Wang and Xian 2008; Bechtold et al. 2010). In addition to detecting protein PTMs, tailored probes can also detect small-molecule thiols. Of particular note are the probes for hydrogen sulfide (H_2S) (Lin et al. 2013; Lippert 2014; Lin et al. 2015a, b) and caged hydrogen sulfide sources (Cerda et al. 2017; Steiger et al. 2017; Zhao et al. 2017a, b), which provide a versatile set of tools to elucidate H_2S's production, trafficking, and contributions to cell signaling.

With the advent of dimedone-based probes as well as Biotin Switch type enrichment strategies, great advances have been made in identifying and characterizing cysteine posttranslational modifications. As with many PTMs, determination of the fractional occupancy of labeling for most thiol modifications, particularly those low-occupancy and unstable modifications, has been difficult. Quantitative proteomics methods, such as isoTOP-ABPP (Weerapana et al. 2010), have proven useful for the identification of high-occupancy labeling sites. However, determining the percentage labeling for lower occupancy sites has remained challenging. To date, the vast majority of identified cysteine PTMs remain functionally uncharacterized. As such, the development of new methods to characterize the contributions of PTMs to protein function would find utility in the identification of new protein regulatory sites.

4 Covalent Inhibitor Discovery at Cysteine Residues

4.1 Strengths and Weaknesses of Covalent Compounds

Covalent compounds, both in general, and specifically those that are cysteine-reactive, offer many desirable attributes for probe and drug development (Johnson et al. 2010). For long-half-life protein targets, covalent compounds can achieve long residence time with less frequent dosing, as the pharmacodynamic effect will outlast the pharmacokinetics of the inhibitor. Covalent compounds are often extremely potent, frequently engaging targets with subnanomolar binding strength (Smith et al. 2009) and can target tough-to-drug or shallow pockets, as exemplified by the compounds that target KRAS (Ostrem et al. 2013; Patricelli et al. 2016; Janes et al. 2018). The ligand efficiency (defined as the average binding energy per non-hydrogen atom) of covalent compounds can far exceed their non-covalent counterparts), allowing low molecular covalent compounds to achieve high target affinity. Given their nonequilibrium binding kinetics, irreversible molecules can compete efficiently against endogenous substrates, particularly those substrates that are found at high concentrations in living systems (e.g., ATP for kinases) or when the mechanism of inhibition leads to a large buildup of substrate (Swinney 2009; Tonge 2018). Covalent compounds are also particularly useful when complete and sustained target engagement is required, exemplified by the covalent kinase inhibitors, which achieve a sustained and durable target inhibition until protein resynthesis occurs. However, this sustained occupancy can be a liability for targets where partial engagement is favored, for some receptors and ion channels, where partial agonism/antagonism or short residence time is preferred (Lipton 2006).

Limitations of covalent compounds include potential toxicity due to irreversible modification of off-target nucleophiles and possible immunogenicity of the corresponding protein adducts (Lavergne et al. 2008; Uetrecht 2009). Most of the safety concerns surrounding covalent compounds stem from idiosyncratic adverse drug reactions (ADRs) caused by compounds known to form reactive metabolites capable of modifying proteins (Evans et al. 2004). Chemical biology methods, particularly the synthesis and application of alkyne-containing chemical probes or "click probes", have provided a deeper understanding of the behavior of drugs and drug-like molecules and their potential to generate reactive metabolites (Bateman et al. 2013; Whitby et al. 2017). Profiling of four structurally dissimilar probes for compounds that generate reactive metabolites (probes for Acetominophen, Tienilic acid, Troglitazone, and Clozapine) revealed that a unifying theme across all four compounds was enrichment of proteins containing functional cysteine and selenocysteine residues (Whitby et al. 2017), suggesting that reactive thiols are primary sites of labeling for reactive drug metabolites.

To evaluate the potency, selectivity and off-target liabilities of covalent compounds, in principle, any covalent compound can be modified with a "clickable" tag to generate an activity-based probe for rapid identification of the on- and off-target(s) of lead compounds in any living system. Exemplifying this probe strategy, Lanning and colleagues applied alkyne-containing "clickable" variants of established covalent kinase inhibitors in conjunction with a chemical proteomics approach, which identified the protein off-targets of these compounds, both those that are saturable and high affinity and lower affinity substoichiometric labeling events (Lanning et al. 2014; Niessen et al. 2017). Notably, despite the safety concerns surrounding covalent compounds, almost 30% of the clinically approved drugs that target enzymes act in an irreversible manner (Robertson 2005). This abundance of clinical compounds corroborates that covalent compounds can be safe and efficacious, and supports the development of new covalent compounds, including those that label reactive cysteine residues.

4.2 *Methods to Develop Covalent Compounds*

Until recently, most covalent compounds were discovered fortuitously, either from natural products (Table 1), or as phenotypic screening hits (Bauer 2015). More recently, structure-guided design, particularly by retrofitting existing non-covalent inhibitors with electrophiles (e.g., covalent kinase inhibitors), has been routinely used as a method to develop covalent inhibitors (Cohen et al. 2005; Pan et al. 2007; Zhang et al. 2009). For example, the covalent kinase inhibitors afatinib and ibrutinib were both developed by retrofitting non-covalent kinase inhibitors with electrophiles (Wissner et al. 2003; Pan et al. 2007). Although retrofitting is useful for cases where an existing inhibitor is found to bind proximal to a suitable cysteine residue, this strategy, by definition, requires an optimized inhibitor and thus precludes the vast majority of proteins for which no small-molecule tool compounds are available. Some excellent reviews have been written on the recent advances in covalent kinase inhibitor development (Muller et al. 2015; Zhao and Bourne 2018). High-throughput screening at reactive cysteine residues (Bachovchin et al. 2009; Dillon et al. 2012), fragment-based tethered screening (Ostrem et al. 2013), in silico screening (Krishnan et al. 2014; London et al. 2014), and unbiased proteome-wide fragment electrophile screening (Backus et al. 2016) also have been successfully applied to identify cysteine-reactive chemical probes and will be discussed in more detail in the next sections.

4.2.1 High-Throughput Screening (HTS)

HTS involves screening large (100,000+) compound libraries against a target and is a relatively unbiased approach that can yield compounds with novel binding modes. HTS requires no structural information about a target, and thus is useful for identifying lead compounds for proteins that lack structural information. For a HT screen at a cysteine residue, a broadly reactive, cysteine-selective chemical probe, such as tetramethylrhodamine iodoacetamide (IA-Rho) or Alexa Fluor maleimide, is often employed in a fluorescence polarization (FluoPol) assay (Fig. 8) (Bachovchin et al. 2009; Lea and Simeonov 2011). Competition or blockade of probe labeling by a hit compound is indicated by a decrease in the polarization of polarized, excited light, caused by the rapid tumbling of the fluorescent probe. HTS screening at cysteine residues identified inhibitors of PAD4 (Knuckley et al. 2010a, b), PRMT1 (Dillon et al. 2012) and GSTO1 (Bachovchin et al. 2009). Despite its utility for discovering probes, HTS is inherently challenged by maintenance of compound libraries, false positives due to compound aggregation, and difficulties in lead compound discovery and optimization. Most HTS campaigns also require a biochemical assay and knowledge of target function, making HTS screens less suitable for uncharacterized proteins.

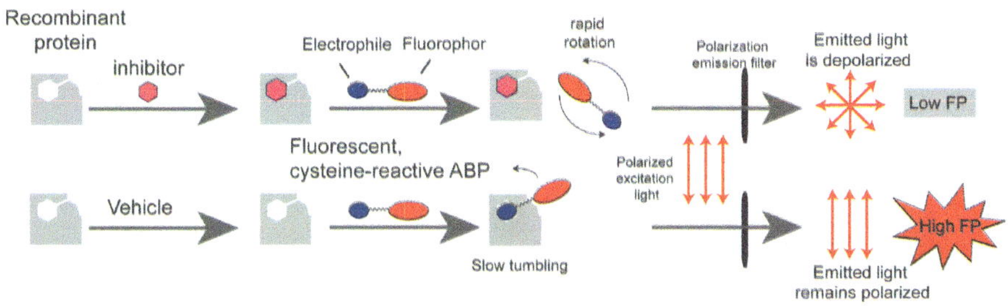

Fig. 8 Fluorescence polarization (FluoPol) high-throughput screening (HTS) at reactive cysteine residues

4.2.2 Covalent Fragment-Based Ligand Discovery

In contrast with HTS, fragment-based ligand discovery (FBLD) (Shuker et al. 1996) screens a small ($\sim$1000 member) library of relatively simple compounds against a target of interest with the aim of identifying weakly potent fragment hits. Fragment electrophile libraries are comprised of compounds that conform to the "rule of three", which recommends a molecular weight <300, number of H-bond acceptors and donors ≤ 3, rotatable bonds ≤ 3, $c\mathrm{Log}P \leq 3$, and the polar surface area to ≤ 60 Å^2 (Congreve et al. 2003; Jhoti et al. 2013). The low molecular weight of the starting fragment allows for rapid hit to lead (H2L) elaboration into probe/drug-like compounds, avoiding issues with chemical integrity that often trouble HTS campaigns (Foloppe 2011). Although most fragment-based screens to date have focused on non-covalent compounds, new screening methods have been developed to conduct covalent FBLD. Disulfide-based reactions, such as disulfide-trapping, or tethering, have found utility in covalent fragment-based screening applications (Erlanson et al. 2000a, b; Sadowsky et al. 2011), including the discovery of small-molecule probes that label G12C KRAS and allosterically modulate GTP affinity (Ostrem et al. 2013). In tethering (Fig. 9), a reactive cysteine-containing protein is allowed to interact simultaneously with a library of disulfide-containing fragments and the most stable compound–protein interactions are detected by mass spectrometry. Lead fragments are then elaborated and the disulfide is typically replaced with other electrophilic moieties. Limitations of tethering include a lack of commercially available disulfide libraries and difficulties replacing the screening disulfide with other electrophiles, while maintaining reactivity with the target protein. Cardoso, Nonoo, and colleagues found that covalent tethering (Cardoso et al. 2012; Nonoo et al. 2012), where libraries of electrophilic compounds are screened in a tethered format, circumvented some of these limitations and enabled the discovery of lead compounds for the cysteine protease papain and the HECT E3 ligase Nedd4-1 (Kathman and Statsyuk 2016). As with most FBLD, tethering still requires purified protein and is therefore difficult to apply broadly across the proteome.

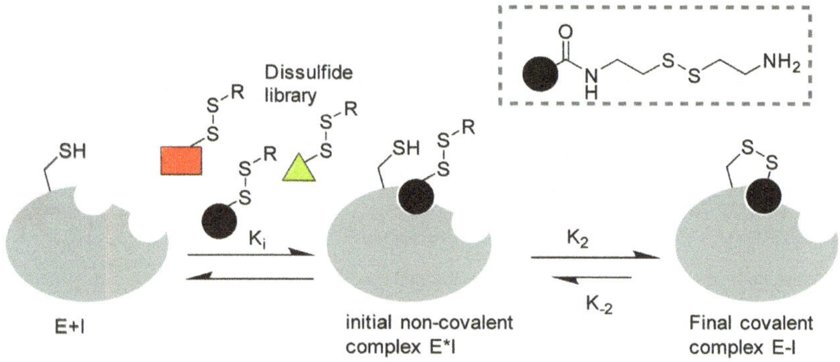

Fig. 9 Covalent Tethering. Libraries of disulfide-containing fragments are incubated with a target protein in the presence of a reducing agent, such as 2-mercaptoethanol. If a library member shows an affinity for the target, the mass action equilibrium will shift towards modified protein, which can be detected by mass spectrometry. Inset shows structure of library members with variable binding groups derived from carboxylic acids (black sphere) coupled to a flexible disulfide linker

Competitive isoTOP-ABPP has been developed to screen covalent fragments in whole proteomes to broadly identify proteins that contain ligandable cysteine residues and corresponding chemical probes for those residues (Backus et al. 2016). Conceptually, covalent fragment screening via isoTOP-ABPP (Fig. 10) operates in an analogous manner to the competitive profiling of redox modified cysteine residues (Fig. 6b). Cells or lysates are treated with compounds or DMSO and then labeled with a broadly reactive, iodoacetamide alkyne probe (IAA). Competition or blockade of probe labeling by pretreatment with compound indicates a saturable ligand binding event, just as competition of probe labeling indicated a cysteine PTM. When competitive isoTOP-ABPP was used to screen a 50+ member cysteine-reactive fragment library against human cancer cells, over 700 cys-ligand interactions were identified, most of which were found in proteins that did not previously have chemical probes (Backus et al. 2016). Among these were novel ligands for many "undruggable" proteins, including proteins involved in tumor metabolism, signaling and epigenetic remodeling. Exemplifying this platform's capacity to discover ligands for tough-to-drug proteins, isoTOP-ABPP identified compounds that selectively target the pro- (inactive) forms of caspase-8 and -10. Competitive isoTOP-ABPP broadly enables the identification of new ligandable cysteine residues in cells or lysates, assaying the potency and selectivity of compounds across thousands of cysteines in parallel. IsoTOP-ABPP requires no protein purification and can assay proteins of unknown function, thus identifying tool compounds early in the discovery process. However, as with other screening methods, optimization of lead compounds identified by isoTOP-ABPP is challenging, particularly for those proteins that lack structural information.

Competitive isoTOP-ABPP has also uncovered new proteins labeled by many other electrophilic compounds, including natural products, herbicides, and clinically approved drugs. Blewett and colleagues used isoTOP-ABPP to identify

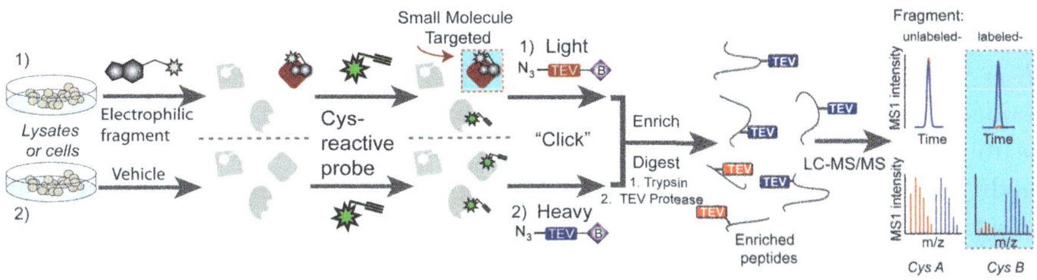

Fig. 10 Competitive isoTOP-ABPP to identify proteins labeled by fragment electrophiles. Cells or lysates are labeled by (1) electrophilic compounds or (2) vehicle, followed by cysteine-reactive probe iodoacetamide alkyne and evaluation by isoTOP-ABPP

cysteines labeled by dimethyl fumarate (DMF) or Tecfidera, a clinically approved therapeutic for the treatment of multiple sclerosis (MS) and found that DMF labeled cysteines in a non-catalytic domain of the kinase PKCθ and that this labeling may contribute to DMF's mechanism of action (Blewett et al. 2016). Competitive isoTOP-ABPP also successfully identified an acrylamide probe for Cys1101 in reticulon 4 (RTN4) (Bateman, Nguyen et al. 2017) and a covalent ligand for the ubiquitin-like modifier activating enzyme (UBA5) (Roberts et al. 2017a, b). Numerous other applications of competitive isoTOP-ABPP include screening chloroacetamide-containing quinazolines (Chen et al. 2017) and natural products withaferin A (Grossman et al. 2017) and licochalcone A (Roberts et al. 2017a, b). In addition to identifying the site(s) labeled by electrophilic compounds, competitive isoTOP-ABPP can also report proteinacious metal binding sites, particularly zinc-binding thiols (Pace and Weerapana 2014a, b), sites modified by lipid-derived electrophiles, particularly 4-hydroxy-2-nonenal (HNE) and prostaglandin J_2 (PGJ2) (Wang et al. 2014) and the identification of putative off-targets of commonly used herbicides glyphosate (Ford et al. 2017) and acetochor (Counihan et al. 2017). Considerations for all isoTOP-ABPP studies include whether matched inactive control compounds were employed to control for off-target effects of compounds, whether functional studies included cysteine mutated control proteins (e.g., cys to ala or cys to ser), and whether competition studies were used to determine the fractional occupancy of compound labeling. For most proteins, gel-based ABPP (Fig. 3b) can be used to further assay the potency and selectivity of lead compounds identified by competitive isoTOP-ABPP.

4.3 Compound Optimization

The main determinants of the potency and selectivity of cysteine-reactive small molecules are: (1) the non-covalent interactions between the compound and the protein (binding mode, kinetics, residence time) that generate a non-covalent complex ($E * I$), positioning the electrophile proximal to the nucleophilic thiol, (2) the relative reactivity and nature of the electrophile, (3) the nucleophilicity of the cysteine thiol, and (4) the positioning of the electrophile proximal to the nucleophile, which together result in bond formation (K_2) to produce a covalent complex (*E-I*), and, for α, β unsaturated electrophiles, (5) the presence of catalytic acidic residues nearby (see Fig. 9 for non-covalent and covalent interactions during disulfide tethering). Covalent compounds can be independently optimized by modifying both the pharmacophore's electrophile and binding elements. As different cysteines tend to exhibit preferences for specific electrophiles (Backus et al. 2016), it is often not straightforward to tune the compound electrophilicity by replacing the initial electrophile with a more attenuated one. Many different cysteine-reactive electrophiles have been developed to address the associated challenges with electrophile optimization. Typically, binding elements are identified and optimized through screening, including in silico screens, and medicinal chemistry efforts, for example, elaborating or connecting lead fragments.

4.3.1 Library Design

Binding element optimization has benefited from the design and synthesis of high-quality libraries of electrophilic compounds, including fragment electrophile libraries, DNA-encoded libraries, and the solid-phase synthesis of combinatorial libraries. One focus across all of these libraries has been to increase library shape diversity by including more SP^3-rich structures to complement the large numbers of flat aromatic compounds that make up many compound libraries. Several solid-phase methods for library synthesis have been developed. Halo-triazines (Banerjee et al. 2013; Shannon et al. 2014) and 4-amino-piperidine scaffolds (Couvertier and Weerapana 2014) allow for modular synthesis. Pels and colleagues reported that the cysteine-reactive, covalent reversible cyanoacrylamide-based compound libraries can be synthesized on resin in two high-yield steps, with minimal off-targets (Pels et al. 2018). DNA-encoded libraries identified covalent JNK-1 inhibitors (Zimmermann et al. 2017) and covalent bromodomain binders (Daguer et al. 2015; Chan et al. 2017). Focused electrophile libraries yielded a potent, selective agonist of TRP1A (Takaya et al. 2015), a transient receptor cation channel. High-quality, focused covalent fragment libraries have identified lead compounds for caspase-8, NR0B1 and HIF-1α (Cardoso et al. 2012; Backus et al. 2016; Bar-Peled et al. 2017), among many others.

Covalent docking methods have recently been put forth as an attractive means to optimize lead molecules, enabling the in silico screening of large libraries of chemical probes. Several software packages, including CovalentDock (Ouyang et al. 2013), Gold (Verdonk et al. 2003), CovDoc (Zhu et al. 2014), and Autodock (Bianco et al. 2016), support covalent docking. Of particular note, London et al. adapted Dock3.6 to screen a 650,000+ compound library for inhibitors of Janus Kinase 3 (JAK3) (London et al. 2014). Typically, in an in silico screen, compounds are tethered via the cysteine of interest to the protein and then the interactions with the surrounding residues are quantified. Although useful for ligand optimization, tethered docking approaches do not account for any non-covalent, pre-alkylation binding poses that contribute to the thiol labeling and typically hold the protein static, and so fail to assay how the dynamics of neighboring residues contribute to binding and reactivity. These docking methods also suffer from many of the same inherent weaknesses of non-covalent docking, namely a lack of accuracy and poor binding scores, a lack of protein flexibility, and a lack of automation.

4.3.2 Electrophiles

Choice of electrophile is a critical step in the optimization of potent, selective covalent probes (see Fig. 11 for a summary of electrophiles). Although some electrophiles, including β-unsubstituted vinyl sulfonamides, vinyl sulfones, many α-haloacetamides, and maleimides are broadly reactive with proteinacious cysteines and glutathione, many electrophiles, including most acrylamides, particularly acrylamides, and vinyl sulfonamides that feature β-substituted acetylenes, as well as 2-chloropropionates react very slowly in the absence of templating by a protein binding site (Ward et al. 2013; Flanagan et al. 2014). Despite their reactivity, α-haloacetamides have been featured widely in cysteine-reactive inhibitors, including inhibitors of nitrilases (Mittag et al. 2016), protein disulfide isomerases (PDIs) (Hoffstrom et al. 2010), glutaredoxin (GRX1) (Gorelenkova Miller et al. 2017), and glutathione S-transferase omega 1 (GSTO1) (Couvertier and Weerapana 2014). Most clinical candidates and clinically approved cysteine-reactive compounds (Carmi et al. 2012; Liu et al. 2013) to date have featured acrylamide warheads, which, given their attenuated proteome-wide reactivity (Backus et al.

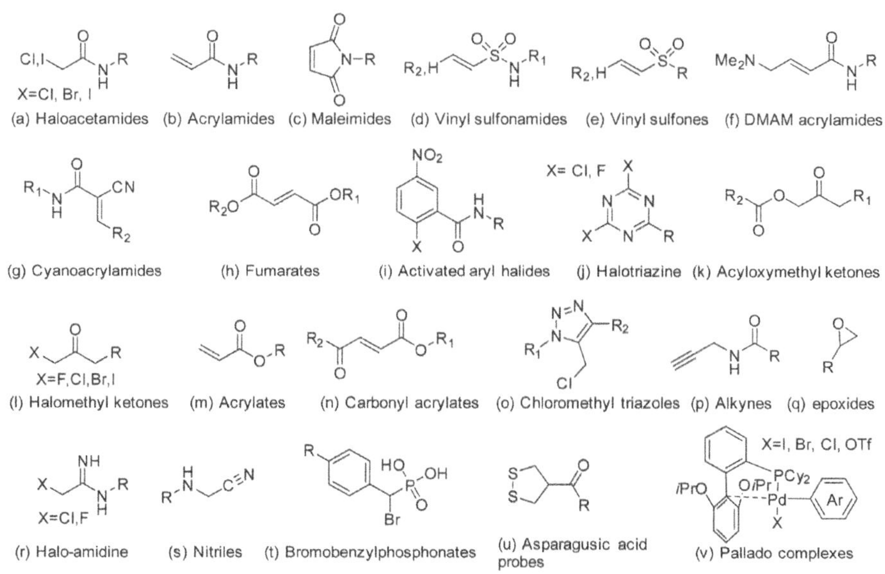

Fig. 11 Chemical structures of selected cysteine-reactive electrophiles

2016) and ease of incorporation, are attractive from a probe development perspective. Dimethylamino-(DMAM) substituted acrylamides are featured in many cysteine-reactive compounds, including afatinib, and have been found in some cases to increase the overall compound reactivity when compared to non-DMAM-containing compounds, either due to increased solubility or by serving as an intramolecular catalytic base (Carmi et al. 2011; Lanning et al. 2014).

Covalent reversible electrophiles have been proposed as a means to reduce the off-target activity and haptogenicity of electrophilic compounds. The cyanoacrylamide electrophile is one example of a covalent reversible scaffold, with β-elimination yielding the starting electrophile (Miller et al. 2013; Krishnan et al. 2014; Bradshaw et al. 2015). Such reversible electrophiles are attractive given that they should mitigate some of the liabilities of covalent compounds. Basu and colleagues demonstrated that cyanoacrylamide afatinib analogues can inhibit EGFR in a covalent reversible manner (Basu et al. 2015). One of the challenges associated with cyanoacrylamide-containing compounds is their short half-life in cells. Substitution of the acrylamide group with bulky amines has been found to dramatically improve the half-life of cyanoacrylamides (Bradshaw et al. 2015). The determination of sites of labeling of reversible cyanoacrylamide products can also be challenging. Pels et al. found that cyanoacrylamide cysteine adducts are stable at low pH (Pels et al. 2018) and Backus et al. found that isoTOP-ABPP could identify the sites of labeling for cyanoacrylamides fragments (Backus et al. 2016). Kinetically controlled electrophiles have been developed as another means to reduce off-target affects. Dimethylfumarate was shown to hydrolyze rapidly in plasma and Zaro and colleagues leveraged this rapid hydrolysis to create kinetically controlled fumarate-based analogues of ibrutinib (Zaro et al. 2016), designed to reduce nonspecific background reactivity that plagues many covalent inhibitors. These compounds rapidly engaged BTK in cells, and subsequent hydrolysis of the fumarate ester to the unreactive free acid blocked time-dependent accumulation of background probe labeling.

In addition to cyanoacrylamides and fumarates, the development of other new cysteine-reactive electrophiles has been another area of considerable interest. Aryl halides, particularly chloro and fluoronitrobenzenes, were found to react selectively with proteinacious thiols, whereas dichlorotriazines labeled lysines preferentially (Shannon et al. 2014). Acyloxymethyl ketones (AOMK) have been widely employed in cysteine protease inhibitors and chemical probes (Pliura et al. 1992; Thornberry et al. 1994), but have found limited utility beyond protease inhibition and are plagued by hydrolytic instability. Fluoromethyl ketones have also found utility as protease inhibitors (Slee et al. 1996). Acrylate (Kathman et al. 2014) and carbonylacrylates are another class of cysteine-reactive electrophiles that have been used for site-selective modifications (Bernardim et al. 2016). Chloromethyltriazoles offer cysteine selectivity and tuneable reactivities (Wang et al. 2016) and have been incorporated into selective inhibitors of the human DNA repair protein MGMT (Wang et al. 2016). Terminal alkynes were found to label cysteines in human deubiquitinases (Ekkebus et al. 2013; Sommer et al. 2013), cytochromes (Wright et al. 2009), and monoamine oxidases (MAOs) (Krysiak et al. 2012). Epoxides, including aziridines, oxiranes, and thiiranes, have also been reported to react with cysteines (Barrett et al. 1982; Pitscheider et al. 2012). However, epoxides generally do not show selectivity for cysteine residues, and also react with other nucleophilic amino acids including histidines (Lowther et al. 1998), lysines (Evans et al. 2007), glutamates, and aspartates (Withers and Aebersold 1995), and even an N-terminal threonine of the active site of the 20S proteasome beta-subunit (Lowther et al. 1998). Fluoro and chloroacetamidine compounds have been shown to irreversibly label a conserved active site cysteine residue in the protein arginine deiminase 2, 3 and 4 (PAD2, PAD3 and PAD4) (Knuckley et al. 2010a, b; Muth et al. 2017). Nitriles have also been developed into potent cysteine-reactive inhibitors (Oballa et al. 2007), including those that selectively label cathepsin K (Gauthier et al. 2008). Biotinylated bromobenzylphosphonate probes react selectively and irreversibly with the catalytic cysteine residues of tyrosine phosphatases (PTPs) (Kumar et al. 2004). Asparagusic acid (AspA) strained disulfide probes have been used as a cysteine-reactive delivery tag to facilitate cellular uptake (Abegg et al. 2017). Organometallic complexes that selectively label cysteine residues have been developed, particularly pallado complexes (Vinogradova et al. 2015), enabled site-specific incorporation of fluorophors, photocrosslinkers, peptides, carbohydrates, PEG, and other biomolecules. These labeling strategies are particularly attractive for protein engineering and the production of antibody-drug conjugates (ADCs).

A risk of any covalent screen is that the most electrophilic compound will be selected as a lead compound rather than the compound with the highest specific binding affinity for the target. Compound reactivity can be measured by thiol consumption assays (e.g., glutathione or N-acetyl cysteine) (Craven et al. 2018) or by chemical proteomic methods (Backus et al. 2016). A particularly elegant solution that has been put forth to this problem of disparate reactivities was to synthesize electrophile fragment libraries of more uniform electrophilicity. In stark contrast to other electrophiles, NMR-based reactivity studies revealed that acrylate-based electrophiles exhibited near uniform reactivity, independent of binding element (Kathman et al. 2014). Krishnan and colleagues screened a panel of electron withdrawing groups to optimize the electrophilicity of cyanoacrylamide-based RSK2 inhibitors (Krishnan et al. 2014), demonstrating that covalent reversible electrophile reactivity is tunable. Despite these advances in electrophile profiling, there still is no systematic means of tuning the reactivity of a specific electrophile (for example, reducing the reactivity of a chloroacetamide to improve compound selectivity, while still maintaining on-target activity). Future work should focus on the systematic development of a reactivity-optimized suite of electrophiles.

5 Conclusions

Cysteine reactivity profiling has uncovered thousands of "hyperreactive," PTM-modified, and ligandable cysteine residues proteome-wide, enabled high-throughput screening at cysteine residues, and facilitated the development of new selective chemical probes and even informed the mechanism of action of clinically approved drugs. Despite the broad utility of cysteine reactivity profiling, the approach still has some inherent limitations. One obvious weakness of reactivity profiling is that, although reactivity and PTMs are widely thought to be good indicators of functional importance, in many cases establishing the functional roles of specific cysteine residues remains difficult. Advances in genome engineering, in particular, CRISPR-Cas9, in part will help assign function through the rapid generation of site-specific point mutations at cysteine residues. The identification of acquired cysteine missense mutations linked to disease through genomics will also inform the functions of reactive cysteines. Alongside functional genomics, the development of other more tailored chemical probes and the combination of cysteine-reactive chemical probes with targeted protein degradation should also help to inform the functions of specific cysteine residues. Cysteine-reactive probe development will benefit from new methods to tune compound reactivity and to reduce off-target activities, including nonspecific alkylation. Another weakness of cysteine profiling is that it, by definition, is restricted to proteins that contain suitable cysteine residues. However, this cysteine focus is beginning to shift and recent studies have accessed other amino acids. IsoTOP-ABPP when coupled with lysine-selective probes can uncover hyperreactive and functional lysine residues (Shannon et al. 2014; Hacker et al. 2017; Ward et al. 2017). Reactivity profiling and proteomics more generally are inherently limited by the stochastic nature of peptide sampling by the mass spectrometer and inherent bias towards more abundant peptides. With advances in proteomic instrumentation and acquisition methods and the design of new chemical probes, including those capable of labeling other amino acids (e.g., histidine, lysine, tyrosine), reactivity profiling will broadly enable the proteome-wide identification of functional and ligandable sites on proteins.

References

Abegg D, Frei R, Cerato L, Prasad Hari D, Wang C, Waser J, Adibekian A (2015) Proteome-wide profiling of targets of cysteine reactive small molecules by using ethynyl benziodoxolone reagents. Angew Chem Int Ed Engl 54(37):10852–10857

Abegg D, Gasparini G, Hoch DG, Shuster A, Bartolami E, Matile S, Adibekian A (2017) Strained cyclic disulfides enable cellular uptake by reacting with the transferrin receptor. J Am Chem Soc 139(1):231–238

Abo M, Weerapana E (2015) A caged electrophilic probe for global analysis of cysteine reactivity in living cells. J Am Chem Soc 137(22):7087–7090

Abo M, Bak DW, Weerapana E (2017a) Optimization of caged electrophiles for improved monitoring of cysteine reactivity in living cells. ChemBioChem 18(1):81–84

Abo M, Li C, Weerapana E (2017) Isotopically-labeled iodoacetamide-alkyne probes for quantitative cysteine-reactivity profiling. Mol Pharm

Alcock LJ, Farrell KD, Akol MT, Jones GH, Tierney MM, Kramer HB, Pukala TL, Bernardes GJL, Perkins MV, Chalker JM (2018) Norbornene probes for the study of cysteine oxidation. Tetrahedron 74(12):1220–1228

Andersson A, Hutlberg B, Lindgren A (2000) Redox status of plasma homocysteine and other plasma thiols in stroke patients. Atherosclerosis 151(2):535–539
Ansbacher T, Chourasia M, Shurki A (2013) Copper-chaperones with dicoordinated Cu(I)–unique protection mechanism. Proteins 81(8):1411–1419
Bachovchin DA, Brown SJ, Rosen H, Cravatt BF (2009) Identification of selective inhibitors of uncharacterized enzymes by high-throughput screening with fluorescent activity-based probes. Nat Biotechnol 27(4):387–394
Backus KM, Correia BE, Lum KM, Forli S, Horning BD, Gonzalez-Paez GE, Chatterjee S, Lanning BR, Teijaro JR, Olson AJ, Wolan DW, Cravatt BF (2016) Proteome-wide covalent ligand discovery in native biological systems. Nature 534(7608):570–574
Bak DW, Weerapana E (2015) Cysteine-mediated redox signalling in the mitochondria. Mol BioSyst 11(3):678–697
Banerjee R, Pace NJ, Brown DR, Weerapana E (2013) 1,3,5-Triazine as a modular scaffold for covalent inhibitors with streamlined target identification. J Am Chem Soc 135(7):2497–2500
Bantscheff M, Lemeer S, Savitski MM, Kuster B (2012) Quantitative mass spectrometry in proteomics: critical review update from 2007 to the present. Anal Bioanal Chem 404(4):939–965
Barglow KT, Cravatt BF (2006) Substrate mimicry in an activity-based probe that targets the nitrilase family of enzymes. Angew Chem Int Ed Engl 45(44):7408–7411
Bar-Peled L, Kemper EK, Suciu RM, Vinogradova EV, Backus KM, Horning BD, Paul TA, Ichu TA, Svensson RU, Olucha J, Chang MW, Kok BP, Zhu Z, Ihle NT, Dix MM, Jiang P, Hayward MM, Saez E, Shaw RJ, Cravatt BF (2017) Chemical proteomics identifies druggable vulnerabilities in a genetically defined cancer. Cell 171(3):696–709 e623
Barrett AJ, Kembhavi AA, Brown MA, Kirschke H, Knight CG, Tamai M, Hanada K (1982) L-trans-Epoxysuccinyl-leucylamido(4-guanidino)butane (E-64) and its analogues as inhibitors of cysteine proteinases including cathepsins B, H and L. Biochem J 201(1):189–198
Basu D, Richters A, Rauh D (2015) Structure-based design and synthesis of covalent-reversible inhibitors to overcome drug resistance in EGFR. Bioorg Med Chem 23(12):2767–2780
Bateman LA, Zaro BW, Miller SM, Pratt MR (2013) An alkyne-aspirin chemical reporter for the detection of aspirin-dependent protein modification in living cells. J Am Chem Soc 135 (39):14568–14573
Bateman LA, Nguyen TB, Roberts AM, Miyamoto DK, Ku WM, Huffman TR, Petri Y, Heslin MJ, Contreras CM, Skibola CF, Olzmann JA, Nomura DK (2017) Chemoproteomics-enabled covalent ligand screen reveals a cysteine hotspot in reticulon 4 that impairs ER morphology and cancer pathogenicity. Chem Commun (Camb) 53(53):7234–7237
Baty JW, Hampton MB, Winterbourn CC (2002) Detection of oxidant sensitive thiol proteins by fluorescence labeling and two-dimensional electrophoresis. Proteomics 2(9):1261–1266
Bauer RA (2015) Covalent inhibitors in drug discovery: from accidental discoveries to avoided liabilities and designed therapies. Drug Discov Today 20(9):1061–1073
Bechtold E, Reisz JA, Klomsiri C, Tsang AW, Wright MW, Poole LB, Furdui CM, King SB (2010) Water-soluble triarylphosphines as biomarkers for protein S-nitrosation. ACS Chem Biol 5(4):405–414
Benitez LV, Allison WS (1974) The inactivation of the acyl phosphatase activity catalyzed by the sulfenic acid form of glyceraldehyde 3-phosphate dehydrogenase by dimedone and olefins. J Biol Chem 249(19):6234–6243
Berg JM (1990) Zinc finger domains: hypotheses and current knowledge. Annu Rev Biophys Biophys Chem 19:405–421
Bernardim B, Cal PMSD, Matos MJ, Oliveira BL, Martínez-Sáez N, Albuquerque IS, Perkins E, Corzana F, Burtoloso ACB, Jiménez-Osés G, Bernardes GJL (2016) Stoichiometric and irreversible cysteine-selective protein modification using carbonylacrylic reagents. Nat Commun 7:13128
Besancon M, Simon A, Sachs G, Shin JM (1997) Sites of reaction of the gastric H, K-ATPase with extracytoplasmic thiol reagents. J Biol Chem 272(36):22438–22446

Bianco G, Forli S, Goodsell DS, Olson AJ (2016) Covalent docking using autodock: two-point attractor and flexible side chain methods. Protein Sci 25(1):295–301

Blackinton J, Lakshminarasimhan M, Thomas KJ, Ahmad R, Greggio E, Raza AS, Cookson MR, Wilson MA (2009) Formation of a stabilized cysteine sulfinic acid is critical for the mitochondrial function of the parkinsonism protein DJ-1. J Biol Chem 284(10):6476–6485

Blewett M, Xie J, Zaro B, Backus KM, Olenchock BA, Patel H, Altman A, Teijaro JR, Cravatt BF (2016) Chemical proteomic map of dimethylfumarate-sensitive cysteine in primary human T cells. Sci Signal 9(445):rs10

Blum G, von Degenfeld G, Merchant MJ, Blau HM, Bogyo M (2007) Noninvasive optical imaging of cysteine protease activity using fluorescently quenched activity-based probes. Nat Chem Biol 3(10):668–677

Böttcher T, Sieber SA (2010) Showdomycin as a versatile chemical tool for the detection of pathogenesis-associated enzymes in bacteria. J Am Chem Soc 132(20):6964–6972

Bradshaw JM, McFarland JM, Paavilainen VO, Bisconte A, Tam D, Phan VT, Romanov S, Finkle D, Shu J, Patel V, Ton T, Li X, Loughhead DG, Nunn PA, Karr DE, Gerritsen ME, Funk JO, Owens TD, Verner E, Brameld KA, Hill RJ, Goldstein DM, Taunton J (2015) Prolonged and tunable residence time using reversible covalent kinase inhibitors. Nat Chem Biol 11(7):525–531

Briggs KJ, Koivunen P, Cao S, Backus KM, Olenchock BA, Patel H, Zhang Q, Signoretti S, Gerfen GJ, Richardson AL, Witkiewicz AK, Cravatt BF, Clardy J, Kaelin WG Jr (2016) Paracrine induction of HIF by glutamate in breast cancer: EglN1 senses cysteine. Cell 166 (1):126–139

Canet-Aviles RM, Wilson MA, Miller DW, Ahmad R, McLendon C, Bandyopadhyay S, Baptista MJ, Ringe D, Petsko GA, Cookson MR (2004) The Parkinson's disease protein DJ-1 is neuroprotective due to cysteine-sulfinic acid-driven mitochondrial localization. Proc Natl Acad Sci U S A 101(24):9103–9108

Cardoso R, Love R, Nilsson CL, Bergqvist S, Nowlin D, Yan J, Liu KK, Zhu J, Chen P, Deng YL, Dyson HJ, Greig MJ, Brooun A (2012) Identification of Cys255 in HIF-1alpha as a novel site for development of covalent inhibitors of HIF-1alpha/ARNT PasB domain protein-protein interaction. Protein Sci 21(12):1885–1896

Carmi C, Lodola A, Rivara S, Vacondio F, Cavazzoni A, Alfieri RR, Ardizzoni A, Petronini PG, Mor M (2011) Epidermal growth factor receptor irreversible inhibitors: chemical exploration of the cysteine-trap portion. Mini Rev Med Chem 11(12):1019–1030

Carmi C, Mor M, Petronini PG, Alfieri RR (2012) Clinical perspectives for irreversible tyrosine kinase inhibitors in cancer. Biochem Pharmacol 84(11):1388–1399

Carrington JC, Dougherty WG (1988) A viral cleavage site cassette: identification of amino acid sequences required for tobacco etch virus polyprotein processing. Proc Natl Acad Sci U S A 85 (10):3391–3395

Cerda MM, Hammers MD, Earp MS, Zakharov LN, Pluth MD (2017) Applications of synthetic organic tetrasulfides as H2S Donors. Org Lett 19(9):2314–2317

Chalker JM, Gunnoo SB, Boutureira O, Gerstberger SC, Fernandez-Gonzalez M, Bernardes GJL, Griffin L, Hailu H, Schofield CJ, Davis BG (2011) Methods for converting cysteine to dehydroalanine on peptides and proteins. Chem Sci 2(9):1666–1676

Chalker JM, Lercher L, Rose NR, Schofield CJ, Davis BG (2012) Conversion of cysteine into dehydroalanine enables access to synthetic histones bearing diverse post-translational modifications. Angew Chem Int Ed Engl 51(8):1835–1839

Chan AI, McGregor LM, Jain T, Liu DR (2017) Discovery of a covalent kinase inhibitor from a DNA-encoded small-molecule library × protein library selection. J Am Chem Soc 139 (30):10192–10195

Chen CA, Wang TY, Varadharaj S, Reyes LA, Hemann C, Talukder MA, Chen YR, Druhan LJ, Zweier JL (2010) S-glutathionylation uncouples eNOS and regulates its cellular and vascular function. Nature 468(7327):1115–1118

Chen YC, Backus KM, Merkulova M, Yang C, Brown D, Cravatt BF, Zhang C (2017) Covalent modulators of the vacuolar ATPase. J Am Chem Soc 139(2):639–642

Chung KK (2006) Say NO to neurodegeneration: role of S-nitrosylation in neurodegenerative disorders. Neurosignals 15(6):307–313

Cohen MS, Zhang C, Shokat KM, Taunton J (2005) Structural bioinformatics-based design of selective, irreversible kinase inhibitors. Science 308(5726):1318–1321

Cohen MS, Hadjivassiliou H, Taunton J (2007) A clickable inhibitor reveals context-dependent autoactivation of p90 RSK. Nat Chem Biol 3(3):156–160

Congreve M, Carr R, Murray C, Jhoti H (2003) A 'rule of three' for fragment-based lead discovery? Drug Discov Today 8(19):876–877

Counihan JL, Duckering M, Dalvie E, Ku WM, Bateman LA, Fisher KJ, Nomura DK (2017) Chemoproteomic profiling of acetanilide herbicides reveals their role in inhibiting fatty acid oxidation. ACS Chem Biol 12(3):635–642

Couvertier SM, Weerapana E (2014) Cysteine-reactive chemical probes based on a modular 4-aminopiperidine scaffold. MedChemComm 5(3):358–362

Craven G, Affron D, Allen C, Matthies S, Greener J, Morgan R, Tate E, Armstrong A, Mann DJ (2018) High-throughput kinetic analysis for target-directed covalent ligand discovery. Angew Chem Int Ed Engl

Cumming RC, Andon NL, Haynes PA, Park M, Fischer WH, Schubert D (2004) Protein disulfide bond formation in the cytoplasm during oxidative stress. J Biol Chem 279(21):21749–21758

Daguer JP, Zambaldo C, Abegg D, Barluenga S, Tallant C, Muller S, Adibekian A, Winssinger N (2015) Identification of covalent bromodomain binders through DNA display of small molecules. Angew Chem Int Ed Engl 54(20):6057–6061

Dayon L, Hainard A, Licker V, Turck N, Kuhn K, Hochstrasser DF, Burkhard PR, Sanchez JC (2008) Relative quantification of proteins in human cerebrospinal fluids by MS/MS using 6-plex isobaric tags. Anal Chem 80(8):2921–2931

De Cesco S, Kurian J, Dufresne C, Mittermaier AK, Moitessier N (2017) Covalent inhibitors design and discovery. Eur J Med Chem 138:96–114

Dennehy MK, Richards KA, Wernke GR, Shyr Y, Liebler DC (2006) Cytosolic and nuclear protein targets of thiol-reactive electrophiles. Chem Res Toxicol 19(1):20–29

Dickens F (1933) Interaction of halogenacetates and SH compounds: the reaction of halogenacetic acids with glutathione and cysteine. the mechanism of iodoacetate poisoning of glyoxalase. Biochem J 27(4):1141–1151

Dillon MB, Bachovchin DA, Brown SJ, Finn MG, Rosen H, Cravatt BF, Mowen KA (2012) Novel inhibitors for PRMT1 discovered by high-throughput screening using activity-based fluorescence polarization. ACS Chem Biol 7(7):1198–1204

Ding Z, Kim S, Dorsam RT, Jin J, Kunapuli SP (2003) Inactivation of the human P2Y12 receptor by thiol reagents requires interaction with both extracellular cysteine residues, Cys17 and Cys270. Blood 101(10):3908–3914

Doerr A (2014) DIA mass spectrometry. Nat Methods 12:35

Drahl C, Cravatt BF, Sorensen EJ (2005) Protein-reactive natural products. Angew Chem Int Ed Engl 44(36):5788–5809

Duan J, Gaffrey MJ, Qian WJ (2017) Quantitative proteomic characterization of redox-dependent post-translational modifications on protein cysteines. Mol BioSyst 13(5):816–829

Dyson HJ, Jeng MF, Tennant LL, Slaby I, Lindell M, Cui DS, Kuprin S, Holmgren A (1997) Effects of buried charged groups on cysteine thiol ionization and reactivity in Escherichia coli thioredoxin: structural and functional characterization of mutants of Asp 26 and Lys 57. Biochemistry 36(9):2622–2636

Eaton P (2006) Protein thiol oxidation in health and disease: techniques for measuring disulfides and related modifications in complex protein mixtures. Free Radic Biol Med 40(11):1889–1899

Ekkebus R, van Kasteren SI, Kulathu Y, Scholten A, Berlin I, Geurink PP, de Jong A, Goerdayal S, Neefjes J, Heck AJ, Komander D, Ovaa H (2013) On terminal alkynes that can react with active-site cysteine nucleophiles in proteases. J Am Chem Soc 135(8):2867–2870

Ellis HR, Poole LB (1997) Novel application of 7-chloro-4-nitrobenzo-2-oxa-1,3-diazole to identify cysteine sulfenic acid in the AhpC component of alkyl hydroperoxide reductase. Biochemistry 36(48):15013–15018

Eng JK, McCormack AL, Yates JR (1994) An approach to correlate tandem mass spectral data of peptides with amino acid sequences in a protein database. J Am Soc Mass Spectrom 5 (11):976–989

Eom KS, Cheong JS, Lee SJ (2016) Structural analyses of zinc finger domains for specific interactions with DNA. J Microbiol Biotechnol 26(12):2019–2029

Erlanson DA, Hansen SK (2004) Making drugs on proteins: site-directed ligand discovery for fragment-based lead assembly. Curr Opin Chem Biol 8(4):399–406

Erlanson DA, Braisted AC, Raphael DR, Randal M, Stroud RM, Gordon EM, Wells JA (2000a) Site-directed ligand discovery. Proc Natl Acad Sci 97(17):9367–9372

Erlanson DA, Braisted AC, Raphael DR, Randal M, Stroud RM, Gordon EM, Wells JA (2000b) Site-directed ligand discovery. Proc Natl Acad Sci U S A 97(17):9367–9372

Erlanson DA, Wells JA, Braisted AC (2004) TETHERING: fragment-based drug discovery. Annu Rev Biophys Biomol Struct 33(1):199–223

Eschenburg S, Priestman M, Schönbrunn E (2005) Evidence that the fosfomycin target Cys115 in UDP-*N*-acetylglucosamine enolpyruvyl transferase (MurA) is essential for product release. J Biol Chem 280(5):3757–3763

Evans DC, Watt AP, Nicoll-Griffith DA, Baillie TA (2004) Drug-protein adducts: an industry perspective on minimizing the potential for drug bioactivation in drug discovery and development. Chem Res Toxicol 17(1):3–16

Evans MJ, Morris GM, Wu J, Olson AJ, Sorensen EJ, Cravatt BF (2007) Mechanistic and structural requirements for active site labeling of phosphoglycerate mutase by spiroepoxides. Mol BioSyst 3(7):495–506

Flanagan ME, Abramite JA, Anderson DP, Aulabaugh A, Dahal UP, Gilbert AM, Li C, Montgomery J, Oppenheimer SR, Ryder T, Schuff BP, Uccello DP, Walker GS, Wu Y, Brown MF, Chen JM, Hayward MM, Noe MC, Obach RS, Philippe L, Shanmugasundaram V, Shapiro MJ, Starr J, Stroh J, Che Y (2014) Chemical and computational methods for the characterization of covalent reactive groups for the prospective design of irreversible inhibitors. J Med Chem 57(23):10072–10079

Foloppe N (2011) The benefits of constructing leads from fragment hits. Future Med Chem 3 (9):1111–1115

Ford B, Bateman LA, Gutierrez-Palominos L, Park R, Nomura DK (2017) Mapping proteome-wide targets of glyphosate in mice. Cell Chem Biol 24(2):133–140

Forrester MT, Thompson JW, Foster MW, Nogueira L, Moseley MA, Stamler JS (2009) Proteomic analysis of S-nitrosylation and denitrosylation by resin-assisted capture. Nat Biotechnol 27(6):557–559

Fox JH, Connor T, Stiles M, Kama J, Lu Z, Dorsey K, Lieberman G, Sapp E, Cherny RA, Banks M, Volitakis I, DiFiglia M, Berezovska O, Bush AI, Hersch SM (2011) Cysteine oxidation within N-terminal mutant huntingtin promotes oligomerization and delays clearance of soluble protein. J Biol Chem 286(20):18320–18330

Frei R, Wodrich MD, Hari DP, Borin PA, Chauvier C, Waser J (2014) Fast and highly chemoselective alkynylation of thiols with hypervalent iodine reagents enabled through a low energy barrier concerted mechanism. J Am Chem Soc 136(47):16563–16573

Gan ZR, Wells WW (1987) Identification and reactivity of the catalytic site of pig liver thioltransferase. J Biol Chem 262(14):6704–6707

Gao J, Mfuh A, Amako Y, Woo CM (2018) Small molecule interactome mapping by photoaffinity labeling reveals binding site hotspots for the NSAIDs. J Am Chem Soc 140(12):4259–4268

Garcia-Pineres AJ, Castro V, Mora G, Schmidt TJ, Strunck E, Pahl HL, Merfort I (2001) Cysteine 38 in p65/NF-kappaB plays a crucial role in DNA binding inhibition by sesquiterpene lactones. J Biol Chem 276(43):39713–39720

Garcia-Santamarina S, Boronat S, Domenech A, Ayte J, Molina H, Hidalgo E (2014) Monitoring in vivo reversible cysteine oxidation in proteins using ICAT and mass spectrometry. Nat Protoc 9(5):1131–1145

Garske AL, Peters U, Cortesi AT, Perez JL, Shokat KM (2011) Chemical genetic strategy for targeting protein kinases based on covalent complementarity. Proc Natl Acad Sci U S A 108 (37):15046–15052

Gartner CA, Elias JE, Bakalarski CE, Gygi SP (2007) Catch-and-release reagents for broadscale quantitative proteomics analyses. J Proteome Res 6(4):1482–1491

Gauthier JY, Chauret N, Cromlish W, Desmarais S, Duong LT, Falgueyret JP, Kimmel DB, Lamontagne S, Leger S, LeRiche T, Li CS, Masse F, McKay DJ, Nicoll-Griffith DA, Oballa RM, Palmer JT, Percival MD, Riendeau D, Robichaud J, Rodan GA, Rodan SB, Seto C, Therien M, Truong VL, Venuti MC, Wesolowski G, Young RN, Zamboni R, Black WC (2008) The discovery of odanacatib (MK-0822), a selective inhibitor of cathepsin K. Bioorg Med Chem Lett 18(3):923–928

Geromanos SJ, Vissers JP, Silva JC, Dorschel CA, Li GZ, Gorenstein MV, Bateman RH, Langridge JI (2009) The detection, correlation, and comparison of peptide precursor and product ions from data independent LC-MS with data dependant LC-MS/MS. Proteomics 9 (6):1683–1695

Gersch M, Kreuzer J, Sieber SA (2012) Electrophilic natural products and their biological targets. Nat Prod Rep 29(6):659–682

Giles GI, Jacob C (2002) Reactive sulfur species: an emerging concept in oxidative stress. Biol Chem 383(3–4):375–388

Gillet LC, Navarro P, Tate S, Rost H, Selevsek N, Reiter L, Bonner R, Aebersold R (2012) Targeted data extraction of the MS/MS spectra generated by data-independent acquisition: a new concept for consistent and accurate proteome analysis. Mol Cell Proteomics 11(6):O111 016717

Go Y-M, Chandler JD, Jones DP (2015) The cysteine proteome. Free Radic Biol Med 84:227–245

Gorelenkova Miller O, Cole KS, Emerson CC, Allimuthu D, Golczak M, Stewart PL, Weerapana E, Adams DJ, Mieyal JJ (2017) Novel chloroacetamido compound CWR-J02 is an anti-inflammatory glutaredoxin-1 inhibitor. PLoS ONE 12(11):e0187991

Grauschopf U, Winther JR, Korber P, Zander T, Dallinger P, Bardwell JC (1995) Why is DsbA such an oxidizing disulfide catalyst? Cell 83(6):947–955

Green NM (1990) Avidin and streptavidin. Meth Enzymol 184:51–67

Grossman EA, Ward CC, Spradlin JN, Bateman LA, Huffman TR, Miyamoto DK, Kleinman JI, Nomura DK (2017) Covalent ligand discovery against druggable hotspots targeted by anti-cancer natural products. Cell Chem Biol 24(11):1368–1376 e1364

Guo CJ, Chang FY, Wyche TP, Backus KM, Acker TM, Funabashi M, Taketani M, Donia MS, Nayfach S, Pollard KS, Craik CS, Cravatt BF, Clardy J, Voigt CA, Fischbach MA (2017) Discovery of reactive microbiota-derived metabolites that inhibit host proteases. Cell 168 (3):517–526 e518

Gupta V, Carroll KS (2016) Rational design of reversible and irreversible cysteine sulfenic acid-targeted linear C-nucleophiles. Chem Commun (Camb) 52(16):3414–3417

Gupta V, Yang J, Liebler DC, Carroll KS (2017) Diverse redoxome reactivity profiles of carbon nucleophiles. J Am Chem Soc 139(15):5588–5595

Gygi SP, Rist B, Gerber SA, Turecek F, Gelb MH, Aebersold R (1999) Quantitative analysis of complex protein mixtures using isotope-coded affinity tags. Nat Biotechnol 17(10):994–999

Hacker SM, Backus KM, Lazear MR, Forli S, Correia BE, Cravatt BF (2017) Global profiling of lysine reactivity and ligandability in the human proteome. Nat Chem 9(12):1181–1190

Hahn Y-I, Kim S-J, Choi B-Y, Cho K-C, Bandu R, Kim KP, Kim D-H, Kim W, Park JS, Han BW, Lee J, Na H-K, Cha Y-N, Surh Y-J (2018) Curcumin interacts directly with the Cysteine 259 residue of STAT3 and induces apoptosis in H-Ras transformed human mammary epithelial cells. Sci Rep 8(1):6409

Han J, Adman ET, Beppu T, Codd R, Freeman HC, Huq LL, Loehr TM, Sanders-Loehr J (1991) Resonance Raman spectra of plastocyanin and pseudoazurin: evidence for conserved cysteine

ligand conformations in cupredoxins (blue copper proteins). Biochemistry 30(45):10904–10913
Hansen KC, Schmitt-Ulms G, Chalkley RJ, Hirsch J, Baldwin MA, Burlingame AL (2003) Mass spectrometric analysis of protein mixtures at low levels using cleavable 13C-isotope-coded affinity tag and multidimensional chromatography. Mol Cell Proteomics 2(5):299–314
Hansen RE, Ostergaard H, Winther JR (2005) Increasing the reactivity of an artificial dithiol-disulfide pair through modification of the electrostatic milieu. Biochemistry 44 (15):5899–5906
Hao G, Derakhshan B, Shi L, Campagne F, Gross SS (2006) SNOSID, a proteomic method for identification of cysteine S-nitrosylation sites in complex protein mixtures. Proc Natl Acad Sci U S A 103(4):1012–1017
Hatahet F, Ruddock LW (2009) Protein disulfide isomerase: a critical evaluation of its function in disulfide bond formation. Antioxid Redox Signal 11(11):2807–2850
Hess DT, Matsumoto A, Kim SO, Marshall HE, Stamler JS (2005) Protein S-nitrosylation: purview and parameters. Nat Rev Mol Cell Biol 6(2):150–166
Hing ZA, Fung HY, Ranganathan P, Mitchell S, El-Gamal D, Woyach JA, Williams K, Goettl VM, Smith J, Yu X, Meng X, Sun Q, Cagatay T, Lehman AM, Lucas DM, Baloglu E, Shacham S, Kauffman MG, Byrd JC, Chook YM, Garzon R, Lapalombella R (2016) Next-generation XPO1 inhibitor shows improved efficacy and in vivo tolerability in hematological malignancies. Leukemia 30(12):2364–2372
Hoffman S, Nolin J, McMillan D, Wouters E, Janssen-Heininger Y, Reynaert N (2015) Thiol redox chemistry: role of protein cysteine oxidation and altered redox homeostasis in allergic inflammation and asthma. J Cell Biochem 116(6):884–892
Hoffstrom BG, Kaplan A, Letso R, Schmid RS, Turmel GJ, Lo DC, Stockwell BR (2010) Inhibitors of protein disulfide isomerase suppress apoptosis induced by misfolded proteins. Nat Chem Biol 6(12):900–906
Holmgren A (1985) Thioredoxin. Annu Rev Biochem 54:237–271
Holmstrom KM, Finkel T (2014) Cellular mechanisms and physiological consequences of redox-dependent signalling. Nat Rev Mol Cell Biol 15(6):411–421
Hsu JL, Huang SY, Chow NH, Chen SH (2003) Stable-isotope dimethyl labeling for quantitative proteomics. Anal Chem 75(24):6843–6852
Iqbalsyah TM, Moutevelis E, Warwicker J, Errington N, Doig AJ (2006) The CXXC motif at the N terminus of an alpha-helical peptide. Protein Sci 15(8):1945–1950
Jacobi A, Huber-Wunderlich M, Hennecke J, Glockshuber R (1997) Elimination of all charged residues in the vicinity of the active-site helix of the disulfide oxidoreductase DsbA. Influence of electrostatic interactions on stability and redox properties. J Biol Chem 272(35):21692–21699
Janes MR, Zhang J, Li LS, Hansen R, Peters U, Guo X, Chen Y, Babbar A, Firdaus SJ, Darjania L, Feng J, Chen JH, Li S, Li S, Long YO, Thach C, Liu Y, Zarieh A, Ely T, Kucharski JM, Kessler LV, Wu T, Yu K, Wang Y, Yao Y, Deng X, Zarrinkar PP, Brehmer D, Dhanak D, Lorenzi MV, Hu-Lowe D, Patricelli MP, Ren P, Liu Y (2018) Targeting KRAS mutant cancers with a covalent G12C-specific inhibitor. Cell 172(3):578–589 e517
Jang BC, Munoz-Najar U, Paik JH, Claffey K, Yoshida M, Hla T (2003) Leptomycin B, an inhibitor of the nuclear export receptor CRM1, inhibits COX-2 expression. J Biol Chem 278 (5):2773–2776
Jeffery DA, Bogyo M (2003) Chemical proteomics and its application to drug discovery. Curr Opin Biotechnol 14(1):87–95
Jeffrey SC, Burke PJ, Lyon RP, Meyer DW, Sussman D, Anderson M, Hunter JH, Leiske CI, Miyamoto JB, Nicholas ND, Okeley NM, Sanderson RJ, Stone IJ, Zeng W, Gregson SJ, Masterson L, Tiberghien AC, Howard PW, Thurston DE, Law CL, Senter PD (2013) A potent anti-CD70 antibody-drug conjugate combining a dimeric pyrrolobenzodiazepine drug with site-specific conjugation technology. Bioconjug Chem 24(7):1256–1263
Jhoti H, Williams G, Rees DC, Murray CW (2013) The 'rule of three' for fragment-based drug discovery: where are we now? Nat Rev Drug Discov 12(8):644–645

Johnson DS, Weerapana E, Cravatt BF (2010) Strategies for discovering and derisking covalent, irreversible enzyme inhibitors. Future Med Chem 2(6):949–964

Kallis GB, Holmgren A (1980) Differential reactivity of the functional sulfhydryl groups of cysteine-32 and cysteine-35 present in the reduced form of thioredoxin from Escherichia coli. J Biol Chem 255(21):10261–10265

Kathman SG, Statsyuk AV (2016) Covalent tethering of fragments for covalent probe discovery. Medchemcomm 7(4):576–585

Kathman SG, Xu Z, Statsyuk AV (2014) A fragment-based method to discover irreversible covalent inhibitors of cysteine proteases. J Med Chem 57(11):4969–4974

Khan S, Vihinen M (2007) Spectrum of disease-causing mutations in protein secondary structures. BMC Struct Biol 7(1):56

Klatt P, Lamas S (2000) Regulation of protein function by S-glutathiolation in response to oxidative and nitrosative stress. Eur J Biochem 267(16):4928–4944

Klomsiri C, Karplus PA, Poole LB (2011) Cysteine-based redox switches in enzymes. Antioxid Redox Signal 14(6):1065–1077

Knuckley B, Causey CP, Jones JE, Bhatia M, Dreyton CJ, Osborne TC, Takahara H, Thompson PR (2010a) Substrate specificity and kinetic studies of PADs 1, 3, and 4 identify potent and selective inhibitors of protein arginine deiminase 3. Biochemistry 49(23):4852–4863

Knuckley B, Jones JE, Bachovchin DA, Slack J, Causey CP, Brown SJ, Rosen H, Cravatt BF, Thompson PR (2010b) A fluopol-ABPP HTS assay to identify PAD inhibitors. Chem Commun (Camb) 46(38):7175–7177

Kolb HC, Finn MG, Sharpless KB (2001) Click chemistry: diverse chemical function from a few good reactions. Angew Chem Int Ed Engl 40(11):2004–2021

Kortemme T, Creighton TE (1995) Ionisation of cysteine residues at the termini of model alpha-helical peptides. Relevance to unusual thiol pKa values in proteins of the thioredoxin family. J Mol Biol 253(5):799–812

Kranz RG, Richard-Fogal C, Taylor J-S, Frawley ER (2009) Cytochrome c biogenesis: mechanisms for covalent modifications and trafficking of heme and for heme-iron redox control. Microbiol Mol Biol Rev: MMBR 73(3):510–528

Krishnan S, Miller RM, Tian B, Mullins RD, Jacobson MP, Taunton J (2014) Design of reversible, cysteine-targeted Michael acceptors guided by kinetic and computational analysis. J Am Chem Soc 136(36):12624–12630

Krysiak JM, Kreuzer J, Macheroux P, Hermetter A, Sieber SA, Breinbauer R (2012) Activity-based probes for studying the activity of flavin-dependent oxidases and for the protein target profiling of monoamine oxidase inhibitors. Angew Chem Int Ed Engl 51 (28):7035–7040

Kumar S, Zhou B, Liang F, Wang WQ, Huang Z, Zhang ZY (2004) Activity-based probes for protein tyrosine phosphatases. Proc Natl Acad Sci U S A 101(21):7943–7948

Kuo YH, Konopko AM, Borotto NB, Majmudar JD, Haynes SE, Martin BR (2017) Profiling protein S-sulfination with maleimide-linked probes. ChemBioChem 18(20):2028–2032

Lanning BR, Whitby LR, Dix MM, Douhan J, Gilbert AM, Hett EC, Johnson TO, Joslyn C, Kath JC, Niessen S, Roberts LR, Schnute ME, Wang C, Hulce JJ, Wei B, Whiteley LO, Hayward MM, Cravatt BF (2014) A road map to evaluate the proteome-wide selectivity of covalent kinase inhibitors. Nat Chem Biol 10(9):760–767

Lapalombella R, Sun Q, Williams K, Tangeman L, Jha S, Zhong Y, Goettl V, Mahoney E, Berglund C, Gupta S, Farmer A, Mani R, Johnson AJ, Lucas D, Mo X, Daelemans D, Sandanayaka V, Shechter S, McCauley D, Shacham S, Kauffman M, Chook YM, Byrd JC (2012) Selective inhibitors of nuclear export show that CRM1/XPO1 is a target in chronic lymphocytic leukemia. Blood 120(23):4621–4634

Lavergne SN, Park BK, Naisbitt DJ (2008) The roles of drug metabolism in the pathogenesis of T-cell-mediated drug hypersensitivity. Curr Opin Allergy Clin Immunol 8(4):299–307

Lea WA, Simeonov A (2011) Fluorescence polarization assays in small molecule screening. Expert Opin Drug Discov 6(1):17–32

Leichert LI, Gehrke F, Gudiseva HV, Blackwell T, Ilbert M, Walker AK, Strahler JR, Andrews PC, Jakob U (2008) Quantifying changes in the thiol redox proteome upon oxidative stress in vivo. Proc Natl Acad Sci U S A 105(24):8197–8202

Levin Y, Bahn S (2010) Quantification of proteins by label-free LC-MS/MS. Methods Mol Biol 658:217–231

Li GZ, Vissers JP, Silva JC, Golick D, Gorenstein MV, Geromanos SJ (2009) Database searching and accounting of multiplexed precursor and product ion spectra from the data independent analysis of simple and complex peptide mixtures. Proteomics 9(6):1696–1719

Lin VS, Lippert AR, Chang CJ (2013) Cell-trappable fluorescent probes for endogenous hydrogen sulfide signaling and imaging H_2O_2-dependent H_2S production. Proc Natl Acad Sci U S A 110 (18):7131–7135

Lin VS, Chen W, Xian M, Chang CJ (2015a) Chemical probes for molecular imaging and detection of hydrogen sulfide and reactive sulfur species in biological systems. Chem Soc Rev 44(14):4596–4618

Lin VS, Lippert AR, Chang CJ (2015b) Azide-based fluorescent probes: imaging hydrogen sulfide in living systems. Methods Enzymol 554:63–80

Lindemann C, Lupilova N, Muller A, Warscheid B, Meyer HE, Kuhlmann K, Eisenacher M, Leichert LI (2013) Redox proteomics uncovers peroxynitrite-sensitive proteins that help *Escherichia coli* to overcome nitrosative stress. J Biol Chem 288(27):19698–19714

Link AJ, Hays LG, Carmack EB, Yates JR 3rd (1997) Identifying the major proteome components of Haemophilus influenzae type-strain NCTC 8143. Electrophoresis 18(8):1314–1334

Lippert AR (2014) Designing reaction-based fluorescent probes for selective hydrogen sulfide detection. J Inorg Biochem 133:136–142

Lipton SA (2006) Paradigm shift in neuroprotection by NMDA receptor blockade: memantine and beyond. Nat Rev Drug Discov 5(2):160–170

Liu CT, Benkovic SJ (2013) Capturing a sulfenic acid with arylboronic acids and benzoxaborole. J Am Chem Soc 135(39):14544–14547

Liu Y, Barrett JE, Schultz PG, Santi DV (1999) Tyrosine 146 of Thymidylate Synthase Assists Proton Abstraction from the 5-Position of 2′-Deoxyuridine 5′-Monophosphate. Biochemistry 38(2):848–852

Liu Q, Sabnis Y, Zhao Z, Zhang T, Buhrlage SJ, Jones LH, Gray NS (2013) Developing irreversible inhibitors of the protein kinase cysteinome. Chem Biol 20(2):146–159

Llopis J, McCaffery JM, Miyawaki A, Farquhar MG, Tsien RY (1998) Measurement of cytosolic, mitochondrial, and Golgi pH in single living cells with green fluorescent proteins. Proc Natl Acad Sci U S A 95(12):6803–6808

Lo Conte M, Carroll KS (2013) The redox biochemistry of protein sulfenylation and sulfinylation. J Biol Chem 288(37):26480–26488

Lo Conte M, Lin J, Wilson MA, Carroll KS (2015) A chemical approach for the detection of protein sulfinylation. ACS Chem Biol 10(8):1825–1830

London N, Miller RM, Krishnan S, Uchida K, Irwin JJ, Eidam O, Gibold L, Cimermancic P, Bonnet R, Shoichet BK, Taunton J (2014) Covalent docking of large libraries for the discovery of chemical probes. Nat Chem Biol 10(12):1066–1072

Lowther WT, McMillen DA, Orville AM, Matthews BW (1998) The anti-angiogenic agent fumagillin covalently modifies a conserved active-site histidine in the <em> *Escherichia coli* </em> methionine aminopeptidase. Proc Natl Acad Sci 95(21):12153–12157

MacKintosh RW, Dalby KN, Campbell DG, Cohen PT, Cohen P, MacKintosh C (1995) The cyanobacterial toxin microcystin binds covalently to cysteine-273 on protein phosphatase 1. FEBS Lett 371(3):236–240

Majmudar JD, Konopko AM, Labby KJ, Tom CTMB, Crellin JE, Prakash A, Martin BR (2016) Harnessing redox cross-reactivity to profile distinct cysteine modifications. J Am Chem Soc 138(6):1852–1859

Marino SM, Gladyshev VN (2010) Cysteine function governs its conservation and degeneration and restricts its utilization on protein surfaces. J Mol Biol 404(5):902–916

Maron BA, Tang SS, Loscalzo J (2013) S-nitrosothiols and the S-nitrosoproteome of the cardiovascular system. Antioxid Redox Signal 18(3):270–287

Martell J, Seo Y, Bak DW, Kingsley SF, Tissenbaum HA, Weerapana E (2016) Global cysteine-reactivity profiling during impaired insulin/IGF-1 signaling in *C. elegans* Identifies uncharacterized mediators of longevity. Cell Chem Biol 23(8):955–966

Matsushita K, Morrell CN, Cambien B, Yang SX, Yamakuchi M, Bao C, Hara MR, Quick RA, Cao W, O'Rourke B, Lowenstein JM, Pevsner J, Wagner DD, Lowenstein CJ (2003) Nitric oxide regulates exocytosis by S-nitrosylation of *N*-ethylmaleimide-sensitive factor. Cell 115 (2):139–150

Maurer A, Zeyher C, Amin B, Kalbacher H (2013) A periodate-cleavable linker for functional proteomics under slightly acidic conditions: application for the analysis of intracellular aspartic proteases. J Proteome Res 12(1):199–207

McDonald WH, Yates JR 3rd (2002) Shotgun proteomics and biomarker discovery. Dis Markers 18(2):99–105

Meissner T, Krause E, Vinkemeier U (2004) Ratjadone and leptomycin B block CRM1-dependent nuclear export by identical mechanisms. FEBS Lett 576(1–2):27–30

Miller RM, Paavilainen VO, Krishnan S, Serafimova IM, Taunton J (2013) Electrophilic fragment-based design of reversible covalent kinase inhibitors. J Am Chem Soc 135(14):5298–5301

Miseta A, Csutora P (2000) Relationship between the occurrence of cysteine in proteins and the complexity of organisms. Mol Biol Evol 17(8):1232–1239

Mittag S, Valenta T, Weiske J, Bloch L, Klingel S, Gradl D, Wetzel F, Chen Y, Petersen I, Basler K, Huber O (2016) A novel role for the tumour suppressor Nitrilase1 modulating the Wnt/β-catenin signalling pathway. Cell Discov 2:15039

Muller S, Chaikuad A, Gray NS, Knapp S (2015) The ins and outs of selective kinase inhibitor development. Nat Chem Biol 11(11):818–821

Muth A, Subramanian V, Beaumont E, Nagar M, Kerry P, McEwan P, Srinath H, Clancy K, Parelkar S, Thompson PR (2017) Development of a selective inhibitor of protein arginine deiminase 2. J Med Chem 60(7):3198–3211

Nelson JW, Creighton TE (1994) Reactivity and ionization of the active site cysteine residues of DsbA, a protein required for disulfide bond formation in vivo. Biochemistry 33(19):5974–5983

Nielsen ML, Vermeulen M, Bonaldi T, Cox J, Moroder L, Mann M (2008) Iodoacetamide-induced artifact mimics ubiquitination in mass spectrometry. Nat Methods 5(6):459–460

Niessen S, Dix MM, Barbas S, Potter ZE, Lu S, Brodsky O, Planken S, Behenna D, Almaden C, Gajiwala KS, Ryan K, Ferre R, Lazear MR, Hayward MM, Kath JC, Cravatt BF (2017) Proteome-wide map of targets of T790M-EGFR-directed covalent inhibitors. Cell Chem Biol 24(11):1388–1400 e1387

Nonoo RH, Armstrong A, Mann DJ (2012) Kinetic template-guided tethering of fragments. ChemMedChem 7(12):2082–2086

Oballa RM, Truchon JF, Bayly CI, Chauret N, Day S, Crane S, Berthelette C (2007) A generally applicable method for assessing the electrophilicity and reactivity of diverse nitrile-containing compounds. Bioorg Med Chem Lett 17(4):998–1002

Ohnishi K, Irie K, Murakami A (2009) In vitro covalent binding proteins of zerumbone, a chemopreventive food factor. Biosci Biotechnol Biochem 73(8):1905–1907

Ohnishi K, Ohkura S, Nakahata E, Ishisaka A, Kawai Y, Terao J, Mori T, Ishii T, Nakayama T, Kioka N, Matsumoto S, Ikeda Y, Akiyama M, Irie K, Murakami A (2013) Non-specific protein modifications by a phytochemical induce heat shock response for self-defense. PLoS ONE 8 (3):e58641

Olejnik J, Sonar S, Krzymańska-Olejnik E, Rothschild KJ (1995) Photocleavable biotin derivatives: a versatile approach for the isolation of biomolecules. Proc Natl Acad Sci 92 (16):7590–7594

Ong SE, Mann M (2007) Stable isotope labeling by amino acids in cell culture for quantitative proteomics. Methods Mol Biol 359:37–52

Ostrem JM, Peters U, Sos ML, Wells JA, Shokat KM (2013) K-Ras(G12C) inhibitors allosterically control GTP affinity and effector interactions. Nature 503(7477):548–551
Ouyang X, Zhou S, Su CT, Ge Z, Li R, Kwoh CK (2013) CovalentDock: automated covalent docking with parameterized covalent linkage energy estimation and molecular geometry constraints. J Comput Chem 34(4):326–336
Pace NJ, Weerapana E (2014a) A competitive chemical-proteomic platform to identify zinc-binding cysteines. ACS Chem Biol 9(1):258–265
Pace NJ, Weerapana E (2014b) Zinc-binding cysteines: diverse functions and structural motifs. Biomolecules 4(2):419–434
Pan Z, Scheerens H, Li S-J, Schultz BE, Sprengeler PA, Burrill LC, Mendonca RV, Sweeney MD, Scott KCK, Grothaus PG, Jeffery DA, Spoerke JM, Honigberg LA, Young PR, Dalrymple SA, Palmer JT (2007) Discovery of selective irreversible inhibitors for Bruton's tyrosine kinase. ChemMedChem 2(1):58–61
Patricelli MP, Janes MR, Li LS, Hansen R, Peters U, Kessler LV, Chen Y, Kucharski JM, Feng J, Ely T, Chen JH, Firdaus SJ, Babbar A, Ren P, Liu Y (2016) Selective inhibition of oncogenic KRAS output with small molecules targeting the inactive state. Cancer Discov 6(3):316–329
Paul BD, Snyder SH (2015) Protein sulfhydration. Methods Enzymol 555:79–90
Paulsen CE, Carroll KS (2013) Cysteine-mediated redox signaling: chemistry, biology, and tools for discovery. Chem Rev 113(7):4633–4679
Paulsen CE, Truong TH, Garcia FJ, Homann A, Gupta V, Leonard SE, Carroll KS (2011) Peroxide-dependent sulfenylation of the EGFR catalytic site enhances kinase activity. Nat Chem Biol 8(1):57–64
Pels K, Dickson P, An H, Kodadek T (2018) DNA-compatible solid-phase combinatorial synthesis of beta-cyanoacrylamides and related electrophiles. ACS Comb Sci 20(2):61–69
Perkins DN, Pappin DJ, Creasy DM, Cottrell JS (1999) Probability-based protein identification by searching sequence databases using mass spectrometry data. Electrophoresis 20(18):3551–3567
Persson C, Sjoblom T, Groen A, Kappert K, Engstrom U, Hellman U, Heldin CH, den Hertog J, Ostman A (2004) Preferential oxidation of the second phosphatase domain of receptor-like PTP-alpha revealed by an antibody against oxidized protein tyrosine phosphatases. Proc Natl Acad Sci U S A 101(7):1886–1891
Pierce A, Unwin RD, Evans CA, Griffiths S, Carney L, Zhang L, Jaworska E, Lee CF, Blinco D, Okoniewski MJ, Miller CJ, Bitton DA, Spooncer E, Whetton AD (2008) Eight-channel iTRAQ enables comparison of the activity of six leukemogenic tyrosine kinases. Mol Cell Proteomics 7(5):853–863
Pitscheider M, Mausbacher N, Sieber SA (2012) Antibiotic activity and target discovery of three-membered natural product-derived heterocycles in pathogenic bacteria. Chem Sci 3 (6):2035–2041
Pliura DH, Bonaventura BJ, Smith RA, Coles PJ, Krantz A (1992) Comparative behaviour of calpain and cathepsin B toward peptidyl acyloxymethyl ketones, sulphonium methyl ketones and other potential inhibitors of cysteine proteinases. Biochem J 288(Pt 3):759–762
Ploger M, Sendker J, Langer K, Schmidt TJ (2015) Covalent modification of human serum albumin by the natural sesquiterpene lactone parthenolide. Molecules 20(4):6211–6223
Ponader S, Chen SS, Buggy JJ, Balakrishnan K, Gandhi V, Wierda WG, Keating MJ, O'Brien S, Chiorazzi N, Burger JA (2012) The Bruton tyrosine kinase inhibitor PCI-32765 thwarts chronic lymphocytic leukemia cell survival and tissue homing in vitro and in vivo. Blood 119:1182–1189. https://doi.org/10.1182/blood-2011-10-38641
Poole LB, Nelson KJ (2008) Discovering mechanisms of signaling-mediated cysteine oxidation. Curr Opin Chem Biol 12(1):18–24
Poole TH, Reisz JA, Zhao W, Poole LB, Furdui CM, King SB (2014) Strained cycloalkynes as new protein sulfenic acid traps. J Am Chem Soc 136(17):6167–6170
Prinarakis E, Chantzoura E, Thanos D, Spyrou G (2008) S-glutathionylation of IRF3 regulates IRF3-CBP interaction and activation of the IFN beta pathway. EMBO J 27(6):865–875

Qian Y, Martell J, Pace NJ, Ballard TE, Johnson DS, Weerapana E (2013) An isotopically tagged azobenzene-based cleavable linker for quantitative proteomics. ChemBioChem 14(12):1410–1414

Ricci G, Bello ML, Caccuri AM, Pastore A, Nuccetelli M, Parker MW, Federici G (1995) Site-directed mutagenesis of human glutathione transferase P1-1: mutation of Cys-47 induces a positive cooperativity in glutathione transferase P1-1. J Biol Chem 270(3):1243–1248

Roberts AM, Miyamoto DK, Huffman TR, Bateman LA, Ives AN, Akopian D, Heslin MJ, Contreras CM, Rape M, Skibola CF, Nomura DK (2017a) Chemoproteomic screening of covalent ligands reveals UBA5 as a novel pancreatic cancer target. ACS Chem Biol 12(4):899–904

Roberts LS, Yan P, Bateman LA, Nomura DK (2017b) Mapping novel metabolic nodes targeted by anti-cancer drugs that impair triple-negative breast cancer pathogenicity. ACS Chem Biol 12(4):1133–1140

Robertson JG (2005) Mechanistic basis of enzyme-targeted drugs. Biochemistry 44(15):5561–5571

Rogers LK, Leinweber BL, Smith CV (2006) Detection of reversible protein thiol modifications in tissues. Anal Biochem 358(2):171–184

Ross PL, Huang YN, Marchese JN, Williamson B, Parker K, Hattan S, Khainovski N, Pillai S, Dey S, Daniels S, Purkayastha S, Juhasz P, Martin S, Bartlet-Jones M, He F, Jacobson A, Pappin DJ (2004) Multiplexed protein quantitation in *Saccharomyces cerevisiae* using amine-reactive isobaric tagging reagents. Mol Cell Proteomics 3(12):1154–1169

Rost HL, Rosenberger G, Navarro P, Gillet L, Miladinovic SM, Schubert OT, Wolski W, Collins BC, Malmstrom J, Malmstrom L, Aebersold R (2014) OpenSWATH enables automated, targeted analysis of data-independent acquisition MS data. Nat Biotechnol 32 (3):219–223

Röst HL, Rosenberger G, Navarro P, Gillet L, Miladinović SM, Schubert OT, Wolski W, Collins BC, Malmström J, Malmström L, Aebersold R (2014) OpenSWATH enables automated, targeted analysis of data-independent acquisition MS data. Nat Biotechnol 32:219

Sadowsky JD, Burlingame MA, Wolan DW, McClendon CL, Jacobson MP, Wells JA (2011) Turning a protein kinase on or off from a single allosteric site via disulfide trapping. Proc Natl Acad Sci U S A 108(15):6056–6061

Sajadimajd S, Khazaei M (2017) Oxidative stress and cancer: the role of Nrf2. Curr Cancer Drug Targets

Segovia-Mendoza M, González-González ME, Barrera D, Díaz L, García-Becerra R (2015) Efficacy and mechanism of action of the tyrosine kinase inhibitors gefitinib, lapatinib and neratinib in the treatment of HER2-positive breast cancer: preclinical and clinical evidence. Am J Cancer Res 5(9):2531–2561

Serafimova IM, Pufall MA, Krishnan S, Duda K, Cohen MS, Maglathlin RL, McFarland JM, Miller RM, Frodin M, Taunton J (2012) Reversible targeting of noncatalytic cysteines with chemically tuned electrophiles. Nat Chem Biol 8(5):471–476

Sethuraman M, McComb ME, Heibeck T, Costello CE, Cohen RA (2004a) Isotope-coded affinity tag approach to identify and quantify oxidant-sensitive protein thiols. Mol Cell Proteomics 3 (3):273–278

Sethuraman M, McComb ME, Huang H, Huang S, Heibeck T, Costello CE, Cohen RA (2004b) Isotope-coded affinity tag (ICAT) approach to redox proteomics: identification and quantitation of oxidant-sensitive cysteine thiols in complex protein mixtures. J Proteome Res 3(6):1228–1233

Shannon DA, Banerjee R, Webster ER, Bak DW, Wang C, Weerapana E (2014) Investigating the proteome reactivity and selectivity of aryl halides. J Am Chem Soc 136(9):3330–3333

Shevchenko A, Jensen ON, Podtelejnikov AV, Sagliocco F, Wilm M, Vorm O, Mortensen P, Shevchenko A, Boucherie H, Mann M (1996) Linking genome and proteome by mass spectrometry: large-scale identification of yeast proteins from two dimensional gels. Proc Natl Acad Sci U S A 93(25):14440–14445

Shuker SB, Hajduk PJ, Meadows RP, Fesik SW (1996) Discovering high-affinity ligands for proteins: SAR by NMR. Science 274(5292):1531–1534

Silva JC, Gorenstein MV, Li GZ, Vissers JP, Geromanos SJ (2006) Absolute quantification of proteins by LCMSE: a virtue of parallel MS acquisition. Mol Cell Proteomics 5(1):144–156

Singh J, Petter RC, Baillie TA, Whitty A (2011) The resurgence of covalent drugs. Nat Rev Drug Discov 10(4):307–317

Slee EA, Zhu H, Chow SC, MacFarlane M, Nicholson DW, Cohen GM (1996) Benzyloxycarbonyl-Val-Ala-Asp (OMe) fluoromethylketone (Z-VAD.FMK) inhibits apoptosis by blocking the processing of CPP32. Biochem J 315(Pt 1):21–24

Smith AJ, Zhang X, Leach AG, Houk KN (2009) Beyond picomolar affinities: quantitative aspects of noncovalent and covalent binding of drugs to proteins. J Med Chem 52(2):225–233

Sommer S, Weikart ND, Linne U, Mootz HD (2013) Covalent inhibition of SUMO and ubiquitin-specific cysteine proteases by an in situ thiol-alkyne addition. Bioorg Med Chem 21 (9):2511–2517

Speers AE, Cravatt BF (2004) Profiling enzyme activities in vivo using click chemistry methods. Chem Biol 11(4):535–546

Speers AE, Cravatt BF (2005) A tandem orthogonal proteolysis strategy for high-content chemical proteomics. J Am Chem Soc 127(28):10018–10019

Steiger AK, Yang Y, Royzen M, Pluth MD (2017) Bio-orthogonal "click-and-release" donation of caged carbonyl sulfide (COS) and hydrogen sulfide (H2S). Chem Commun (Camb) 53 (8):1378–1380

Sun J, Xin C, Eu JP, Stamler JS, Meissner G (2001) Cysteine-3635 is responsible for skeletal muscle ryanodine receptor modulation by NO. Proc Natl Acad Sci U S A 98(20):11158–11162

Sussman D, Westendorf L, Meyer DW, Leiske CI, Anderson M, Okeley NM, Alley SC, Lyon R, Sanderson RJ, Carter PJ, Benjamin DR (2018) Engineered cysteine antibodies: an improved antibody-drug conjugate platform with a novel mechanism of drug-linker stability. Protein Eng Des Sel 31(2):47–54

Swinney DC (2009) The role of binding kinetics in therapeutically useful drug action. Curr Opin Drug Discov Devel 12(1):31–39

Szychowski J, Mahdavi A, Hodas JJL, Bagert JD, Ngo JT, Landgraf P, Dieterich DC, Schuman EM, Tirrell DA (2010) Cleavable biotin probes for labeling of biomolecules via azide–alkyne cycloaddition. J Am Chem Soc 132(51):18351–18360

Takaya J, Mio K, Shiraishi T, Kurokawa T, Otsuka S, Mori Y, Uesugi M (2015) A potent and site-selective agonist of TRPA1. J Am Chem Soc 137(50):15859–15864

Tanaka H, Nishida K, Sugita K, Yoshioka T (1999) Antitumor efficacy of hypothemycin, a new Ras-signaling inhibitor. Jpn J Cancer Res 90(10):1139–1145

The M, MacCoss MJ, Noble WS, Käll L (2016) Fast and accurate protein false discovery rates on large-scale proteomics data sets with percolator 3.0. J Am Soc Mass Spectrom 27(11):1719–1727

Thompson A, Schafer J, Kuhn K, Kienle S, Schwarz J, Schmidt G, Neumann T, Johnstone R, Mohammed AK, Hamon C (2003) Tandem mass tags: a novel quantification strategy for comparative analysis of complex protein mixtures by MS/MS. Anal Chem 75(8):1895–1904

Thornberry NA, Peterson EP, Zhao JJ, Howard AD, Griffin PR, Chapman KT (1994) Inactivation of interleukin-1 beta converting enzyme by peptide (acyloxy)methyl ketones. Biochemistry 33 (13):3934–3940

Tian C, Sun R, Liu K, Fu L, Liu X, Zhou W, Yang Y, Yang J (2017) Multiplexed thiol reactivity profiling for target discovery of electrophilic natural products. Cell Chem Biol 24(11):1416–1427 e1415

Tonge PJ (2018) Drug-target kinetics in drug discovery. ACS Chem Neurosci 9(1):29–39

Truong TH, Ung PM, Palde PB, Paulsen CE, Schlessinger A, Carroll KS (2016) Molecular basis for redox activation of epidermal growth factor receptor kinase. Cell Chem Biol 23(7):837–848

Uetrecht J (2009) Immune-mediated adverse drug reactions. Chem Res Toxicol 22(1):24–34

Uzozie AC, Aebersold R (2018) Advancing translational research and precision medicine with targeted proteomics. J Proteomics

Verdonk ML, Cole JC, Hartshorn MJ, Murray CW, Taylor RD (2003) Improved protein-ligand docking using GOLD. Proteins 52(4):609–623

Vinogradova EV, Zhang C, Spokoyny AM, Pentelute BL, Buchwald SL (2015) Organometallic palladium reagents for cysteine bioconjugation. Nature 526(7575):687–691

Visscher M, Arkin MR, Dansen TB (2016) Covalent targeting of acquired cysteines in cancer. Curr Opin Chem Biol 30:61–67

Wachnowsky C, Wesley NA, Fidai I, Cowan JA (2017) Understanding the molecular basis of multiple mitochondrial dysfunctions syndrome 1 (MMDS1)-impact of a disease-causing Gly208Cys substitution on structure and activity of NFU1 in the Fe/S cluster biosynthetic pathway. J Mol Biol 429(6):790–807

Wang H, Xian M (2008) Fast reductive ligation of S-nitrosothiols. Angew Chem Int Ed Engl 47 (35):6598–6601

Wang Z, Udeshi ND, O'Malley M, Shabanowitz J, Hunt DF, Hart GW (2010) Enrichment and site mapping of O-Linked *N*-Acetylglucosamine by a combination of chemical/enzymatic tagging, photochemical cleavage, and electron transfer dissociation mass spectrometry. Mol Cell Proteomics 9(1):153–160

Wang C, Weerapana E, Blewett MM, Cravatt BF (2014) A chemoproteomic platform to quantitatively map targets of lipid-derived electrophiles. Nat Meth 11(1):79–85

Wang C, Abegg D, Hoch DG, Adibekian A (2016) Chemoproteomics-enabled discovery of a potent and selective inhibitor of the DNA repair protein MGMT. Angew Chem Int Ed Engl 55 (8):2911–2915

Wang H, Chen X, Li C, Liu Y, Yang F, Wang C (2018) Sequence-based prediction of cysteine reactivity using machine learning. Biochemistry 57(4):451–460

Ward RA, Anderton MJ, Ashton S, Bethel PA, Box M, Butterworth S, Colclough N, Chorley CG, Chuaqui C, Cross DA, Dakin LA, Debreczeni JE, Eberlein C, Finlay MR, Hill GB, Grist M, Klinowska TC, Lane C, Martin S, Orme JP, Smith P, Wang F, Waring MJ (2013) Structure- and reactivity-based development of covalent inhibitors of the activating and gatekeeper mutant forms of the epidermal growth factor receptor (EGFR). J Med Chem 56(17):7025–7048

Ward CC, Kleinman JI, Nomura DK (2017) NHS-esters As versatile reactivity-based probes for mapping proteome-wide ligandable hotspots. ACS Chem Biol 12(6):1478–1483

Washburn MP (2015) The H-index of 'an approach to correlate tandem mass spectral data of peptides with amino acid sequences in a protein database'. J Am Soc Mass Spectrom 26 (11):1799–1803

Weerapana E, Speers AE, Cravatt BF (2007) Tandem orthogonal proteolysis-activity-based protein profiling (TOP-ABPP)—a general method for mapping sites of probe modification in proteomes. Nat Protoc 2(6):1414–1425

Weerapana E, Simon GM, Cravatt BF (2008) Disparate proteome reactivity profiles of carbon electrophiles. Nat Chem Biol 4(7):405–407

Weerapana E, Wang C, Simon GM, Richter F, Khare S, Dillon MB, Bachovchin DA, Mowen K, Baker D, Cravatt BF (2010) Quantitative reactivity profiling predicts functional cysteines in proteomes. Nature 468(7325):790–795

Whitby LR, Obach RS, Simon GM, Hayward MM, Cravatt BF (2017) Quantitative chemical proteomic profiling of the in vivo targets of reactive drug metabolites. ACS Chem Biol 12 (8):2040–2050

Wissner A, Overbeek E, Reich MF, Floyd MB, Johnson BD, Mamuya N, Rosfjord EC, Discafani C, Davis R, Shi X, Rabindran SK, Gruber BC, Ye F, Hallett WA, Nilakantan R, Shen R, Wang Y-F, Greenberger LM, Tsou H-R (2003) Synthesis and structure–activity relationships of 6,7-disubstituted 4-Anilinoquinoline-3-carbonitriles. The design of an orally active, irreversible inhibitor of the tyrosine kinase activity of the epidermal growth factor receptor (EGFR) and the human epidermal growth factor receptor-2 (HER-2). J Med Chem 46 (1):49–63

Withers SG, Aebersold R (1995) Approaches to labeling and identification of active site residues in glycosidases. Protein Sci 4(3):361–372

Witt AC, Lakshminarasimhan M, Remington BC, Hasim S, Pozharski E, Wilson MA (2008) Cysteine pKa depression by a protonated glutamic acid in human DJ-1. Biochemistry 47 (28):7430–7440

Wolters DA, Washburn MP, Yates JR 3rd (2001) An automated multidimensional protein identification technology for shotgun proteomics. Anal Chem 73(23):5683–5690

Wong HL, Liebler DC (2008) Mitochondrial protein targets of thiol-reactive electrophiles. Chem Res Toxicol 21(4):796–804

Wood ZA, Schroder E, Robin Harris J, Poole LB (2003) Structure, mechanism and regulation of peroxiredoxins. Trends Biochem Sci 28(1):32–40

Wright AT, Song JD, Cravatt BF (2009) A Suite of Activity-based probes for human cytochrome P450 enzymes. J Am Chem Soc 131(30):10692–10700

Wright FA, Lu JP, Sliter DA, Dupre N, Rouleau GA, Wojcikiewicz RJ (2015) A point mutation in the ubiquitin ligase RNF170 that causes autosomal dominant sensory ataxia destabilizes the protein and impairs inositol 1,4,5-Trisphosphate receptor-mediated Ca^{2+} signaling. J Biol Chem 290(22):13948–13957

Wu C, Parrott AM, Liu T, Jain MR, Yang Y, Sadoshima J, Li H (2011) Distinction of thioredoxin transnitrosylation and denitrosylation target proteins by the ICAT quantitative approach. J Proteomics 74(11):2498–2509

Wulff JE, Siegrist R, Myers AG (2007) The natural product avrainvillamide binds to the oncoprotein nucleophosmin. J Am Chem Soc 129(46):14444–14451

Xia YF, Ye BQ, Li YD, Wang JG, He XJ, Lin X, Yao X, Ma D, Slungaard A, Hebbel RP, Key NS, Geng JG (2004) Andrographolide attenuates inflammation by inhibition of NF-kappa B activation through covalent modification of reduced cysteine 62 of p50. J Immunol 173 (6):4207–4217

Xu T, Park SK, Venable JD, Wohlschlegel JA, Diedrich JK, Cociorva D, Lu B, Liao L, Hewel J, Han X, Wong CCL, Fonslow B, Delahunty C, Gao Y, Shah H, Yates JR 3rd (2015) ProLuCID: an improved SEQUEST-like algorithm with enhanced sensitivity and specificity. J Proteomics 129:16–24

Yang Y, Verhelst SHL (2013) Cleavable trifunctional biotin reagents for protein labelling, capture and release. Chem Commun 49(47):5366–5368

Yang J, Gupta V, Tallman KA, Porter NA, Carroll KS, Liebler DC (2015) Global, in situ, site-specific analysis of protein S-sulfenylation. Nat Protoc 10(7):1022–1037

Yang J, Carroll KS, Liebler DC (2016) The expanding landscape of the thiol redox proteome. Mol Cell Proteomics 15(1):1–11

Yi EC, Li XJ, Cooke K, Lee H, Raught B, Page A, Aneliunas V, Hieter P, Goodlett DR, Aebersold R (2005) Increased quantitative proteome coverage with (13)C/(12)C-based, acid-cleavable isotope-coded affinity tag reagent and modified data acquisition scheme. Proteomics 5(2):380–387

Yu L-R, Conrads TP, Uo T, Issaq HJ, Morrison RS, Veenstra TD (2004) Evaluation of the acid-cleavable isotope-coded affinity tag reagents: application to camptothecin-treated cortical neurons. J Proteome Res 3(3):469–477

Yver A (2016) Osimertinib (AZD9291)-a science-driven, collaborative approach to rapid drug design and development. Ann Oncol 27(6):1165–1170

Zaro BW, Whitby LR, Lum KM, Cravatt BF (2016) Metabolically labile fumarate esters impart kinetic selectivity to irreversible inhibitors. J Am Chem Soc 138(49):15841–15844

Zhang R, Sioma CS, Wang S, Regnier FE (2001) Fractionation of isotopically labeled peptides in quantitative proteomics. Anal Chem 73(21):5142–5149

Zhang J, Yang PL, Gray NS (2009) Targeting cancer with small molecule kinase inhibitors. Nat Rev Cancer 9(1):28–39

Zhao Z, Bourne PE (2018) Progress with covalent small-molecule kinase inhibitors. Drug Discov Today 23(3):727–735

Zhao Y, Bolton SG, Pluth MD (2017a) Light-activated COS/H_2S donation from photocaged thiocarbamates. Org Lett 19(9):2278–2281

Zhao Y, Henthorn HA, Pluth MD (2017b) Kinetic insights into hydrogen sulfide delivery from caged-carbonyl sulfide isomeric donor platforms. J Am Chem Soc 139(45):16365–16376

Zhou H, Ranish JA, Watts JD, Aebersold R (2002) Quantitative proteome analysis by solid-phase isotope tagging and mass spectrometry. Nat Biotechnol 20(5):512–515

Zhou W, Ercan D, Chen L, Yun CH, Li D, Capelletti M, Cortot AB, Chirieac L, Iacob RE, Padera R, Engen JR, Wong KK, Eck MJ, Gray NS, Janne PA (2009) Novel mutant-selective EGFR kinase inhibitors against EGFR T790 M. Nature 462(7276):1070–1074

Zhou WY, Zheng H, Du XL, Yang JL (2016) Characterization of FGFR signaling pathway as therapeutic targets for sarcoma patients. Cancer Biol Med 13(2):260–268

Zhu K, Borrelli KW, Greenwood JR, Day T, Abel R, Farid RS, Harder E (2014) Docking covalent inhibitors: a parameter free approach to pose prediction and scoring. J Chem Inf Model 54 (7):1932–1940

Zimmermann G, Rieder U, Bajic D, Vanetti S, Chaikuad A, Knapp S, Scheuermann J, Mattarella M, Neri D (2017) A specific and covalent JNK-1 ligand selected from an encoded self-assembling chemical library. Chemistry 23(34):8152–8155

The manufacturer's authorised representative in the EU is Springer Nature Customer Service Centre GmbH, Europaplatz 3, 69115 Heidelberg, Germany. If you have any concerns regarding our products, please contact ProductSafety@springernature.com

Printed and bound by CPI Group (UK) Ltd, Croydon, CR0 4YY

30/09/2026

02225228-0016